ADVANCES IN NUMERICAL METHODS AND APPLICATIONS

Proceedings of the Third International Conference

ADVANCES IN NUMERICAL METHODS AND APPLICATIONS

$\mathcal{O}(h^3)$

Sofia, Bulgaria 21 – 26 August 1994

Editors

I. T. Dimov

Bl. Sendov

P. S. Vassilevski

Bulgarian Academy of Sciences

World Scientific

Singapore • New Jersey • London • Hong Kong

Published by

World Scientific Publishing Co Pte Ltd
5 Toh Tuck Link, Singapore 596224
USA office: 27 Warren Street, Suite 401-402, Hackensack, NJ 07601
UK office: 57 Shelton Street, Covent Garden, London WC2H 9HE

British Library Cataloguing-in-Publication Data
A catalogue record for this book is available from the British Library.

ADVANCES IN NUMERICAL METHODS AND APPLICATIONS

ISBN-13 978-981-02-1926-0
ISBN-10 981-02-1926-1

Preface

The new advanced numerical methods together with the increasing speed and expanded storage capacity of modern high-performance computers, have greatly improved the ability to solve large-scale computational problems. At present, the revolutionary impact of the next generations of computers are fully recognized by many scientists. New parallel and vector machines form the basic of computing power needed to address the large-scale problems which engineers faced today. The new machines not only can tackle more sizable problems, but also employ a different computational approach to obtain the most efficient results. Moreover, the new computers and new computational techniques stimulate new searches and solutions in the field of parallel algorithms. On the other hand, the progress in numerical analysis and approximation theory is conducive to the creation of new numerical algorithms. Now it is commonplace that the new results on error estimation and convergence theory in Finite Element and Boundary Element Methods as well as in Monte-Carlo Methods lead to efficient numerical algorithms suitable for modern parallel machines. Development on the domain decomposition and Multigrid approaches opens up new possibilities for formulating efficient parallel algorithms. New theoretical findings on overconvergent Monte-Carlo methods also encourage the use of new parallel computers for solving complex physical and technical problems of real-world computing.

The book contains refereed and edited versions of the papers presented at the Third International Conference on Numerical Methods and Applications to be held in Sofia, Bulgaria, during 21–26 August 1994. The conference was organized by the Bulgarian Academy of Sciences (B A S), Sofia, Bulgaria. The University of Sofia, Bulgaria is a co-organizer. The book presents and discusses advances in the numerical methods and applications.

The conference follows the traditions of the First and Second conferences on Numerical Methods and Applications held in Sofia in 1984 and 1988. The subjects of the conference range from basic research to applications in physics, mechanics, engineering, environmental science and other areas, including the following topics:

- Numerical Methods (Finite Difference, Finite Element and Boundary Element Methods; Numerical Methods of Approximation Theory; Monte-Carlo Methods; Preconditioning Methods);
- Parallel Algorithms;
- Applications of Numerical Methods.

I am grateful to all the authors for their contributions and in particular to the members of the International Program Committee for their help.

The International Conference on Numerical Methods and Applications, Sofia, 21–26 August and the present volume are the outcome of the joint effort of many colleagues from various institutes and organizations, among whom I would like to mention academicians Bl. Sendov, Dr. P. Vassilevski, Dr. S. Margenov, Dr. K. Georgiev from the Center for Informatics and Computer Technology of B A S, Prof. Z. Zlatev from the National Environmental Research Institute, Roskilde, Denmark, Prof. M. Kaschiev from the Institute of Mathematics of the B A S, Prof. B. Boyanov from the University of Sofia, Prof. S. Radev from the Institute of Mechanics of B A S.

The hard work related to the every-day technical and scientific organization has been shared by my close associates: A. Karaivanova, T. Gurov, S. Fidanova, P. Marinov and I. Lirkov.

Special thanks are due to Anneta Karaivanova who took the responsibility of solving many organizing problems of the conference. Without her effort it would have been impossible to organize this conference.

Support for the conference from the EVRICA foundation, National Academic Foundation and St. Kiril i Methodii Foundation is gratefully acknowledged.

I hope that this volume will increase the awareness of the numerical methods amongst scientists and engineers.

Ivan Dimov
Chairman of the Organizing Committee, NM&A 94
Sofia, June, 1994.

Contents

II Parallel Algorithms

III Applications of Numerical Methods

Part I

Numerical Methods

Computer-Aided Interval Interpolation

Y. Akyildiz, S. Markov*, E. Popova* and J. Schulze

Abstract

We consider interpolation involving interval input data for the dependent variable using interpolating functions which are linear in the parameters. Computer-aided modelling involving uncertain data is demonstrated using the computer-algebra system *Mathematica*.

1 Introduction

A natural setting for practical interpolation problems especially when arising from biology should be based on the assumption that certain input values are given in the form of intervals [6]. A further assumption is that the modelling functions are generalized polynomials, that is they are linear in the parameters.

The last version 2.2 of the computer-algebra system *Mathematica* [7] supports interval arithmetic [4] providing thus an easy access to such a software, a representation of the usefulness of interval methods and a possibility for combining computer-algebraic and interval-analytical methods. The ability to realize arbitrary functions and expressions easily and to manipulate them in different forms make *Mathematica* suitable for the development of a system for computer modelling powerful enough for exploration, validation and documentation of a model under consideration.

The main purpose of the present paper is to demonstrate the possibilities of Mathematica 2.2 to compute and visualize interval generalized polynomials in relation to a particular interval interpolation problem. In the next section we briefly formulate the problem, as it has been discussed in [5]. In section 3 we present some numerical experiments obtained with the program system *Mathematica* 2.2. We also include some basic *Mathematica* code helping the reader to run his own examples.

2 An Interval Interpolation Problem

The problem can be formulated as follows. Given are

*This work has been partially supported by the National Foundation for Scientific Research under contract No MM10/91.

i) a class of generalized polynomials defined in $D \subseteq R^k$:

$$\eta(\lambda; \xi) = \sum_{i=1}^{m} \lambda_i \varphi_i(\xi) = \varphi(\xi)^{\mathsf{T}} \lambda, \quad \xi \in D, \tag{1}$$

where $\varphi(\cdot) = (\varphi_1(\cdot), \ldots, \varphi_m(\cdot))^{\mathsf{T}}$ is a Chebishev system of m continuous functions on D and $\lambda = (\lambda_1, \ldots, \lambda_m)^{\mathsf{T}} \in R^m$ is a vector of m unknown parameters;

ii) $n \geq m$ input data $x_j \in D$, $j \in J = \{1, \ldots, n\}$, such that $x_i \neq x_j$, $i \neq j$, and n interval measurements $Y_j = [y_j^-, y_j^+]$, $j \in J$, which contain the correct values of the corresponding measured quantities.

Consider the problem of finding the envelope of the set of functions η defined by (1) which interpolate (pass through) the vertical segments (x_j, Y_j), $j \in J$, whenever this set is not empty. For a fixed $\xi \in D$ denote the set of values at ξ of all functions η interpolating the segments (x_j, Y_j), $j \in J$ by

$$\{\eta = \eta(\lambda; \xi) \mid \eta \text{ is such that } \eta(\lambda; x_j) \in Y_j,\ j \in J\}, \quad \xi \in D. \tag{2}$$

The above set will be denoted by $\eta(\mathbf{x}, Y; \xi)$, with $\mathbf{x} = (x_1, x_2, \ldots, x_n)^{\mathsf{T}} \in R^{n \times k}$, $Y = (Y_1, \ldots, Y_n)^{\mathsf{T}} \in IR^n$, where IR^n is the set of n-dimensional interval vectors [1]. Formula (2) defines an interval-valued function $\eta(\mathbf{x}, Y; \xi)$ on D.

The requirement that the values of η at x_j range in the corresponding intervals Y_j leads to a system of inequalities for λ

$$\eta(\lambda; x_j) = \varphi(x_j)^{\mathsf{T}} \lambda \in Y_j, \quad j \in J, \tag{3}$$

which can be written in matrix form as $\Phi(\mathbf{x})\lambda \in Y$, where $\Phi(\mathbf{x})$ is the following $(n \times m)$-matrix of full rank:

$$\Phi(\mathbf{x}) = \begin{pmatrix} \varphi_1(x_1) & \ldots & \varphi_m(x_1) \\ \vdots & \ddots & \vdots \\ \varphi_1(x_n) & \ldots & \varphi_m(x_n) \end{pmatrix} = \begin{pmatrix} \varphi(x_1)^{\mathsf{T}} \\ \vdots \\ \varphi(x_n)^{\mathsf{T}} \end{pmatrix}. \tag{4}$$

The data $\mathbf{x}$ and Y are known, and the parameter vector λ is unknown. Thus (3) is a system of n algebraic inclusions for the m-dimensional parameter λ. Every parameter λ satisfying the system generates a solution function $\eta(\lambda; \cdot)$. The set $\eta(\mathbf{x}, Y; \xi)$ defined by (2) is an interval for any fixed $\xi \in D$. Thus (2) defines an interval-valued function (briefly, interval function) on D. We are concerned with the computation of the boundary lower and upper functions $\eta^-(\mathbf{x}, Y; \cdot)$, $\eta^+(\mathbf{x}, Y; \cdot)$ of the interval function (2). The latter are also called *enveloping functions* of the family of modeling functions interpolating $(\mathbf{x}, Y)$.

Proposition 1 *For $m = n$ we have (see [3], [5])*

$$\eta(\mathbf{x}, Y; \xi) = \left(\varphi(\xi)^{\mathsf{T}} \Phi^{-1}(\mathbf{x})\right) Y. \tag{5}$$

If Y is a degenerated interval vector, that is a numeric (real-valued) vector of the form $Y = y = (y_1, ..., y_n)$, then (5) is the well-known generalized interpolation polynomial for the data (x, y) in Lagrange form. In the nondegenerate case we call (5) an *interval generalized interpolation polynomial*, or briefly a *sausage-function* (because of its form; see Fig.1 - Fig.4).

Proposition 1 says that the sausage-function can be computed for every ξ by multiplying the real-valued vector $\varphi(\xi)^{\mathsf{T}}\Phi^{-1}(\mathbf{x})$ and the interval vector Y in the sense of interval arithmetic [1]. It can be efficiently computed by any programming system supporting matrix, vector and interval computations.

We now consider the case $m < n$. Let the index set Q be a subset of the index set J with m elements: $Q \subseteq J$, $card(Q) = m$. Assume that Q is ordered in increasing order and let $q(i)$ be the i-th element of Q. Denote by $\mathbf{x}^Q = \left(x_{q(1)}, \ldots, x_{q(m)}\right)^{\mathsf{T}}$ the matrix $\mathbf{x}$ reduced to the index set Q. Analogously $Y^Q = \left(Y_{q(1)}, \ldots, Y_{q(m)}\right)^{\mathsf{T}}$ is the vector Y reduced to Q.

To find the set of functions interpolating a reduced set of m data $(\mathbf{x}^Q, Y^Q)$, we consider the corresponding system $\Phi(\mathbf{x}^Q)\lambda \in Y^Q$, which is a system of m algebraic inclusions for m unknown parameters and applying (5), obtain the sausage function $\eta(\mathbf{x}^Q, Y^Q; \xi) = \left(\varphi(\xi)^{\mathsf{T}}\Phi^{-1}(\mathbf{x}^Q)\right) Y^Q$ corresponding to the data $(\mathbf{x}^Q, Y^Q)$.

Proposition 2 *[5] The value of $\eta(\mathbf{x}, Y; \cdot)$ at a point ξ is given by*

$$\eta(\mathbf{x}, Y; \xi) = \bigcap_{Q \subseteq J} \eta(\mathbf{x}^Q, Y^Q; \xi) = \bigcap_{Q \subseteq J} (\varphi(\xi)^{\mathsf{T}}\Phi^{-1}(\mathbf{x}^Q))Y^Q. \tag{6}$$

Proposition 2 says that the value of $\eta(\mathbf{x}, Y; \cdot)$ at ξ can be determined by an intersection the values at ξ of $\binom{n}{m}$ sausage-functions.

3 Numerical examples using *Mathematica*

We shall next demonstrate the computation and visualization of the expressions given by Propositions 1 and 2 by means of the computer-algebra system *Mathematica* 2.2.

We first consider the one-dimensional case $k = 1$.

In our first example the basic functions are polynomials of $(i-1)$-th degree of the form $\varphi_i(x) = x^{i-1}$, $i = 1, ..., m$. We take nine uniformly distributed mesh-points in the interval $[-1, 1]$ and constant values for the measurement intervals $Y_i = [-0.5, 0.5]$. Below we give a *Mathematica* code for computation and visualization of the resulting sausage function defined by (5).

The code is self-explanatory. Inputs *In[1]-In[3]* enter the values of n and m, the mesh x and the interval vector Y. *In[4]* enters the basic functions. *In[5]* defines a *Mathematica* function "sausage" which computes the value of the sausage function $\left(\varphi(\xi)^{\mathsf{T}}\Phi^{-1}(\mathbf{x}^Q)\right) Y^Q$ at ξ for a particular index set Q. In our case we have $m = n$,

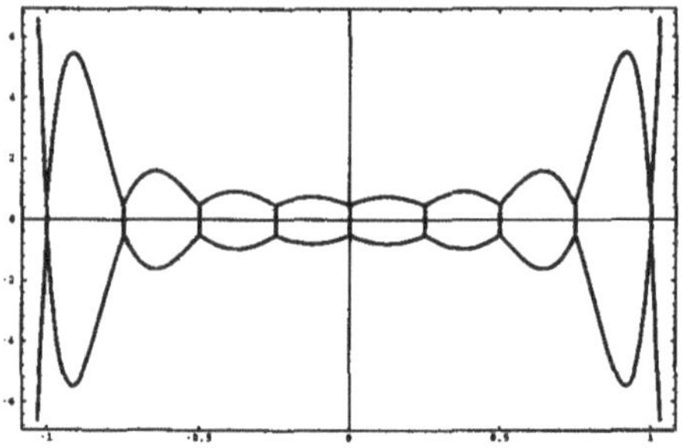

Figure 1: Uniform Mesh

resp. $Q = J$. *In[6]* plots the vertical segments (x, Y). *In[7]* graphs the interval solution function, which in this case is the sausage function. Both the segments and the sausage function are displayed by *Show* (see *In[8]*).

$In[1] := n = 9; m = 9;$

$In[2] := Do[x[i] = -1 + (2 * (i-1))/(n-1), \{i, n\}];$

$In[3] := Do[y[i] = Interval[\{-2\hat{}(-1), 1/2\}], \{i, n\}];$

$In[4] := Do[fi[i, t\_] = t\hat{}(i-1), \{i, m\}];$

$In[5] := sausage[ksi\_, set\_] := (fi[\#1, ksi]\&)/@Range[m].$
$\quad Inverse[Table[fi[i, x[set[[j]]]], \{j, m\}, \{i, m\}]].$
$\quad (z[\#1]\&)Range[m]/.Table[z[i]-> y[set[[i]]], \{i, m\}];$

$In[6] := sampleDraw := Module[\{segment\}, segment[x\_, y\_] :=$
$\quad Line[\{\{x, Min[y]\}, \{x, Max[y]\}\}];$
$\quad Graphics[\{(segment[x[\#1], y[\#1]]\&)/@Range[n]\}]];$

$In[7] := iplot[y\_, r\_, opt\_\_] :=$
$\quad Plot[\{Min[y], Max[y]\}, r, AspectRatio-> 1, opt];$

$In[8] := Show[sampleDraw, iplot[sausage[ksi, Range[n]], \{ksi, -1.03, 1.03\},$
$\quad DisplayFunction-> Identity],$
$\quad DispalyFunction :> \$DispalayFunction,$
$\quad Frame-> True, Axes-> True,$
$\quad PlotRange-> \{-6, 6\},$
$\quad FrameLabel-> \{"Fig.1.UniformMesh", " "\}];$

The Chebishev mesh minimizes the maximum of the width of the interval solution function on $[-1, 1]$. This is a special case of the problem of planning of experiment under interval data (see [2]).

Fig. 3 presents the sausage function for the uniform mesh (same as in Fig. 1) and the interval "step" data $Y_i = Y(x_i) = \{[-1-\varepsilon, -1+\varepsilon]$, if $x_i < 0$; $[-1, 1]$, if $x_i = 0$;

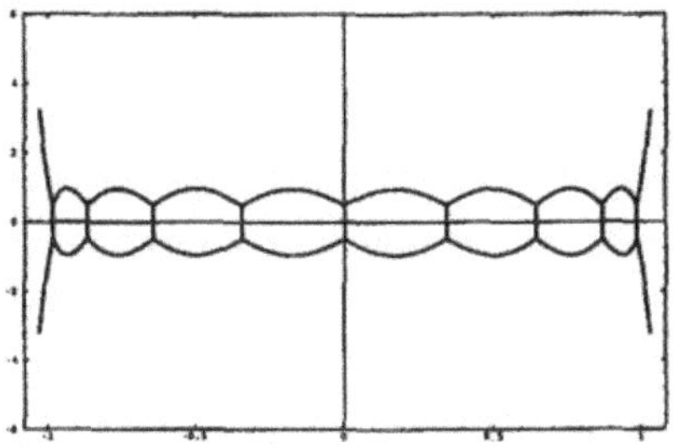

Figure 2: Chebishev Mesh

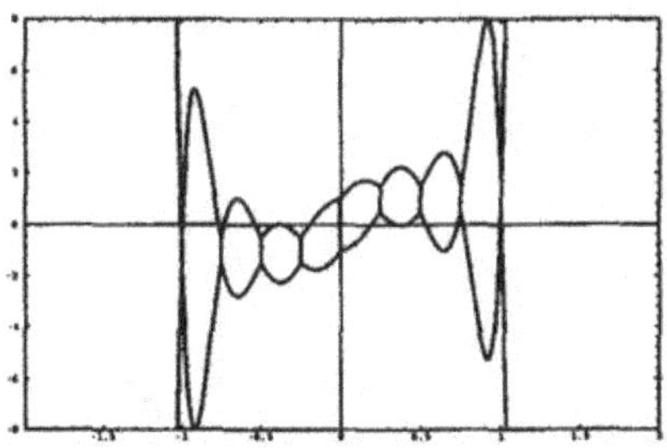

Figure 3: Step Function; Uniform Mesh

$[1-\varepsilon, 1+\varepsilon]$, if $x_i > 0\}$, $\varepsilon = 0.5$.

The effect of exploding of the interpolation error when the argument moves from the center of the domain comprising the mesh points towards outside this domain is clearly seen on Fig. 1 and Fig. 3. As we have seen (Fig. 2) this effect can be controlled by using special mesh points. This effect can be also diminished by considering the interpolation problem with $m < n$ parameters using Proposition 2. Fig. 4 presents the envelop of the set of all interpolating polynomials of degree 7.

Our last example is two-dimensional. The mesh points are three: $(0,0)$, $(0,1)$, $(1,0)$ and the interval values are constant: $[1,2]$. The basic functions are polynomial: $\phi_1(\xi_1,\xi_2) = 1, \phi_2(\xi_1,\xi_2) = \xi_1, \phi_3(\xi_1,\xi_2) = \xi_2$. The graph is presented on Fig. 5.

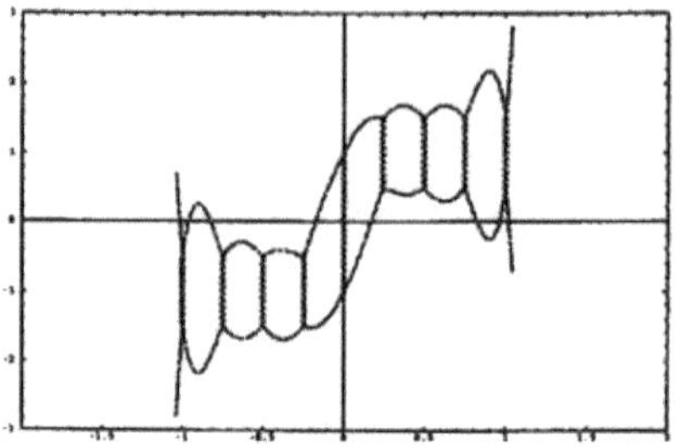

Figure 4: Step Function; $m = 7$

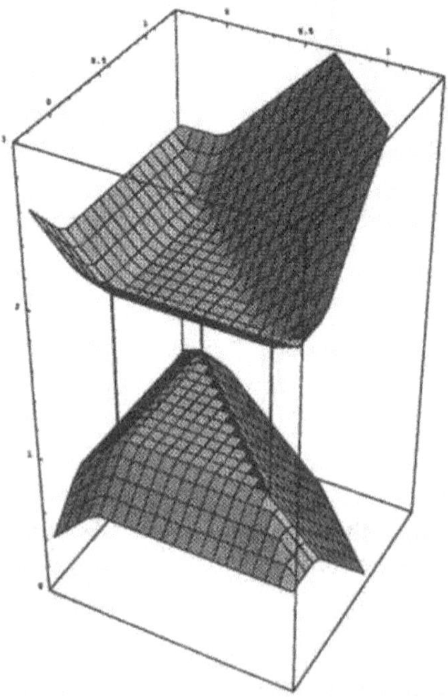

Figure 5: Two-Dimensional Case; $m = n = 3$

4 Conclusion

Our interval approach to interpolation problems can be very useful in certain modelling situations. Suppose that the experimentally obtained measurement intervals Y contain the true values of the measured quantities and that we have computed the interval solution function $\eta(x, Y; \cdot)$ from these measurements under the assumption that the model function η belongs to the class (1). Suppose further that a *new* experiment provides us with a *new* measurement (x_N, Y_N) such that $\eta(x, Y; x_N) \cap Y_N = \emptyset$. The only conclusion then can be that the chosen class of model functions is inadequate for the description of the experimental data.

Therefore our approach and programing techniques can be used from experimental scientists, for checking hypotheses with respect to the type of the modeling functions. A next task is to design a program package generating in a user friendly way interval solution functions corresponding to given interval experimental data. If such solution set does not exist in particular cases then the package will offer to the user a variety of classes of modeling function (possibly involving more parameters or other basic functions). The package will provide tools for solving related problems like choosing suitable mesh-points (planing of experiment), choosing certain individual solution satisfying additional extremal properties or additional restrictions etc. *Mathematica* seems to provide a suitable environment for the development of such a package. Extensive graphics capabilities of *Mathematica* allow the user to generate two- and three-dimensional graphics, including contour graphics and density plots which can be useful in the process of mathematical modelling.

References

[1] Alefeld, G.; Herzberger, J.: *Introduction to Interval Computations*, Academic Press, 1983.

[2] Dyvak, N. P.: *Design of Saturated Experiment in Interval Model Building*, Proc. Int. Conf. on Interval and Stochastic Methods in Science and Engineering (INTERVAL-92), 1992, Moscow, v. I (Ed. A. P. Voschinin), Publ. House of the Moscow Power Eng. Inst. (in Russian).

[3] Herzberger, J.: *Note on a Bounding Technique for Polynomial Functions*, SIAM J. Appl. Math., **34**, No.4, 1978.

[4] J. Keiper: *Interval Arithmetic in Mathematica*, Proc. of Int. Conf. on Numerical Analysis with Automatic Result Verification, Lafayette, March 1993, to appear in Interval Computations.

[5] Markov, S.; Popova, E.; Schneider, U.; Schulze, J.: *On Linear Interpolation under Interval Data*, Submitted to Mathematics and Computers in Simulation, 1993.

[6] Milanese, M.: *Estimation and Prediction in the Presence of Unknown but Bounded Uncertainty: A Survey.* In: Milanese, M.; Tempo, R. and Vicino, A. (Eds.): *Robustness in Identification and Control.* Plenum Press, New York, 1989, pp. 3–24.

[7] Wolfram, S.: *Mathematica: A System for Doing Mathematics by Computer*, Second Edition. Addison-Wesley, 1991.

Y. Akyildiz
Mathematics Department
Bosphorus University
80815 Bebek
Istanbul
Turkiye

S.M. Markow & E.D. Popova
Division of Mathematical Modelling
Central Laboratory of Biophysics
Bulgarian Academy of Science
Acad. G. Bonchev str., bldg. 21
1113 Sofia, Bulgaria

J. Schulze
Breite Str. 33/34
Luckenwalde
Germany 14943

Lumped Mass Approximation for the Mixed Finite Element Method

A.B. Andreev and T.T. Dimov*

Abstract

Lumped mass approximation for the mixed finite element solution of the homogeneous Dirichlet problem on rectangular domain is considered. Optimal error estimate for this type approximation is derived.

1 Introduction

Let $\Omega \subset \mathbb{R}^2$ be a bounded domain with boundary $\partial\Omega$. We consider the homogeneous Dirichlet problem:

$$\left| \begin{array}{l} -\mathrm{div}(a(x)\underline{\nabla} p) = f(x), \text{ in } \Omega \\ p = 0, \text{ on } \partial\Omega \end{array} \right. \tag{1}$$

where $\underline{\nabla} w$ denotes the gradient of a scalar function w, $div\underline{v}$ denotes the divergence of vector function $\underline{v}$ and $a(x)$ is diagonal matrix which elements satisfy the requirements $a_i(x) \geq a_0 > 0,\ i = 1, 2$. We set

$$\underline{u} \equiv (u_1, u_2) = a(x)\underline{\nabla} p, \alpha_i(x) = a_i(x)^{-1}, i = 1, 2.$$

We consider the spaces $\underline{V}$ and W defined by

$$\begin{aligned} \underline{V} &= \underline{H}(div; \Omega) = \{\underline{v} \in L^2(\Omega)^2 \,:\, div\underline{v} \in L^2(\Omega)\}, \\ W &= L^2(\Omega) \end{aligned}$$

provided with the norms

$$\begin{aligned} \|\underline{v}\|_{\underline{V}} &\equiv \|\underline{v}\|_{\underline{H}(div;\Omega)} = \left(\|\underline{v}\|_{0,\Omega}^2 + \|div\underline{v}\|_{0,\Omega}^2\right)^{1/2} \text{ and} \\ \|w\|_W &= \|w\|_{L^2(\Omega)} = \|w\|_{0,\Omega} \end{aligned}$$

respectively.

---

*This work was supported in part by the Ministry of Science and Education of Bulgaria under grants MU MSA-3/94 and MU 20-IN/93

Then the mixed variational formulation of the problem (1) is given by characterizing the pair $(\underline{u}, p)$, as the solution of

$$\left| \begin{array}{ll} a(\underline{u},\underline{v}) + b(\underline{v},p) = 0, & \forall \underline{v} \in \underline{V}; \\ b(\underline{u},w) = -(f,w), & \forall w \in W, \end{array} \right. \tag{2}$$

where

$$a(\underline{u},\underline{v}) = (\alpha u_1, v_1) + (\alpha u_2, v_2), \quad b(\underline{u},w) = (div\underline{u}, w) \tag{3}$$

and $(\cdot,\cdot)$ indicated the inner product in $L^2(\Omega)$.

Optimal error estimate of the mixed finite element solution $(\underline{u}, p)$, using Raviart-Thomas (RT) spaces have been established in [4] and [5]. Superconvergence along the Gauss lines for this type of approximation has been proved in [7]. Effects of numerical integration in the mixed finite element method have been discussed in [8].

The diagonalization of the mass matrix (lumped mass approximation) leads to more efficient algorithm for solving the finite element system. In the finite element method this phenomena have been analyzed in [1], [2], [3]. The purpose of the present paper is to construct lumped mass approximation having the optimal rate of convergence of the mixed finite element solution. In our approach the lumping of the mass matrix results from the use of Gauss-Lobatto quadrature formula for the integrals over rectangular RT finite elements of index r.

2 Mixed finite element approximation

We shall use Sobolev space of functions $H^m(\Omega)$ with the norm

$$\|w\|_{m,\Omega} = \left(\sum_{i=0}^{m} |w|_{i,\Omega}^2 \right)^{1/2},$$

where

$$|w|_{i,\Omega} = \left(\int_\Omega \sum_{|\alpha|=i} |D^\alpha w|^2 \, dx \right)^{1/2}.$$

For the sake of simplicity we take the domain Ω to be the square$(0,1)^2$. The results that we derive in this paper can be extended to more general domains.

Let D_1 and D_2 be partition of $I = [0,1]$:

$$\begin{aligned} D_1 &= \{0 = x_{1,0} < x_{1,1} < \ldots < x_{1,N} = 1\} \\ D_2 &= \{0 = x_{2,0} < x_{2,1} < \ldots < x_{2,M} = 1\} \end{aligned}$$

Denote

$$h_{i-1} = x_{1,i} - x_{1,i-1}, \; i = 1, \ldots N, \quad h'_{j-1} = x_{2,j} - x_{2,j-1}, \; j = 1, \ldots M$$

and $h = \max_{ij}(h_i, h'_j)$.

Thus the rectangle $\overline{\Omega} = [0,1]^2$ can be partitioned by the Cartesian product of D_1 and D_2:

This partition τ_h is defined by

$$\tau_h = \{T \equiv T_{ij} \subset \Omega : T_{ij} = [x_{1,i}-x_{1,i-1}] \times [x_{2,j}-x_{2,j-1}],\ i = 1,\dots,N,\ j = 1,\dots,M\}.$$

We suppose that the partition t_h is regular [6], i.e. there exist constants C_1 and C_2 such that

$$C_1 h^2 \le \text{meas} T \le C_2 h^2 \quad \text{for all } T \in \tau_h\,.$$

Denote the following polynomials of two variables

$$\begin{aligned} P(n) &= \{p(x_1,x_2) : p(x_1,x_2) = \sum_{0\le i+j\le n} a_{ij} x_1^i x_2^j\} \\ Q(n,m) &= \{q(x_1,x_2) : q(x_1,x_2) = \sum_{i=0}^{n}\sum_{j=0}^{m} a_{ij} x_1^i x_2^j\}\,. \end{aligned}$$

Define the Raviart-Thomas spaces [4] $\underline{V_h}$ and W_h of index r as follows:

$$\begin{aligned} \underline{V_h} &= \{\underline{v}_h \in V : \underline{v}_h \in Q(r+1,r) \times Q(r,r+1) \text{ on any } T \in \tau_h\}; \\ W_h &= \{w_h \in W : w_h \in Q(r,r) \text{ on any } T \in \tau_h\}. \end{aligned}$$

Introduce nodal bases for these spaces. Consider the Gauss and Lobato points in the interval $[-1,1]$ as follows:

$$\begin{aligned} \hat{G} &= \{\hat{g}_i : L_{r+1}(\hat{g}_i) = 0,\ i = 1,\dots,r+1\} \\ \hat{L} &= \{\hat{l}_j : L'_{r+1}(\hat{l}_j) = 0,\ j = 1,\dots,r,\ l_0 = -1,\ l_{r+1} = 1\}, \end{aligned}$$

where $L_{r+1}(\xi)$ is the Legendre polynomial of degree $r+1$.

Denote the reference element $\hat{T} = [-1,1]^2$ and the spaces $\underline{\hat{V}}$ and $\hat{W}$ defined on $\hat{T}$:

$$\begin{aligned} \underline{\hat{V}}(\hat{x}_1,\hat{x}_2) &= \{\underline{\hat{v}} \in V : \underline{\hat{v}} \in Q(r+1,r) \times Q(r,r+1) \text{ on } \hat{T}\}; \\ \hat{W}(\hat{x}_1,\hat{x}_2) &= \{\hat{w} \in W : \hat{w} \in Q(r,r) \text{ on } \hat{T}\}. \end{aligned}$$

Then the function $\underline{\hat{u}} \in \underline{\hat{V}}$ is uniquely determined by the values of the points $\hat{S}_1 \times \hat{S}_2$ and the function $w \in W_h$ at points $\hat{S}_0$, where

$$\begin{aligned} \hat{S}_0 &= \{(\hat{g}_i,\hat{g}_j),\ i,j = 1,\dots,r+1\}; \\ \hat{S}_1 &= \{(\hat{l}_j,\hat{g}_i),\ j = 0,\dots,r+1,\ i = 1,\dots r+1\}; \\ \hat{S}_2 &= \{(\hat{g}_i,\hat{l}_j),\ i = 1,\dots,r+1,\ j = 0,\dots,r+1\}. \end{aligned}$$

Let $\hat{w}$ and $\underline{\hat{v}} \equiv (\hat{v}_1,\hat{v}_2)$ be continuous function defined on $\hat{T}$. Their interpolants $\hat{w}^I, \hat{v}_1^I, \hat{v}_2^I$ are polynomials in $Q(r,r), Q(r+1,r)$ and $Q(r,r+1)$ respectively.

Obviously,

$$\hat{w}^I \in \hat{W};\ \hat{\underline{v}}^I \equiv (\hat{v}_1^I, \hat{v}_2^I) \in \hat{\underline{V}}. \tag{4}$$

For each element $T \in \tau_h$ there exists the unique affine invertible mapping F_T : $\hat{T} \mapsto T$ such that

$$\hat{x} \mapsto x = F_T(\hat{x}) = B_T\hat{x} + \underline{b_T},$$

where $B_T \in \mathbb{R}(2 \times 2)$, $\underline{b_T} \in R(2 \times 1)$ and $\det B_T = \frac{1}{4}\text{meas}(T)$.

In what follows for any scalar function w on T we set

$$\hat{w}(\hat{x}) = w(x),\ x = F_T(\hat{x}),\ \forall \hat{x} \in \hat{T}. \tag{5}$$

The mixed finite element approximation of (1) is defined by determining a pair $(\underline{u_h}, p_h) \in \underline{V_h} \times W_h$ such that

$$\left| \begin{array}{ll} a_h(\underline{u_h}, \underline{v_h}) + b(\underline{v_h}, p_h) = 0, & \forall \underline{v_h} \in \underline{V_h}; \\ b(\underline{u_h}, w_h) = -(f, w_h), & \forall w_h \in W_h\,. \end{array} \right. \tag{6}$$

Optimal error estimate for (2) is given in [4], i. e.

$$\|\underline{u} - \underline{u_h}\|_{\underline{H}(div;\Omega)} + \|p - p_h\|_{0,\Omega} \le C\, h^{r+1}\|p\|_{r+3}\,.$$

where C is a positive constant.

In this paper, C denotes a generic positive number, taking different values at different occurrences depending only on a_i, $i = 1,2$ but it is independent of h.

3 Lumped mass approximation using numerical integration

We evaluate the bilinear form $a(\cdot,\cdot)$ in (2) with $a_h(\cdot,\cdot)$ using numerical integration. Since the integration is performed elementwise, it is enough to define it on the reference finite element $\hat{T}$.

We put

$$\hat{a}(\hat{\underline{u}}, \hat{\underline{v}}) = \hat{a}_1(\hat{u}_1, \hat{v}_1) + \hat{a}_2(\hat{u}_2, \hat{v}_2) = \int_{\hat{T}} \hat{\alpha}_1\hat{u}_1\hat{v}_1\, d\hat{x} + \int_{\hat{T}} \hat{\alpha}_2\hat{u}_2\hat{v}_2\, d\hat{x}, \tag{7}$$

$$\hat{a}_h(\hat{\underline{u}}, \hat{\underline{v}}) = \hat{a}_{h_1}(\hat{u}_1, \hat{v}_1) + \hat{a}_{h_2}(\hat{u}_2, \hat{v}_2), \tag{8}$$

where

$$\hat{a}_{h_1}(\hat{u}_1, \hat{v}_1) = \sum_{j=0}^{r+1}\sum_{i=1}^{r+1} \hat{\omega}_{ij}\hat{\alpha}_1\hat{u}_1\hat{v}_1(\hat{l}_j, \hat{g}_i),\ (\hat{l}_j, \hat{g}_i) \in \hat{S}_1,\ \hat{\omega}_{ij} > 0,$$

$$\hat{a}_{h_2}(\hat{u}_2, \hat{v}_2) = \sum_{i=1}^{r+1}\sum_{j=0}^{r+1} \hat{\omega}_{ij}\hat{\alpha}_2\hat{u}_2\hat{v}_2(\hat{g}_i, \hat{l}_j),\ (\hat{g}_i, \hat{l}_j) \in \hat{S}_2,\ \hat{\omega}_{ij} > 0.$$

From the results for the one-dimensional case it is easy seen that this quadrature formulas are exact for polynomials of $Q(2r+1, 2r+1)$.

We define $a_h(\cdot,\cdot)$ according to (8) and (5). Note that the considered partition and quadrature formulas give a diagonal mass matrix.

We replace the problem (2) by the following problem: find $(\underline{u}_h^\star, p_h^\star) \in \underline{V_h} \times W_h$ satisfying

$$\left| \begin{array}{ll} a_h(\underline{u}_h^\star, \underline{v}_h) + b(\underline{v}_h, p_h^\star) = 0, & \forall \underline{v}_h \in \underline{V_h}; \\ b(\underline{u}_h^\star, w_h) = -(f, w_h), & \forall w_h \in W_h\,. \end{array} \right. \tag{9}$$

Lemma 1. *There exist constants C_1 and C_2 such that*

$$a_h(\underline{u}_h, \underline{v}_h) \leq C_1 \|\underline{u}_h\|_{\underline{V}} \|\underline{v}_h\|_{\underline{V}}, \ \forall \underline{u}_h \in \underline{V_h}, \ \forall \underline{v}_h \in \underline{V_h}\,. \tag{10}$$

$$a_h(\underline{v}_h, \underline{v}_h) \geq C_2 \|\underline{v}_h\|^2_{\underline{V}}\,, \quad \forall \underline{v}_h \in \underline{V}_h^\star\,, \tag{11}$$

where

$$\underline{V}_h^\star = \{\underline{v}_h \in \underline{V_h} : (\mathrm{div}\underline{v}_h, w_h) = 0, \quad \forall w_h \in W_h\}\,.$$

Proof. Let $a: \underline{V} \times \underline{V} \mapsto \mathbb{R}$ be bounded bilinear form given by (3)

We recall that $\underline{v}_h = (v_{1,h}, v_{2,h})$ and $(\cdot,\cdot)_h$ is the approximation of the inner product using the quadrature formulas according to (8) and (5) giving the lumped mass.

First we prove that the mapping

$$\hat{v}_1 \in Q(r+1, r) \ \mapsto \ (\hat{a}_{h_1}(\hat{v}_1, \hat{v}_1))^{1/2}$$

is a norm on $Q(r+1, r)$. It is sufficient to note that $\hat{v}_1 = 0$ in $(r+2) \times (r+1)$ points implies that $\hat{v}_1 = 0$, since the set of Lobatto-Gauss points is $Q(r+1, r)$- unisolvent on $\hat{T}$. From the equivalence of all norms on the finite-dimensional space we infer the existence of constants $\hat{C}_1, \hat{C}_2 > 0$, such that

$$\hat{C}_1\, \hat{a}_{h_1}(\hat{v}_1, \hat{v}_1) \leq \|\hat{v}_1\|^2_{0,\hat{T}} \leq \hat{C}_2\, \hat{a}_{h_1}(\hat{v}_1, \hat{v}_1)$$

Taking into account

$$a_{h_1,T}(v_1, v_1) = (\det B_T)\, \hat{a}_{h_1}(\hat{v}_1, \hat{v}_1)$$

and

$$\|w\|_{0,T} = (\det B_T)^{1/2} \|\hat{w}\|_{0,\hat{T}}, \quad \forall w \in L^2(T),$$

we get

$$C_3\, a_{h_1,T}(v_1, v_1) \leq \|v_1\|^2_{0,T} \leq C_4\, a_{h_1,T}(v_1, v_1)$$

A similar argument gives the corresponding relation for v_2 and $a_{h_2,T}(v_2, v_2)$.

Using the following relations for all $\underline{v} \in (L^2(\Omega))^2$

$$\|\underline{v}\|_{0,T}^2 = \|v_1\|_{0,T}^2 + \|v_2\|_{0,T}^2$$

$$\|v_{1,h}\|_{0,\Omega}^2 = \sum_{T\in\tau_h} \|v_{1,h}\|_{0,T}^2$$

$$a_h(\underline{v},\underline{v}) = \sum_{T\in\tau_h} a_{h,T}(\underline{v},\underline{v})$$

we obtain

$$C_5(a_h(\underline{v_h},\underline{v_h})) \le \|\underline{v_h}\|_{0,\Omega}^2 \le C_6(a_h(\underline{v_h},\underline{v_h})) \tag{12}$$

We note that for any $\underline{u_h},\underline{v_h} \in \underline{V_h}$

$$a_h(\underline{u_h},\underline{v_h}) \le C\,(a_h(\underline{u_h},\underline{u_h}))^{1/2}(a_h(\underline{v_h},\underline{v_h}))^{1/2} \tag{13}$$

and for any $\underline{v_h} \in \underline{V_h^*}$

$$\|\underline{v_h}\|_{0,\Omega} = \|\underline{v_h}\|_{\underline{V}} \tag{14}$$

since $\|\mathrm{div}\underline{v_h}\|_{0,\Omega} = 0$

The inequalities (10) and (11) follow from (12), (13) and (14).

Lemma 2. *Let $(\underline{u},p) \in V \times W$ be the solution of (2), $p \in H^{r+3}(\Omega)$ and $\underline{u}^I$ be the interpolant of $\underline{u}$ defined according to (4) and (5). Then there exists a number C_3 such that:*

$$\sup_{\underline{t_h}\in\underline{V_h}} \frac{a(\underline{u}^I,\underline{t_h}) - a_h(\underline{u}^I,\underline{t_h})}{\|\underline{t_h}\|_{\underline{V}}} \le C_3 h^{r+1}\|p|_{r+3,\Omega}$$

Proof. For $r \ge 0$ the following relation is satisfied: $H^{r+2}(\hat{T}) \hookrightarrow C(\hat{T})$. Then for all $\underline{t_h} \in \underline{V_h}$

$$|a(\underline{u}^I,\underline{t_h}) - a_h(\underline{u}^I,\underline{t_h})| \le \sum_{T\in\tau_h} |E_{1,T}(\alpha_1 u_1^I t_{1,h})| + |E_{2,T}(\alpha_2 u_2^I t_{2,h})|,$$

where

$$E_{i,T}(\alpha_i u_i^I t_{i,h}) = a_i(u_i^I,t_{i,h}) - a_{i,h}(u_i^I,t_{i,h}),\ i = 1,2.$$

More detailed but simple analysis shows that

$$|E_{i,T}(\alpha_i u_i^I t_{i,h})|_{0,T} \le C h^{r+1}\|u_i^I\|_{r+1,T}\|t_{i,h}\|_{0,T},\ i = 1,2 \tag{15}$$

and

$$\|u_i^I\|_{r+1,T} \le C\|u_i\|_{r+2,T},\ i = 1,2 \tag{16}$$

Then using Cauchy-Schwarz inequality and relation (15) and (16) we complete the proof.

The results of Lemma 1 give us possibility to apply the abstract error estimate [8] for the problem (9):

$$\|\underline{u} - \underline{u}_h^*\|_{\underline{V}} + \|p - p_h^*\|_W \leq$$

$$C\{ \inf_{\underline{v}_h \in \underline{V}_h} \|\underline{u} - \underline{v}_h\|_{\underline{V}} + \sup_{\underline{t}_h \in \underline{V}_h} \frac{a(\underline{v}_h, \underline{t}_h) - a_h(\underline{v}_h, \underline{t}_h)}{\|\underline{t}_h\|_{\underline{V}}} + \inf_{w_h \in W_h} \|p - w_h\|_W \}.$$

Then in the choice $\underline{v}_h = \underline{u}^I \in \underline{V}_h$ and Lemma 2 we derive the main result.

Theorem. *Let $(\underline{u}, p)$ be a solution of (2) and $(\underline{u}_h^*, p_h^*)$ be a solution of (9). Assume that $\underline{u} \in H^{r+2}(\Omega)$ and $p \in H^{r+3}(\Omega)$. Then there exists a constant C such that*

$$\|\underline{u} - \underline{u}_h^*\|_{\underline{H}(div;\Omega)} + \|p - p_h^*\|_{0,\Omega} \leq C\, h^{r+1} \|p\|_{r+3},$$

4 Matrix problem

We introduce vector notations U_1, U_2, P for the approximate unknown values $u_{1,h}^*, u_{2,h}^*, p_h^*$ respectively and V_1, V_2, W for $v_{1,h}, v_{2,h}, w_h$ respectively as follows:

$$\begin{aligned} a_h(\underline{u}_h^*, \underline{v}_h) &= \sum_{i=1}^{2} U_i^T M_i V_i \\ b(\underline{v}_h, p_h^*) &= \sum_{i=1}^{2} V_i^T N_i P \\ b(\underline{u}_h, w_h) &= \sum_{i=1}^{2} W^T N_i^T U_i \end{aligned}$$

Than the matrix form of the approximate solution will be:

$$\begin{pmatrix} M_1 & 0 & N_1 \\ 0 & M_2 & N_2 \\ N_1^T & N_2^T & 0 \end{pmatrix} \begin{pmatrix} U_1 \\ U_2 \\ P \end{pmatrix} = \begin{pmatrix} 0 \\ 0 \\ -F \end{pmatrix}$$

Lumped mass procedure presented above leads to diagonal matrices $M_i, i = 1, 2$. Then this problem can be easily reduced to the problem:

$$(N_1^T M_1^{-1} N_1 + N_2^T M_2^{-1} N_2) P = F,$$

where $U_i = -M_i^{-1} N_i P, i = 1, 2$.

References

[1] U. Banerjee, J. E. Osborn, *Estimation of the Effect of Numerical Integration in Finite Element Eigenvalue Approximation*, Numer. Math. **56**, 1990, pp. 735–762

[2] C. M. Chen, V. Thomée, *The Lumped Mass Finite Element Method for a Parabolic Problem*, J. Austral. Math. Soc. Ser. B **26**, 1985, pp. 329–354.

[3] A. B. Andreev, V. A. Kascieva, M. Vanmaele, *Some Results in Lumped Mass Finite Element Approximation of Eigenvalue Problems using Numerical Quadrature Formulas*, J. Comp. Appl. Math. **43**, 1992, pp. 291–311.

[4] P. A. Raviart, J. M. Thomas, *A Mixed Finite Element Method for Second Order Elliptic Problems*, Mathematical Aspects of the Finite Element Method, Lect. Notes Math. **606**, Springer Verlag, Berlin, 1977, pp. 292–315.

[5] J. Douglas Jr., J. E. Roberts, *Global Estimates for Mixed Methods for Second Order Elliptic Problems*, Math. Comp. **44**, 1985, pp. 39–52.

[6] P. G. Ciarlet, *Finite Element Methods for Elliptic Problems*, North-Holland, Amsterdam, 1978.

[7] R. E. Ewing, R. D. Lazarov, J. Wang, *Superconvergence of the Velocity along the Gauss Lines in Mixed Finite Element Methods*, SIAM J. Numer. Anal. **28**, 1991, pp. 1015–1029.

[8] J. E. Roberts, J. M. Thomas, *Mixed and Hybrid Methods, Handbook of Numerical Analysis*, North Holland, 1991.

A. Andreev
High Institute of Machine
and Electrical Engineering
H. Dimitar str.4
5300 Gabrovo
Bulgaria

T. Dimov
CICT - Bulg. Acad. of Sciences
Akad. G Bonntcev str. Bl. 25-A
1113 Sofia
Bulgaria

Time Integration of Free Boundary Problems in Fluid Mechanics

Ivan Bazhlekov* and Ludmil Zickatanov†

Abstract

An estimation of the error arising at the numerical investigation of unsteady free boundary problems in fluid mechanics is performed. It leads to the condition of the form $\Delta t = O(h)$ at which a numerically stable solution could be obtained. An iterative scheme for a time integration is proposed whose application might reduce the computational time.

1 Introduction

In many free boundary problems in fluid mechanics there are subdomains where the velocity changes very rapidly [1]. In order to guarantee a satisfactory approximation at the numerical simulation of such problems, it is necessary to use fine grids in these subdomains, e.g. in some cases the dimensionless space step must be of order 10^{-10} [2].

For unsteady free boundary problems it is shown numerically (see [3]) that the maximum value of time step Δt is limited by numerical unstability and decreases with decreasing space step h becoming prohibitively small in many cases. Hence the following questions exist:

(a) What is the reason of the numerical unstability mentioned above?

(b) What is the relation between the optimal time step and the space step?

(c) How can the time integration schemes be chosen or modified to provide reduction of the computational time?

In the next two sections we try to answer these questions considering a concrete problem formulated in Section 2. In Section 3 the error behaviour and a time integration scheme are discussed.

*This author is partially supported by Grant MM-6/91 with Bulgarian Ministry of Education and Science

†This author is partially supported by Grant MM-207/92 with Bulgarian Ministry of Education and Science

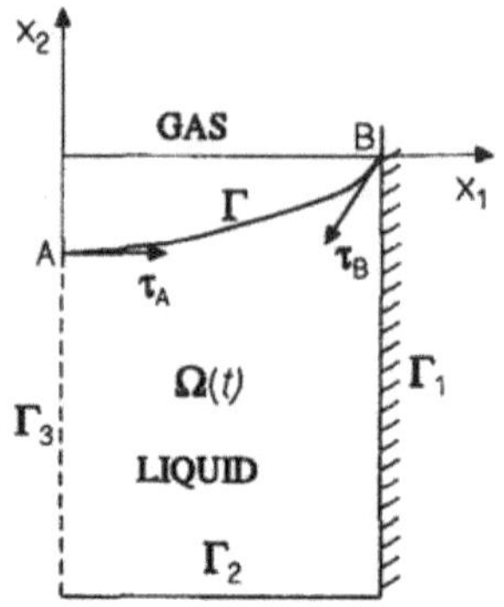

Figure 1: *Schematic diagram of a free boundary problem*

2 Formulation of the problem

Consider an incompressible viscous Newtonian liquid with constant density and viscosity occupying domain $\Omega(t)$, where the part Γ of the $\partial\Omega(t)$ is a gas–liquid free surface (see Fig. 1). For the sake of simplicity, we assume that the effects of gravity and inertia are negligible and the problem considered is two dimensional. These assumptions are not imposed by the techniques used in the next section. They make the analysis more understandable and do not restrict the applicability of the results obtained.

Also we suppose that there exists a unique sufficiently smooth solution (all norms used through the next section exist).

The problem is governed by the dimensionless creeping flow approximation of the Navier–Stokes equation as follows:

$$\nabla \cdot \mathbf{T}(\mathbf{x}, t) = 0, \quad \mathbf{x} = (x_1, x_2) \in \Omega(t) \tag{2.1}$$

and the continuity equation

$$\nabla \cdot \mathbf{u} = 0. \tag{2.2}$$

Here $\mathbf{T}$ is the stress tensor with components

$$T_{ij} = -p\delta_{ij} + \frac{1}{2}((u_i)_{,j} + (u_j)_{,i}), \tag{2.3}$$

where $(\cdot)_{,i}$ denotes the co–variant derivative with respect to x_i; δ_{ij} are the components of the unit tensor; p is the pressure and $\mathbf{u} = \begin{pmatrix} u_1 \\ u_2 \end{pmatrix}$ is the velocity of the liquid.

We suppose that the gas–liquid free surface $\Gamma(t)$ can be described by the equation

$$x_2 = S(x_1,t) \ ,x_1 \in I, I := [0, x_{\max}] \,, \tag{2.4}$$

where S is a bounded differentiable function with bounded first derivatives. The free surface evolution can be written in terms of standard kinematic condition.

$$\frac{\partial S(x_1,t)}{\partial t} = \mathbf{N}\cdot\mathbf{u}\,, \tag{2.5}$$

where $\mathbf{N}$ denotes the normal vector to Γ defined with the relation $\mathbf{N} = (-S_{x_1}, 1)$.

The balance of the stress and the capillary pressure on the gas–liquid free surface Γ is a non–homogenous Neumann boundary condition:

$$\mathbf{T}\cdot\boldsymbol{\nu} - p_g\cdot\boldsymbol{\nu} = \frac{1}{Ca}\frac{1}{R}\cdot\boldsymbol{\nu}. \tag{2.6}$$

Here Ca denotes the capillary number, R is the main radii of the curvature of Γ; $\boldsymbol{\nu} := \mathbf{N}/|\mathbf{N}|$–unit outward normal vector; p_g is the pressure in the gas and it is assumed to be a constant with respect to space variables. This kind of boundary condition is used in the free boundary problems in fluid mechanics.

The main goal of the presented investigation is concerned with the behaviour of the evolution of the numerical error.

As attention is concentrated on the boundary condition (2.6), the conditions on the other boundaries are written only for completeness and their effects on the accuracy and the approximation are not studied.

Therefore on the other boundaries we have:

- vanishing normal component of the fluid velocity and the tangential stress on the axis of symmetry ($\Gamma_3 :\ x_1 = 0$)

$$u_1 = 0 \text{ on } x_1 = 0 \tag{2.7}$$

$$T_{12} = 0 \text{ on } x_1 = 0\,; \tag{2.8}$$

- prescribed velocity at some distance from the free surface

$$\mathbf{u} = \mathbf{u}_d \text{ on } \Gamma_2\,; \tag{2.9}$$

- no–slip condition on the solid wall

$$\mathbf{u} = \mathbf{u}_w \text{ on } \Gamma_1\,, \tag{2.10}$$

  where $\mathbf{u}_w$ is the velocity of the wall.

In order to determine the free surface position the tangential vectors $\boldsymbol{\tau}$ at its ends have to be known:

- at the point A lying on the axis of symmetry

$$\boldsymbol{\tau}_A = (0,1)\,; \tag{2.11}$$

- at the wall

$$\tau_B = \tau_w \,, \tag{2.12}$$

where τ_w is known.

The initial conditions for the velocity and the free surface position are:

$$\mathbf{u}(\mathbf{x},t)|_{t=0} = \mathbf{u}_0(\mathbf{x}); \quad S(x_1,t)|_{t=0} = S_0(x_1)\,. \tag{2.13}$$

The problem is an unsteady one in the cases when the initial conditions (2.13) do not correspond to the steady solution or when the right–hand sides of (2.9), (2.10), (2.12) are time-dependent.

In the next section attention will be concentrated on the spreading of the numerical error between two joined time steps.

3 Error behaviour

For the numerical solving of the problem (2.1-13) a finite element discretization is used. After eliminating the pressure the following continuous problem is under consideration: Find $\mathbf{u} \in \mathbf{V}$ such that

$$a(\mathbf{u},\mathbf{v}) = f(\mathbf{v}) \text{ for all } \mathbf{v} \in \mathbf{V}\,, \tag{3.1}$$

where $a(\mathbf{u},\mathbf{v})$ is the continuous $\mathbf{V}$–elliptic bilinear form corresponding to the Laplace operator. For $f(\mathbf{v})$ we have

$$f(\mathbf{v}) = \int_{\Gamma(t)} \frac{1}{Ca}\frac{1}{R}\boldsymbol{\nu}\cdot\mathbf{v}d\gamma\,, \tag{3.2}$$

and $\mathbf{V} = \{\mathbf{v} \in (H^1_E(\Omega(t)))^2 \;:\; \operatorname{div}\mathbf{v} = 0\}$. As the problem is an unsteady one, a discrete form of (2.5) is introduced as follows (i.e. a space continuous problem which is discrete in time is considered):

$$\Gamma(t+\Delta t) \;:\quad S(t+\Delta t, x_1) := S(t,x_1) + \Delta t\boldsymbol{\nu}\cdot\mathbf{u}\,. \tag{3.3}$$

To discretize the problem (3.1-3) let $\Gamma_h(t)$ be an approximation of the boundary $\Gamma(t)$ (for instance such an approximation might be the piecewise quadratic interpolant of $\Gamma(t)$ for $t = 0$). Let us consider the partition $\mathcal{J}$ consisting of curve side quadrilaterals cover $\tilde{\Omega}(t) = \Omega(t)\cup\Omega_h(t)$, where $\Omega_h(t)$ is the obtained approximating domain. We suppose that the partition $\mathcal{J}_h$ of $\Omega_h(t)$ is a subset of $\mathcal{J}$. The partition $\mathcal{J}$ enables us to introduce a finite dimensional subspace $\mathbf{V}_h$ of $\mathbf{V}$ (finite element space) (see [4], [5]).

Now the projection of the solution of the problem (3.1) could be defined: Find $\tilde{\mathbf{u}}_h \in \mathbf{V}_h$ such that

$$a(\tilde{\mathbf{u}}_h,\mathbf{v}_h) = f(\mathbf{v}_h) \text{ for all } \mathbf{v}_h \in \mathbf{V}_h\,. \tag{3.4}$$

Following Strang and Fix [6] we consider an approximation of (3.1) in the form:

Find $\mathbf{u}_h \in \mathbf{V}_h$ such that

$$a_h(\mathbf{u}_h, \mathbf{v}_h) = f_h(\mathbf{v}_h) \text{ for all } \mathbf{v}_h \in \mathbf{V}_h . \tag{3.5}$$

The new bilinear form $a_h(\cdot,\cdot)$ is obtained from $a(\cdot,\cdot)$ in (3.4) but the integration is over $\Omega_h(t)$ instead of over $\Omega(t)$. The linear form $f_h(\cdot)$ reads as follows:

$$f_h(\mathbf{v}_h) = \int_{\Gamma_h(t)} \frac{1}{Ca}\frac{1}{R}\nu \cdot \mathbf{v}_h d\gamma + \omega_h(\mathbf{v}_h) . \tag{3.6}$$

Here $\omega_h(\mathbf{v}_h) := \sum_i \mathbf{v}_h(i)\tau(-i) - \mathbf{v}_h(i)\tau(+i)$ and this sum includes all grid points lying on $\Gamma_h(t)$. As the $S_h(t)$ has discontinuous derivatives between these grid points with $\tau(-i)$ and $\tau(+i)$ we denote the left and right unit tangential vector to $\Gamma_h(t)$ at node i respectively.
Finally the equation (3.3) is approximated with

$$\Gamma_h(t+\Delta t) \ : \quad S_h(x_1, t+\Delta t) := S_h(x_1, t) + \Delta t \nu_h \cdot \mathbf{u}_h \tag{3.7}$$

Remark 3.1 Note that this construction looks like a numerical integration for stationary problem– the integrals in (3.4) are not exactly calculated because $\Omega_h(t)$ is different from $\Omega(t)$ in general. In addition we should note that the term $\omega_h(\mathbf{v}_h)$ in (3.6) makes true the equality

$$f_h(\mathbf{v}_h) = \int_0^{x_{\max}} \frac{1}{Ca}\tau(x_1)\cdot\frac{\partial \mathbf{v}_h}{\partial x_1}dx_1 ,$$

which holds for $\Gamma(t)$ and $f(\mathbf{v}_h)$ (if $\Gamma(t)$ is sufficiently smooth). Without adding $\omega_h(\mathbf{v}_h)$ the difference between the right-hand sides of (3.5) and (3.4) depends on the jumps of the first derivatives of $S_h(x_1)$ at the grid points on the $\Gamma_h(t)$ which produces an worse approximation.

It is shown in [6] that the error produced by this "numerical integration" is proportional to the $[\text{meas}(\Omega(t)) - \text{meas}(\Omega_h(t))]$. Using the technique developed in [6] it is easy to obtain the following estimates.

Theorem 3.1 *For* $\mathbf{u}_h^k = \mathbf{u}_h(k\Delta t)$, $\tilde{\mathbf{u}}_h^k = \tilde{\mathbf{u}}_h(k\Delta t)$ *and* $S_h^k = S_h(k\Delta t)$, $S^k = S(k\Delta t)$ *the following inequalities hold:*

$$|\tilde{\mathbf{u}}_h^k - \mathbf{u}_h^k|_{1,\tilde{\Omega}(k\Delta t)} \leq C\left(|S^k - S_h^k|_{0,I} + \frac{1}{Ca\sqrt{h}}|S^k - S_h^k|_{1,I}\right), \tag{3.8}$$

$$\begin{aligned} |S^k - S_h^k|_{1,I} \ &\leq \ C\Big((1+\frac{\Delta t}{h})|S^{k-1} - S_h^{k-1}|_{1,I} \\ &+ \ \frac{1}{\sqrt{h}}|\tilde{\mathbf{u}}_h^{k-1} - \mathbf{u}_h^{k-1}|_{1,\tilde{\Omega}((k-1)\Delta t)}\Big) + O(h^2). \end{aligned} \tag{3.9}$$

From this theorem it immediately follows that

$$|S^k - S_h^k|_{1,I} \leq C\left((1+\frac{\Delta t}{h}+\frac{\Delta t}{Ca\,h})|S^{k-1} - S_h^{k-1}|_{1,I}\right) + O(h^2), \tag{3.10}$$

i.e. that if Δt is properly chosen ($\Delta t \leq \min\{h, Ca\, h\}$) then the error on level k depends on the error on level $(k-1)$ and it is of the same order.

Remark 3.2 It is important to note that $\tilde{\mathbf{u}}_h$ can not be found in general, because the domain $\Omega(t)$ is not known. But this projection makes applicable the inequality of the form $|\mathbf{v}_h|_{1,\Gamma} \leq \frac{1}{\sqrt{h}}|\mathbf{v}_h|_{1,\hat{\Omega}}$ with $\mathbf{v}_h := \mathbf{u}_h - \tilde{\mathbf{u}}_h$, because it is an element of a finite dimensional space (see [5]). Moreover the boundary condition (2.6) is of the Neumann type (i.e. there are no restrictions for the functions from $\mathbf{V}_h$ on Γ and Γ_h).

It can be easily observed that the second term on the right–hand side of (3.8) is "larger" then the first one, which comes from the difference between $a(\cdot,\cdot)$ and $a_h(\cdot,\cdot)$ (see [6]). This makes the assumption reasonable that the main error term for $\mathbf{u}$ comes from the approximation of the right–hand side $f(\mathbf{v})$. Using this intuitive guess as a basis, it could be thought that two different time steps might be used– the larger one (Δt_a) for the stiffness matrices and the smaller (Δt_f) for the calculation of right–hand sides. In this way $f_h(\mathbf{v}_h)$ is calculated at each small time step and the stiffness matrix– at the larger one. That means to solve linear systems of algebraic equations in the time interval $(t, t+\Delta t_a)$ changing only the right–hand sides. Thus, using the *LU* decomposition, the numerical solution could be obtained at little extra computational cost at each time step Δt_f. Therefore, the proposed numerical scheme for calculation of the FE solution is as follows:

1. Set $l = 0$; Choose $\Delta t_f : \Delta t_f \leq Ca\, h$ and $\Delta t_a : \Delta t_a > \Delta t_f$
2. Set $l = l + 1$. Calculate the *LU* decomposition of the stiffness matrix $\mathbf{K}$ at level l.
3. Set $k = 0$.
4. Set k = k+1. Calculate the right–hand side $\mathbf{b}$ and find the solution $\mathbf{u}_h^{(l+k)}$.
5. Calculate the new boundary position from (3.7) with $\Delta t := \Delta t_f$
6. **If** $k.\Delta t_f < \Delta t_a$ **then** go to 4 **else** go to 7.
7. **If** $l.\Delta t_a \leq 1$ **then** go to 2 **else** STOP.

4 Conclusions

The presented analysis of the error behaviour for free boundary problems provides a relation between the time and space steps and leads to a numerical scheme which might reduce the computational time. The further investigations for real problems derived from fluid mechanics will be concerned with the numerical tests using this scheme.

References

[1] R.J. Hansen, T.Y. Toong, *Dynamic Contact Angle and Its Relationship to Forces of Hydrodynamic Origin*, J. Colloid and Interface Sci., v. 37, 1971, pp. 196–207.

[2] J. Lowndes, *The Numerical Simulation of the Steady Movement of a Fluid Maniscus in a Capillary Tube* , J. Fluid Mech., v. 101, 1980, pp. 631–646.

[3] I. Bazhlekov, A. Chesters, *The Unsteady Wetting of Solid Surfaces*, Project 8144 S & T Cooperation with Eastern and Central European Countries, Final Report, 1993.

[4] P. Shopov, P. Minev, I. Bazhlekov, *Numerical Method for Unsteady Viscous Hydrodynamical Problem with Free Boundaries* , Int. J. Num. Meth. Fluids, v. 14, 1992, pp. 681–705.

[5] P.G. Ciarlet, The Finite Element Method for Elliptic Problems, North-Holand, Amsterdam, 1978.

[6] G. Strang, G.J. Fix, An Analysis of the Finite Element Method, Prentice Hall, Englewood Cliffs, NJ, 1973.

Address of Authors:
Institute of Mathematics,
Acad. G.Bonchev str. bl. 8
1113 Sofia
BULGARIA

MILU factorizations for 2-processor orderings

Robert Beauwens Laurent Dujacquier Samuel Hitimana
Magolu monga-Made

Abstract

We review here a few families of MILU preconditionings that have the unusual feature that the use of appropriate 2-processor orderings *improves* their convergence behaviour. Besides block factorizations, these methods include point factorizations with specific orderings such as S/P consistent and Kutcherov-Makarov orderings.

Key words : Iterative methods, preconditioning, incomplete factorizations, M-matrices, matrix graphs, parallel processing.

AMS(MOS) : 65F10, 65N20, 65F35, 15A23, 15A48.

1 Introduction

We consider here MILU factorizations of Stieltjes matrices. Such factorizations play a basic role in a wide class of preconditioning techniques used to solve finite element or finite difference approximations to elliptic problems.

Usually, they need be coupled with a *reduction* technique, by which the given matrix is first preconditioned by a spectrally equivalent *Stieltjes* one but we shall not address this problem here (there are by now a growing set of techniques for doing this), assuming that this step has already been done or is not required.

Our concern here will be to review the conditioning properties of a few families of MILU factorizations that appear as particularly performant for 2-processor orderings of the matrix graph. Generally, multiprocessor orderings lead to decreased performances. However, the particular feature of the methods analyzed here is that using appropriate 2-processor orderings *improves* their convergence behaviour.

4-processor and 8-processor orderings are also under study; for larger parallel architectures, these basic units have however to be replicated, leading then to the usual trade-off between parallelism and conditioning.

2 Conditioning analysis

Let A be a Stieltjes matrix and x a positive vector such that $Ax \geq 0$. A matrix B of the form

$$B = U^T P^{-1} U \tag{1}$$

where U is upper triangular with positive diagonal entries and $P = diag(U)$, is a MILU factorization of A if

$$offdiag(U^T + U) = offdiag(A - \beta * ((U^T - P)P^{-1}(U - P))) \tag{2}$$

where β is a symmetric $(0,1)$ matrix and $*$ denotes the Hadamard or elementwise matrix multiplication, and if

$$Bx = Ax + \Lambda Dx \tag{3}$$

where $D = diag(A)$ and $\Lambda = (\lambda_i \delta_{ij}) \geq 0$. Λ is called the *perturbation* matrix; the method is said *unperturbed* when $\Lambda = 0$.

There is now a fairly well developed conditioning analysis of such factorizations, which may be summarized by the following Lemma and Theorem. Proofs are not reported here because these results can easily be deduced from the more general ones to be proved in the next section.

Lemma 1 *Let U be an upper triangular M-matrix with $P = diag(U)$, let x be a positive vector and define*

$$t = \max_{i \in Pc(U)} \frac{((P - U)x)_i}{(Px)_i} \, . \tag{4}$$

Let further the matrices B_1 and B_2 be defined by

$$\begin{aligned} offdiag(B_1) &= offdiag(U^T + U) \ , \\ offdiag(B_2) &= -offdiag((U - P)^T P^{-1}(U - P)) \ , \\ B_1 x &= B_2 x = 0 \ . \end{aligned}$$

Then, the matrix $\tau B_1 - B_2$ is positive semidefinite for any $\tau \geq t$.

Remark. t is called the dominance ratio of U with respect to x and $Pc(U)$; $Pc(U)$ denotes the set of common precursors in the (ordered undirected) graph of U, i.e. the set of nodes with at least two successors; a precursor (successor) of i is a neighbour j of i such that $j < i$ $(j > i)$.

Theorem 1 *Let A be a Stieltjes matrix, U an upper triangular M-matrix, $P = diag(U)$, $B = U^T P^{-1} U$ and assume that*

$$-\, offdiag((U - P)^T P^{-1}(U - P)) \leq offdiag(U^T + U - A) \leq 0 \ . \tag{5}$$

Let x be a positive vector, define t by (4) and assume further that $t < 1$ and that

$$(1-\tau)Ax \;\leq\; Bx \;\leq\; Ax \tag{6}$$

for some τ such that

$$t \leq \tau < 1 \,. \tag{7}$$

Then, for any $z \neq 0$,

$$1 \;\leq\; \frac{(z,Az)}{(z,Bz)} \;\leq\; \frac{1}{1-\tau}\,, \tag{8}$$

whence $\kappa(B^{-1}A) \leq 1/(1-\tau)$.

The main bearing of this result is that diagonal dominance of the upper triangular factor (with respect to x and $Pc(U)$) *controls* an upper bound on the resulting spectral condition number, a fact which is at the root of the development of *dynamic* factorization algorithms [1, 2, 10].

3 Block methods

The MILU factorizations considered in the former § may be generalized by reinterpreting the triangular and diagonal factors U and P as block triangular and block diagonal with respect to some given partitioning, the perturbation matrix Λ remaining pointwise diagonal.

The matrix β still controls the sparsity of P; however, the block diagonal entries of P^{-1} are needed in (2) and these entries are dense. This is why (2) is further generalized to

$$\mathit{offdiag}(U^T + U) = \mathit{offdiag}(A - \beta * ((U^T - P)K(U - P))) \tag{9}$$

where K is some sparse approximation to P^{-1}; we assume here that, when P is monotone, then K satisfies $0 \leq K \leq P^{-1}$.

In the following results, which generalize those of the former section and are largely inspired by Notay's work [9], no mention is made of any partitioning, nor even of the structure of the approximate factors S^T and S. The block case is covered however, as it is readily seen, by *adding* the assumptions that S is block upper triangular with P as block diagonal. Nevertheless, these assumptions have been omitted because they are not needed in the proofs. In this sense, these results generalize those of Notay.

Lemma 2 *Let P be a Stieltjes matrix, S an M-matrix such that*

$$\mathit{offdiag}(S - P) \leq 0\,,$$

let x be a positive vector, $T = (t_i\delta_{ij})$ the diagonal matrix defined by

$$Tx = P^{-1}(P - S)x \tag{10}$$

and t the parameter defined by

$$t = \max_i t_i \,. \tag{11}$$

Let further the matrices B_1 and B_2 be defined by

$$\begin{aligned} \mathit{offdiag}(B_1) &= \mathit{offdiag}(S^T - P + S - P) \;, \\ \mathit{offdiag}(B_2) &= -\mathit{offdiag}((S-P)^T P^{-1}(S-P) + TPT) \;, \\ B_1 x &= B_2 x = 0 \;. \end{aligned}$$

Then, the matrix $\tau B_1 - B_2$ is positive semidefinite for any $\tau \geq t$.

Proof: Since B_1 is positive semidefinite, it is sufficient to prove that $tB_1 - B_2$ is positive semidefinite.

For this purpose, let B' be

$$B' = (T + (S-P)^T P^{-1})P(T + P^{-1}(S-P)) \tag{12}$$

and note that the definition of T implies that $B'x = 0$.

We have

$$\begin{aligned} &\mathit{offdiag}(tB_1 - B_2 - B') \\ &\quad = \mathit{offdiag}((S-P)^T(tI - T) + (tI - P)(S-P)) \leq 0 \,. \end{aligned}$$

and

$$(tB_1 - B_2 - B')x = 0 \,.$$

Therefore, $tB_1 - B_2 - B'$ is positive semidefinite, thus also $tB_1 - B_2$ since B' is positive semidefinite too. ∎

Theorem 2 *Let A and P be Stieltjes matrices and S an M-matrix, such that*

$$\mathit{offdiag}(S - P) \leq 0 \,, \tag{13}$$

and

$$\mathit{offdiag}((S-P)^T P^{-1}(S-P)) \leq \mathit{offdiag}(S^T + S - P - A) \leq 0 \,, \tag{14}$$

and let $B = S^T P^{-1} S$.

Let x be a positive vector and define T and t by (10) and (11). Assume further that $t < 1$ and that

$$(1-\tau)Ax \leq Bx \leq Ax \tag{15}$$

for some τ such that

$$t \leq \tau < 1 \,. \tag{16}$$

Then, for any $z \neq 0$,

$$1 \leq \frac{(z, Az)}{(z, Bz)} \leq \frac{1}{1-\tau} \tag{17}$$

whence $\kappa(B^{-1}A) \leq 1/(1-\tau)$.

Proof: Letting the matrices B_1 and B_2 be defined as in Lemma 2, it is readily checked that

$$\begin{aligned} & \mathit{offdiag}(B-(1-\tau)A-(\tau B_1-B_2)) \\ &\quad = \mathit{offdiag}((1-\tau)(S^T+S-P-A)+\tau P-TPT) \\ &\quad = \mathit{offdiag}((1-\tau)(S^T+S-P-A)+(\tau I-T)P+TP(I-T)) \leq 0 \end{aligned}$$

while

$$(B-(1-\tau)A-(\tau B_1-B_2))x = Bx-(1-\tau)Ax \geq 0$$

implying that $B-(1-\tau)A-(\tau B_1-B_2)$ is positive semidefinite. Since $\tau B_1 - B_2$ is positive semidefinite by Lemma 2, the right inequality (17) follows.

The left inequality follows in a similar way, by checking that $\mathit{offdiag}(A-B) \leq 0$ with $(A-B)x \geq 0$. ∎

To give a concrete illustration of the bearing of these results, let us consider the 5-point finite difference Poisson problem on a $N \times M$ grid with line partitioning. As shown in [9] on the basis of [11], the former results lead in this case to

$$\kappa(B^{-1}A) \leq M \quad . \tag{18}$$

One could similarly consider a 3D grid with plane partitioning, leading again to (18) when the Poisson problem is considered on a grid with M planes.

A simple way to introduce a 2-processor ordering here consists in choosing as last numbered line (or plane) the one in the middle of the mesh. Clearly, both half of the mesh can then be treated concurrently (with one communication needed to complete the forward step for the last line or plane). The nice feature of this ordering is that the conditioning properties are improved. Thus, for the Poisson problem, we still get (18), but M is now understood as the number of levels of the level structure $\mathcal{L} = (L_k)_{0 \leq k \leq \ell}$ generated by defining L_ℓ as the middle line (or plane), so that its actual value is halved.

The main drawback of the method is that it requires exact plane factorizations for 3D problems. This is the main motivation for turning back in the next sections to point factorization methods which, thanks to specific ordering techniques, recover the useful properties of the block factorizations.

4 Specific orderings

For some orderings, it is possible to refine Lemma 1, and therefore also Theorem 1.

4.1 S/P consistent orderings

According to [4], an ordered undirected graph is called S/P consistently ordered if whenever a pair of nodes $i < j$ has a common neighbour k with $k < i$, then either it also has a common neighbour ℓ with $j < \ell$ or neither i has any neighbour ℓ with $\ell > i$ nor j any neighbour ℓ with $\ell > j$.

An S/P consistently ordered matrix is a matrix whose graph is S/P consistently ordered and it turns out that modified incomplete factorizations with S/P consistently ordered M-factors have attractive conditioning properties, provided that the submatrix determined by the nodes without successors is factored exactly [3].

It is appropriate to recall here that a graph is consistently ordered if there exists a partitioning $\mathcal{L} = (L_k)_{0 \le k \le \ell}$ of its node set such that

$$P(L_k) \subset L_{k-1} \tag{19}$$

for $0 \le k \le \ell$ (being understood that $L_{-1} = \emptyset$), and to note that an S/P consistently ordered graph is indeed consistently ordered [4]. (We recall here that, for a set of nodes L in a graph G, $P(L)$ denotes the set of the precursors of the nodes of L, while $S(L)$ similarly denotes the set of the successors of the nodes of L). An example is shown on Fig. 1 where the white nodes define the last level while the other nodes are numbered according to the rule

$$L_k = P(L_{k+1}) \tag{20}$$

for $0 \le k < \ell$.

Further, for an upper triangular M-matrix U, S/P consistency can be turned into a quantified notion trough a parameter called *maximal reduction ratio*, denoted by $\kappa_t(U, x)$, which depends on a given positive vector x (used to measure diagonal dominance) and a given positive number t (a prescribed level of diagonal dominance) [4].

The relevant fact is that, if U is an upper triangular S/P consistently ordered M-matrix, then without any other additional assumption, the conclusion of Lemma 1 improves to: the matrix $\tau(1 - \frac{\kappa}{M})B_1 - B_2$ is positive semidefinite for any $\tau \ge t$ and $\kappa \le \kappa_t(U, x)$, where M denotes the number of levels of the associated level structure.

Similarly, the conclusion of Theorem 1 improves to

$$\kappa(\widetilde{B}^{-1}A) \le 1/(1 - \tau + \tau\frac{\kappa}{M}) \tag{21}$$

provided that B is modified into $\widetilde{B}$ by accepting all fill-in in the last level [3].

There is no difficulty to also introduce here a 2-processor ordering, again by putting the last level in the middle of the mesh; this has already been used on the example shown on Fig. 1, and it leads to similar advantages as in the block case.

On the other hand, the particular graph G_a shown on this figure was obtained by starting with a 2D grid to which a red/black reduction step was first applied. The resulting matrix is spectrally equivalent to the original one [5, 7].

Extensions of this technique to 3D problems are under development.

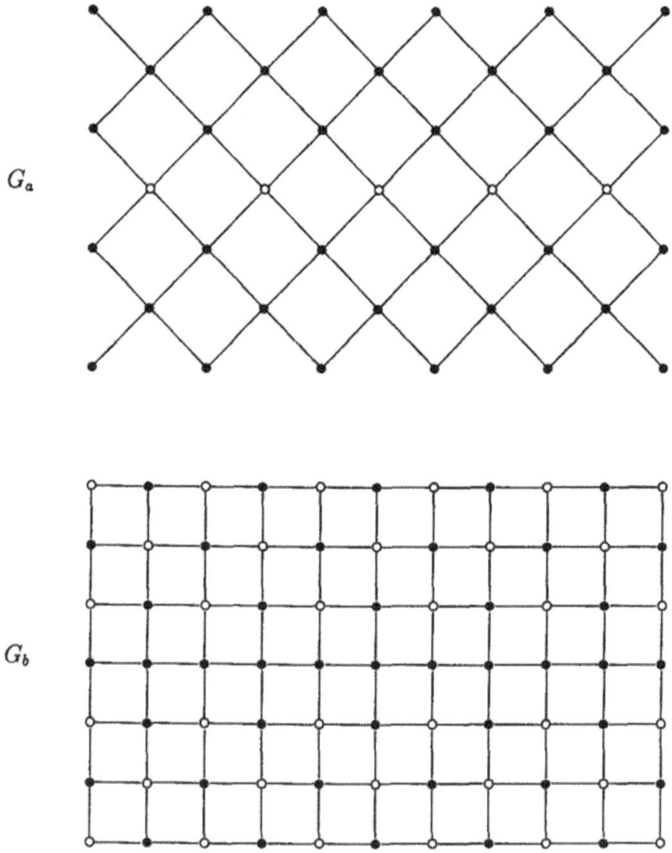

Figure 1: S/P consistent (G_a) and Kutcherov-Makarov (G_b) orderings.

4.2 The Kutcherov-Makarov orderings

Kutcherov and Makarov have introduced in [8] an ordering for 2D rectangular grids which leads to similar properties as S/P consistent orderings and which may even be easier to implement. On the other hand, their present results with this ordering are limited to 2D grids.

The Kutcherov-Makarov ordering consists in ordering the lines of the grid successively but using a red-black ordering inside each line in such a way that, if the first node of some line is red, then the first node of the next line is black; these rules are further altered for the last line which is ordered by the natural (i.e. lexicographic) ordering.

The Kutcherov-Makarov analysis leads to the same improved version of Lemma 1 as obtained for S/P consistently ordered M-matrices, with the conclusion that $\tau(1-\frac{\kappa}{M})B_1 - B_2$ is positive semidefinite for any $\tau \geq t$, M being the number of grid lines and κ a positive number (for which an algorithmic determination can be deduced from the results reported in [8]).

Here, no modification of the factorization is required to get the corresponding improved version of Theorem 1, but the particular ordering used for the last line actually implies that this last line is factored exactly.

Putting the last line in the middle of the mesh can again be used in conjunction with the Kutcherov-Makarov ordering and this is the example shown on Fig. 1 where the red nodes are represented by white circles; here again, this produces a 2-processor ordering with conditioning properties twice better than the original Kutcherov-Makarov ordering.

The main drawback of the Kutcherov-Makarov ordering lies in its limitation to 2D grids. A generalized definition has however recently been proposed [6] under the following form: a Kutcherov-Makarov ordering of a graph G is an ordering for which there exists a level structure $\mathcal{L} = (L_k)_{0\leq k\leq \ell}$ and a red/black partitioning of each level $k < \ell$, $L_k = (L_k^r, L_k^b)_{0\leq k<\ell}$, such that

$$\begin{aligned} S(L_k^r) &\subset L_k^b \cup L_{k+1} \\ S(L_k^b) &\subset L_{k+1} \\ \forall i \in L_k^b \cup (L_k^r \cap Pc(G)) &: \; Card(S(i) \cap L_{k+1}) = 1 \end{aligned}$$

Work is in progress to show that these properties suffice to get the same results as with the original definition, provided that the last level is factored exactly.

For 3D problems, this would reduce the exact factorization requirement (with respect to block methods with plane partitioning) to only one plane. One may further hope that in actual practice, a few iterations with the (2-processor) 2D Kutcherov-Makarov ordering may reveal sufficient.

Acknowledgements

This work presents research results of the Belgian Incentive Program "Information Technology" - Computer Science of the future, initiated by the Belgian State - Prime Minister's Service - Science Policy Office (Contract No. IT/IF/14). The scientific responsibility is assumed by their authors.

This work was also supported by IBM trough a research contract between ULB and IBM.

References

[1] O. Axelsson and V. Barker, *Finite Element Solution of Boundary Value Problems. Theory and Computation*, Academic Press, New York (1984).

[2] R. Beauwens, *Modified incomplete factorization strategies*, in Preconditioned Conjugate Gradient Methods, edited by O. Axelsson and L. Kolotilina, pages 1–16, Lectures Notes in Mathematics No. 1457, Springer-Verlag, 1990.

[3] R. Beauwens, Num. Lin. Alg. with Appl. **1**, 3 (1994).

[4] R. Beauwens, Y. Notay, and B. Tombuyses, Num. Lin. Alg. with Appl. **1**, 19 (1994).

[5] C. W. Brand, Numer. Math. **61**, 433 (1992).

[6] L. Dujacquier, Memoir, in progress, 1994.

[7] S. Hitimana, 2-processor and 4-processor S/P consistent orderings, in progress, 1994.

[8] A. Kutcherov and M. Makarov, J. Numer. Lin. Alg. Appl. **1**, 1 (1992).

[9] Y. Notay, Lin. Alg. Appl. **154-156**, 711 (1991).

[10] Y. Notay, DRIC : a dynamic version of the RIC method, to appear, 1994.

[11] M. M. Magolu, Appl. Numer. Math. **8**, 25 (1991).

Service de Métrologie Nucléaire,
Université Libre de Bruxelles (C.P. 165)
50, av. F.D. Roosevelt,
B-1050 Brussels,
Belgium

A numerical scheme for a quadratic porous media equation

G. Bourdin * J.M. Greenberg † A.Y. Le Roux *

Abstract

A numerical scheme of the Godunov type is built for a quadratic porous media equation. The properties of the scheme are studied and the convergence is proved.

1 Introduction

We consider the following equation, for x $\in$ **R**, t>0,

$$u_t + \lambda u u_x = u_x^2 \tag{1}$$

for $\lambda \in \mathbf{R}$, $\lambda \neq 0$.
Such an equation occurs in some model of flow in a porous media. In this paper, we are concerned with the convergence of numerical schemes for (1). The usual hyperbolic methods, for instance TVD schemes, cannot allow to go to the limit on the quadratic term. In practice, u corresponds to a non negative parameter.
This equation can be read as

$$u_t + w u_x = 0 \tag{2}$$

with the flux w given by

$$w = \lambda u - u_x. \tag{3}$$

Then we can derive

$$w_t + (2w - \lambda u) w_x = 0. \tag{4}$$

*CeReMaB, Université Bordeaux I, 33405 Talence, France

†Department of Mathematics and Statistics, University of Maryland, Baltimore County, Baltimore, Maryland 21228

Now, by considering (4), we can derive a numerical scheme which will be stable in both L^∞ and BV norms, for w. Then, from (3) we get the same estimates for u_x and this will be enough to handle the quadratic term.

2 The numerical scheme

We denote by h the space meshsize and set

$$\Delta t = rh$$

as the time increment. The parameter r will be used to control the stability through a so-called CFL condition. We shall use different cells for the numerical values of u and w. We shall use two indices $i \in \mathbf{Z}$ and $j = i + \frac{1}{2}$. This way, the approximate values of w will be w_i^n on the cell

$$M_i =](j-1)h, jh[$$

at time $t_n = n\Delta t$, $n \in \mathbf{N}$, and the approximate values of u will be u_j^n on $M_i =](j-1)h, jh[$ at time t_n.
We suppose that the values w_i^n are given at some time t_n. Since from (2), we can expect a bounded solution, so we build the u_j^n by interpolating the unique bounded solution of (3), which is seen as a differential equation for u, with given bounded data w_i^n. We get

$$u_{j-1}^n = e^{-\lambda h} u_j^n + (1 - e^{-\lambda h}) \frac{w_i^n}{\lambda}. \tag{5}$$

It is now sufficient to know one $u_{j_0}^n$ to be able to derive all the u_j^n. This can be done for example by assuming that the support of u_0 is bounded in one side.
Now (4) can be seen as a quasilinear equation for w since u is known. We can compute the interface values

$$w_j^n = \begin{cases} w_i^n & if\ w_i^n > \frac{\lambda u_j^n}{2} \quad , \quad w_i^n + w_{i+1}^n > \lambda u_j^n \\ \frac{\lambda u_j^n}{2} & if\ w_i^n < \frac{\lambda u_j^n}{2} \quad , \quad w_{i+1}^n > \frac{\lambda u_j^n}{2} \\ w_{i+1}^n & if\ w_{i+1}^n < \frac{\lambda u_j^n}{2} \quad , \quad w_i^n + w_{i+1}^n \le \lambda u_j^n \end{cases} \tag{6}$$

This correspond to the Riemann problem solution for (4). (See for example [1])
Next, we compute the projection of the exact solution of (4) on the cell M_i at time

t_{n+1}, which can be easily performed as long as the waves coming from the two interfaces cannot meet. This is true provide the condition on r

$$rMax(|2w_i^n - \lambda u_j^n|, |2w_i^n - \lambda u_{j-1}^n|) \leq \frac{1}{2} \tag{7}$$

is fulfilled. We have got the following scheme

$$w_i^{n+1} = w_i^n - r(w_j^{n^2} - \lambda u_j^n w_j^n - w_{j-1}^{n^2} + \lambda u_{j-1}^n w_{j-1}^n) - \lambda r w_i^n(u_j^n - u_{j-1}^n). \tag{8}$$

3 The convergence theorem

We first derive some estimates on w.

Proposition 3.1 : *Under the condition (7), we have*

$$Min(w_{i-1}^n, w_i^n, w_{i+1}^n) \leq w_i^{n+1} \leq Max(w_{i-1}^n, w_i^n, w_{i+1}^n) \tag{9}$$

and

$$\Sigma_i |w_{i+1}^{n+1} - w_i^{n+1}| \leq \Sigma_i |w_{i+1}^n - w_i^n|. \tag{10}$$

The inequalities (9) tell us that w is bounded and that the scheme will preserve the positiveness of any positive initial data. The proof of proposition 3.1 is a technical but straightforward one.
We now need a L^∞-estimate for u, which will allow us to use a constant parameter r for the whole computation. We have

Proposition 3.2 : *Under the condition (7), we have*

$$Min\left(u_j^n, \frac{w_i^n}{\lambda}\right) \leq u_{j-1}^n \leq Max\left(u_j^n, \frac{w_i^n}{\lambda}\right) \tag{11}$$

$$\left|\frac{u_j^n - u_{j-1}^n}{h}\right| \leq M|\lambda u_j^n - w_i^n| \tag{12}$$

for some constant M which does not depend on h.

From these estimates, and by assuming that the support of w is bounded from above, we get

Proposition 3.3 : *From the previous estimates , one can extract a sequence w_{h_m}, with h_m going to zero, which converges in L^1-norm towards a bounded and BV function w, whose support is bounded from above.*
The corresponding sequence u_{h_m} converges in C^0-norm (uniform convergence) towards a lipschitz continous function u whose second derivative is bounded, and such that the sequence $\frac{\partial}{\partial x}u_{h_m}$ converges in L^1-norm towards u_x.

It remains to prove that w and u satisfy (3) and (4) in some weak sense which is compatible with the previous results.
From (8), by multiplying by some test function ϕ with a compact support in $\mathbf{R}\times(0,T)$, we get

$$\int\int(w\phi_t + (w^2 - \lambda uw)\phi_x + \lambda wu_x\phi)dxdt = 0.$$

So, we get

$$w_t + (w^2 - \lambda uw)_x = \lambda wu_x$$

which makes sense in $L^1(\mathbf{R}\times(0,T))$
By using the same way, we get

$$\lambda u - u_x = w$$

which also has a sense in $L^1(\mathbf{R}\times(0,T))$.
Now, by noticing that, at any time t

$$(\lambda I - \frac{\partial}{\partial x})(u_t + \lambda uu_x - u_x^2) = 0$$

with $u_t + \lambda uu_x - u_x^2$ bounded, we get that

$$u_t + \lambda uu_x - u_x^2 = 0$$

is the unique bounded solution of the differential equation

$$\lambda y - y_x = 0.$$

Uniqueness is obviously derived since $u = u_1 - u_2$ is the solution of a linear equation with bounded and BV coefficients. This implies that in proposition 3. 3, the whole sequences will converge.
Our result is now

Theorem 3.1 : *Let us suppose that the initial data u_0 are such that the support of u_0 is bounded from above, and*

$$u_0 \in C^0(\mathbf{R}), u_{0,x} \in BV(\mathbf{R}),$$

then the Godunov scheme (5), (6), (8) converges under the CFL condition (7), towards the unique bounded continuous solution of (1).

Remarks : The CFL condition (7) can be taken up to one in practice. This scheme has the convergence property, but presents some drawbacks. For instance, a compact support of u_0 will not be preserved from (5) which builds a fast decreasin g function near $-\infty$. Another drawback is that a nonconstant stationary solution will not be preserved through the computation. Some other scheme with the properties of preservation (compact support and stationary solutions) does exit; however the one we have build is sometimes not a TVD scheme for w.

The authors thank E. Sibe and B. Capbern from CEA-CESTA who suggested to study some physical models from which this one was derived. (See [2]).

References

[1] A.Y. LeRoux, *Convergence of the Godunov scheme for first order quasilinear equation*, Proc. Japan Acad.,52 (1976), pp. 217-243.

[2] K. Moncapjuzan, *Stage de DEA*, CEA-CESTA (1993).

Computing Extreme Eigenvalues of Large Sparse Symmetric Problems

Clemens W. Brand   Svetozara Petrova *

Abstract

Iterative algorithms to calculate a few extreme eigenvalues of a large, sparse, symmetric matrix are Lanczos method, power iterations or Davidson's method and its generalizations. Their common characteristic is to build a sequence of subspaces, which contains in the limit the desired eigenvectors. All these methods converge at a rate proportional to the the so-called *gap ratio*, which shows the separation of the extreme eigenvalues from the rest of the spectrum. With a preconditioning matrix we improve the separation of the eigenvalues. Our iterative algorithm is related to Davidson's method and its generalizations but differs in the way of preconditioning and orthogonalizing. Davidson's method uses diagonal preconditioning and has been very successful for problems in quantum chemistry. We have proved global convergence of the proposed iterative procedure. Moreover, for a class of problems of dimension n we prove that the number of iterations to reduce the error of an initial approximation by some fixed order of magnitude is $O(\sqrt{n})$. The numerical results show the good performance of our algorithm.

1 Introduction

Eigenvalue problems arise, among others, in the description of vibrations of objects like strings, plates or buildings. For a structure with complex shape, a discrete representation of the problem on a fine grid yields a large, sparse matrix. Typically just a few of the lowest eigenvalues and corresponding eigenvectors of this matrix describe the practically important modes of vibration.

In this paper, we present an iterative method for finding the lowest eigenvalue and corresponding eigenvector of a symmetric positive definite matrix. We analyze the convergence of this method and compare its computational efficiency with the Lanczos method. The results favor the new method. We discuss also how the rate of convergence of iterative methods to find the lowest eigenvalue λ_1 of a symmetric

*On leave from Center of Informatics and Computer Technology, Bulgarian Academy of Sciences, "Acad. G. Bonchev" str., block 25A, 1113 Sofia, Bulgaria

positive definite matrix A is related to the relative separation of λ_1 from the rest of the spectrum. Our idea is to replace the eigenvalue problem $A\mathbf{u}_1 = \lambda_1\mathbf{u}_1$ by a preconditioned problem, where the eigenvalue we are interested in is separated much better from the others. Section 2 describes in detail the iterative method to calculate the lowest eigenvalue of a symmetric positive matrix and analyzes the convergence. The main result is, that—under some assumptions on the matrix and the preconditioner—the number of iterations required to get an approximate eigenvalue and eigenvector, so that $\|A\mathbf{x} - \lambda\mathbf{x}\| < \varepsilon$ with some (fixed) accuracy threshold ε, is of order $O(\sqrt{n})$ for an $n \times n$ matrix. Section 3 presents numerical results.

Our method can be compared with Davidson's method [2] and its generalizations. Morgan and Scott [5, 6] apply preconditioning to Davidson's method and the Lanczos algorithm. Their matrix operator $(\tilde{A} - \lambda^{(k)}I)^{-1}(A - \lambda^{(k)}I)$ differs from the one we use as the basis of our method, see Equation (3). As a consequence, modified incomplete factorizations like MIC(p,q) or MINV(p) [1] become unstable when used in their method but work very well in our algorithms.

To introduce the basic concepts of our method, consider a symmetric positive definite matrix $A \in \mathbb{R}^{n\times n}$. Let $0 < \lambda_1 < \cdots \leq \lambda_n$ be the eigenvalues of A, and denote the corresponding eigenvectors by $\mathbf{u}_1, \ldots, \mathbf{u}_n$.

A simple, though inefficient, method to calculate $\mathbf{u}_1$ is the power method. We briefly comment on the convergence properties of this method in comparison to the Lanczos method. For a detailed discussion of the two methods see e.g. Golub and Van Loan[4]. To find the smallest eigenvalue by the power method, form a matrix $B = sI - A$, where $s > 0$ is a suitably chosen shift parameter. The eigenvalues of this matrix B are $s - \lambda_n \leq \cdots \leq s - \lambda_2 < s - \lambda_1$. For a shift parameter $s \geq \lambda_n$, the eigenvalue $s - \lambda_1$ is a dominant eigenvalue of B. The power method applied to B therefore converges. At the k–th iteration, it finds an approximation $\lambda_1^{(k)}$ to λ_1, and [4]

$$|\lambda_1^{(k)} - \lambda_1| = O(\mathcal{R}^k), \quad \text{where } \mathcal{R} = \frac{s - \lambda_2}{s - \lambda_1} < 1. \tag{1}$$

Call $\delta = (\lambda_2 - \lambda_1)/(\lambda_n - \lambda_1)$ the relative separation of λ_1 from the rest of the spectrum. The number of iterations to make $\mathcal{R}^k < \varepsilon$, for some accuracy threshold ε, is approximately $\log\varepsilon/\log\mathcal{R}$. For small δ, it is of order $O(\delta^{-1})$, since $\log\mathcal{R} \approx \delta$. For practical problems with large n, typically $\lambda_n/\lambda_1 = O(n)$ and $\lambda_1/\lambda_2 = O(1)$. The relative separation δ therefore is $O(1/n)$. So we conclude that this form of the power method needs $O(n)$ iterations. For the Lanczos method, the Kaniel-Paige theory estimates the number of iterations to be $O(\delta^{-1/2}) = O(\sqrt{n})$. This is, of course, an estimate much better than the one for the power method.

We consider now how to precondition an eigenvalue problem. Let A be the matrix corresponding to the discrete Laplace operator on a square grid with the lowest eigenvalue λ_1 and its corresponding eigenvector $\mathbf{u}_1$. Let $\tilde{A}$ be a preconditioner for A

and we consider the matrix

$$M = \tilde{A}^{-1}(A - \lambda_1 I). \tag{2}$$

Clearly, if $\mathbf{u}_1$ is an eigenvector of A, it is an eigenvector of M as well, and its eigenvalue is 0.

Typically, preconditioners result from some variant of incomplete Cholesky factorization of A. A simple preconditioner, called DKR in the following, is a modified pointwise incomplete factorization first introduced by Dupont, Kendall and Rachford[3]. It is equivalent to the method MIC(1,1) in the terminology of Concus, Golub and Meurant [1]. Their paper investigates (among many others) also a block incomplete method denoted INV(p).

2 The Preconditioned Power Method

The *preconditioned power method* generates a sequence of pairs $(\lambda^{(k)}, \mathbf{x}^{(k)})$ as follows: Choose a relaxation parameter $\omega > 0$ and an $\mathbf{x}^{(0)} \in \mathbb{R}^n$ with unit 2-norm, not orthogonal to $\mathbf{u}_1$.

$$\begin{array}{l} \text{Set } \lambda^{(0)} \leftarrow \mathbf{x}^{(0)T} A \mathbf{x}^{(0)} \\ \textbf{for } \; k \leftarrow 0, 1, 2, \ldots \\ \quad M^{(k)} \leftarrow \tilde{A}^{-1}(A - \lambda^{(k)} I) \\ \quad \mathbf{y}^{(k+1)} \leftarrow (I - \omega M^{(k)})\mathbf{x}^{(k)} \\ \quad \mathbf{x}^{(k+1)} \leftarrow \mathbf{y}^{(k+1)}/\|\mathbf{y}^{(k+1)}\| \\ \quad \lambda^{(k+1)} \leftarrow \mathbf{x}^{(k+1)T} A \mathbf{x}^{(k+1)} \end{array} \tag{3}$$

Remark. The condition that $\mathbf{x}^{(0)}$ must not be orthogonal to $\mathbf{u}_1$ has rather theoretical significance. If indeed $\mathbf{x}^{(0)}$ had no component in direction $\mathbf{u}_1$, then rounding errors during the iteration would introduce some.

For $\tilde{A} = I$, i.e. without preconditioning, the method essentially reduces to the shifted power method considered in Section 1: For $\omega = 1/\lambda_n$ and $\mathbf{x}^{(0)}$ not orthogonal to $\mathbf{u}_1$, the iteration (3) converges to $(\lambda_1, \mathbf{u}_1)$. For a non-trivial $\tilde{A}$ we therefore call Method (3) the *preconditioned* power method.

For $\tilde{A} = A$ and $\omega = 1$, (3) is equivalent to inverse iteration. Actually, it is just the power method applied with A^{-1}. Convergence is determined (see e.g. [4]) by the ratio of λ_1 (the dominant eigenvalue of A^{-1}) to λ_2: $\|(A - \lambda^{(k)} I)\mathbf{x}^{(k)}\| = O\left[(\lambda_1/\lambda_2)^k\right]$. The number of iterations to decrease an initial error by some constant is $O(1)$ for large n. However, each iteration involves solving a linear system with A. This costs a large amount of computation.

Interpolating between the two extreme cases $\tilde{A} = I$ and $\tilde{A} = A$ and using a well chosen preconditioner the rate of convergence will be better than for the power method, but not as good as for inverse iteration. We show for a class of preconditioners that the number of iterations is of the order of the condition number $\kappa(\tilde{A}^{-1}A)$. We have proved that the only fixed points of the mapping associated with the procedure (3) are $\mathbf{u}_i$, $i = 1, \ldots, n$; show that $\mathbf{u}_1$ is the only point of attraction, while the

$\mathbf{u}_i$ for $i = 2, \ldots, n$ are repelling fixed points; show that for a suitable choice of ω the procedure (3) converges for any choice of the initial guess $\mathbf{x}^{(0)}$.

We derive now a parameter-free version of the preconditioned power method. In Method (3), $\mathbf{y}^{(k+1)}$ lies in the subspace spanned by $\mathbf{x}^{(k)}$ and $M^{(k)}\mathbf{x}^{(k)}$. Instead of using a fixed value for ω throughout the iterations, consider selecting it at each step so that $\mathbf{y}^{(k+1)} = \mathbf{x}^{(k)} - \omega^{(k)} M^{(k)}\mathbf{x}^{(k)}$ minimizes the Rayleigh quotient. That is,

$$\mathbf{y}^{(k+1)} \text{ minimizes } \frac{\mathbf{y}^T A \mathbf{y}}{\mathbf{y}^T \mathbf{y}} \text{ over all } \omega \in \mathbb{R}, \quad \mathbf{y} = \mathbf{x}^{(k)} - \omega M^{(k)}\mathbf{x}^{(k)}.$$

To do so, get an orthonormal basis $(\mathbf{q}_1, \mathbf{q}_2)$ of the two-dimensional subspace spanned by $\mathbf{x}^{(k)}$ and $M^{(k)}\mathbf{x}^{(k)}$. Define $Q \in \mathbb{R}^{n\times 2} = (\mathbf{q}_1, \mathbf{q}_2)$. Finding a linear combination $\mathbf{y} = a_1\mathbf{q}_1 + a_2\mathbf{q}_2$ that minimizes the Rayleigh quotient is equivalent to the problem: find the smaller eigenvalue β_1 of the 2-by-2 matrix $Q^T A Q$ and a corresponding eigenvector $\mathbf{b}_1$. For $\mathbf{y} = Q\mathbf{b}_1$, the Rayleigh quotient $\mathbf{y}^T A\mathbf{y}/\mathbf{y}^T\mathbf{y}$ assumes its minimum value β_1 among all linear combinations of $\mathbf{q}_1$ and $\mathbf{q}_2$.

The following method carries out preconditioned power iterations and minimizes at each step the Rayleigh quotient over a two-dimensional subspace. As in Method (3) let A and $\tilde{A} \in \mathbb{R}^{n\times n}$ be symmetric positive definite and $0 < \lambda_1 < \lambda_2 \leq \cdots \leq \lambda_n$. Start with an $\mathbf{x}^{(0)} \in \mathbb{R}^n$ with unit 2-norm, not orthogonal to $\mathbf{u}_1$.

$$\begin{array}{l}
\text{Set } \lambda^{(0)} \leftarrow \mathbf{x}^{(0)T} A \mathbf{x}^{(0)} \\
\textbf{for } k \leftarrow 0, 1, 2, \ldots \\
\quad M^{(k)} \leftarrow \tilde{A}^{-1}(A - \lambda^{(k)} I) \\
\quad \mathbf{y}^{(k)} \leftarrow M^{(k)}\mathbf{x}^{(k)} \\
\quad \mathbf{y}^{(k)} \leftarrow \mathbf{y}^{(k)} - (\mathbf{x}^{(k)T}\mathbf{y}^{(k)})\mathbf{x}^{(k)} \text{ (orthogonalize)} \\
\quad \mathbf{y}^{(k)} \leftarrow \mathbf{y}^{(k)}/\|\mathbf{y}^{(k)}\| \text{ (normalize)} \\
\quad B \in \mathbb{R}^{2\times 2} \leftarrow (\mathbf{x}^{(k)}, \mathbf{y}^{(k)})^T A (\mathbf{x}^{(k)}, \mathbf{y}^{(k)}) \\
\quad \text{get } \beta_1, \text{ the minimum eigenvalue of } B, \\
\qquad \text{and its eigenvector } \mathbf{b}_1 \in \mathbb{R}^2, \|\mathbf{b}_1 \ | = 1 \\
\quad \mathbf{x}^{(k+1)} \leftarrow (\mathbf{x}^{(k)}, \mathbf{y}^{(k)})\mathbf{b}_1, \ (\mathbf{x}^{(k)}, \mathbf{y}^{(k)}) \in \mathbb{R}^{n\times 2} \\
\quad \lambda^{(k+1)} \leftarrow \beta_1
\end{array} \tag{4}$$

For estimating the rate of convergence, we have to be more specific about the preconditioner $\tilde{A}$. Assume $\lambda_1 < \lambda_2 \leq \cdots \leq \lambda_n$ for the eigenvalues of a symmetric positive definite matrix $A \in \mathbb{R}^{n\times n}$. Consider a splitting $A = \tilde{A} + E$, where $\tilde{A}$ is symmetric positive definite and E is nonnegative definite. Modified incomplete Cholesky factorizations of M-matrices generate splittings of this kind. The simplest examples are (using the terminology of Concus, Golub and Meurant [1]) DKR, MIC(p, q) or MINV(p). (A matrix $A = \{a_{ij}\}$ is an M-matrix if $a_{ij} \leq 0$ for $i \neq j$, A is nonsingular and $A^{-1} \geq 0$.)

We have proved that the number of iterations to compute an approximate eigenvector of A is of order $O(\sqrt{n})$. This is the same order of convergence that Lanczos methods typically achieve. But one iteration in Method (3) involves more computations than one Lanczos step—how much more, depends mostly on the preconditioner.

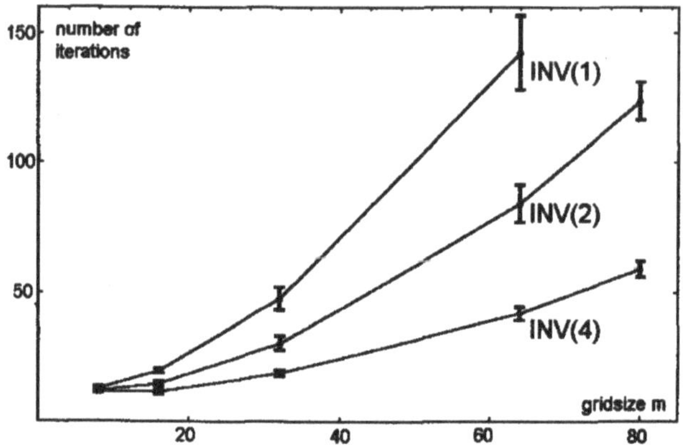

Figure 1:

3 Numerical Results

This section presents numerical comparisons for the iterative procedures (3) and (4) described in Section 2. An extensive screening of preconditioners is beyond the scope of this paper; the DKR and INV(p) are simple, but typical examples. The results of the numerical experiments clearly indicate that Method (4) is a powerful algorithm for large problems. All algorithms were programmed in FORTRAN and executed on a HP 750 Workstation and a Convex C-220.

We apply the iterative procedures (3) and (4) to find the smallest eigenvalue of a model problem $A\mathbf{x} = \lambda\mathbf{x}$. This problem is the discrete version of the eigenvalue problem for the two-dimensional Laplace operator on the unit square with Dirichlet boundary conditions. The initial approximation for the iterations, $\mathbf{x}^{(0)}$, is a vector with elements randomly chosen in $(0, 1]$ and then normalized so that $\|\mathbf{x}^{(0)}\| = 1$. We use a stopping criterion $\left\|(A - \lambda^{(k)} I)x^{(k)}\right\| \leq 10^{-5}$.

Figure 1 presents the number of iterations for Method (3) versus gridsize m. The grids range from 8×8 to 80×80. Note that for an $m \times m$ grid the corresponding matrix A is in $\mathbb{R}^{n \times n}$, where $n = m^2$. We choose the relaxation parameter $\omega = 1$. Preconditioners are INV(1), INV(2), INV(4). (DKR does not converge for $\omega = 1$). The number of iterations to reach the accuracy threshold depends also on the initial guess. The values given here are an average for ten different seed values in the random number generator. The variance of the data is indicated by error bars.

Figure 2 presents the number of iterations for Method (3) versus relaxation parameter ω. The preconditioner here is INV(2), and we compare results for a 16×16 and a 32×32 grid. The figure illustrates a general trend that we observed in cal-

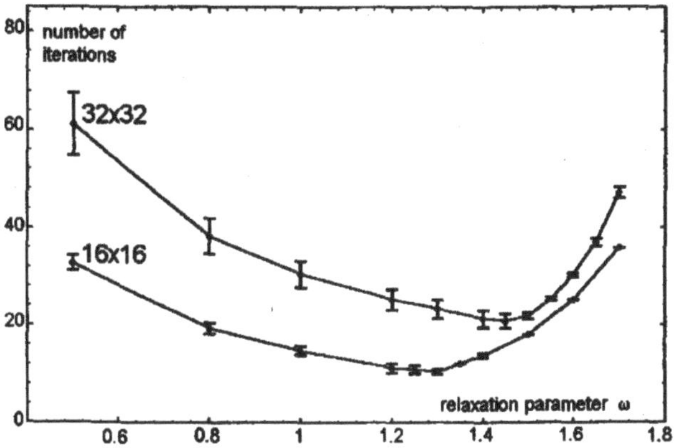

Figure 2:

culations with other preconditioners and gridsizes: for the INV(p) preconditioners, the minimum number of iterations is for an ω_{opt} somewhere between 1 and 2. The minimum is rather shallow, i.e. the number of iterations is not very sensitive to small errors in ω_{opt}. The optimum value ω_{opt} increases slightly for larger grids. It seems that ω_{opt} is also smaller for higher p in INV(p). For the DKR preconditioner, our computations showed that ω_{opt} is always smaller than 1 and of order $O(1/m)$.

Figure 3 presents the number of iterations for the parameter-free Method (4) versus gridsize m. As for Figure 1, the grids range from 8×8 to 80×80. Preconditioners are DKR and INV(1), INV(2), INV(4). For the parameter-free method, the number of iterations does not depend as strongly on the initial guess as for Method (3). The values given here are an average over ten runs, but no error bars are drawn (they are just too small). Note that the number of iterations increases almost strictly linear for DKR, which is in accordance with theoretical predictions. For the INV(p) preconditioners, the tendency is less clear. The graph for INV(2) seems to grow superlinearly, while for INV(4) it stays almost constant.

All preconditioners need significantly less iterations than for Method (3) with $\omega = 1$. Even for an optimum choice of the relaxation parameter ω, the parameter-free method needs less iterations (no figures are presented for this case). This is not too big a surprise, since Method (4) is essentially Method (3), augmented with one-dimensional optimization at each step.

We have compared also the number of iterations for the Lanczos method and the parameter-free Method (4) using DKR as a preconditioner. Although it is (as far as the number of iterations is concerned) the least powerful preconditioner among the ones tested here, it always needs only about half the number of iterations.

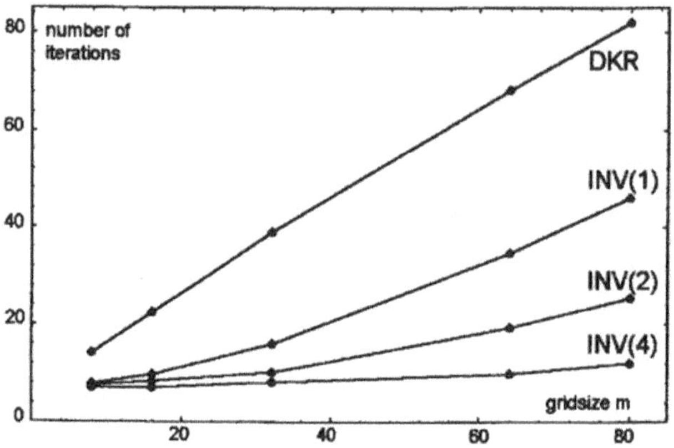

Figure 3:

Conclusion. The numerical tests and the considerations regarding numerical stability and memory requirement show clearly that Method (4) is an efficient algorithm for large, sparse eigenvalue problems. At least in terms of memory requirements it is definitely superior to the Lanczos algorithm.

References

[1] P. Concus, G. H. Golub, and G. Meurant. Block preconditioning for the conjugate gradient method. *SIAM J. Sci. Stat. Comput.*, 6:220–252, 1985.

[2] E. R. Davidson. The iterative calculation of a few of the lowest eigenvalues and corresponding eigenvectors of large real-symmetric matrices. *J. Comp. Phys.*,17:87–94, 1975.

[3] T. Dupont, R. P. Kendall and H. H. Rachford. An approximate factorization procedure for solving self-adjoint elliptic difference equations. *SIAM J. Numer. Anal.*, 5:559–573, 1968.

[4] G. Golub and C. Van Loan. *Matrix Computations.* 2 ed., Johns Hopkins Univ. Pr., 1989.

[5] R. B. Morgan and D. S. Scott. Generalizations of Davidson's method for computing eigenvalues of sparse symmetric matrices. *SIAM J. Sci. Stat. Comp.*, 7(3):817–825, 1986.

[6] R. B. Morgan and D. S. Scott. Preconditioning the Lanczos algorithm for sparse symmetric eigenvalue problems. *SIAM J. Sci. Comp.*, 14(3):585–593, 1993.

[7] J. Ortega. *Numerical Analysis: A Second Course.* SIAM Classics in applied mathematics, Vol. 3, 1990.

Institut für Angewandte Mathematik
Montanuniversität Leoben
Franz-Josef-Str. 18
A-8700 Leoben, Austria

Stability of the Krylov basis and subspaces*

J-F. Carpraux, S.K. Godunov, S.V. Kuznetsov

1 Introduction

The problems of numerical analysis with large sparse matrices involve often a projection of this matrix onto a Krylov subspace to obtain a smaller matrix which is used to solve the initial problem. The subspace depends on the matrix and on an arbitrary vector. We consider, in this paper, a method to study the stability of the Krylov basis and subspace through a data perturbation. This method includes a definition of the condition numbers for the computation of the Krylov basis and subspace. A practical method for estimating these numbers is provided. It is based on the solution of a large triangular system.

2 Theory

Let $A \in \mathrm{R}^{n\times n}$ and $f \in \mathrm{R}^n$, $\|f\|_2 = 1$. The Krylov subspaces are, for $1 \le k \le n$, the subspaces $\mathcal{K}_k(A,f) = span[f, Af, A^2f, \dots, A^{k-1}f]$ of dimension $\le k$. Let l be the dimension of $\mathcal{K}_n(A,f) = span[f, Af, \dots, A^{n-1}f]$: in fact, we have $\mathcal{K}_n(A,f) = span[f, Af, \dots, A^{l-1}f]$.

Definition 1 *For $1 \le k \le l$, the natural orthonormal Krylov basis of $\mathcal{K}_k(A,f)$ is an orthonormal basis $F_k = \{f_1, f_2, \dots, f_k\}$ such that for $1 \le j \le k$, F_j is a basis of $\mathcal{K}_j(A,f)$. F_k is unique up to the sign.*

For $1 \le k \le l$, let

- $\mathcal{K} = \mathcal{K}_k(A,f)$, and F its natural orthonormal Krylov basis. Since $k \le l$, F is of dimension k.
- $\tilde{\mathcal{K}} = \mathcal{K}_k(A+\Delta, f)$, and $\tilde{F}$ its natural orthonormal Krylov basis.

*This work was supported in part by Russian Fund of Fundamental Researches (93-011-1515) and DRET

We assume that $\|\Delta\|$ is small enough to ensure that $\tilde{F}$ is also of dimension k. We give in Definition 3 the definition of the condition number for the Krylov subspace and basis computation. These definitions come from the usual definition of condition number where the metric in the set of subspaces is defined in Definition 2 and we choose the Frobenius norm in the space of matrices.

Definition 2 *At the first order in* $\|X\|$, $d(F,\tilde{F}) = \min_{X\in\mathcal{X}} \frac{1}{\sqrt{2}}\|X\|_F$, *where*

$$\mathcal{X} = \left\{X \text{ such that } \|X\| \ll 1, X^* = -X \text{ and } \tilde{F} = \left(I + X + O(\|X\|^2)\right) F\right\}.$$

The distance between $\mathcal{K}$ *and* $\tilde{\mathcal{K}}$ *is given by* $d(\mathcal{K},\tilde{\mathcal{K}}) = \min_{F,\tilde{F}} d(F,\tilde{F})$ *where* F *and* $\tilde{F}$ *are respectively two orthonormal bases of* $\mathcal{K}$ *and* $\tilde{\mathcal{K}}$.

Definition 3 *The condition number* $\mu\{\mathcal{K}_k(A,f)\}$ *and* $\mu_b\{\mathcal{K}_k(A,f)\}$ *for, respectively, the Krylov subspace and basis are given by :*

$$\mu\{\mathcal{K}_k(A,f)\} = \inf_{\epsilon>0}\left\{\sup_{\|\Delta\|_F\le\epsilon}\left(\frac{d(\mathcal{K},\tilde{\mathcal{K}})}{\|\Delta\|_F}\|A\|_F\right)\right\}$$

and

$$\mu_b\{\mathcal{K}_k(A,f)\} = \inf_{\epsilon>0}\left\{\sup_{\|\Delta\|_F\le\epsilon}\left(\frac{d(F,\tilde{F})}{\|\Delta\|_F}\|A\|_F\right)\right\}$$

Remarks :

- It is easy to see that $\mu\{\mathcal{K}_1(A,f)\} = \mu_b\{\mathcal{K}_1(A,f)\} = 0$.
- If $l = n$, then $\mu\{\mathcal{K}_n(A,f)\} = 0$ and $\mu_b\{\mathcal{K}_n(A,f)\} = \mu_b\{\mathcal{K}_{n-1}(A,f)\}$ because in this case, the last vector is uniquely defined up to the sign.

We can show that, for A and f given, we can find a basis in which this matrix is an Heisenberg matrix and $f = (1,0,\ldots,0)^T$. We also show that the stability of the Krylov basis and subspaces through a vector perturbation can be deduced from the computation of the stability through a matrix perturbation. Therefore, we only consider in the following of this paper the stability of the Krylov basis and subspaces constructed from an Heisenberg matrix and the vector $f = (1,0,\ldots,0)^T$, through a matrix perturbation.

Let us suppose now that $2 \le k \le \min(l, n-1)$, A is an Heisenberg matrix and $f = (1,0,\ldots,0)^T$.
Then the natural orthonormal basis of $\mathcal{K}_k(A,f)$ is $F = \begin{pmatrix} I_k \\ 0 \end{pmatrix}$.

Let $\tilde{F} = [\tilde{f}_1, \tilde{f}_2, \dots, \tilde{f}_k]$ be the natural orthonormal basis of $\mathcal{K}_k(\tilde{A}, f)$ where $\tilde{A} = A + \Delta + O(\|\Delta\|^2)$ $(\|\Delta\| \ll 1)$ and $\tilde{f}_1 = f = (1, 0, \dots, 0)^T$. We are searching $X \in \mathrm{R}^{n \times n}$ $(\|X\| \ll 1)$ such that $\tilde{F} = (I + X + O(\|\Delta\|^2))\, F$ with $X^* = -X$.

Remark : If Δ is an Heisenberg matrix for the $k-1$ first columns then $\tilde{F} = F$. Therefore, the supremum in Definition 3 is attained for some

$$\Delta = \begin{pmatrix} 0 & \cdots\cdots & & & & 0 \\ 0 & & & & & \\ \delta_{3,1} & \ddots & & 0 & & \\ \vdots & \ddots & \ddots & & & \\ \vdots & & \delta_{k+1,k-1} & 0 & \dots & 0 \\ \vdots & & \vdots & \vdots & & \vdots \\ \delta_{n,1} & \dots & \delta_{n,k-1} & 0 & \dots & 0 \end{pmatrix}.$$

Let

$$m = (k-1)n + 1 - k(k+1)/2,$$

$$x^{(k)} = (x_{3,2}, \dots, x_{n,2} \,|\, x_{4,3}, \dots \,|\, \dots \,|\, x_{k+1,k}, \dots, x_{n,k})^T \in \mathrm{R}^m$$

$$\text{and} \quad \delta^{(k)} = (\delta_{3,1}, \dots, \delta_{n,1} \,|\, \dots \,|\, \delta_{k+1,k-1}, \dots, \delta_{n,k-1})^T \in \mathrm{R}^m$$

Then, we can prove that

$$X \text{ such that } \left\{ \begin{array}{c} \tilde{F} = (I + X + O(\|\Delta\|^2))F \\ X^* = -X \end{array} \right. \iff x^{(k)} \text{ solution of } \quad B^{(A,k)} x^{(k)} = \delta^{(k)}$$

where $B^{(A,k)}$ is the following triangular matrix built with the elements of A :

$$\begin{pmatrix} a_{2,1} I_{n-2} & & & & \\ a_{2,2} J^{n-2}_{n-3} - A^{(4)} & a_{3,2} I_{n-3} & & & \\ a_{2,3} J^{n-2}_{n-4} & \ddots & \ddots & 0 & \\ \vdots & \ddots & \ddots & \ddots & \\ a_{2,k-1} J^{n-2}_{n-k} & \dots & a_{k-2,k-1} J^{n-k+2}_{n-k} & a_{k-1,k-1} J^{n-k+1}_{n-k} - A^{(k+1)} & a_{k,k-1} I_{n-k} \end{pmatrix}$$

Finally, we find

Theorem 1 *Let l be the dimension of $\mathcal{K}_n(A, f)$, then for $2 \le k \le \min(l, n-1)$, the matrix $C^{(A,k)} = \left(B^{(A,k)}\right)^{-1}$ exists, and then :*

The condition number of the natural orthonormal basis of $\mathcal{K}_k(A, f)$ is

$$\mu_b\{\mathcal{K}_k(A, f)\} = \|C^{(A,k)}\|_2 \|A\|_F$$

The condition number of $\mathcal{K}_k(A, f)$ *is*

$$\mu\{\mathcal{K}_k(A, f)\} = \|\hat{C}^{(A,k)}\|_2 \|A\|_F$$

where $\hat{C}^{(A,k)}$ *is the matrix composed by a few rows of* $C^{(A,k)}$ *such that*

$$\hat{x}^{(k)} = (x_{k+1,2}, \ldots, x_{n,2} \mid \ldots \mid x_{k+1,k}, \ldots, x_{n,k})^T = \hat{C}^{(A,k)} \delta^{(k)}.$$

3 Algorithm and examples

Let $A \in \mathbb{R}^{n \times n}$ and $f \in \mathbb{R}^n$. We propose here an algorithm to compute the condition numbers of the subspace of Krylov $\mathcal{K}_k(A, f)$ and of its basis.

1. Choose a basis such that A is in an Heisenberg form and $f = (1, 0, \ldots, 0)^T$
2. Compute $C^{(A,k)} = \left(B^{(A,k)}\right)^{-1}$
3. Estimate the condition numbers thanks to Theorem 1

The difficulty is to compute precisely $C^{(A,k)}$. Indeed, it often happens, when the condition number is high, that the matrix $B^{(A,k)}$ is near from a singular matrix. Therefore, we use a special technique for the computation of scalar products. We also study the stability of the algorithm for the computation of $C^{(A,k)}$, and give bounds for the estimation of $\|C^{(A,k)}\|_2$.

We give now some examples of computation of the condition numbers of Krylov basis and subspaces. Let $B = B^{(A,k)}$ and $C = \left(C^{(A,k)}\right)_{mach}$, then we have to be sure that the computed matrix C is not too far from B^{-1}. The study of the stability of this algorithm allows us to estimate an upper bound of $\|\Omega\|_F$ where $BC = I + \Omega$. Then, we can give a lower and upper bound of the condition number of the basis :

$$\|C\|_2 \frac{1 - 2\|\Omega\|_F}{1 - \|\Omega\|_F} \|A\|_F \leq \mu_b\{\mathcal{K}_k(A, f)\} \leq \|C\|_2 \frac{1}{1 - \|\Omega\|_F} \|A\|_F$$

For each example, we give a table in which we can see the lower and upper bounds of the condition numbers of the basis. Indeed, these bounds allow us to validate the result when they are close enough from each other. We also can validate the result, by computing $\|BC - I\|_2$.

Let us consider the following matrix $A \in \mathbb{R}^{16 \times 16}$ and $f \in \mathbb{R}^{16}$:

$$A = \begin{pmatrix} -7 & 36 & & \\ -1 & 0 & \ddots & 0 \\ & \ddots & \ddots & 36 \\ 0 & & -1 & 0 \end{pmatrix} \qquad f = \begin{pmatrix} 1 \\ 0 \\ \vdots \\ 0 \end{pmatrix}$$

k	lower bound	$\mu_b\{\mathcal{K}_k(A,f)\}$	upper bound	$\mu\{\mathcal{K}_k(A,f)\}$	$\|BC - I\|_2$
2	1.397e+02	1.397e+02	1.397e+02	1.397e+02	-0.0e+00
3	5.158e+03	5.158e+03	5.158e+03	5.158e+03	-0.0e+00
4	1.856e+05	1.856e+05	1.856e+05	1.856e+05	-0.0e+00
5	6.671e+06	6.671e+06	6.671e+06	6.671e+06	-0.0e+00
6	2.395e+08	2.395e+08	2.395e+08	2.395e+08	-0.0e+00
7	8.556e+09	8.573e+09	8.589e+09	8.573e+09	-0.0e+00
8	2.851e+11	3.045e+11	3.238e+11	3.045e+11	-0.0e+00
9	4.393e+11	4.924e+11	5.455e+11	4.304e+11	-0.0e+00
10	4.322e+11	4.924e+11	5.526e+11	1.691e+10	-0.0e+00
11	4.260e+11	4.924e+11	5.588e+11	5.755e+08	-0.0e+00
12	4.208e+11	4.924e+11	5.640e+11	1.847e+07	-0.0e+00
13	4.169e+11	4.924e+11	5.679e+11	5.737e+05	-0.0e+00
14	4.144e+11	4.924e+11	5.704e+11	1.746e+04	-0.0e+00
15	4.132e+11	4.924e+11	5.716e+11	5.225e+02	-0.0e+00

The condition number of the Krylov basis increases with its size, and becomes quickly bad. It is interesting to see that the condition number of the Krylov subspace increases first up to a large value then decreases.

Let us now take the transposed previous matrix and the same vector :

$$A = \begin{pmatrix} -7 & -1 & & \\ 36 & 0 & \ddots & 0 \\ & \ddots & \ddots & -1 \\ 0 & & 36 & 0 \end{pmatrix} \in \mathbb{R}^{16\times 16} \quad \text{and} \quad f = \begin{pmatrix} 1 \\ 0 \\ \vdots \\ 0 \end{pmatrix} \in \mathbb{R}^{16}$$

k	lower bound	$\mu_b\{\mathcal{K}_k(A,f)\}$	upper bound	$\mu\{\mathcal{K}_k(A,f)\}$	$\|BC-I\|_2$
2	3.879e+00	3.879e+00	3.879e+00	3.879e+00	-0.0e+00
3	6.349e+00	6.349e+00	6.349e+00	6.348e+00	-0.0e+00
4	8.856e+00	8.856e+00	8.856e+00	8.851e+00	-0.0e+00
5	1.138e+01	1.138e+01	1.138e+01	1.136e+01	3.4e-21
6	1.389e+01	1.389e+01	1.389e+01	1.386e+01	5.5e-21
7	1.641e+01	1.641e+01	1.641e+01	1.633e+01	7.7e-21
8	1.892e+01	1.892e+01	1.892e+01	1.875e+01	9.8e-21
9	2.144e+01	2.144e+01	2.144e+01	2.101e+01	1.1e-20
10	2.391e+01	2.391e+01	2.391e+01	2.269e+01	1.2e-20
11	2.642e+01	2.642e+01	2.642e+01	2.344e+01	1.2e-20
12	2.881e+01	2.881e+01	2.881e+01	2.319e+01	1.2e-20
13	3.130e+01	3.130e+01	3.130e+01	2.187e+01	1.2e-20
14	3.343e+01	3.343e+01	3.343e+01	1.922e+01	1.2e-20
15	3.586e+01	3.586e+01	3.586e+01	1.452e+01	1.2e-20

The only difference between this example and the previous is that we consider here the transposed matrix. But, as we can see in the tables the results are quite different. Here, the condition numbers of the Krylov basis and subspaces are always good.

4 Conclusion

We provide an algorithm to measure the sensitivity of the Krylov subspace and basis to a data perturbation. Moreover, the examples show surprising and, at this time, unexplained results. Therefore, this tool will be very useful to understand instabilities of a Krylov subspace or basis. It will be also interesting to see if there are some links between the instability of the Krylov subspace and the convergence of iterative methods in linear algebra using these subspaces.

J-F Carpraux
IRISA/INRIA - Campus de Beaulieu
35042 Rennes Cedex, France

S.K. Godunov, S.V. Kuznetsov
Institute of Mathematics, Universitetski pr. 4
Novosibirsk, 630090, Russia

Multigrid and Domain Decomposition Methods for Unstructured Meshes

Tony F. Chan* Barry Smith * Jun Zou †

Abstract

We summarize some recent work of ours on the design and analysis of multigrid and overlapping domain decomposition algorithms for solving elliptic problems on unstructured meshes in two and three dimensions. We describe a class of algorithms which are based on the construction of a coarse grid hierarchy, and the associated standard finite element interpolation operators, from the given fine grid. We develop a theory which shows that our domain decomposition algorithms have the same optimal convergence rate as the usual algorithms for structured meshes. Very general meshes and subdomains are allowed: neither the fine mesh nor the coarse mesh need to be quasi-uniform, the subdomains can be of arbitrary shapes and sizes, and the coarse mesh need not be nested to, or cover the same physical domain as, the fine mesh. Some numerical results will be presented to demonstrate both the theory and the algorithms.

Key Words. Unstructured meshes, multigrid, domain decomposition.

1 Introduction

Unstructured meshes have become increasing popular in scientific computing [1, 12]. Compared to structured meshes, they have increased flexibility in adapting to complicated geometries as well as to large changes in the solution. However, traditional solvers have to be modified so that their efficiency will not be adversely affected by the lack of structure in the mesh.

In this short paper, we summarize some recent work of ours on the design and analysis of multigrid and overlapping domain decomposition algorithms for solving elliptic problems on unstructured meshes in two and three dimensions. For this class

*The work of these two authors was partially supported by the National Science Foundation under contract ASC 92-01266, the Army Research Office under contract DAAL03-91-G-0150, and ONR under contract ONR-N00014-92-J-1890.

†The work of this author was partially supported by the National Science Foundation under contract ASC 92-01266, the Army Research Office under contract DAAL03-91-G-0150, and subcontract DAAL03-91-C-0047.

of algorithms, unstructured meshes present several additional practical and theoretical difficulties: the absence of a natural coarse grid hierarchy, the need to construct well-balanced subdomains, the typically highly non-uniformity of the mesh, the non-matching of the fine grid boundary and the coarse grid boundary and the non-nestedness of the finite element spaces on the fine mesh and the coarse mesh. We shall describe some approaches we have taken to overcome these problems.

We describe a class of algorithms which are based on the construction of a coarse grid hierarchy, and the associated standard finite element interpolation operators, from the given fine grid. We use recursive spectral graph partitioning techniques to construct the subdomains. We develop a convergence theory for overlapping domain decomposition algorithms which allows very general meshes and subdomains: neither the fine mesh nor the coarse mesh need to be quasi-uniform, the subdomains can be of arbitrary shapes and sizes, and the coarse mesh need not be nested to, or cover the same physical domain as, the fine mesh. The theory shows that our algorithms have the same optimal convergence rate as the usual algorithms for structured meshes.

There has been relatively little work on domain decomposition and multigrid algorithms for unstructured meshes. Some references are: [12, 11, 5, 4, 14].

The present paper is essentially a summary of our results in the papers [8, 9, 7]. More details can be found in those papers.

2 Construction of coarse grid hierarchy

In both multigrid and domain decomposition algorithms, a coarse grid hierarchy is need. Unlike for a structured mesh, this is not naturally given as part of the refinement procedure. Our approach is to construct the coarse grid hierarchy, as well as the associated interpolation and restriction operators, directly from the given unstructured fine mesh. It suffices to describe this for one coarse level because the procedure can be recursively applied to obtain all the coarse meshes. Typically, the whole coarse grid hierarchy is used by a multigrid algorithm but only one of the coarser grids is used by a standard two-level domain decomposition algorithm.

We shall need the notion of a *maximal independent set* of the vertices of a graph. A subset of vertices V of a graph G is said to be *independent* if no two vertices of V are connected by an edge. V is said to be *maximally independent* if adding any additional vertex to it makes it dependent. Note that maximal independent sets of vertices of a graph are generally not unique.

Our procedure has four steps:

1. Form a maximally independent set of the boundary vertices and from these construct a set of coarse boundary edges,

2. Form a maximally independent set of the interior vertices,

3. Apply a Cavendish type algorithm [6] to triangulate the resulting collection of coarse boundary edges and coarse interior vertices,

4. Construct the interpolation and restriction operators.

Step (i) is fairly straightforward. For each disjoint boundary segment, the boundary vertices are ordered say in a clockwise direction, starting with a random vertex. Then every other vertex is thrown out and the remaining ones are connected with new coarse boundary edges. This forms a coarse representation of the boundary segment. After several coarsenings, one may find that the boundary is no longer qualitatively similar to the original boundary. This may be prevented by simply retaining some of the vertices in the coarse grid boundary that would normally be dropped.

Step (ii) uses a greedy wavefront type algorithm. A random interior vertex is selected for inclusion in the maximally independent set. Then every interior vertex connected to it is eliminated from consideration for inclusion in the maximally independent set. Next, one of the interior vertices connected to the newly eliminated vertices is selected for inclusion and the procedure repeats until all interior vertices have been considered. An algorithm similar to this has been used by Barnard and Simon [2] in designing graph partitioning algorithms, also see Guillard [11]. This procedure can be implemented in linear time, i.e. proportional to the total number of interior vertices. An alternative to step (ii) is to construct a maximal independent set of the *dual* graph of the mesh and use the center of the remaining elements as the coarse grid vertices.

The input to Step (iii) is thus a collection of coarse boundary edges and coarse interior vertices. A version of Cavendish's algorithm [6] is then applied to triangulate this collection. This algorithm is an advancing front technique and "grows" new triangles from those already built by selecting an interior vertex to be "mated" to an existing edge. In doing so, it tries to optimize the aspect ratio of the new triangle formed, preferring those that are close to being equilateral. It is possible to implement this algorithm in linear time, i.e. proportional to the number of interior vertices, but our current implementation is not optimal.

Finally, in Step (iv), the interpolation operator is constructed in the form of a sparse matrix and stored. To determine the entries of this interpolation matrix, the coarse triangles are taken in sequence and the entries corresponding to all the fine grid vertices within the coarse triangle are then computed using the standard piecewise linear interpolation. This procedure can also be implemented in linear time because the fine grid triangles close to the vertices of the coarse triangle (which are also fine grid vertices as well) can be found by a local search. We emphasize that this is not possible if the coarser grids are generated completely independently. The restriction matrix is then just the transpose of the interpolation matrix.

3 Convergence analysis of Two Level Schwarz Methods

We will develop a theory to make the two level Schwarz overlapping domain decomposition methods applicable to unstructured meshes and keep the same optimal convergence. We give only the results for the model problem with Dirichlet boundary condition. For more general problems and boundary conditions, we refer to Chan, Smith and Zou [9] [7].

3.1 Uniformly elliptic problems with non-matching boundary

We consider the following model problem: Find $u \in H_0^1(\Omega)$ such that

$$a(u,v) \equiv \int_\Omega \rho(x)\,\nabla u\,\nabla v\,dx = f(v),\ \forall\, v \in H_0^1(\Omega) \tag{1}$$

where $\Omega \subset R^d$ (d=2, 3) and $\rho(x) \geq \alpha > 0$, in Ω.

Suppose we are given a family of triangulations $\{\mathcal{T}^h\}$ on Ω, consisting of simplices. We will not discuss the effects of approximating Ω but always assume in the paper that the triangulations $\{\mathcal{T}^h\}$ of Ω are exact. So we have $\Omega = \Omega^h \equiv \cup_{\tau\in\mathcal{T}^h}\tau$. Let $h_\tau =$ diam τ and $h = \overline{h} = \max_{\tau\in\mathcal{T}^h} h_\tau$. Let V^h be a piecewise linear finite element subspace of $H_0^1(\Omega)$ defined on $\mathcal{T}^h$ with its basis denoted by $\{\phi_i^h\}_{i=1}^n$, and O_i =supp ϕ_i^h. As usual, we decompose the domain Ω into p nonoverlapping subdomains Ω_i such that $\overline{\Omega} = \cup_{i=1}^p \overline{\Omega}_i$, then extend each subdomain Ω_i to a larger one Ω_i' such that the distance between $\partial\Omega_i$ and $\partial\Omega_i'$ is bounded from below by $\delta_i > 0$. We denote the minimum of all δ_i by δ. We assume that $\partial\Omega_i'$ does not cut through any element $\tau \in \mathcal{T}^h$. For the subdomains meeting the boundary we cut off the part of Ω_i' which is outside of $\overline{\Omega}$. No other assumptions will be made on $\{\Omega_i\}$ except that any point $x \in \Omega$ belongs only to a finite number of subdomains $\{\Omega_i'\}$. Therefore, we allow each Ω_i to be of quite different size and of quite different shape from other subdomains. We define the subspaces of V^h corresponding to the subdomains $\{\Omega_i'\}$, $i = 1, 2, \cdots, p$ by

$$V_i^h = V^h \cap H_0^1(\Omega_i'). \tag{2}$$

In order to get an optimal preconditioner even when the number of subdomains increases greatly, we introduce a coarse grid $\mathcal{T}^H$ which can be completely independent of $\mathcal{T}^h$, i.e., none of the nodes of $\mathcal{T}^H$ need to be nodes of $\mathcal{T}^h$. In general, $\Omega^H \neq \Omega$. Let H be the maximum diameter of the elements of $\mathcal{T}^H$, and $\Omega^H = \cup_{\tau^H\in\mathcal{T}^H}\tau^H$. By V^H we denote a subspace of $H_0^1(\Omega^H)$ consisting of piecewise polynomials defined on $\mathcal{T}^H$, by $\{\psi_i^H\}_{i=1}^m$ we denote its basis functions related to the nodes $\{q_i^H\}_{i=1}^m$. Let O_i^H = supp ψ_i^H. We note that V^H need not necessarily be piecewise linear as V^h. Thus we do not necessarily have the usual condition: $V^H \subset V^h$.

For technical reasons, we make two additional reasonable assumptions on the coarse grid:

(A1): $\tau^H \cap \Omega \neq \emptyset$ for all $\tau^H \in \mathcal{T}^H$,

that is, no coarse grid element lies completely outside the fine grid. For the complement set $\Omega \setminus \Omega^H$, let S be the set of all vertices q_i^H of Ω^H which belongs to $\overline{\Omega \setminus \Omega^H}$, and $B_p(r)$ be a ball at the point p with radius r. We assume that

(A2): $\Omega \setminus \Omega^H \subset \cup_{q_i^H \in S} B_{q_i^H}(\text{diam } O_i^H)$,

that is, the coarse grid must cover a significant part of the fine grid.

To overcome the difficulty that $V^H \not\subset V^h$, in both the theory and the algorithms, we need a way of mapping values from V^H to V^h. For the coarse space to be effective, this mapping must possess the properties of H^1-stability and L^2 optimal approximation. We will show that the standard finite element interpolant Π_h is one of such choices.

3.2 Two level overlapping Schwarz algorithms

Based on the finite element spaces V_i^h and V^H given previously, we formulate the two level overlapping Schwarz methods for nonnested grids. Schwarz methods are techniques to construct preconditioners for the stiffness matrix $A = (a_{ij})_{i,j=1}^n$ with $a_{ij} = a(\phi_i^h, \phi_j^h)$. These preconditioners are built using local and coarse grid solves.

The local solves are defined as in Dryja and Widlund [10], and Bramble, Pasciak, Wang, and Xu [3]. Define the H^1-projection operators $P_i : V^h \rightarrow V_i^h$, $i = 1, \cdots, p$ such that for any $u \in V^h$, $P_i u \in V_i^h$ satisfies

$$a(P_i u, v_i) = a(u, v_i), \qquad \forall\, v_i \in V_i^h. \tag{3}$$

There are two ways to define the coarse grid projection, in order to handel the non-nestedness of the coarse grid space. Let $\mathcal{I}_($ be any linear operator which maps V^H into a subspace $\mathcal{I}_h V^H$ of V^h. But in the rest of the paper, we only discuss the choice $\mathcal{I}_h = \Pi_h$-the standard finite element interpolant.

In **Method 1** we define $\tilde{P}_0$ by first defining $P_H u \in V^H$ on the original coarse grid space by

$$a(P_H u, v) = a(u, \mathcal{I}_h v), \qquad u \in V^h, \quad \forall v \in V^H \tag{4}$$

and then define $\tilde{P}_0 = \mathcal{I}_h P_H$: $V^h \rightarrow V_0^h$. Here the subspace $V_0^h = \mathcal{I}_h V^H \subset V^h$.

In **Method 2** we define P_0 by calculating the projection directly onto the subspace V_0^h: For any $u \in V^h$, $P_0 u \in V_0^h$ such that

$$a(P_0 u, v) = a(u, v), \quad \forall v \in V_0^h. \tag{5}$$

We now derive the matrix representation of the operators P_i and $\tilde{P}_0$. Next, we use u^h to denote finite element functions and u its coefficient vector, that is $u^h = \sum u_k \phi_k$.

Let $\{\phi_{i,j}^h\}_{j=1}^{n_i} \subset \{\phi_i^h\}_{i=1}^{n}$ be the set of nodal basis functions of V_i^h, $i = 1, 2, \cdots, p$. For each i, we define a matrix extension operator R_i^T as follows: For any $u_i^h \in V_i^h$, we denote by u_i the coefficient vector of u_i^h in the basis $\{\phi_{i,j}^h\}_{j=1}^{n_i}$, and we define that $R_i^T u_i$ to be the coefficient vector of u_i^h in the basis $\{\phi_i^h\}_{i=1}^{n}$.

Since $\{\psi_i^H\}_{i=1}^m$ is the set of basis functions of V^H, then $\{\mathcal{I}_h \psi_i^H\}_{i=1}^m$ is the set of basis functions of V_0^h. We define a matrix extension operator R_0^T as follows: For any $u_0^h \in V_0^h$, we denote by u_0 the coefficient vector of u_0^h in the basis $\{\mathcal{I}_h \psi_i^H\}_{i=1}^m$, define $R_0^T u_0$ to be the coefficient vector of u_0^h in the basis $\{\phi_j^h\}_{j=1}^n$. Let A_H be the stiffness matrix corresponding to the original coarse space V^H.

Then it is straightforward to derive that for any $u^h \in V^h$, the matrix representation of $\tilde{P}u^h$ with $\tilde{P} = \tilde{P}_0 + \sum_{i=1}^p P_i$ is

$$M_1 Au = R_0^T A_H^{-1} R_0 Au + \sum_{i=1}^{p} R_i^T A_i^{-1} R_i Au, \tag{6}$$

and the matrix representation of Pu^h with $P = \sum_{i=0}^p P_i$ is

$$M_2 Au = \sum_{i=0}^{p} R_i^T A_i^{-1} R_i Au \tag{7}$$

where $A_i = R_i A R_i^T$ for $i = 0, 1, \cdots, p$ are the stiffness matrices corresponding to the subspaces V_i^h. The preconditioners M_1 and M_2 may be also thought of as an overlapping block Jacobi method with the addition of a coarse grid correction. The multiplicative Schwarz method is the Gauss-Seidel version of the additive algorithm. We write down the symmetrized version, using Method 1 as,

$$\begin{aligned} M &= (I - (I - R_1^T A_1^{-1} R_1 A) \ldots (I - R_p^T A_p^{-1} R_p A)(I - R_0^T A_H^{-1} R_0 A) \\ &\quad (I - R_p^T A_p^{-1} R_p A) \ldots (I - R_1^T A_1^{-1} R_1 A))A^{-1}. \end{aligned} \tag{8}$$

In practice the application of the multiplicative Schwarz preconditioner is done directly, not as given in (8).

For **Method 1** and **Method 2**, we have the following estimates for their condition numbers, see Chan, Smith and Zou [7] for details:

Theorem 3.1 *Suppose that both triangulations $\mathcal{T}^h$ and $\mathcal{T}^H$ are shape regular (not necessarily quasi-uniform), and satisfy Assumptions* **(A1)** *and* **(A2)**. *Then, with $\mathcal{I}_h = \Pi_h$, we have*

$$\kappa(M_1 A),\ \kappa(M_2 A) \le C\,(1 + \frac{H}{\delta})^2. \tag{9}$$

We remark that the ratio H/δ in (9) can be replaced by a completely localized ratio.

4 Numerical results

In this section we present convergence results for two basic algorithms: two level overlapping multiplicative Schwarz methods and standard V-cycle multigrid. The fine grid is as depicted in Figure 1. In Figure 2 we depict the three coarser grids which have been generated automatically.

For our computations we have used Laplace's equation with homogeneous boundary conditions. This problem was discretized with the usual piecewise linear finite elements. In our calculations the discrete right hand side was chosen to be the vector of all 1's. Other numerical examples may be found in Chan and Smith [8] and Chan, Smith and Zou [7].

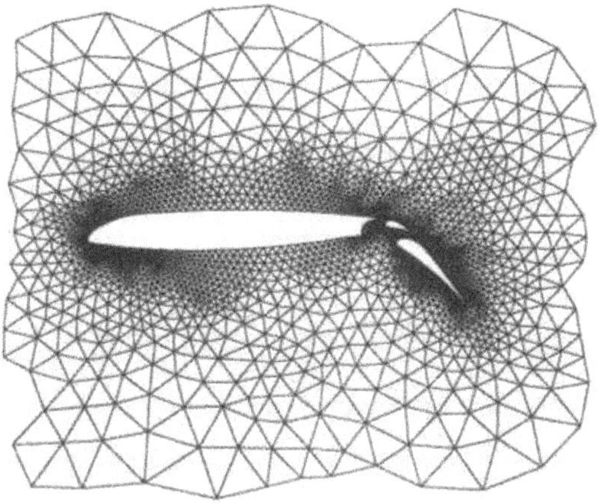

Figure 1: The *Airfoil* mesh: 4253 nodes

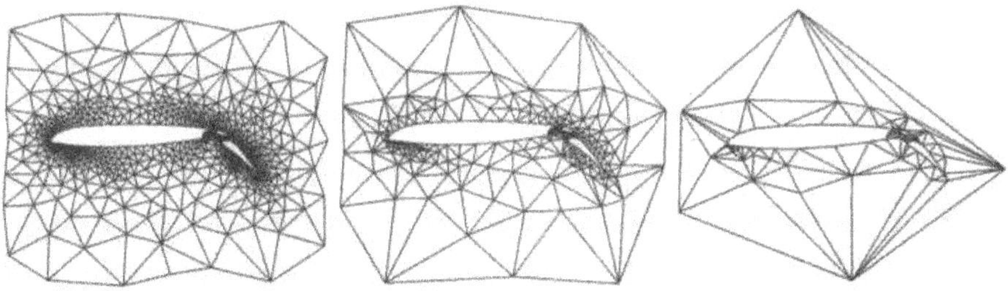

Figure 2: The coarser *Airfoil* meshes

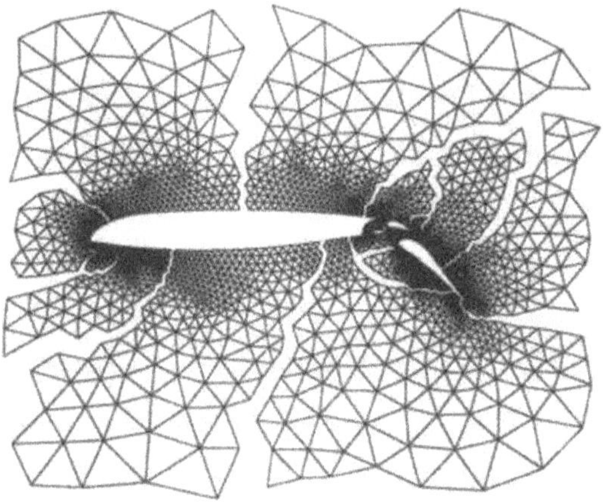

Figure 3: The *Airfoil* mesh: 16 subdomains computed by RSB

4.1 Multiplicative Schwarz results

We have partitioned the original fine grid into 16 pieces using the *recursive spectral bisection* method, [13]. These are given in Figure 3. In Table 1 we give the number of iterations required to achieve a relative decrease in the discrete two norm of 10^{-6}. Exact solvers (sparse LU factorization) were used to solve both the local problems and the coarse grid problem.

Table 1: Iterations for the *Airfoil* mesh. Multiplicative Schwarz

Overlap (no. elements)	Level of coarse grid	Regular coarsening	Dual graph coarsening
0	None	56	56
0	4	21	22
0	3	15	12
1	None	16	16
1	4	10	10
1	3	7	7
2	None	14	14
2	4	8	8
2	3	5	5

4.2 Multigrid results

In Table 2 we give the convergence results for standard V-cycle multigrid. We have used 2 pre and 2 post smoothing steps of symmetric pointwise Gauss-Seidel on each level. These results are comparable to those obtained on a uniformly refined mesh.

Table 2: Multigrid iterations for the *Airfoil* mesh

	Regular coarsening		Dual graph coarsening	
MG Levels	Nodes	Dir. B.C. .	Nodes	Dir. B.C.
2	1180	4	1507	4
3	518	4	328	4
4	89	4	171	5

References

[1] In H. Deconinck and Tim Barth, editors, *Special course on unstructured grid methods for advection dominated flows*, March 1992. AGARD REPORT 787, Special course at the VKI, Belgium.

[2] S.T. Barnard and H.D. Simon. A fast multilevel implementation of recursive spectral bisection. *Proc. 6th SIAM Conf. Parallel Proc. for Sci. Comp.*, pages 711 -718, 1993.

[3] James H. Bramble, Joseph E. Pasciak, Junping Wang, and Jinchao Xu. Convergence estimates for product iterative methods with applications to domain decomposition. *Math. Comp.*, 57(195):1–21, 1991.

[4] Xiao-Chuan Cai. A non-nested coarse space for Schwarz type domain decomposition methods. Technical report, Department of Computer Science, University of Colorado at Boulder, November, 1993.

[5] Xiao-Chuan Cai and Youcef Saad. Overlapping domain decomposition algorithms for general sparse matrices. Technical Report Tech. Rep. 93-027, Army High Performance Comp. Res. Center, Univ. of Minn., March 1993.

[6] J.C. Cavendish. Automatic triangulation of arbitrary planar domains for the finite element method. *Int'l J. Numer. Meth. Engineering*, 8:679–696, 1974.

[7] Tony F. Chan, , Barry Smith, and Jun Zou. Overlapping schwarz methods on unstructured meshes using non-matching coarse grids. Technical Report 94-8, Department of Mathematics, University of California at Los Angeles, February 1994.

[8] Tony F. Chan and Barry Smith. Domain decomposition and multigrid methods for elliptic problems on unstructured meshes. Proceedings of the 7th SIAM Conference on Domain Decomposition Methods, SIAM, Philadelphia, 1994.

[9] Tony F. Chan and Jun Zou. Additive Schwarz domain decomposition methods for elliptic problems on unstructured meshes. Technical Report 93-40, Department of Mathematics, University of California at Los Angeles, December 1993.

[10] Maksymilian Dryja and Olof B. Widlund. An additive variant of the Schwarz alternating method for the case of many subregions. Technical Report 339, also Ultracomputer Note 131, Department of Computer Science, Courant Institute, 1987.

[11] H. Guillard. Node-nested multi-grid method with Delaunay coarsening. Technical report, INRIA, Sophia Antipolis, France, 1993.

[12] D.J. Mavriplis. Unstructured mesh algorithms for aerodynamic calculations. Technical Report 92-35, ICASE, NASA Langley, Virginia, July 1992.

[13] A. Pothen, H. Simon, and K.P. Liou. Partitioning sparse matrices with eigenvector of graphs. *SIAM J. Mat. Anal. Appl.*, 11(3):430–452, 1990.

[14] L. R. Scott and S. Zhang. Higher-dimensional nonnested multigrid methods. *Math. Comp.*, 58:457–466, 1992.

Tony F. Chan and Barry Smith
Department of Mathematics
University of California at Los Angeles
Los Angeles, CA 90024-1555
E-mail: chan@math.ucla.edu, bsmith@math.ucla.edu

Jun Zou
Department of Mathematics
University of California at Los Angeles
Los Angeles, CA 90024-1555
and
Computing Center the Chinese Academy of Sciences
Beijing 100080, P. R. China
E-mail: zou@math.ucla.edu.

Superconvergence of the mixed finite element approximations to parabolic equations

Hongsen Chen* Richard Ewing* Raytcho Lazarov †

Abstract

Semidiscrete mixed finite element approximation to parabolic initial-boundary value problems is introduced and analyzed. Superconvergence estimates for both pressure and velocity are obtained. The estimates for the errors in pressure and velocity depend on the smoothness of the initial data including the limiting cases of data in L^2 and data in H^r, for r sufficiently large. Because of the smoothing properties of the parabolic operator these estimates for large time levels essentially coincide with the estimates obtained earlier for smooth solutions. However, for small time intervals we obtain the correct convergence orders for nonsmooth data.

1 Introduction

Since the pioneering work of Raviart and Thomas [14] the mixed finite element approximations to second order elliptic problems have drawn the attention of many specialists on numerical partial differential equations. This method provides direct approximation of the physical quantities such as fluxes and velocities and leads to schemes that are locally conservative. Due largely to Babuška[1] and Brezzi [2], it is now well understood that the finite element spaces approximating different physical quantities (pressure and velocity, temperature and flux, etc.) cannot be chosen independently. Then the so-called *inf-sup* condition of Babuška–Brezzi is essential if one wants to construct unconditionally stable schemes with optimal convergence rates.

Superconvergence results are important from an application point of view since under reasonable assumptions on the grid and with additional smoothness of the solution, they provide higher order accuracy. To our knowledge the first superconvergence results in the mixed method for second order elliptic problems were proven by

*Department of Mathematics, and Institute for Scientific Computation, Texas A&M University, College Station, TX 77840

†Department of Mathematics, Texas A&M University, College Station, TX 77840, and Institute of Mathematics, Bulgarian Academy of Sciences, Sofia

Douglas and Milner in [5] and by Nakata, Weiser and Wheeler in [12]. Since then a wide variety of results has been obtained by many other authors, see, e.g., Duran [7], Douglas and Wang [6], Wang [18], and Ewing, Lazarov and Wang [9].

The corresponding error analysis of mixed finite element methods applied to time-dependent problems was developed by Johnson and Thomée in [10]. Then, Squeff in [15] using the quasi-projections of Douglas, Dupont and Wheeler, and the averaging method of Bramble and Schatz obtained superconvergence rates for parabolic mixed finite element solutions in R^2.

All known superconvergence results for mixed parabolic finite element equations require that the solutions be sufficiently smooth, uniformly in time. However, many practical problems involve irregularities, such as, in the initial data. This may results in a break-down of the uniform regularity of the solutions and, therefore, a non-availability of the error estimates obtained before. Fortunately, the linear parabolic operators have the so-called *smoothing property*. Namely, if the initial data given at $t = 0$ is nonsmooth or does not satisfy the compatibility conditions on the boundary of the domain then the solution is sufficiently smooth away from $t = 0$ but has a singularity of the form $t^{-\beta}$ with some β positive. This kind of smoothing property has been proved also for the standard Galerkin parabolic finite element equations and used to derive optimal error estimates for problems with rough initial data, see, e.g., Luskin and Rannacher [11], Thomée [16] and Rannacher [13]. Furthermore, superconvergence results for the gradient for the standard Galerkin finite element methods with initial data of low regularity were obtained by Thomée, Xu and Zhang in [17] and by Chen in [4]. The main goal of this paper is to establish superconvergence estimates for the mixed finite element methods when applied to parabolic problems with rough initial data.

2 Problem Formulation

Let $\Omega \in R^2$ be a bounded domain with boundary $\partial\Omega$. We consider the following parabolic problem:

$$\begin{cases} p_t - \nabla\cdot(a\nabla p) = f & \text{in } \Omega\times(0,\infty), \\ p = 0, \quad \text{on } \partial\Omega\times(0,\infty), & p|_{t=0} \text{ is given }, \end{cases} \tag{1}$$

where $a = a(x,t)$ is a sufficiently smooth function that is bounded below by a positive constant on $\Omega\times(0,\infty)$. Our purpose is to solve problem (1) using mixed finite element methods. To describe the mixed variational form for (1), as usual, we introduce two Hilbert spaces. Let

$$W = L^2(\Omega), \qquad V = \left\{\varphi \in L^2(\Omega)^2,\ \nabla\cdot\varphi \in L^2(\Omega)\right\},$$

and the space V be equipped with the norm $\|\varphi\|_V = (\|\varphi\|^2 + \|\nabla\cdot\varphi\|^2)^{1/2}$. The inner product and the norm in $L^2(\Omega)$ are denoted by $(\cdot,\cdot)$ and $\|\cdot\|$, respectively. For the sake of simplicity, $(\cdot,\cdot)$ and $\|\cdot\|$ are also used as the inner product and norm, respectively, in the product space $L^2(\Omega)^2$.

Let $\boldsymbol{u} = a\nabla p$; then the pair $(p, \boldsymbol{u}) \in W \times \boldsymbol{V}$ satisfies the following mixed variational equation:

$$\begin{cases} (p_t, \psi) - (\nabla \cdot \boldsymbol{u}, \psi) = (f, \psi), & \forall\, \psi \in W, t \in (0,\infty), \\ (\alpha \boldsymbol{u}, \boldsymbol{\varphi}) + (\nabla \cdot \boldsymbol{\varphi}, p) = 0, & \forall\, \boldsymbol{\varphi} \in \boldsymbol{V}, t \in (0,\infty), \end{cases} \tag{2}$$

where $p_t = \partial p/\partial t$, $\alpha = a^{-1}$ and $p(0)$ is given. We note that the boundary condition $p = 0$ on $\partial\Omega$ is implicitly contained in (2).

Given the finite-dimensional spaces $W_h \subset W$ and $\boldsymbol{V}_h \subset \boldsymbol{V}$, $0 < h < 1$, the so-called mixed finite element approximation $(p_h, \boldsymbol{u}_h) \in W_h \times \boldsymbol{V}_h$ to the pair $(p, \boldsymbol{u}) \in W \times \boldsymbol{V}$ is the solution of the following problem:

$$\begin{cases} (p_{h,t}, \psi_h) - (\nabla \cdot \boldsymbol{u}_h, \psi_h) = (f, \psi_h), & \forall\, \psi_h \in W_h, t \in (0,\infty), \\ (\alpha \boldsymbol{u}_h, \boldsymbol{\varphi}_h) + (\nabla \cdot \boldsymbol{\varphi}_h, p_h) = 0, & \forall\, \boldsymbol{\varphi}_h \in \boldsymbol{V}_h, t \in (0,\infty), \end{cases} \tag{3}$$

where, $p_h(0) \in W_h$ is given. We note that $\boldsymbol{u}_h(0)$ is determined by $p_h(0)$ through the second equation of (3).

To ensure the existence and convergence of the solution of the above formulation, we assume that $\nabla \cdot \boldsymbol{V}_h \subset W_h$ and there exists a linear operator $\boldsymbol{\Pi}_h$: $\boldsymbol{V} \to \boldsymbol{V}_h$ such that $\nabla \cdot \boldsymbol{\Pi}_h = Q_h \nabla \cdot$. Here, the operator $Q_h : W \to W_h$ is the L^2-projection. The classical inf-sup condition is then satisfied. Further, we assume that there exists an integer $r \geq 0$ such that the following approximation properties are satisfied:

$$\|\boldsymbol{\varphi} - \boldsymbol{\Pi}_h \boldsymbol{\varphi}\| \leq ch^i \|\boldsymbol{\varphi}\|_{\boldsymbol{H}^i(\Omega)}, \quad \forall\, \boldsymbol{\varphi} \in \boldsymbol{H}^i(\Omega),\ 1 \leq i \leq r+1, \tag{4}$$

$$\|\psi - Q_h \psi\| \leq ch^i \|\psi\|_{H^i(\Omega)}, \quad \forall\, \psi \in H^i(\Omega),\ 0 \leq i \leq r+1. \tag{5}$$

Here, and throughout the paper, the letter c is used as a generic constant, which is independent of h, p, $\boldsymbol{u}$, etc. $H^i(\Omega)$ denotes the standard Sobolev space $W^{i,2}(\Omega)$ with $H^0(\Omega) = L^2(\Omega)$. Examples of spaces of piece-wise polynomials that satisfy the conditions stated above are the triangular and rectangular Raviat-Thomas elements from [14] (for other examples see Brezzi and Fortin [3]).

3 L^2-error estimates

This section is designed to provide the preliminaries for deriving the superconvergence results obtained in the next section. The main results are contained in Theorem 1, where L^2- estimates for $(p-p_h)^{(k)}(t)$ and $(\boldsymbol{u}-\boldsymbol{u}_h)^{(k)}(t)$ are derived for homogeneous equations with initial data from $H^i(\Omega)$, $0 \leq i \leq k+1$, including the case of data just in $L^2(\Omega)$.

Lemma 1 *Let $(p, \boldsymbol{u}) \in W \times \boldsymbol{V}$ and $(p_h, \boldsymbol{u}_h) \in W_h \times \boldsymbol{V}_h$ be the solutions of problems (2) and (3), respectively. If $p_h(0) = Q_h p(0)$, then*

$$\| (Q_h p - p_h)(t) \|^2 + \int_0^t \| \boldsymbol{u} - \boldsymbol{u}_h \|^2 \, ds \leq c \int_0^t \| \boldsymbol{u} - \boldsymbol{\Pi}_h \boldsymbol{u} \|^2 \, ds. \tag{6}$$

$$\| (p - p_h)(t) \|_{H^{-j}(\Omega)} \leq ch^{i+j} \| p(0) \|_{H^i(\Omega)}, \quad 0 \leq i, j \leq r+1. \tag{7}$$

$$\int_0^t \| p - p_h \|^2 \, ds \leq ch^2 \| p(0) \|^2 . \tag{8}$$

$$\| (p - p_h)(t) \| \leq cht^{-1/2} \| p(0) \| . \tag{9}$$

Based on the above lemmas we prove the main result of this section stated in the following theorems. Below, for a function $\theta(t)$, we denote its k-th derivative with respect to t as $\theta^{(k)}(t)$.

Theorem 1 *Let $(p, \boldsymbol{u}) \in W \times \boldsymbol{V}$ and $(p_h, \boldsymbol{u}_h) \in W_h \times \boldsymbol{V}_h$ be the solutions of problems (2) and (3), respectively. If $f = 0$ and $q_h(0) = Q_h q(0)$, then,*

$$\| (p - p_h)^{(k)}(t) \| \leq ch^m t^{-(m-i)/2-k} \| p(0) \|_{H^i(\Omega)}, \tag{10}$$

$$\| (\boldsymbol{u} - \boldsymbol{u}_h)^{(k)}(t) \| \leq ch^m t^{-(m+1-i)/2-k} \| p(0) \|_{H^i(\Omega)}, \tag{11}$$

where $0 \leq i \leq m \leq r+1$, $k \geq 0$ and $p(0) \in H_0^i(\Omega)$.

Proof : This is the most technical result. We prove it by induction and using the results of the lemmas and theorems stated above.

4 Superconvergence

The estimate obtained in the previous sections can be interpreted in the following way: the maximum rate of convergence for $p - p_h$ is $O(h^{r+1}t^{-(r+1-i)/2})$ while for $\boldsymbol{u} - \boldsymbol{u}_h$ is $O(h^{r+1}t^{r+2-i)/2})$ for initial data in $H^i(\Omega)$. Obviously, for any fixed $t > 0$, the convergence is asymptotically $O(h^{r+1})$, due to the smoothing properties of the parabolic operator. For smooth initial data, i.e., $p(0)$ from $H^{r+1}(\Omega)$, the error for p is asymptotically $O(h^{r+1})$ for any t. It should be pointed out that the estimates (10) and (11) cover the whole range of smoothness of the initial data, including the worst case of data only in $L^2(\Omega)$.

The estimates (10) and (11) are of optimal type and cannot be improved in terms of the norms involved even if the solutions are smoother. However, in the case of a smooth solution one can select special points or postprocess the finite element solution in order to obtain higher order convergence. Such estimates, called superconvergence estimates, have been obtained for the standard Galerkin method for parabolic problems by Thomée, Xu and Zhang [17].

For Raviart-Thomas finite elements of order $r+1$, which fit into the framework of this paper, Ewing and Lazarov in [8] derive $O(h^{r+2})$ error estimates for both $p - p_h$ and $\boldsymbol{u} - \boldsymbol{u}_h$ at the Guass points in the case of sufficiently smooth solution for all $t \geq 0$. Our goal in this section is to obtain superconvergent type estimates for more realistic situations, namely, for data in $H^i(\Omega)$, for $0 \leq i \leq r+2$, including the case of data in $L^2(\Omega)$.

In order to bound the temporal derivatives of $Q_h p - p_h$ we we need estimates for $Q_h p - p_h$ and $\boldsymbol{\Pi}_h \boldsymbol{u} - \boldsymbol{u}_h$ in the L^2-norm. We begin with an estimate for $Q_h p - p_h$ and then we derive estimates for the derivatives of $\boldsymbol{\Pi}_h \boldsymbol{u} - \boldsymbol{u}_h$ and $Q_h p - p_h$ with respect to time. Next, the duality argument for the error estimates discussed in Theorem 2, requires estimates in negative Sobolev norm. These estimates are obtained in the following lemma.

Lemma 2 *Let $(p, \boldsymbol{u}) \in W \times \boldsymbol{V}$ and $(p_h, \boldsymbol{u}_h) \in W_h \times \boldsymbol{V}_h$ be the solutions of problems* (2) *and* (3), *respectively. If $f = 0$ and $p_h(0) = Q_h p(0)$, then*

$$\| (Q_h p - p_h)(t) \|_{H^{-j}(\Omega)} \leq c h^{r+2+j} \, \| p(0) \|_{H^{r+2}(\Omega)}, \quad 0 \leq j \leq r-1, \tag{12}$$

$$\| (Q_h p - p_h)(t) \|_{H^{-r}(\Omega)} \leq c h^{2r+2} \, | \log h | \, \| p(0) \|_{H^{r+2}(\Omega)} \, . \tag{13}$$

Now, we are ready to obtain the main results of this section, namely, estimates of $Q_h p - p_h$ for problems with nonsmooth initial data.

Theorem 2 *Let $(p, \boldsymbol{u}) \in W \times \boldsymbol{V}$ and $(p_h, \boldsymbol{u}_h) \in W_h \times \boldsymbol{V}_h$ be the solutions of problems* (2) *and* (3), *respectively. If $f = 0$ and $p_h(0) = Q_h p(0)$, and $k \geq 0$, then*

$$\| (Q_h p - p_h)^{(k)}(t) \| \leq c h^{r+2} t^{-(r+2-i)/2-k} \, \| p(0) \|_{H^i(\Omega)}, \tag{14}$$

$$\| (\boldsymbol{\Pi}_h \boldsymbol{u} - \boldsymbol{u}_h)^{(k)}(t) \| \leq c h^{r+2} t^{-(r+3-i)/2-k} \, \| p(0) \|_{H^i(\Omega)}, \tag{15}$$

where $1 \leq i \leq r+2$.

Now we present one possible way of using the results of Theorem 2 for superconvergent recovery of the solution p and $\boldsymbol{u}$ from their finite element approximations p_h and $\boldsymbol{u}_h$ in the case of Raviart-Thomas rectangular elements in rectangular domains Ω. Let T_h be partition of the domain Ω into rectangles with sides parallel to the coordinate axes.

Let $P_r(e)$ be the restriction of the polynomials of total degree r to the set e and $P_{r,l}(e)$ the restriction of $P_r(R^1) \times P_l(R^1)$ to element e. We set

$$\boldsymbol{V}(e) = P_{r+1,r}(e) \times P_{r,r+1}(e), \quad W(e) = P_{r,r}(e),$$

$$\boldsymbol{V}_h = \{ \boldsymbol{v} \in \boldsymbol{V} : \boldsymbol{v}|_e \in \boldsymbol{V}(e), e \in T_h \}, \quad W_h = \{ w \in W : w|_e \in W(e), e \in T_h \}.$$

The spaces $\boldsymbol{V}_h$ and W_h defined above satisfy the approximation properties (4) and (5) and the projection operators $\boldsymbol{\Pi}_h$ and Q_h are defined element by element.

Next, we introduce some discrete seminorms in W and $\boldsymbol{V}$. Let $\hat{g}_i$, $i = 1, ..., r+1$ be the Gauss points in the interval $[-1,1]$, i.e., $L_{r+1}(\hat{g}_i) = 0$, where L_{r+1} is the Legendre polynomial of degree $r+1$ orthogonal on the interval $[-1,1]$. Each finite element $e \in T_h$ can be transferred by an affine mapping F to the reference element $\hat{e} = [-1,1] \times [-1,1]$. Then, the inverse mapping F^{-1} introduces in e the set of points $G(e) = \{F^{-1}(\hat{g}_i, \hat{g}_j), i, j = 1, \cdots, r+1\}$. The seminorms in W and $\boldsymbol{V}$ are defined by

$$|||w|||= \left(\sum_{e\in T_h} \sum_{x\in G(e)} h^2 w^2(x) \right)^{1/2}, \quad |||\boldsymbol{v}|||=|||v_1||| + |||v_2|||, \quad \boldsymbol{v} = (v_1, v_2).$$

It is easy to see that there is a constant $c > 0$ such that

$$|||w_h|||\leq c \, \| w_h \|, \qquad |||\boldsymbol{v}_h|||\leq c \, \| \boldsymbol{v}_h \|, \quad \forall \, w_h \in W_h, \; \boldsymbol{v}_h \in \boldsymbol{V}_h.$$

Now, we bound the errors for $p - p_h$ and $\boldsymbol{u} - \boldsymbol{u}_h$ in the discrete seminorms defined above, which arte the desired superconvergence estimates.

Theorem 3 *Let $(p, \boldsymbol{u}) \in W \times \boldsymbol{V}$ and $(p_h, \boldsymbol{u}_h) \in W_h \times \boldsymbol{V}_h$ be the solutions of problems* (2) *and* (3), *respectively. If $f = 0$ and $p_h(0) = Q_h p(0)$, then*

$$|||(p - p_h)(t)|||\leq ch^{r+2} t^{-(r+2-i)/2} \, \| p(0) \|_{H^i(\Omega)}, \tag{16}$$

$$|||(\boldsymbol{u} - \boldsymbol{u}_h)(t)|||\leq ch^{r+2} t^{-(r+3-i)/2} \, \| p(0) \|_{H^i(\Omega)}, \tag{17}$$

where $1 \leq i \leq r+2$.

References

[1] I. Babuška, *The finite element method with Lagrange multipliers*, Numer. Math., 20, 179–192 (1973).

[2] F. Brezzi, *On existence, uniqueness and approximation of saddle-point problems arising from Lagrange multipliers*, RAIRO Anal. Numer., 2, 129–151 (1974).

averaging in

elements

[3] F. Brezzi, M. Fortin, *Mixed and Hybrid Finite Element Methods*, Springer-Verlag New York Inc., 1991, 350 p.

[4] H. Chen, *An L^2- and L^∞-error analysis for the parabolic finite element equations with applications to superconvergence and error expansions*, Ph.D Thesis, Univ. Heidelberg (1993).

[5] J. Douglas and F. Milner, *Interior and superconvergence estimates for mixed methods for second order elliptic equations*, RAIRO Model. Math. Anal. Numer., 19, 297–328 (1985).

[6] J. Douglas, J. Wang, *Superconvergence of mixed finite element methods on rectangular domains*, Calcolo, 26, 121-134 (1989).

[7] R. Duran, *Superconvergence for rectangular mixed finite elements*, Numer. Math., 58, 287-298 (1990).

[8] R. E. Ewing, R. D. Lazarov, *Superconvergence of the mixed finite element approximations of parabolic problems using rectangular finite elements*, East-West J. Numer. Math., 1, 199-212 (1993).

[9] R. E. Ewing, R. D. Lazarov, J. Wang, *Superconvergence of the velocity along the Gauss lines in mixed finite element methods*, SIAM J. Numer. Anal., 28(4), 1015-1029 (1991).

RAIRO Numer.

[10] C. Johnson, V. Thomée, *Error estimates for some mixed finite element methods for parabolic type problems*, RAIRO Anal. Numer., 15, 41-78 (1982)

[11] M. Luskin, R. Rannacher, *On the smoothing property of the Galerkin method for parabolic equations*, SIAM J. Numer. Anal., 19, 93-113 (1981).

[12] M. Nakata, A. Weiser, M. F. Wheeler, *Some superconvergence results for mixed finite element methods for elliptic problems on rectangular domains*, in the mathematics of finite elements and applications, MAFELAP 1985, J. R. Whiteman, ed., 1985.

[13] R. Rannacher, *Finite element solution of diffusion problems with irregular data*, Numer. Math., 43, 309-327 (1984).

[14] P. A. Raviart, J. M. Thomas, *A mixed finite element method for second order elliptic problems*, Mathematical Aspect of the Finite Element Method. Lecture Notes Math., 606, 292-315 (1977).

[15] M. C. Squeff, *Superconvergence of mixed finite element methods for parabolic problems*, M^2AN Math. Model. Numer. Anal., 21,327-352 (1987).

[16] V. Thomée, *Galerkin finite element methods for parabolic problems*, Lecture Notes in Mathematics, 1054, Springer-Verlag, 1984

[17] V. Thomée, J. Xu, N. Zhang, *Superconvergence of the gradient in piecewise linear finite element approximation to a parabolic problem*, SIAM J. Numer. Anal. , 26(3),553-573 (1989).

[18] J. Wang, *Superconvergence and extrapolation for mixed finite element methods on rectangular domains*, Math. Comp., 56, 477–503 (1991).

EFFICIENT DIFFERENCE SCHEMES FOR SOLVING CONVECTION PROBLEMS IN THE STREAM FUNCTION – VORTICITY FORMULATION

V.V. Chudanov A.G. Churbanov P.N. Vabishchevich

1 Introduction

The stream function–vorticity variables are in general use for the numerical simulation of incompressible flows [1]. There do exist pro and contra arguments in using such a formulation, but evidently that this approach seems to be rather suitable and efficient for two-dimensional flow predictions in simply connected calculation domains. At the present time, there is a variety of numerical studies on the basis of this formulation with emphasis on the approximation of convective terms and the boundary conditions for the vorticity.

In the present work, a new efficient class of difference schemes is developed for two-dimensional time-dependent problems of a viscous incompressible fluid in the stream function–vorticity formulation. The schemes are unconditionally stable (in linear sense), i.e. the time-step is practically independent on a spatial grid. The only restriction do exist due to the nonlinearity of convection problems. Transition to the next time level in the schemes involves the solution of linear grid elliptic equations for the stream function and vorticity. For any time-step used, an a priori estimate is valid for the discrete solution, similar to that obtained for the original continuous differential problem.

2 Governing equations

Let us consider a time-dependent incompressible flow in a 2D rectangular cavity $\Omega = \{x \mid x = (x_1, x_2), \quad 0 < x_\alpha < l_\alpha, \quad \alpha = 1, 2\}$. Such a flow is governed by the following momentum equation:

$$\frac{\partial \mathbf{v}}{\partial t} + \mathrm{V}(\mathbf{v})\mathbf{v} + \mathrm{grad} p - \nu \Delta \mathbf{v} = \mathbf{f}(x,t), \quad x \in \Omega, \quad 0 < t \leq T, \tag{1}$$

and the incompressibility constraint:

$$\mathrm{div}\ \mathbf{v} = 0, \quad x \in \Omega, \quad 0 < t \leq T. \tag{2}$$

Here $\mathbf{v} = (v_1, v_2)$ is the velocity vector, p – the pressure normalized with respect to density, $\mathrm{V} = \mathrm{V}(\mathbf{v})$ – the convective transport operator, ν – the cinematic viscosity, Δ – the Laplace operator and $\mathbf{f}(x,t)$ – the vector of volumetric forces.

The cavity boundary $\partial\Omega$ is assumed to be rigid and fixed, so the no-slip and no-permeability conditions are imposed:

$$\mathbf{v}(x,t) = 0, \quad x \in \partial\Omega, \quad 0 < t \leq T. \tag{3}$$

The quiescent state is considered as the initial condition for the time evolution:

$$\mathbf{v}(x,0) = 0, \quad x \in \Omega. \tag{4}$$

The problem (1)–(4) provides the complete description of fluid motion in Ω at any time moment $t > 0$.

To reduce the number of the dependent variables, let us introduce the new ones. The velocity components can be expressed using the stream function $\psi(x,t)$ as follows:

$$v_1 = \frac{\partial \psi}{\partial x_2}, \quad v_2 = -\frac{\partial \psi}{\partial x_1}, \tag{5}$$

and hence, the incompressibility constraint (2) identically holds. Next, for the vorticity ω we have:

$$\omega = \frac{\partial v_2}{\partial x_1} - \frac{\partial v_1}{\partial x_2}. \tag{6}$$

Taking into account definitions (5) and (6), the momentum equation (1) can be rewritten as the vorticity transport equation:

$$\frac{\partial \omega}{\partial t} + \mathrm{V}(\mathbf{v})\omega - \nu \Delta \omega = \varphi(x,t), \quad x \in \Omega, \quad 0 < t \leq T, \tag{7}$$

where $\varphi = \partial f_2/\partial x_1 - \partial f_1/\partial x_2$. For the stream function we obtain the Poisson equation:

$$-\Delta \psi = \omega, \quad x \in \Omega, \quad 0 < t \leq T. \tag{8}$$

This system of equations is complemented by the boundary and initial conditions:

$$\psi(x,t) = 0, \quad \frac{\partial \psi}{\partial \mathbf{n}}(x,t) = 0, \quad x \in \partial\Omega, \quad 0 < t \leq T, \tag{9}$$

$$\psi(x,0) = 0, \quad x \in \Omega, \tag{10}$$

where $\mathbf{n}$ stands for the outer normal vector.

In equation (7) the convective terms can be written in different ways. For example, using definition (5) we can introduce the following divergence form of the convective terms:

$$\mathrm{V}(\mathbf{v}) = \sum_{\alpha=1}^{2} \mathrm{V}_\alpha(v_\alpha), \quad \mathrm{V}_\alpha(v_\alpha)\omega = \frac{\partial}{\partial x_\alpha}(v_\alpha \omega), \quad \alpha = 1,2, \tag{11}$$

$$\mathrm{V}_1(v_1)\omega = \frac{\partial}{\partial x_1}(\frac{\partial \psi}{\partial x_2}\omega), \quad \mathrm{V}_2(v_2)\omega = -\frac{\partial}{\partial x_2}(\frac{\partial \psi}{\partial x_1}\omega). \tag{12}$$

For this definition of the convective terms we have the following fundamental property:

$$(\mathrm{V}(\mathbf{v})\omega, \psi) = 0. \tag{13}$$

For the problem (7)–(10) using (13) we can derive a simple a priori estimate. Let $\mathcal{H} = L_2(\Omega)$ be a Hilbert space with the norm $\|\cdot\|$ and scalar product $(\cdot,\cdot)$. Multiplying equation (7) by ψ term-by-term, we get

$$\frac{1}{2}\frac{d}{dt}\|\mathrm{grad}\psi\|^2 \leq \nu\|\omega\|^2 + 2\|\mathbf{f}\|\|\mathrm{grad}\psi\|.$$

Hence the inequality

$$\|\mathrm{grad}\psi(x,t)\| \leq \|\mathrm{grad}\psi(x,0)\| + 2\int_0^t \|\mathbf{f}(x,s)\|ds. \tag{14}$$

holds. The estimate (14), which expresses boundedness of the solution of (7)–(10), will serve as a check point in the construction of difference schemes.

3 Difference scheme

To solve equations (7)–(10) approximately, difference methods are used. Let us introduce in Ω an uniform grid $\bar{\omega}_h = \omega_h \cup \partial\omega_h$ with steps h_1 and h_2. Let ω_h be the set of the internal points, i.e.

$$\begin{aligned} \omega_h &= \{x \mid x = (x_1, x_2), \quad x_\alpha = i_\alpha h_\alpha, \\ & \quad i_\alpha = 1, 2, \dots, N_\alpha - 1, \quad N_\alpha h_\alpha = l_\alpha, \quad \alpha = 1, 2\}, \end{aligned}$$

and $\partial\omega_h$– the set of the boundary points. We shall use the common index-free notation of the theory of difference schemes [2]. For the forward, backward and central difference derivatives we use respectively the following notations $w_{x_\alpha}, w_{\bar{x}_\alpha} w_{\overset{\circ}{x}_\alpha}$ for any grid function w. Let us define the discrete Laplace operator on the set of grid functions $y \in H$, which are equal to zero at the boundary:

$$\Lambda y = \sum_{\alpha=1}^{2} \Lambda_\alpha y, \quad \Lambda_\alpha y = -y_{\bar{x}_\alpha x_\alpha}, \quad \alpha = 1, 2. \tag{15}$$

The scalar product in the finite-dimensional Hilbert space H is defined as

$$(y, z) = \sum_{x \in \omega_h} y(x) z(x) h_1 h_2.$$

The operator Λ in H is selfadjoint and positive definite [2], i.e. $\Lambda = \Lambda^* > 0$.

Approximation of convective terms should be given a special consideration [3]. Aimed at using linearized schemes, we can employ neither the divergence form (11), (12) nor the advective (non-divergence) form due to the fact that the fulfillment of a discrete analog of (13) is hardly expected.

Using equations (5), (6) we can rewrite the convective terms as follows

$$\begin{aligned} \mathrm{V}(\mathbf{v})\omega &= \frac{\partial}{\partial x_1}\left(\frac{\partial \psi}{\partial x_2}\omega\right) - \frac{\partial}{\partial x_2}\left(\frac{\partial \psi}{\partial x_1}\omega\right) = -\frac{\partial}{\partial x_1}\left(\frac{\partial \omega}{\partial x_2}\psi\right) + \frac{\partial}{\partial x_2}\left(\frac{\partial \omega}{\partial x_1}\psi\right) \\ &= -\frac{1}{2}\left(\frac{\partial \omega}{\partial x_2}\frac{\partial \psi}{\partial x_1} + \frac{\partial}{\partial x_1}\left(\frac{\partial \omega}{\partial x_2}\psi\right)\right) + \frac{1}{2}\left(\frac{\partial \omega}{\partial x_1}\frac{\partial \psi}{\partial x_2} + \frac{\partial}{\partial x_2}\left(\frac{\partial \omega}{\partial x_1}\psi\right)\right). \end{aligned}$$

Introducing the vector $\mathbf{w}$,

$$\mathbf{w} = \{w_1, w_2\}, \quad w_1 = \frac{\partial \omega}{\partial x_2}, \quad w_2 = -\frac{\partial \omega}{\partial x_1}, \tag{16}$$

we can rewrite the convective transport of the vorticity as

$$\mathrm{V}(\mathbf{v})\omega = -\tilde{\mathrm{V}}(\mathbf{w})\psi \equiv -\frac{1}{2}(\mathrm{V}(\mathbf{w}) - \mathrm{V}^*(\mathbf{w}))\psi, \tag{17}$$

i.e., we treat the convective transport of the vorticity as an effective transport of the stream function! In light of (11), the adjoint operator in (17) has the form

$$\mathrm{V}^*(\mathbf{v}) = \sum_{\alpha=1}^{2} \mathrm{V}_\alpha^*(v_\alpha), \quad \mathrm{V}_\alpha^*(v_\alpha)\omega = -v_\alpha \frac{\partial}{\partial x_\alpha}\omega, \quad \alpha = 1, 2$$

and corresponds to the non-divergence form of convective terms. Now with the operator formulation (16), (17), the fundamental constraint (13) is fulfilled for all vectors $\mathbf{w}$.

In accordance with (16), (17), using central difference approximations for derivatives, we approximate the convective terms as follows

$$\tilde{V}(\mathbf{w}) = \sum_{\alpha=1}^{2} \tilde{V}_\alpha(w_\alpha), \tag{18}$$

where, in light of (16),

$$\begin{aligned} \tilde{V}_1(w_1)\psi &= \tfrac{1}{2}(\omega_{\overset{\circ}{x}_2}\psi_{\overset{\circ}{x}_1} + (\omega_{\overset{\circ}{x}_2}\psi)_{\overset{\circ}{x}_1}), \\ \tilde{V}_2(w_2)\psi &= -\tfrac{1}{2}(\omega_{\overset{\circ}{x}_1}\psi_{\overset{\circ}{x}_2} + (\omega_{\overset{\circ}{x}_1}\psi)_{\overset{\circ}{x}_2}). \end{aligned} \tag{19}$$

For this approximation of convective transfer, we have

$$(\tilde{V}(\mathbf{w})\psi, \psi) = 0, \tag{20}$$

which means that the operator $\tilde{V}$ is skew-symmetric in H ($\tilde{V} = -\tilde{V}^*$). Moreover, both operators $\tilde{V}_\alpha(w_\alpha), \alpha = 1, 2$, are also skew-symmetric.

In the sequent, we shall use for approximate solutions the same notation as for the exact ones. Boundary condition (9) is treated using the common Thom approximation [1] for the vorticity so that, for example, we have on the left boundary

$$\omega(0, x_2) = -\frac{2}{h_1^2}\psi(h_1, x_2).$$

The discretized problem is formulated for the internal grid points, hence the spatial discretization of (7), (8) leads to the semidiscrete equation (discrete in space and continuous in time)

$$\frac{d}{dt}\Lambda\psi - \tilde{V}(\mathbf{w})\psi + \nu\Lambda^2\psi + \nu\rho(x)\psi = \varphi(x,t), \quad x \in \omega_h, \quad 0 < t \leq T. \qquad (21)$$

In (21), the grid function $\rho(x)$ in the Thom boundary conditions for the vorticity is defined as

$$\rho(x) = \textstyle\sum_{\alpha=1}^{2} \rho_\alpha(x_\alpha),$$
$$\rho_\alpha(x_\alpha) = \begin{cases} 0, & h_\alpha < x_\alpha < l_\alpha - h_\alpha, \\ 2h_\alpha^{-4}, & x_\alpha = h_\alpha, l_\alpha - h_\alpha, \end{cases} \quad \alpha = 1, 2.$$

We shall write (21) in the following form, convenient for computational realization:

$$\frac{d}{dt}\Lambda\psi + (A_1 + A_2)\psi = \varphi, \qquad (22)$$

where

$$A_1 = \nu\Lambda, \quad A_2 = -\tilde{V}(\mathbf{w}) + \nu\rho(x)E. \qquad (23)$$

This implies $A_1 = A_1^* > 0$ and $A_2 \geq 0$ for any $\mathbf{w}$.

To construct discretization in time, let us introduce a uniform grid in time with step $\tau > 0$:

$$\omega_\tau = \{t | t = t_n = n\tau, \quad n = 0, 1, \ldots, N_\tau, \quad N_\tau \tau = T\}$$

and let $y_n = y(t_n)$.

The simplest schemes for (22), (23) are based on a natural linearization, when the operator $\tilde{V}(\mathbf{w})$ is determined by the solution at the previous time level. Due to the non-negativeness of operators $A_\alpha, \alpha = 1, 2$, various difference schemes based on operator-splitting technique [2] can be considered. When a purely implicit factorized scheme (an analog of the Douglas- Rachford scheme [4]) is used, we have on the first stage, in view of (23), the difference equation

$$\frac{\Lambda\psi_{n+1/2} - \Lambda\psi_n}{\tau} - \tilde{V}(\mathbf{w}_n)\psi_n + \Lambda^2\psi_{n+1/2} + \rho(x)\psi_n = \varphi_n. \qquad (24)$$

Thus, the convective transfer, as well as the data for the boundary condition for the vorticity, are taken from the previous time level. Using the notation $\omega_{n+1/2} = \Lambda\psi_{n+1/2}$, equation (24) can be written as an equation for the vorticity:

$$\frac{\omega_{n+1/2} - \omega_n}{\tau} - \tilde{V}(\mathbf{w}_n)\psi_n + \Lambda\omega_{n+1/2} + \rho(x)\psi_n = \varphi_n. \qquad (25)$$

The second stage consists in the correction in convective transfer and the boundary condition for the vorticity:

$$\frac{\Lambda\psi_{n+1} - \Lambda\psi_{n+1/2}}{\tau} + (\rho(x) - \tilde{V}(\mathbf{w}_n))(\psi_{n+1} - \psi_n) = 0. \tag{26}$$

In going to the next time level, the vorticity is found as a solution of the selfadjoint elliptic problem (25), while for the stream function we have the non-selfadjoint problem (26).

Using the above mentioned properties of the grid operators, it is easy to obtain the following estimate for the scheme (24)-(26):

$$\|\psi_{n+1}\|_* \leq \|\psi_n\|_* + \tau\|\varphi_n\|_{\Lambda^{-1}}, \tag{27}$$

where

$$\|\psi\|_*^2 = \|\psi\|_\Lambda^2 + \tau^2\nu^2\|\rho\psi\|_{\Lambda^{-1}}^2.$$

The estimate (27) was obtained without any restrictions on grid parameters, and can be regarded as a difference analog of a priori estimate (14).

4 Numerical results

To study the features of the developed schemes, the following problems have been studied numerically: a lid-driven cavity flow [5] and a buoyancy-driven flow in a square cavity with side walls of different temperatures. The sequence of uniform grids derived by division in two ($21 \times 21, 41 \times 41, 81 \times 81$ and 161×161) was employed in our calculations on IBM PC AT-486 to verify the accuracy and stability of the schemes. For the first problem the following values of Reynolds number has been considered Re=100, 400, 1000 and 3200. For free convection problem the considered Rayleigh number were $10^4, 10^5$ and 10^6. The predictions indicate that the scheme provides the results with the second-order accuracy in space. Next, a great deal of predictions shows that the time-step is practically the same for various spatial grids.

5 Resume

This work is concerned with the development, numerical implementation and comparative study of finite-difference schemes for fluid flow problems in the stream function-vorticity formulation. Special approximations are developed for convective terms. Linearized additive schemes are constructed. Unconditional apriori estimates for the difference solution are derived. It was shown that they are consistent with the corresponding estimates for the solution of the differential problem. Difference schemes presented here have demonstrated high efficiency in solving well-known benchmark problems of hydrodynamics and convective heat transfer.

References

[1] P.J. Roache, *Computational Fluid Dynamics*, Hermosa, Albuquerque, 1982.

[2] A.A. Samarskii, *Theory of Difference Schemes*, Nauka, Moscow, 1989 (in Russian).

[3] A. Arakawa, Computational design of longterm numerical integration of the equations of fluid motion. 1. Two-dimensional incompressible flow. *J. Comput. Phys.* **1**, 119 (1966).

[4] J. Douglas and H.H. Rachford, On the numerical solution of heat conduction problems in two and three spaces, *Trans. Amer. Math. Soc.* **82**, 421 (1956).

[5] U. Ghia, K.N. Ghia and C.T. Shin, High-Re solutions for incompressible flow using the Navier-Stokes equations and a multigrid method, *J. Comput. Phys.* **48**, 387 (1982).

V.V. Chudanov
Nuclear Safety Institute RAS, Moscow, Russia

A.G. Churbanov, P.N. Vabishchevich
Institute for Mathematical Modeling RAS, Moscow, Russia

On a Parallel Method for Enclosing Real Roots of Nonlinear Equations*

N. S. Dimitrova

Abstract

Considered is an interval iterative method for enclosing all real zeroes of a nonlinear equation $f(x) = 0$ in a given interval. An algorithm with result verification in terms of computer arithmetic operations is proposed. The implementation of the algorithm on a parallel processor is discussed.

1 Introduction

Many problems of applied mathematics and mathematical models lead to the problem of finding all real roots of a nonlinear equation $f(x) = 0$ in a given interval X. In the last two or three decades algorithms with result verification for solving this or other related mathematical problems have been developed. Such algorithms provide full control over all kind of input-, truncation- and computational errors. These methods can be successfully combined with various parallel techniques.

Let IR be the set of all compact intervals on the real line R. Denote by x^- and x^+, $x^- \leq x^+$, the end-points of $X \in IR$, i.e. $X = [x^-, x^+]$. The width of X is defined by $\omega(X) = x^+ - x^-$. For $x_1, x_2 \in R$, where $x_1 \leq x_2$ is not required, the interval X with end-points x_1 and x_2 will be written as $X = [x_1 \vee x_2] = \{[x_1, x_2], \text{ if } x_1 \leq x_2; \quad [x_2, x_1], \text{ if } x_1 \geq x_2\}$. By X^{+0} and X^{-0} we denote the end-points $X^{+0} = \{x^-, \text{ if } |x^-| \leq |x^+|; \quad x^+, \text{ otherwise}\}, X^{-0} = \{x^+, \text{ if } |x^-| \leq |x^+|; \quad x^-, \text{ otherwise}\}$, which satisfy $|X^{+0}| \leq |X^{-0}|$. Thus we can write $X = [X^{+0} \vee X^{-0}]$. Let be $X, Y \in IR$, $X = [x^-, x^+]$, $Y = [y^-, y^+]$. Denote $X \vee Y = [\min\{x^-, y^-\}\ ,\ \max\{x^+, y^+\}]$.

Let $f : D \to R$, $D \subseteq R$, be a continuously differentiable function. Denote by $ID = \{X : X \in IR, X \subseteq D\}$. For $X \in ID$ let $f(X)$ be the range of f on the interval X, i.e. $f(X) = \{f(x) : x \in X\}$. Similarly, $f'(X) = \{f'(x) : x \in X\}$ denotes the range of the derivative f' on X. Define an interval extension F' of f' as any interval function, $F' : ID \to IR$, satisfying the conditions:

$$\begin{cases} f'(X) \subseteq F'(X) \text{ for } X \in ID; \\ F'(X) \subseteq F'(Y), \text{ whenever } X \subseteq Y, \quad X, Y \in ID. \end{cases} \tag{1}$$

*This work has been partially supported by the Ministry of Education and Sciences - National Science Fund under contract No. MM-10/91

Denote by $F'^{+0}(X)$ and $F'^{-0}(X)$ the end-points of $F'(X)$, i.e. $F'(X) = [F'^{+0}(X) \vee F'^{-0}(X)]$. Consider the equation $f(x) = 0$. Let $x_1^*, x_2^*, \dots, x_p^*$ be the real roots in the interval X_0, i.e. $x_i^* \in X_0$, $i = 1, 2, \dots, p$. W. l. g. we can assume that $x_1^* < x_2^* < \dots < x_p^*$. Denote $X^* = [x_1^*, x_p^*]$; obviously, $X^* \subseteq X_0$ holds true. In [2] the following (interval) iterative procedure for enclosing the set X^* is proposed and studied:

$$\begin{cases} X_0 = [x_0^-, x_0^+] \in ID; \\ x_{n+1}^- = x_n^- + |f(x_n^-)|/|F'^{-0}(X_n)|, \\ x_{n+1}^+ = x_n^+ - |f(x_n^+)|/|F'^{-0}(X_n)|; \\ n = 0, 1, \dots. \end{cases} \tag{2}$$

The iteration procedure (2) always generates two sequences $\{x_n^-\}$, $\{x_n^+\}$ such that $x_0^- \geq x_1^- \geq \dots$ and $x_0^+ \leq x_1^+ \leq \dots$. The following statement is valid.

Theorem 1 [2]. *Let $f : D \to R$, $D \subseteq R$, be a continuously differentiable function on D, whose derivative f' has an interval extension F' satisfying* (1). *Then:*

i) if $f(x) = 0$ has (at least one) solution in the initial interval X_0, then $x_n^- \leq x_n^+$ for all $n = 1, 2, \dots$, holds, both sequences $\{x_n^-\}$ and $\{x_n^+\}$ are convergent and $\lim_{n\to\infty} x_n^- = x_1^*$, $\lim_{n\to\infty} x_n^+ = x_p^*$;

ii) if there is an index m such that $x_{m-1}^- \leq x_{m-1}^+$ but $x_m^- > x_m^+$ hold, then the equation $f(x) = 0$ does not possess solutions in the initial interval X_0. □

In the first case *i)* we shall say that the sequence of intervals $\{X_n\}$, $X_n = [x_n^-, x_n^+]$, converges to the interval X^*. In the second case *ii)* we say that the m-th interval X_m becomes empty and write $X_m = \emptyset$.

In the special case when $X^* = [x^*, x^*] = x^*$, i.e. when $f(x)$ has an unique zero $x^* \in X_0$, $f'(x) \neq 0$ for all $x \in X_0$ is satisfied. Assume further that $0 \notin F'(X_0)$. Then $0 \notin F'(X)$ for all $X \subseteq X_0$ holds true [1]. It is a well known fact that $0 \in f(X_0)$ (resp. $0 \notin f(X_0)$) is equivalent to $x^* \in X_0$ (resp. $x^* \notin X_0$). Since f is assumed monotone on X_0, the range $f(X_0)$ can be presented by $f(X_0) = [f(x_0^-) \vee f(x_0^+)]$.

Theorem 2 [2]. *Let $f : D \to R$, $D \subseteq R$, be a continuously differentiable function on D, whose derivative f' has an interval extension F' satisfying (1) and $0 \notin F'(X_0)$. Then:*

i) if $0 \notin f(X_0)$, the equation does not possess solution in X_0;

ii) if $0 \in f(X_0)$, the iterations (2) *generate an interval sequence $\{X_n\}$, $X_n = [x_n^-, x_n^+]$, such that*

a) $X_0 \supseteq X_1 \supseteq \dots \supseteq X_n \supseteq X_{n+1} \supseteq \dots$;

b) $x^* \in X_n$ for $n = 1, 2, \dots$ and $\lim_{n\to\infty} X_n = x^*$;

c) if additionally $\omega(F'(X)) \leq L\omega(X)$, $L = const > 0$, for all $X \in ID$ is assumed, then $\omega(X_{n+1}) \leq c\omega^2(X_n)$, $c = const > 0$, holds, i.e. the interval sequence $\{X_n\}$ converges quadratically to x^*. □

In the present paper we propose a method for enclosing all real zeroes $x_1^*, x_2^*, \dots, x_p^*$ in separate subintervals $X_1^*, X_2^*, \dots, X_l^*$ of X_0, such that each interval X_i^*, $i = 1, 2, \dots, l$, bounds a simple or a multiple zero. The method is based on procedure (2) and will be considered in the next section. Section 3 presents a parallel algorithm with result verification using computer arithmetic.

2 Formulation of the Method for Enclosing All Real Roots

Let $f : D \to R$, $D \subseteq R$, be a continuously differentiable function in D. Denote by $f(X)$ the range of f on $X \in ID$ and by F' an interval extension of the first derivative, satisfying (1). Let $x_1^*, x_2^*, \ldots, x_p^*$ be the real zeroes of $f(x)$ in an interval $X_0 \in ID$. Our goal is to split X_0 into subintervals $X_1^*, X_2^*, \ldots, X_l^*$, $l \leq p$, such that $X_1^* \bigcup X_2^* \bigcup \ldots \bigcup X_l^* \subseteq X_0$, each interval X_i^* containing a zero; a multiple zero will be enclosed in a single interval.

Let $x \in X_0$ be an interior point; for example x may be the mid-point of X_0, $x = (x_0^- + x_0^+)/2$. Consider the subintervals $X_{1,0} = [x_0^-, x]$, $X_{2,0} = [x, x_0^+]$ and do one step according (2) with each of $X_{1,0}$ and $X_{2,0}$ using $F'(X_0)$:

$$\begin{cases} x_{1,1}^- = x_0^- + |f(x_0^-)|/|F'^{-0}(X_0)|, \\ x_{1,1}^+ = x - |f(x)|/|F'^{-0}(X_0)|; \end{cases}$$

$$\begin{cases} x_{2,1}^- = x + |f(x)|/|F'^{-0}(X_0)|; \\ x_{2,1}^+ = x_0^+ - |f(x_0^+)|/|F'^{-0}(X_0)|. \end{cases}$$

Compose the intervals $X_{1,1} = [x_{1,1}^-, x_{1,1}^+]$, $X_{2,1} = [x_{2,1}^-, x_{2,1}^+]$. If both intervals are empty then the equation $f(x) = 0$ has no solutions in X_0. If one of them is empty, say $X_{2,1}$ then it does not contain any zero of $f(x)$; thus $X_{2,1}$ will be no more processed. All zeroes are then enclosed in $X_{1,1}$. We apply the above procedure to $X_{1,1}$. If both subintervals $X_{1,1}$, $X_{2,1}$ are nonempty, then x_1^*, x_2^*, $\ldots$ $x_p^* \in X_{1,1} \bigcup X_{2,1}$ is valid. Both intervals are further processed separately. In this way the initial interval X_0 will be split in several, perhaps many (but finite number of) subintervals. Denote by L the list of subintervals obtained after some steps. Let $X \in L$ be the current interval to be processed. We have to compute first $F'(X)$. If $0 \notin F'(X)$ it follows that $f(x)$ has at most one zero in X. As mentioned in the previous section, we can test this fact by computing $f(X) = [f(x^-) \vee f(x^+)]$. In the situation when $0 \notin f(X)$, the interval X will be deleted from the list L; if $0 \in f(X)$ then $f(x)$ has an unique zero in X. We can enclose it iterating according (2). As mentioned above, $0 \notin F'(X)$ implies $0 \notin F'(Y)$ for any $Y \subseteq X$; thus there is no need to test whether $0 \notin F'(X)$ on each following iteration step. However, if $0 \in F'(X)$ then $0 \notin F'(Y)$ or $0 \in F'(Y)$, $Y \subseteq X$, is possible. Therefore when $0 \in F'(X)$ holds and X results in some interval Y, $Y \subseteq X$, we have to test on the next iteration step if $F'(Y)$ contains or does not contain zero.

3 A Parallel Algorithm with Result Verification for Enclosing the Zeroes

By practical computations, computer arithmetic should be used to obtain guaranteed bounds for the solutions. Let S be a floating point system [3] and IS be the set of all intervals with end-points from S. Two kinds of directed roundings $\nabla, \Delta : R \rightarrow S$ of real numbers are used: $\nabla a = \max\{x \in S : x \leq a\}$, $\Delta a = \min\{x \in S : x \geq a\}$. They generate the computer arithmetic operations $\mathbin{\nabla\!\!\!\!{\scriptstyle *}}$, $\mathbin{\Delta\!\!\!\!{\scriptstyle *}}$: $x \mathbin{\nabla\!\!\!\!{\scriptstyle *}} y = \nabla(x * y)$, $x \mathbin{\Delta\!\!\!\!{\scriptstyle *}} y = \Delta(x * y)$ for $x, y \in R$ and for any operation $* \in \{+, -, \times, /\}$ (by division $y \neq 0$ is assumed).

Let $f : D \rightarrow R$ be the function considered in the previous section. Let be $D_S = \{x \in S : x \in D\}$ and $ID_S = \{X \in IS : X \in ID\}$. For $x \in D_S$ denote by $\Diamond f(x) = [\nabla f(x), \Delta f(x)]$ the interval resulting from the computation of $f(x)$. Obviously, $f(x) \in \Diamond f(x)$ for all $x \in D_S$ holds. For $X \in ID_S$ let $\Diamond F'(X)$ be the computed interval for $F'(X)$; $F'(X) \subseteq \Diamond F'(X)$ is valid for all $X \in ID_S$. For the sake of brevity we shall use the notations $\Diamond f(x) = [f^{+0}(x) \vee f^{-0}(x)]$, $\Diamond F'(X) = [F'^{+0}(X) \vee F'^{-0}(X)]$.

Consider the following computer arithmetic procedure related to (2):

$$\begin{cases} X_0 = [x_0^-, x_0^+] \in ID_S; \\ x_{n+1}^- = x_n^- \mathbin{\nabla\!\!\!\!{\scriptstyle +}} (|f^{+0}(x_n^-)| \mathbin{\nabla\!\!\!\!{\scriptstyle /}} |F'^{-0}(X_n)|), \\ x_{n+1}^+ = x_n^+ \mathbin{\Delta\!\!\!\!{\scriptstyle -}} (|f^{+0}(x_n^+)| \mathbin{\nabla\!\!\!\!{\scriptstyle /}} |F'^{-0}(X_n)|); \\ n = 0, 1, 2, \ldots. \end{cases} \tag{3}$$

Stopping criteria of algorithms written in terms of computer arithmetic operations are discussed in detail in [2]. We shall briefly present them in relation with our algorithm. Due to Moore's finite convergence principle [4] the interval iterative procedure (3) delivers a finite sequence of intervals in the sense that there is an index N such that $X_N = X_{N+i}$ for all $i = 1, 2, \ldots$ holds. The resulting interval for the solution (set) is then X_N, $X^* \subseteq X_N$ is valid and X_N is optimal in the sense that it can not be made smaller in the prescribed precision and by the accuracy used to compute $\Diamond f(x)$ and $\Diamond F'(X)$. Other stopping criteria come from the fact that the optimal roundings $\nabla f(x)$ and $\Delta f(x)$, $x \in D_S$, can not be realized in practice. In general we can only compute a lower resp. upper bound for $f(x)$ instead of $\nabla f(x)$ resp. $\Delta f(x)$. This leads to some additional tests by practical implementation of the algorithm. If for example for some $x \in D_S$, $0 \in \Diamond f(x)$ is obtained, then x is a zero of $f(x)$ or else "near" to one. There are two (practically artificial) situations, which we will consider below. Let on some step of the algorithm the interval $X = [x^-, x^+]$ is to be processed.

i) $0 \notin \Diamond F'(X)$, $0 \in \Diamond f(x^-)$ and/or $0 \in \Diamond f(x^+)$.

This is the case of enclosing a simple zero in X. If X is starting interval for (3), then the algorithm can not determine existence/nonexistence of a solution in X. If X is obtained after several steps according (3), then X contains a solution, but it can not be done better; further iterations are then useless. X is displayed as a resulting interval.

ii) $0 \in \Diamond F'(X)$, $0 \in \Diamond f(x^-)$ and/or $0 \in \Diamond f(x^+)$.

This situation may occur if a multiple zero of $f(x)$ is to be enclosed or $f(x)$ has more zeroes (simple and/or multiple) in X. If for any machine number $x \in X$, $0 \in \Diamond f(x)$ holds, then the algorithm can not determine existence/nonexistence of a solution in X. If there is a $x \in X$ with $0 \notin \Diamond f(x)$, improvements of $X_1 = [x^-, x]$, $X_2 = [x, x^+]$ only at these end-points are expected, where $\Diamond f$ does not contain zero; the other end-points should not be processed.

We describe below the steps of our algorithm. The steps are executed consecutively except the corresponding indications in the branches.

Steps of the algorithm

1. Put the initial interval X_0 in a list L of intervals to be processed.

2. If the list L is empty then stop; otherwise read an interval $X = [x^-, x^+]$ from L. Delete X from L.

3. Compute $\Diamond F'(X)$. If $0 \notin \Diamond F'(X)$, go to step 4; else go to step 5.

4. Compute $\Diamond f(x^-)$, $\Diamond f(x^+)$. If $0 \in \Diamond f(x^-)$ and/or $0 \in \Diamond f(x^+)$, print X; go to step 2. If $0 \notin \Diamond f(x^-)$ and $0 \notin \Diamond f(x^+)$ then compose $\Diamond f(X) = [\Diamond f(x^-) \vee \Diamond f(x^+)]$. If $0 \notin \Diamond f(X)$, go to step 2, else do one step of the iterations (3). Denote the interval obtained by X_1. If $X_1 \subset X$ holds, then set $X := X_1$ and go to step 3; if $X_1 = X$, print X and go to step 2.

5. Compute $\Diamond f(x^-)$ and $\Diamond f(x^+)$. If $0 \in \Diamond f(x^-)$ and/or $0 \in \Diamond f(x^+)$, go to step 6. If $0 \notin \Diamond f(x^-)$ and $0 \notin \Diamond f(x^+)$, choose $x \in X$ with $0 \notin \Diamond f(x)$. Do one step with $X_{1,0} = [x^-, x]$ according (3); let $X_{1,1}$ be the resulting interval. Apply (3) for one iteration to $X_{2,0} = [x, x^+]$ and denote by $X_{2,1}$ the interval obtained. If both intervals $X_{1,1}$ and $X_{2,1}$ are empty, go to step 2. Let the result be a single interval, say $X_{1,1}$ (i.e. $X_{2,1} = \emptyset$). If $X_{1,1} \subset X_{1,0}$ then rename it X and go to step 3. If $X_{1,1} = X_{1,0}$, print $X_{1,1}$ and go to step 2. If the iteration results in two intervals, then put the one in the list, call the other X and go to step 3.

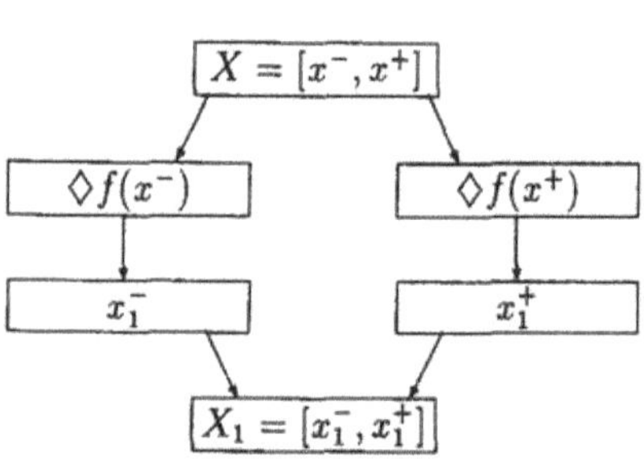

Fig.1. Parallel execution of step 4 of the algorithm.

6. If $0 \in \lozenge f(x^-)$, $0 \in \lozenge f(x^+)$ and $0 \in \lozenge f(x)$ for any machine number $x \in X$ are simultaneously true, then print X and go to step 2.

If $0 \in \lozenge f(x^-)$, $0 \notin \lozenge f(x^+)$, $0 \notin \lozenge f(x)$ for some $x \in X$, then apply (3) for one step to the interval $X_{2,0} = [x, x^+]$. Denote the result by $X_{2,1}$. Apply (3) only to the second end-point x of the interval $X_{1,0} = [x^-, x]$. Denote the result by $X_{1,1} = [x^-, x_{2,1}^+]$. If both $X_{1,1}$ and $X_{2,1}$ are empty, go to step 2. If the result is one interval, say $X_{1,1}$ and if $X_{1,1} = X_{1,0}$ then print $X_{1,1}$; otherwise rename it X and go to step 3. In case of two resulting intervals put the one in the list, call the other X and go to step 3.

If $0 \notin \lozenge f(x^-)$, $0 \in \lozenge f(x^+)$, $0 \notin \lozenge f(x)$ for some $x \in X$, then proceed analogously as described above in this step.

When the algorithm stops all intervals enclosing the zeroes of f in X_0 will have been printed in step 4, step 5 or step 6. It is possible to obtain additional intervals printed (with the necessary information about them), in which the algorithm can not determine existence or nonexistence of a solution. The user is then advised to apply the algorithm to each of these intervals using higher precision and/or computing the values of f and F' more accurately.

Starting with one interval, step 4 of the algorithm results always in one interval. On this step independent computation of two functional values and the resulting end-points of the new iterate is included. It can be implemented using two parallel processors as it is shown in Fig.1. Fig.2 presents the parallel execution of Step 5 of the algorithm, using three and four processors.

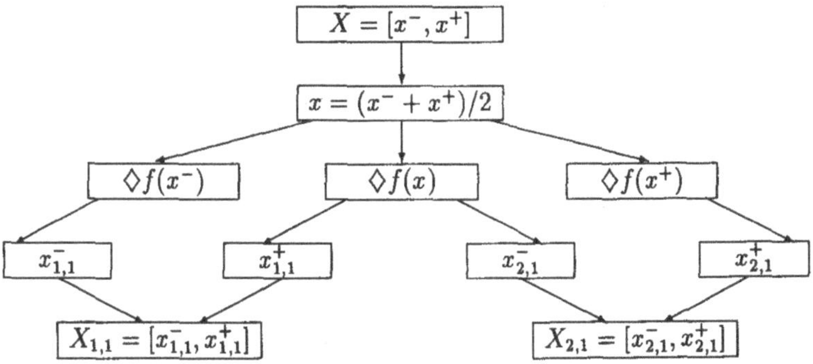

Fig.2. Parallel execution of step 5 of the algorithm.

References

[1] Alefeld, G., Herzberger, J.: *Introduction to Interval Computations*, Academic Press, New York 1983.

[2] Dimitrova, N., Markov, S.: *A Validated Newton Type Method for Nonlinear Equations*, Interval Computations (in press).

[3] Kulisch, U., Miranker, W.: *Computer arithmetic in theory and practice*, Academic Press 1980.

[4] Moore, R. E.: *Interval Analysis*, Prentice-Hall, Englewood Cliffs, N. J. 1966.

Address of the Author:
Division for Mathematical Modeling
Institute of Biophysics
Bulgarian Academy of Sciences
Acad. G. Bonchev str., bldg. 21
BG-1113 Sofia, Bulgaria

Overconvergent Monte Carlo Methods for Density-function Modelling Using B-Splines *

Ivan Dimov and Aneta Karaivanova

Abstract

The aim of this work is to show possibilities for applying spline-functions in creating overconvergent Monte Carlo methods for densities modelling using N realizations of a random variable. It is shown that B-splines are a good tool for construction of efficient methods for modelling . The error analysis is realized. It is proved that when B-splines with defects 1 of degree $k \geq 2$ are used in density modelling the method is overconvergent. For $k = 2$ the convergence is $N^{-3/2}$ while for $k = 3$ the convergence is $N^{-5/2}$.

The problem for balancing both systematic and stochastic errors is studied. The dependence between the number of the realizations of the random variable and the step of the mesh on which B-splines are defined is obtained.

1 Introduction

The density-function modelling is a very important approach in solving many problems of environmental mathematics, probabilistic theory and physics. The problems of this type arise when it is necessary to estimate the unknown density of a random variable using a given number of its realizations.

The aim of this work is to solve this problem using a specially created overconvergent Monte Carlo method for estimation the coefficients in the B-spline approximation of the unknown density. The idea of the method is to express the density of the random variable defined on the interval $[a, b]$ as a linear form of B-splines of degree k with defects 1. It is proved that when B- splines with defects 1 of degree $k \geq 2$ are used in density modelling the proposed method has convergence $N^{-3/2}$, while, as it is shown in [1, 2], the usual Monte Carlo method has convergence $N^{-1/2}$.

The error of the method consists of two components: systematic and stochastic one. The idea of balancing is used for obtaining the optimal ratio between the number of realizations and the size of the mesh on which B-splines are defined.

*This work was supported in part by the Ministry of Science and Education of Bulgaria, Grant $I210-92$ and by the NATO Science under Grant ENVIR.CRG 930449.

The paper is organized as follows: in section 2 the formulation of the problem is given; section 3 contains the description of the methods; in this section two cases ($k \geq 2$ and $k \geq 3$) are considered and error analysis is realised; in section 4 the problem of error balancing is considered; section 5 containes the conclusion.

2 Formulation

Consider a set of points

$$w_m = \{a = x_0 < x_1 < \ldots x_m = b\}, \tag{1}$$

in the interval $[a, b]$.

Let $k \geq 0$ be an integer number and $\{l_i\}$, $1 \leq l_i \leq k$, $i = 2, 3, \ldots, m$ be a sequence of integers. Let the set T_n be introduced in the following way:

(i) $2(k+1)$ new nodes are added to the set w_m:

$$T_n = \{t_1 \leq t_2 \leq \ldots t_{k+1} = x_0 < x_1 < \ldots < x_m = t_{n-k} \leq t_{n-k+1} \leq \cdots \leq t_n\}; \tag{2}$$

(ii) for $i = 2, 3, \ldots, m$ $\quad x_i$ is repeated l_i times in the set T_n such that

$$n = 2(k+1) + \sum_{i=2}^{m} l_i. \tag{3}$$

Denote by k and l_i the degree and defects of the spline $B_{i,k}(x)$.

Define $i - th$ B-spline of k-th degree as

$$B_{i,k}(x) = (t_{i+k+1} - t_i)[t_i, \ldots, t_{i+k+1}](. - x)_+^k, \quad i = 1, 2, \ldots, n-k-1, \tag{4}$$

where $[t_i, \ldots, t_{i+k}](. - x)_+^k$ is the $(k+1)$-th divided difference with respect to t of the function

$$(t - x)_+^k = [max\{0, (t - x)\}]^k \tag{5}$$

where x is a fixed point ([3]).

Consider a density-function $f(x)$, defined on the interval $[a, b]$. Suppose $f(x) \in C^{k+1}[a, b]$. Then, an approximation g_L of f has the following representation ([3]):

$$g_L(x) = \sum_{i=1}^{L} c_i B_{i,k}(x), \quad x \in [a, b], \quad L = n - k - 1, \tag{6}$$

with an error $O(h^{k+1})$ (c_i are constants). Denote by $B_k(x)$ B-spline of k-th degree with defects $l_i = 1$ in the point $t = 0$ with integer nodes $t_j = j - 1$, $j = 1, 2, \ldots, k+1$. In this case the set T_n has $n = 2k + m$ nodes. Let $\gamma_j \in [0, 1]$ be a random number (continuous uniformly distributed random variable with mathematical expectation $E\gamma_j = 1/2$ and dispersion $D\gamma_j = \sigma^2\gamma_j = 1/12$).

Let $p_k(x)$ be a density-function of the random variable

$$\xi_k = \sum_{j=1}^{k} \gamma_j. \tag{7}$$

Then, as it is shown in [4]

$$p_k(x) = B_k(x), \quad x \in [0, k]. \tag{8}$$

Denote by θ_k the following random variable:

$$\theta_k = \sqrt{12/k}(\xi_k - k/2). \tag{9}$$

One can obtain an algorithm for modelling random variables with densities $B_{i,k}(x)$ on a regular grid with step h:

$$\theta_{i,k} = [(i-1) + \theta_k]h. \tag{10}$$

3 The Methods

Consider the problem of error estimation of the unknown density-function $f(x)$ using N realizations $\{\xi_i\}_{i=1}^{N}$ of a random variable with density $f(x)$.

Consider the approximation (6) of $f(x)$. The coefficient c_j in (6) are obtained using the following representation:

$$\sum_{j=1}^{L}(B_{i,k}(x), B_{j,k}(x))c_j = (f, B_{i,k}(x)). \tag{11}$$

Obviously, the functional $(f, B_{i,k}(x))$ is the mathematical expectation of $B_{i,k}(x)$ with a density-function $f(x)$:

$$(f, B_{i,k}(x)) = \int_a^b B_{i,k}(x)f(x)\,dx = E_\xi B_{i,k}(\xi), \tag{12}$$

where the random point ξ has a density $f(x)$. The inner product $(f, B_{i,k}(x))$ is estimated using Monte Carlo method:

$$(f, B_{i,k}(x)) \approx \frac{1}{N}\sum_{j=1}^{N}\theta_j = \hat{\theta}_N. \tag{13}$$

The constants c_j are calculated and an estimation of the unknown density

$$q_f(x) = \sum_{j=1}^{L}\hat{c}_j B_{j,k}(x) \tag{14}$$

is obtained. The error of the method consists of two components:

$$||q_f - f||^2 \leq ||f - g_L||^2 + ||g_L - q_f||^2 = \delta_t^2 + \delta_{L,N}^2, \tag{15}$$

where

$$||f|| = (\int_a^b f^2(x)\, dx)^{1/2}. \tag{16}$$

The first component δ_t^2 is the systematic error while the second one $\delta_{L,N}^2$ is the stochastic error. A method is considered good when $\delta_{L,N}$ is approximately equal to δ_t.

Estimate the stochstic error:

$$\delta_{L,N}^2 = \int_a^b [\sum_{j=1}^{L}(\hat{c}_j - c_j)B_{j,k}(x)]^2\, dx = \sum_{i=1}^{L}\sum_{j=1}^{L}(\hat{c}_i - c_i)(\hat{c}_j - c_j)a_{i,j}, \tag{17}$$

where

$$a_{i,j} = \int_a^b B_{i,k}(x)B_{j,k}(x)\, dx \tag{18}$$

Let $\mathbf{r} = (r_1, r_2, \ldots, r_L)$, $r_i = \hat{c}_i - c_i$, $i = 1, \ldots, L$, and $A = \{a_{i,j}\}_{i,j=1}^{L}$.

Consider the product $A\mathbf{r} = (\varepsilon_1, \varepsilon_2, \ldots, \varepsilon_L)^T = \boldsymbol{\varepsilon}$.

The stochastic error $\delta_{L,N}$ depends on the given sampling of realizations $\{\xi_i\}_{i=1}^{N}$. Consider the estimation for the mathematical expectation of this error. Since the matrix A is positive defined, it is possible to express the expectation of the stochastic error as follows:

$$E(A\mathbf{r}, \mathbf{r}) = E(\boldsymbol{\varepsilon}, A^{-1}\boldsymbol{\varepsilon}) \leq ||A^{-1}||E||\boldsymbol{\varepsilon}||^2. \tag{19}$$

The errors ε_i, $i = 1, \ldots, L$ depend on the error of Monte Carlo method for evaluation of integrals (13). It is so, because the system (11) for finding coefficients c_j can be solved using direct methods. (The matrix A is positive defined and diagonal dominated.) Thus, the problem of increasing the accuracy of the method is reduced to the problem of obtaining a good Monte Carlo method for evaluation integrals (13).

The error is studied for two special cases. The first one is when splines of degree $k \geq 2$ are used for modelling densities.

The second one is when splines of degree $k \geq 3$ are used.

Case 1. When B-spline of degree 2 or more with defects 1 are used it is useful to take into account that the splines $B_{i,2}(x)$ have continuous derivative. In fact, any spline of degree k with defects l_i has continuous derivatives of degree $r = k - l_i$. In our case $r \geq 1$.

Since the first derivative of $B_{i,k}(x)$ is continuous there exists a constant $L_{i,k}$ such that

$$L_{i,k} \geq \left|\frac{dB_{i,k}(x)}{dx}\right|, \quad x \in [a, b]. \tag{20}$$

For the construction of Monte Carlo method with high rate of convergence consider a sequence of random points $\{\xi_j\}_{j=1}^{N}$, $(\xi_1 < \xi_2 < \ldots < \xi_N)$

Define a set of points $\{y_j\}_{j=0}^N$ such that

$$y_j = \frac{1}{2}(\xi_j + \xi_{j+1}), \quad j = 1, 2, \ldots, N-1 \text{ and } y_0 = a; \quad y_L = b. \tag{21}$$

The integral (13) can be represented as

$$J_{i,k} = \int_a^b B_{i,k}(x) f(x)\, dx = \sum_{j=1}^N \int_{y_{j-1}}^{y_j} B_{i,k}(x) f(x)\, dx. \tag{22}$$

Let

$$J_{i,k}^j = \int_{y_{j-1}}^{y_j} B_{i,k}(x) f(x)\, dx \text{ and } f^j = \int_{y_{j-1}}^{y_j} f(x)\, dx. \tag{23}$$

Obviously

$$\sum_{j=1}^N J_{i,k}^j = J_{i,k}, \text{ and } \sum_{j=1}^N f^j = 1, \text{ for } k \geq 3. \tag{24}$$

In this case the random variable $\hat{\theta}_N$ (see, (13)) is

$$\hat{\theta}_N = \sum_{j=1}^N f^j B_{i,k}^j(\xi^j) \tag{25}$$

There exist constants c_1 and c_2 such that

$$f^j \leq c_1/N \text{ and } \Delta x_j = y_j - y_{j-1} \leq c_2/N. \tag{26}$$

For example, one can see that when $f(x) = const$ and $\Delta x_j = \Delta x_{j+1}$ for $j = 1, 2, \ldots, N-1$ both constants c_1 and c_2 are equal to 1.

The probable error of evaluating integrals is defined [7] as a value r_n for which the following equalities hold:

$$Pr\{|J_{i,k} - \Theta_N| \leq r_N\} = Pr\{|J_{i,k} - \Theta_N| > r_N\} = \frac{1}{2}.$$

For the usual Monte Carlo method [7]

$$r_N = c_{0.5}\sigma(\theta)N^{-1/2}, \tag{27}$$

where $c_{0.5} \approx 0.6745$; $\sigma(\theta)$ is the standard deviation.

An overconvergent Monte Carlo method is a method for which

$$r_N = cN^{-1/2-\psi}, \tag{28}$$

where c is a constant and $\psi > 0$ [1,2].

Theorem 1. Let $B_{i,k}(x)$ be a B-spline of degree $k \geq 2$ with defects $l_i = 1$.

Then the probable error r_N for Monte Carlo method (25) for evaluating integrals (13) is

$$r_N \le \sqrt{2} c_1 c_2 L_{i,k} N^{-3/2}.$$

Proof. Consider B-splines $B_{i,k}(x)$, where $k \ge 3$. For a fixed point $s_j \in \Delta x_j = [y_{j-1}, y_j]$ we have

$$B_{i,k}(x) = B_{i,k}(s_j) + B'_{i,k}(\eta_j)(x - s_j), \tag{29}$$

where $\eta_j \in \Delta x_j$.

Since $|B'_{i,k}(\eta_j)| \le L_{i,k}$ we have:

$$\begin{aligned} DB_{i,k}(\xi_j) &\le EB^2_{i,k}(\xi_j) \le L^2_{i,k} E(\xi_j - s_j)^2 \\ &\le L^2_{i,k}(\sup_{x_{1_j}, x_{2_j} \in \Delta x_j} |x_{1_j} - x_{2_j}|)^2 \le L^2_{i,k} c_2^2 / N^2. \end{aligned} \tag{30}$$

Now we can estimate $D\theta^*_N$. Since in our case $\theta_N = \sum_{j=1}^N f_j B^j_{i,k}(\xi^j)$ we have

$$D\theta_N = \sum_{j=1}^N p_j^2 DB^j_{i,k}(\xi^j) \sum_{j=1}^N c_1^2 N^{-2} L^2_{i,k} c_2^2 N^{-2}) = (c_1 c_2 L_{i,k})^2 N^{-3}. \tag{31}$$

To estimate the probable error one can apply the Chebishev's inequality:

$$\Pr\{|\theta_N - E\theta_N| < h\} \ge 1 - (D\theta_N / h^2), \tag{32}$$

where $h > 0$.

Let $h = 1/\epsilon (D\theta_N)^{1/2}$, where ϵ is a positive number. Then

$$\Pr\{|\theta_N - J_{i,k}| < 1/\epsilon c_1 c_2 L_{i,k} N^{-3/2}\} \ge 1 - \epsilon^2. \tag{33}$$

For $\epsilon = 1/\sqrt{2}$ one can get:

$$\Pr\{|\theta_N - J_{i,k}| < \sqrt{2} c_1 c_2 L_{i,k} N^{-3/2}\} \ge 1/2. \tag{34}$$

The last inequality proves the theorem.

This result defines an overconvergent Monte Carlo method, because in this case the rate of convergence is 3/2 ($\psi = 1$).

Case 2. Consider the case when B-splines $B_{i,k}$ are splines of degree $k \ge 3$ with defects 1 . In this case $B_{i,k}(x)$ have continuous and bounded second derivative (since $B_{i,k}(x)$ are defined on finite interval). In fact, $r = k - l_i \ge 2$. Since the second derivative of $B_{i,k}(x)$ is bounded there exists constant $M_{i,k}$ such that

$$M_{i,k} \ge \left| \frac{d^2 B_{i,k}(x)}{dx^2} \right|, \quad x \in [a, b]. \tag{35}$$

Consider a sequence of random points $\{\xi_j\}_{j=1}^N$, $(\xi_1 < \xi_2 < \ldots < \xi_N)$.

Define a set of points $\{y_j\}_{j=0}^N$ such that $y_j = 1/2(\xi_j + \xi_{j+1}),\ j = 1, 2, \dots, N-1$ and $y_0 = a;\ y_L = b$.

Now consider the following Monte Carlo method for estimation of integrals (13) (it is assumed that $\xi_0 = a$ and $\xi_{N+1} = b$:

$$\hat{\theta}_N = -\frac{1}{2}(B_{i,k}(\xi_0) + B_{i,k}(\xi_1))\xi_0$$

$$+\frac{1}{2}\sum_{j=1}^{N}(B_{i,k}(\xi_{j-1}) - B_{i,k}(\xi_{j+1}))\xi_j + \frac{1}{2}(B_{i,k}(\xi_N) + B_{i,k}(\xi_{N+1}))\xi_{N+1}. \tag{36}$$

Obviously, there exist constants c_1 and c_2, such that

$$f^j = \int_{y_{j-1}}^{y_j} f(x)\,dx \leq c_1/N \text{ and } \Delta x_j = y_j - y_{j-1} \leq c_2/N. \tag{37}$$

In this case the following theorem holds:

Theorem 2. Let $B_{i,k}(x)$ be a B-spline of degree $k \geq 3$ with defects $l_i = 1$.

Then the probable error r_N for Monte Carlo method (36) for evaluation integrals (13) is

$$r_N \leq \frac{\sqrt{2}}{4} c_1 c_2^2 M_{i,k} N^{-5/2}.$$

Sketch of the Proof. For B-splines $B_{i,k}(x)$, where $k \geq 3$ and for a fixed point $s_j \in \Delta x_j = [y_{j-1}, y_j]$ we have

$$B_{i,k}(x) = B_{i,k}(s_j) + B'_{i,k}(s_j)(x - s_j), + \frac{1}{2}B''_{i,k}(\eta_j)(x - s_j)^2, \tag{38}$$

where $\eta_j \in \Delta x_j$.

Since $B''_{i,k}(\eta_j) \leq M_{i,k}$ it is easy to get

$$D\theta_N = \left(\frac{1}{2}c_1 c_2^2 M_{i,k}\right)^2 N^{-5}. \tag{39}$$

Using the Chebishev's unequality: $\Pr\{|\theta_N - J_{i,k}| < h\} \geq 1 - (D\theta_N/h^2)$, for $h = 1/\epsilon(D\theta_N)^{1/2}$, we get $r_N \leq \frac{\sqrt{2}}{4} c_1 c_2^2 M_{i,k} N^{-5/2}$.

In this case we have an overconvergent Monte Carlo method, because the convergence is $N^{-5/2}$ ($\psi = 2$).

4 Balancing of errors

There are two errors in our Monte Carlo methods: systematic and stochastic one. The problem of balancing is very important. It is clear that for obtaining good results the stochastic error must be approximately equal to the systematic one. (It is not necessary to use "a large" number of realizations N in order to have "a small" stochastic error if the systematic error is "big".) So, it means that $\delta_{L,N} \approx \delta_t$.

The problem of balancing is closely connected with the problem of obtaining optimal ratio between the number of realizations of the random variable and the step of the mesh on which B- splines are defined.

The systematic error is

$$\delta_t \leq Ch^k, \tag{40}$$

where h is the step of the mesh w_m (see formula (1)) and k is a constant.

For the first Monte Carlo method (case 1) we have:

$$E||\varepsilon||^2 = C_1^2 N^3, \tag{41}$$

where k_1 is a constant.

Thus $\delta_{L,N}^2 \leq ||A^{-1}||C_1^2 N^{-3}$.

Since for the balancing of error it is necessary to have $C^2h^{2m} = C_1^2||A^{-1}||N^{-3}$ one can get the optimal relation between N and h:

$$N = C^{(1)}h^{-\frac{2}{3}k}, \tag{42}$$

where cond(A) is the conditioning number of the matrix A.

It is easy to find the corresponding ratio for the second algorithm (case 2):

$$N = C^{(2)}h^{-\frac{2}{5}k}. \tag{43}$$

Let us note that these results are much more better than the result of Mahotkin [4] , obtained for the usual Monte Carlo method, because for the usual method $E||\varepsilon||^2 = C_0N^{-1}$ and

$$N = C^{(0)}h^{-2k}. \tag{44}$$

5 Conclusion

Overconvergent Monte Carlo methods for modelling densities by using B-splines are proposed. It is proved that when B-splines of degree $k \geq 2$ with defects 1 are used it is possible to construct methods with high rate of convergence. Two methods are proposed and studied. The first one can be used for $k \geq 2$, while the second one - for $k \geq 3$. The optimal relation between the number of realizations N and the step of the mesh is obtained using the idea of balancing.

References

[1] Ivan Dimov and Aneta Karaivanova. Parallel Monte Carlo Algorithms. Proc. of the Third Intern. Conference on Application of Supercomputers in Engineering

ASE/93, **Comp. Mech. Publ., Elsevier Appl. Sci., London, New York,** 1993, pp. 479–495.

[2] Ivan Dimov, Efficient and Overconvergent Monte Carlo Methods. Parallel algorithms., **Addvances in Parallel Algorithms**, I. Dimov, O. Tonev (Eds.), **Amsterdam, IOS Press** (1994), pp. 100–111.

[3] C . De Boor, A Practical Guide to splines.Applied Mathematical Sciences, vol. 27, **Springer-Verlag**, 1978.

[4] O.A.Mahotkin, M.I.Pirimkulov, Application of splines for some problems of statistical modelling, Novosibirsk, Computing Center. **Theory and Practice of Statistical Modelling**, Mikhailov (Ed.) (1984), pp. 43–53.

[5] V. Dupach, Stochasticke pocetni metody, **Cas. pro pest. mat., 81**, No 1 (1956) 55–68.

[6] I.M. Sobol. Monte Carlo numerical methods. **"Nauka", Moscow**, 1973.

[7] I. Dimov, O. Tonev. Random walk on distant mesh points Monte Carlo methods, **J. of Statistical Physics, vol. 70, Nos. 5/6**, 1993, pp. 1333 –1342.

[8] Dimov I.T., Tonev O.I. Monte Caralo algorithms: performance analysis for some computer architectures, **J. of Computational and Applied Methematics, vol. 48** (1993), pp.253– 277.

Address of the Authors:
Bulgarian Academy of Sciences
Acad.G.Bonchev str. Bl.25-A, 1113 Sofia
Bulgaria

FINITE-DIFFERENCE APPROXIMATION TO A CONTROLLED BOUNDARY VALUE PROBLEM

Asen L. Dontchev*

Abstract

We estimate the Hausdorff distance between sets of solutions of a boundary value problem with a control and of its approximation provided by the Euler scheme.

1 Introduction

We consider the following problem: find a pair of functions (y, u), y absolutely continuous in [0,1] with values in $\mathbf{R}^n$, u bounded and measurable with values in $\mathbf{R}^m$, such that

$$\dot{y}(t) = f(y(t), u(t)) \text{ and } u(t) \in U \text{ for all } t \in [0,1], y(0) = a, y(1) \in K, \qquad (1)$$

where $f : \mathbf{R}^n \times \mathbf{R}^m \to \mathbf{R}^n$, $U \subset \mathbf{R}^m, K \subset \mathbf{R}^n$, and $a \in \mathbf{R}^n$ are given.

If $y(t)$ represents the state of a control system and $u(t)$ is the control, then the problem is to find a feasible state-control pair, that is, a function u and a corresponding trajectory y with $y(0) = a$ which reaches the target set K at time $t = 1$. Problem (1) can be also viewed as an estimation problem where one has to determine a solution y of a boundary value problem as well as a unknown (uncertain) time-varying parameter u from a given set of admissible parameters U. If one is interested in finding a solution y for an arbitrary u then (1) reduced to a boundary value problem for a differential inclusion

$$\dot{y}(t) \in f(y(t), U) \text{ for all } t \in [0,1], y(0) = a, y(1) \in K,$$

which in turn can be obtained from a differential equation with discontinuous right-hand side.

*Mathematical Reviews, Ann Arbor, MI 48107. On leave from the Institute of Mathematics, Bulgarian Academy of Sciences, Sofia, Bulgaria.

We examine the convergence of Euler method applied to problem (1). Given a natural number N let $h = 1/N$ be the mesh spacing and let $t_0 = 0, t_{i+1} = t_i + h, i = 1, \cdots, N-1$. The Euler integration scheme applied to (1) results in

$$y_{i+1} = y_i + hf(y_i, u_i), u_i \in U, i = 0, 1, \cdots, N-1, y_0 = a, y_N \in K. \tag{2}$$

Our purpose is to estimate the error of the discretization, measured by an appropriately defined distance between the set of solutions of the finite-dimensional problem (2) and the set of solutions of the original problem (1). Given a feasible pair $x^* = (y^*, u^*)$ of problem (1) we estimate the l^∞ distance between the sequence of values of x^* at $t_i, i = 0, 1, \cdots, N-1$, and the set of solutions of (2). Next, we take a sequence of solutions $x^N = (y^N, u^N), N = 1, 2, \cdots$, of the discrete problem (2), consider y^N as a piecewise linear and continuous function and u^N as a step function across t_i, and estimate the $C \times L^\infty$ distance between x^N and the set of solutions of (1).

The results presented here take an intermediate place between the rather modest research on numerical analysis of differential inclusions (for a survey see [3]) and much richer but less homogeneous literature on discrete approximations in optimal control. Recently Veliov [11] studied discrete approximations to control systems in a fairly general framework; however, he considers problem linear in control u and estimates the error for the trajectory y only.

The error estimates are obtained in Section 3 and are based on an abstract result of inverse function type for a sequence of set-valued maps given in Section 2. We also present a sequential form of the open mapping principle for convex set-valued maps. To avoid assumptions concerning smoothness of the right-hand side of the differential equation we employ the recently developed technique of averaged moduli of smoothness [9]. Let f be a bounded function defined on the segment [0, 1]. The local modulus of continuity of f is defined by

$$\omega(f; t, h) = \sup \{|f(s) - f(q)| : s, q \in [t - h/2, t + h/2] \cap [0, 1]\},$$

while the averaged modulus of smoothness of f is

$$\tau(f; h) = \int_0^1 \omega(f; t, h) dt.$$

2 A sequential inverse mapping theorem

Throughout this section we use the following notation. The distance from a point x in a metric space (X, ρ) to a set $A \subset X$ is

$$\operatorname{dist}(x, A) = \inf\{\rho(x, y) : y \in A\};$$

and the excess from the set A to the set C is defined by

$$e(C, A) = \sup\{\mathrm{dist}(x, A) : x \in C\}.$$

Let Γ be a set-valued map from X to the subsets of Y. We denote by graphΓ the set $\{(x, y) \in X \times Y : y \in \Gamma(x)\}$. The inverse map Γ^{-1} is defined by $\Gamma^{-1}(y) = \{x \in X : y \in \Gamma(x)\}$. $B_r(a)$ is the closed ball centered at a with radius r and B is the closed unit ball.

Definition 1. Let Γ_N be a sequence of set-valued maps from a metric space (X, ρ) to subsets of a metric space Y and let $y_N \in \Gamma(x_N)$ for all N. We say that the sequence of mappings Γ_N is *uniformly pseudo-Lipschitz around* (x_N, y_N) if there exist positive constants α, β and γ such that

$$e(\Gamma_N(u) \cap B_\alpha(y_N), \Gamma_N(v)) \leq \gamma\rho(u, v)$$

for all $u, v \in B_\beta(x_N)$.

If $\alpha = +\infty$ the sequence Γ_N is *uniformly Lipschitz* (with respect to the Hausdorff metric) in $B_\beta(y_N)$.

When the maps $\Gamma_N = \Gamma$ this definition reduces to the definition of pseudo-Lipschitz continuity introduced in [1]. We note that the latter turns out to be a fairly important concept in set-valued and nonsmooth analysis, see [2], [4], [6], [8]. If Γ_N are single-valued, then we obtain uniform Lipschitz continuity for a sequence of functions.

Definition 2. Let f_N be a sequence of functions from a metric space (X, ρ) to a linear normed space $(Y, \|\cdot\|)$ and let $x_N \in X, N = 1, 2, \cdots$. We say that f_N is *strictly stationary at* x_N if for every $\epsilon > 0$ there exist $\delta > 0$ and a natural number N_0 such that for every $N > N_0$

$$\| f_N(u) - f_N(v) \| \leq \epsilon\rho(u, v)$$

whenever $u, v \in B_\delta(x_N)$.

Given a sequence of functions $f_N, N = 1, 2, \cdots$, from a linear normed space to a linear normed space, if each f_N is strictly differentiable at x_N, uniformly in N, and $\nabla f_N(x_N) = 0$, then f_N is strictly stationary at x_N.

The following theorem can be viewed as a version of a basic results in numerical analysis, sometimes called Dahlquist theorem or Lax principle. It has been established originally for initial value problems and has the following short formulation: consistency plus stability implies convergence.

Theorem 1. *Let X be a complete metric space and Y be a linear normed space. Let x_N be a sequence in X, let the sequence of functions $f_N : X \to Y$ be strictly stationary at x_N, and let F_N be a sequence of maps from X to Y with $(x_N, y_N) \in$ graph F_N. Suppose that the following two conditions hold:*

(Consistency) $\lim_{N\to\infty} \| f_N(x_N) \| = 0$.

(Stability) *For each N the inverse map F_N^{-1} is closed-valued and the sequence of inverse maps F_N^{-1} is uniformly pseudo-Lipschitz around (y_N, x_N) with constant γ.*

Then for every $\gamma^+ > \gamma$ there exists N^+ such that for every $N > N^+$ there exists an x_N^* satisfying

$$x_N^* \in (F_N + f_N)^{-1}(y_N) \text{ and } \rho(x_N^*, x_N) \leq \gamma \parallel f_N(x_N) \parallel . \tag{3}$$

The following theorem gives a sufficient condition for uniform pseudo-Lipschitz continuity of a sequence of convex maps. It is a version of the open mapping principle for set-valued maps; the difference to the classical result, due to Robinson [7] and Ursescu [10], is that we consider a sequence of not necessarily closed-valued maps satisfying an interior point condition uniformly in N, and the spaces X and Y may be not complete.

Recall that a map F is called *convex* if graph$F$ is a convex set.

Theorem 2. Let X and Y be linear normed spaces, let $\mathcal{F}_N : X \to Y, N = 1, 2, \cdots$, be a sequence of convex set-valued maps, and let $y_N \in \mathcal{F}_N(x_N)$ for every N. Suppose that there exist positive constants m and M such that

$$y_N + 2mB \subset \mathcal{F}_N(x_N + MB), N = 1, 2, \cdots .$$

Then the sequence $\mathcal{F}_N^{-1}$ is uniformly pseudo-Lipschitz around (y_N, x_N) with a Lipschitz constant $2M/m$.

Combining theorems 1 and 2 we obtain the following

Corollary 1. *Let X and Y be linear normed spaces, let Ω_N be a sequence of convex and closed sets in X, and let Λ_N be a sequence of convex sets in Y. Let $x_N \in \Omega_N$ and $z_N \in \Lambda_N$, and let ϕ_N be a sequence of functions which are strictly differentiable at x_N, uniformly in N, and such that $\parallel \phi_N(x_N) \parallel \to 0$ as $N \to \infty$. Suppose that there exist positive constants τ and Θ such that for all sufficiently large N*

$$z_N + \tau B \subset \nabla\phi(x_N)(\Theta B \cap \Omega_N) + \Lambda_N. \tag{4}$$

Then there exists a constant γ such that for every sufficiently large N there exists $x_N^* \in X$ with

$$x_N^* \in \Omega_N, z_N \in \phi_N(x_N^*) + \Lambda_N \text{ and } \parallel x_N^* - x_N \parallel \leq \gamma \parallel \phi_N(x_N) \parallel .$$

3 Estimates

First we find an estimate for the appropriately defined distance between a fixed feasible pair of the continuous problem (1) and the set of feasible pairs of the discretized problem (2).

Let $x^* = (y^*, u^*)$ be a feasible pair of (1). We assume that the sets U and K are closed and convex and f is continuously differentiable in a convex and open set in

$\mathbf{R}^{n+m}$ containing $(y^*(t), u^*(t))$ for all $t \in [0,1]$. Along with (1) we consider the linear problem

$$\dot{y}(t) = A(t)y(t) + B(t)u(t), y(0) = a, \tag{5}$$
$$u(t) \in U \text{ for all } t \in [0,1], y(1) \in K,$$

where $A = \nabla_y f, B = \nabla_u f$, both computed at (x^*, u^*). Denote by $\mathcal{R}$ the reachable set at $t = 1$ of (5), that is the set of all points in $\mathbf{R}^n$ that are values at $t = 1$ of solutions of (5) and corresponding to some feasible control u, i.e. u bounded and measurable, $u(t) \in U$ for all $t \in [0,1]$. In the sequel we denote by c a generic constant independent of N.

Theorem 3. *Let $\dot{y}^*, A$ and B be Riemann integrable and suppose that*

$$(int\mathcal{R} \cap K) \cup (intK \cap \mathcal{R}) \neq \emptyset. \tag{6}$$

Then for N sufficiently large there exists a feasible pair (y^N, u^N) for the discretized problem (2) satisfying

$$\max\nolimits_{0 \leq i \leq N-1}(|y^*(t_i) - y_i^N| + |u^*(t_i) - u_i^N|) \leq c\tau(\dot{y}^*; h),$$

From the properties of the averaged modulus of smoothness we obtain

Corollary 2. *Suppose that the conditions of Theorem 3 hold and let $\dot{y}^*$ be of bounded variation. Then the l^∞ distance from (y^*, u^*) to the set of feasible pairs of the discrete problem (2) is $O(h)$.*

Remark 1. Condition (6) cannot be dropped, even for linear systems. Suppose that the equation (1) is linear and let (6) be violated, e.g. K is a point on the boundary of the reachable set $\mathcal{R}$. The reachable set of the discretized problem (2) will be close to $\mathcal{R}$ but may not contain K, that is, the feasible set of the discrete problem may be empty.

Consider now a sequence of feasible pairs $x_N = (y^N, u^N)$ for the discrete problem (2) for $N = 1, 2, \cdots$. Below we give conditions under which for sufficiently large N there exists a solution of the continuous problem (1) the $C \times L^\infty$ distance from which to x_N is $O(h)$.

We assume that the sets U and K are closed and convex, the vectors $(y_i^N, u_i^N), i = 0, 1, \cdots, N, N = 1, 2, \cdots$, are contained in a bounded set Z in $\mathbf{R}^{n+m}$, and the function f is continuous differentiable in a convex and open set containing Z. Suppose that u^N is a piecewise constant function across the grid points t_i which is right-continuous on $[0,1)$,$u^N(t_i) = u_i^N$, and let y^N be piecewise linear and continuous, $y^N(t_i) = y_i^N, i = 1, 2, \cdots, N$. Then (y^N, u^N) satisfies

$$\dot{y}(t) = f(y(t_i), u(t)) \text{ for all } t \in (t_i, t_{i+1}), i = 0, 1, \cdots, N-1.$$

We apply Corollary 1 taking

$$X = C([0,1], \mathbf{R}^n) \times L^\infty([0,1], \mathbf{R}^m), Y = L^1([0,1], \mathbf{R}^n), z_N = 0, \Lambda_N = \{0\},$$

$$\phi_N(x) = \phi(x) = \dot{y}(t) - f(y(t), u(t)),$$

and

$$\Omega_N = \Omega = \{x = (y, u) \in X : y(0) = a, y(1) \in K, u(t) \in U \text{ for all } t \in [0,1]\}.$$

Then

$$\phi(x^N)(t) = -f(y^N(t), u^N(t_i)) + f(y^N(t_i), u^N(t_i)) \text{ for } t \in (t_i, t_{i+1}), i = 0, 1, \cdots, N-1.$$

Since $\dot{y}^N$ is uniformly bounded we have

$$\| \phi(x^N) \|_{L^1} \leq c \sum_{i=0}^{N-1} \int_{t_i}^{ti+1} |y^N(t) - y^N(t_i)| ds \leq ch,$$

The derivative of ϕ at x^N is

$$\nabla\phi(x^N)(t) = \dot{y}(t) - A_N(t)y(t) - B_N(t)u(t)$$

where $A_N = \nabla_y f(y^N, u^N), B^N = \nabla_u f(y^N, u^N)$. Let $\mathcal{R}_N$ be the reachable set of

$$\dot{y}(t) = A_N(t)y(t) + B_N(t)u(t), y(0) = a, u(t) \in U, u \in L^\infty([0,1], \mathbf{R}^m). \tag{7}$$

In the sequel we use the following notation: If C is a set, then

$$C^{-\alpha} = \{x \in C : x + \alpha B \subset C\}.$$

Theorem 4. *Suppose that there exist a constant $\alpha > 0$ such that*

$$(\mathcal{R}_N^{-\alpha} \cap K) \cup (K^{-\alpha} \cap \mathcal{R}_N) \neq \emptyset \tag{8}$$

for all large N. Then for every sufficiently large N there exists a feasible pair $(\bar{y}^N, \bar{u}^N)$ of the continuous problem (1) such that

$$\| y^N - \bar{y}^N \|_C + \| u^N - \bar{u}^N \|_{L^\infty} \leq c'h.$$

Remark 2. If $U = \mathbf{R}^m$, then condition (10) holds when the pair (A_N, B_N) is controllable for large N.

Remark 3. Assume that the matrices A_N, B_N converge to A_0, B_0 weakly in L^1 and let $\mathcal{R}_0$ be the reachable set of the linear system (5) with matrices A_0 and B_0. Let $z \in \mathcal{R}_0$ and u is a corresponding feasible control. Applying u to the system (9) we obtain that, if y_N is the corresponding trajectory, then $y_N(1) \to z$ as $N \to \infty$. This means that every point of $\mathcal{R}_0$ is a limit of a sequences of points of $\mathcal{R}_N$. Then the set $\mathcal{R}_N$ in condition (10) can be replaced by $\mathcal{R}_0$.

References

[1] J.- P. Aubin, Lipschitz behaviour of solutions to convex minimization problems,*Math. Oper. Res.*, 9 (1984) 87-111.

[2] J. M. Borwein, Stability and regular points of inequality systems, *J. Opt. Theory and Appl.* 48 (1986) 9-52.

[3] A.L. Dontchev, F. Lempio, Difference methods for differential inclusions - a survey, *SIAM Review* 34 (1992) 263-294.

[4] A. L. Dontchev, W. W. Hager, An inverse mapping theorem for set-valued maps, *Proc. Amer. Math. Soc.* (to appear).

[5] A. L. Dontchev, T. Zolezzi, *Well-posed optimization problems,* Lecture Notes in Mathematics, vol. 1543, Springer 1993.

[6] J.-P. Penot, Metric regularity, openness and Lipschitz multifunctions, *Nonlinear Analysis* 13 (1989) 629-643

[7] S. M. Robinson, Stability theory for systems of inequalities, Part II:Differentiable nonlinear systems, *SIAM J. Num. Analysis* 13 (1976) 497-513.

[8] R. T. Rockafellar, Lipschitz properties of multifunctions, *Nonlinear Analysis* 9 (1985) 867-885.

[9] Bl. Sendov, V. A. Popov,*The averaged moduli of smoothness*, J. Wiley and Sons, New York,1988.

[10] C. Ursescu, Multifunctions with closed convex graphs, *Czech. Math. J.* 25 (1975) 438-441.

[11] V. M. Veliov, Best Approximations of Control/Uncertain Differential Systems by Means of Discrete-Time Systems, Preprint, IIASA WP-91-45, November 1991.

PRECONDITIONING OF NONCONFORMING FINITE ELEMENT APPROXIMATIONS OF SECOND ORDER ELLIPTIC PROBLEMS

R.E. Ewing, Yu. Kuznetsov, R.D. Lazarov, and S. Maliassov

Abstract

We consider the finite element approximation of Poisson equation in a parallelepiped using linear tetrahedral nonconforming Crouzeix-Raviart elements. Using the idea of substructuring we eliminate most of the unknowns and precondition the obtained Schur complement by a spectrally equivalent very sparse matrix. The numerical experiments show that this solution procedure is very efficient, robust and is a good preconditioner for approximations of general elliptic equations of second order on domains, topologically equivalent to a parallelepiped.

1. Introduction

Let Ω be a convex polyhedral domain in $\mathbb{R}^3$, $f(x) \in L^2(\Omega)$ and $K(x)$ be a three by three symmetric matrix-valued function on $\bar{\Omega}$ satisfying the uniform positive definiteness condition: there exists $\alpha > 0$ such that

$$\alpha^{-1}\xi^T\xi \le \sum_{i,j} \xi^T K(x)\xi \le \alpha\xi^T\xi\,, \qquad \forall x \in \bar{\Omega}\,, \forall \xi \in \mathbb{R}^3\,. \tag{1.1}$$

We consider the Dirichlet boundary value problem:

$$\mathbf{q} + K\nabla u = 0 \quad \text{in} \;\; \Omega\,, \qquad \nabla\cdot\mathbf{q} = f \quad \text{in} \;\; \Omega\,, \qquad u|_{\partial\Omega} = 0, \qquad (1.2),$$

where $\partial\Omega$ is the boundary of Ω. In applications of fluid flow in porous media, $u(x)$ is referred to as pressure and $\mathbf{q}$ a to Darcy velocity vector.

This problem can be discretized in various ways. Among the most popular and frequently used methods of approximation are the finite volume method, the Galerkin finite element method and the mixed finite element method. Each of these methods has its advantages and disadvantages when applied to particular engineering problem. For example, for reservoir problems in geometrically simple domains and heterogeneous media the finite volume method has proven to be reliable, accurate and mass conserving cell-by-cell. Many engineering problems, e.g. petroleum recovery, groundwater contamination, seismoic exploration etc. need very accurate velocity (flux) determination in the presence of heterogeneities, anisotropy and large jumps in the coefficient matrix $A(x)$. More accurate approximation of the velocity can be achieved through the use of mixed finite element method. As shown by Wiser and Wheeler in [13] the lowest order mixed finite element method with special quadratures on rectangular grids is equivalent to the finite volume method. Based on that equivalence Bramble et al. in [3] have developed efficient multigrid solution procedure for structured grids. However, in general the mixed finite element method leads to an algebraic saddle point problem that is more difficult and more expensive to solve. Although some reliable preconditioning algorithms for these saddle point problems have been proposed and studied (see, e.g. [4, 9, 11]), their efficiency depends strongly on the geometry of the domain, on the coefficient matrix $A(x)$ and on the type of the finite elements used.

An alternative approach can be taken by developing of hybrid methods. This approach has been studied by Arnold and Brezzi [2] where the continuity of the velocity vector normal to the boundary of each element is enforced by Lagrange multipliers. This idea can be explained in the following way: first, define the spaces

$$\mathbf{V} = H(div;\Omega) = \left\{\mathbf{q} \in \left(L^2(\Omega)\right)^3, \nabla\cdot\mathbf{q} \in L^2(\Omega)\right\}, \qquad W = L^2(\Omega);$$

then the weak formulation of the system (1.2) is: find a pair $(\mathbf{q}, u) \in \mathbf{V} \times W$ such that

$$(\nabla\cdot\mathbf{q}, w) + (K^{-1}\mathbf{q}, \mathbf{p}) - (u, \nabla\cdot\mathbf{p}) = (f, w), \qquad \forall\, w \in W, \quad \forall\, \mathbf{p} \in \mathbf{V}. \tag{1.3}$$

The standard mixed finite element approximation to (1.5) reads as follows: let $\bar{\mathbf{V}}_h \times W_h \subset \mathbf{V} \times W$ be a finite element space over the partition $\mathcal{T}_T$ of Ω into tetrahedra (or over the partition $\mathcal{T}_C$ into cubes) (see [6]). The requirement $\bar{\mathbf{V}}_h \subset \mathbf{V}$ implies that the normal component of the vector $\mathbf{q}$ is continuous across the interelement boundaries $\partial\mathcal{T}_T$. The construction of Arnold and Brezzi [2] is based on the idea of backing off this continuity requirement and defining the space $\mathbf{V}_h = \left\{\mathbf{q} \in (L_h^2(\Omega))^3 : \mathbf{q}|_T \in \bar{\mathbf{V}}_h,\ T \in \mathcal{T}_T\right\}$. In order to introduce the interelement continuity of the normal component of $\mathbf{q}$ we introduce the space of the Lagrange multipliers

$$L_h = \left\{\lambda \in L^2(\partial\mathcal{T}_T) : \lambda|_{\partial T} \in \bar{\mathbf{V}}_h \cdot \nu \ \text{ for each } T \in \mathcal{T}_T\right\},$$

where ν is the normal to ∂T vector.

Now the approximation to (1.3) using Lagrange multipliers is formulated for the unknown triple $(\mathbf{q}_h, u_h, \lambda_h) \in \mathbf{V}_h \times W_h \times L_h$. We skip the details of the weak formulation of (1.3) over $\mathbf{V}_h \times W_h \times L_h$ referring to [1, 5, 7]. If the vectors $\mathbf{Q}$, $\mathbf{U}$ and

$\mathbf{\Lambda}$ correspond to the representation of $\mathbf{q}_h$, u_h and λ_h with respect to the bases in $\mathbf{V}_h$, W_h and L_h, respectively, the algebraic form of this approximation is (see Brezzi and Fortin [6])

$$\begin{pmatrix} M & B & C \\ B^T & 0 & 0 \\ C^T & 0 & 0 \end{pmatrix} \begin{pmatrix} \mathbf{Q} \\ \mathbf{U} \\ \mathbf{\Lambda} \end{pmatrix} = \mathbf{F}, \tag{1.4}$$

where M is a symmetric and positive definite matrix. Important feature of the matrices M and B is that they are block diagonal since the unknown nodal values of $\mathbf{q}_h$ and u_h over a given finite element T are related to the nodal values on the adjacent element only through the Lagrange multipliers. Therefore, using element-by-element elimination we can reduce this system to the form

$$S\,\mathbf{\Lambda} = \mathbf{\Phi}. \tag{1.5}$$

For the description of the structure of the Shur complement S for particular finite element spaces we refer to [1, 5, 7].

The important discovery of Arnold and Brezzi [2] is that the system (1.5) can be obtained also from application to (1.2) the Galerkin method using nonconforming elements. Namely, in [2] is shown that the lowest-order Raviart-Thomas mixed element approximations are equivalent to the usual P_1-nonconforming finite element approximations when the classical P_1-nonconforming space is augmented with P_3-bubbles. Such a relationship has been studied recently for a large variety of mixed finite element spaces [1, 5, 7].

This equivalence between the hybrid mixed and the nonconforming finite element methods establishes a framework for preconditioning and/or solving the algebraic problem and for postprocessing of the finite element solution. This framework includes the following three steps: (1) forming the reduced algebraic problem for the Lagrange multipliers, which is equivalent to the nonconforming problem; (2) construction and study of efficient methods, based on multigrid, multilevel or domain decomposition, for solving or preconditioning of the reduced problem; (3) recovery of the solution $u(x)$ and the velocity $\mathbf{q}$ from the Lagrange multipliers that were already found.

The recent progress in each of these steps described above (see, e.g. [8, 10, 12]) gives us an indication that the mixed finite element method can be used as an accurate and efficient method for solving general elliptic problems of second order in domains with complicated geometry.

The goal of this paper is to construct, study and implement efficient preconditioners for the nonconforming finite element approximations of problem (1.2) on tetrahedral meshes.

2. Problem Formulation

To explain our approach we consider the model case when Ω is a unit cube in $\mathbb{R}^3$ and $K(x) = a(x)\,I$. Let $\mathcal{T}_T$ be a regular partitioning of Ω on tetrahedra T with a characteristic size $h = diam(T)$ (see Fig. 1).

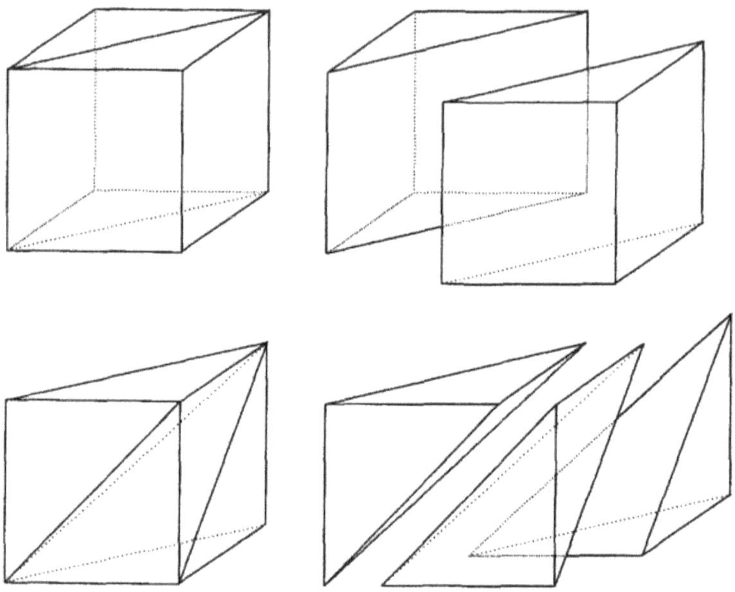

Figure 1: Partition of cube into prisms and tetrahedra.

We introduce the set of barycenters of all faces of the tetrahedral partition of Ω, and the set Q_h of those barycenters that are strictly inside Ω. The Crouzeix-Raviart nonconforming finite element space V_h consists of all piecewise linear functions on $\mathcal{T}_T$ that vanish at barycenters of the boundary faces and are continuous at the barycenters of Q_h. Note that the space V_h is not a subspace of $H_0^1(\Omega)$.

Now we define the bilinear form on V_h by

$$a_h(u,v) = \sum_{T\in\mathcal{T}_T} \int_T a(x)\,\nabla u \cdot \nabla v dx, \qquad \forall\, u,v \in V_h. \tag{2.1}$$

Thus the nonconforming discretization of problem (1.2) is given by seeking $u_h \in V_h$ such that

$$a_h(u_h, v) = (f, v), \qquad \forall\, v \in V_h. \tag{2.2}$$

The natural degrees of freedom of Crouzeix-Raviart nonconforming elements are the values at the barycenters of the faces of the tetrahedral elements. Denote the vector of the unknown values corresponding to a function $v_h \in V_h$ by $\mathbf{v}$ and assume that its dimension is N, i.e., $\mathbf{v} \in \mathbb{R}^N$. Note, that all unknowns on faces on the boundary with Dirichlet data are excluded.

Let $(\mathbf{u}, \mathbf{v})$ and $\|\mathbf{v}\|$ be a bilinear form and the corresponding norm, defined on $\mathbb{R}^N$ by

$$(\mathbf{u},\mathbf{v})_N = h^3 \sum_{x\in Q_h} u(x)v(x), \qquad \|\mathbf{v}\| = (\mathbf{v},\mathbf{v})_N^{1/2}, \qquad u,v \in V_h. \tag{2.3}$$

Then the discretization operator $A : \mathbb{R}^N \to \mathbb{R}^N$, which is symmetric and positive definite is defined by

$$(A\mathbf{v}, \mathbf{w})_N = a_h(u, w), \qquad u, v \in V_h. \tag{2.4}$$

Similarly, we introduce the vector $\mathbf{F}$ as $(f, v) = (\mathbf{F}, \mathbf{v})_N \quad \forall\, v \in V_h$. Now, the problem (2.2) can be rewritten in a matrix form

$$A\mathbf{u} = \mathbf{F}. \tag{2.5}$$

3. Matrix Formulation and Its Properties

Our goal is to introduce an algebraic formulation of the approximate problem using a type of static condensation that eliminates some of the unknowns. In this way we can reduce substantially the size of the problem. For this approach we need a special partitioning of the domain into tetrahedra that have some regularity and preserve the simplicity of the algebraic problem.

First, we partition Ω into cubes with size of the edges $h = 1/n$ and denote them by $C = C^{(i,j,k)}$ where (x_{1i}, x_{2j}, x_{3k}) is the right back upper corner of the cube. This partitioning is denoted by $\mathcal{T}_C$. Next, we divide each cube $C = C^{(i,j,k)}$ into two prisms $P_1 = P_1^{(i,j,k)}$ and $P_2 = P_2^{(i,j,k)}$ as shown in Fig. 1 and denote this partitioning of Ω by $\mathcal{T}_P$. Finally, we divide each prism into three tetrahedra (see Fig. 1) and denote this partitioning of Ω into tetrahedra by $\mathcal{T}_T$.

Let $P = P^{(i,j,k)} \in \mathcal{T}_P$ be a particular prism of the partition $\mathcal{T}_P$. Denote by V_h^P the subspace of restrictions of the functions in V_h onto P. These restrictions define vectors $\mathbf{u}_P$ that are restrictions of a vector $\mathbf{u} \in \mathbb{R}^N$. The dimension of V_h^P we denote by N^P. Obviously, for prisms with no faces on $\partial\Omega$ the dimension $N^P = 10$.

Local stiffness matrices A^P on prisms $P \in \mathcal{T}_P$ is defined by

$$(A^P \mathbf{u}_P, \mathbf{v}_P)_N = \sum_{T \subset P} \int_T a(x)\, \nabla u_h \cdot \nabla v_h \, dx\,, \tag{3.1}$$

for any $P \in \mathcal{T}_P$. Then the global stiffness matrix A is determined by assembling the local stiffness matrices.

Now we consider a prism P of an arbitrary cube that has no face on the boundary $\partial\Omega$ and enumerate its faces s_j, $j = 1, \ldots, 10$. Then the local stiffness matrix of this prism for the case $a(x) \equiv 1$ has the following form:

$$A^P = \frac{3h}{2}\left[\begin{array}{cccccc|cccc} 1 & 0 & 0 & 0 & 0 & 0 & 0 & 0 & -1 & 0 \\ 0 & 1 & 0 & 0 & 0 & 0 & 0 & 0 & 0 & -1 \\ 0 & 0 & 1 & 0 & 0 & 0 & -1 & 0 & 0 & 0 \\ 0 & 0 & 0 & 1 & 0 & 0 & 0 & -1 & 0 & 0 \\ 0 & 0 & 0 & 0 & 1 & 0 & 0 & 0 & -1 & 0 \\ 0 & 0 & 0 & 0 & 0 & 1 & 0 & 0 & 0 & -1 \\ \hline 0 & 0 & -1 & 0 & 0 & 0 & 2 & 0 & -1 & 0 \\ 0 & 0 & 0 & -1 & 0 & 0 & 0 & 2 & 0 & -1 \\ -1 & 0 & 0 & 0 & -1 & 0 & -1 & 0 & 4 & -1 \\ 0 & -1 & 0 & 0 & 0 & -1 & 0 & -1 & -1 & 4 \end{array}\right] \equiv \frac{3h}{2}\begin{bmatrix} A_{11} & A_{12} \\ A_{21} & A_{22} \end{bmatrix}, \tag{3.2}$$

where

$$A_{22} = \begin{bmatrix} 2 & 0 & -1 & 0 \\ 0 & 2 & 0 & -1 \\ -1 & 0 & 4 & -1 \\ 0 & -1 & -1 & 4 \end{bmatrix}.$$

Along with matrix A^P we introduce the following matrix B^P defined on the same space V_h^P:

$$B^P = \frac{3h}{2}\begin{bmatrix} A_{11} & A_{12} \\ A_{21} & B_{22} \end{bmatrix}, \qquad B_{22} = \begin{bmatrix} 3 & -1 & -1 & 0 \\ -1 & 3 & 0 & -1 \\ -1 & 0 & 3 & 0 \\ 0 & -1 & 0 & 3 \end{bmatrix}. \tag{3.3}$$

It easy to show that $ker A^P = ker B^P$.

Then we define the $N \times N$ matrix B by the following equality:

$$(B\mathbf{u}, \mathbf{v})_N = \sum_{P \in \mathcal{T}_P} (B^P \mathbf{u}_P, \mathbf{v}_P)_N \qquad \forall\, \mathbf{u}, \mathbf{v} \in \mathbb{R}^N, \tag{3.4}$$

and prove that $\text{cond}(B^{-1}A) \leq (2+\sqrt{3})^2$.

Now we divide all unknowns in the system into two groups: (1) the first group consists of all unknowns corresponding to faces of the prisms in the partition $\mathcal{T}_P$, excluding, of course, the faces on $\partial\Omega$: (2) the second group consists of the unknowns corresponding to the faces of the tetrahedra that are internal for each prism.

This splitting of the space $\mathbb{R}^N$ induces the following presentation of the vectors $\mathbf{v}^T = (\mathbf{v}_1^T, \mathbf{v}_2^T)$, where $\mathbf{v}_1 \in \mathbb{R}^{N_1}$ and $\mathbf{v}_2 \in \mathbb{R}^{N_2}$. Obviously, $N_2 = 4n^3$. Then matrix B can be presented in the following block form:

$$B = \begin{bmatrix} B_{11} & B_{12} \\ B_{21} & B_{22} \end{bmatrix}, \qquad \text{dim} B_{11} = N_1. \tag{3.5}$$

Denote now by $\hat{B}_{11} = B_{11} - B_{12}B_{22}^{-1}B_{21}$ the Schur complement of B obtained by elimination of the vector $\mathbf{v}_2$. Then $B_{11} = \hat{B}_{11} + B_{12}B_{22}^{-1}B_{21}$, so the matrix B has the form

$$B = \begin{bmatrix} \hat{B}_{11} + B_{12}B_{22}^{-1}B_{21} & B_{12} \\ B_{21} & B_{22} \end{bmatrix}. \tag{3.6}$$

Note that for each prism $P \in \mathcal{T}_P$ the unknowns on the internal for P faces are connected through the equation $B\mathbf{v} = \mathbf{F}$, only with the unknowns associated with this prism, i.e., the matrix B_{22} is block diagonal with 2×2 blocks and can be inverted locally (prism by prism). Thus matrix $\hat{B}_{11}$ is easily computable.

4. Three-Level Preconditioner

Next, we replace the matrix B of (3.6) by

$$\tilde{B} = \begin{bmatrix} \tilde{B}_1 + B_{12}B_{22}^{-1}B_{21} & B_{12} \\ B_{21} & B_{22} \end{bmatrix}. \tag{4.1}$$

This is done in the following way: take the stiffness matrix of any prism and eliminate the unknowns associated with the internal for that prism faces of the three tetrahedra. Using certain ordering of the unknowns we get the matrix:

$$\hat{B}_{1,P} = \left(\frac{3h}{2}\right) \left[\begin{array}{rr|rrrrrr} 8/3 & -1 & -1 & 0 & -1/3 & 0 & -1/3 & 0 \\ -1 & 8/3 & 0 & -1/3 & 0 & -1 & 0 & -1/3 \\ \hline -1 & 0 & 1 & 0 & 0 & 0 & 0 & 0 \\ 0 & -1/3 & 0 & 2/3 & 0 & 0 & 0 & -1/3 \\ -1/3 & 0 & 0 & 0 & 2/3 & 0 & -1/3 & 0 \\ 0 & -1 & 0 & 0 & 0 & 1 & 0 & 0 \\ -1/3 & 0 & 0 & 0 & -1/3 & 0 & 2/3 & 0 \\ 0 & -1/3 & 0 & -1/3 & 0 & 0 & 0 & 2/3 \end{array}\right]. \tag{4.2}$$

Then, replace this prism element matrix by the following spectrally equivalent matrix (they have the same kernel):

$$\tilde{B}_{1,P} = \left(\frac{3h}{2}\right) \left[\begin{array}{rr|rrrrrr} 8/3 & -1 & -2/3 & 0 & -2/3 & 0 & -1/6 & -1/6 \\ -1 & 8/3 & 0 & -2/3 & 0 & -2/3 & -1/6 & -1/6 \\ \hline -2/3 & 0 & 2/3 & 0 & 0 & 0 & 0 & 0 \\ 0 & -2/3 & 0 & 2/3 & 0 & 0 & 0 & 0 \\ -2/3 & 0 & 0 & 0 & 2/3 & 0 & 0 & 0 \\ 0 & -2/3 & 0 & 0 & & 2/3 & 0 & 0 \\ -1/6 & -1/6 & 0 & 0 & 0 & 0 & 1/3 & 0 \\ -1/6 & -1/6 & 0 & 0 & 0 & 0 & 0 & 1/3 \end{array}\right]. \tag{4.3}$$

The matrix $\tilde{B}$ is obtained by assembling these prismatic element stiffness matrices. Again, using superelement analysis, we prove that $\text{cond}(\tilde{B}^{-1}A) \leq 3(2+\sqrt{3})^2$.

Now consider the linear system with matrix $\tilde{B}_1$: $\tilde{B}_1\mathbf{u} = \mathbf{f}$. Matrix $\tilde{B}_1$ can be represented in the block form

$$\tilde{B}_1 = \begin{bmatrix} C_{11} & C_{12} \\ C_{21} & C_{22} \end{bmatrix}, \tag{4.4}$$

where the matrix C_{22} is diagonal. After eliminating the unknowns of the second block we split the reduced system to one linear system of dimension n^3 and n linear systems each of dimension n^2. Thus, we have constructed a spectrally equivalent sparse preconditioner for the Schur compliment after the elimination of more than 80 % of the original unknowns.

5. Results of the Numerical Experiments

The method of preconditioning on the basis of multilevel substructuring as discussed above was tested on the model problem with diagonal matrix K:

$$Lu \equiv -div\, K\, grad\, u = f \quad \text{in } \Omega = (0,1)^3 \subset \mathbb{R}^3, \qquad u|_{\partial\Omega} = 0\,,$$

using described above Crouzeix-Raviart nonconformal finite elements.

The domain was divided into n^3 cubes (n in each direction) and each cube was partitioned into 6 tetrahedra. The total dimension of the original algebraic system was $N = 12n^3 - 6n^2$.

The right hand side was generated randomly. The original algebraic problem has been solved by the conjugate gradient method with the preconditioner in the form of (4.1) with accuracy $\varepsilon = 10^{-6}$. For comparison that problem has been solved by the same method without preconditioning. The condition number of matrix $B^{-1}A$ was calculated from the relation between conjugate gradients and Lanczos algorithm.

Two cases were considered:

1. $K = I$, where I is unity matrix; the results are shown in Table 1;
2. $K = \text{diag}\{a_1, a_2, a_3\}$, where a_i, $i = 1, 2, 3$, are piece-wise constant functions; results are shown in Table 2.

Table 1

		CG Without Preconditioning			CG With Preconditioning		
n	N	n_{iter}	cond	time (sec)	n_{iter}	cond	time (sec)
4	672	40	66	0.18	22	9.84	0.22
8	5760	73	265	2.18	24	10.7	1.27
16	47616	130	1062	49.2	24	11.94	15.7
32	387072	200<	—	1248	25	12.2	163
40	758400				25	12.26	376
50	1485000				25	12.33	771

Table 2

a_1	a_2	a_3	$N = 47,616$ Iter	Cond	$N = 387,072$ Iter	Cond
1	1	1	21	11.	21	11.1
1	1	0.1	16	5.3	16	5.4
1	1	0.01	14	4.5	14	4.5
10	1	1	18	7	18	7.1
1	10	1				
100	1	1	16	5.7	16	5.8
1	100	1				
1	1	10	54	125	55	130
1	1	100	100	1266	102	1270
0.1	1	1	43	50	43	50.5
0.01	1	1	88	330	88	335

References

[1] T. Arbogast and Z. Chen, *On the implementation of mixed methods as nonconforming methods for second order elliptic problems*, IMA Preprint #1172, 1993. (Submitted to Math. Comp.)

[2] D.N. Arnold and F. Brezzi, *Mixed and nonconforming finite element methods: implementation, postprocessing and error estimates*, RAIRO, Model. Math. Anal. Numer., **19** (1985), pp. 7–32.

[3] J. Bramble, R. Ewing, J. Pasciak, and J. Shen, *Analysis of the multigrid algorithms for cell-centered finite difference approximations*, Preprint, 1994.

[4] J.H. Bramble and J.E. Pasciak, *A preconditioning technique for indefinite systems resulting from mixed approximations of elliptic problems*, Math. Comp., **50** (1988), pp. 1–17.

[5] S.C. Brenner, *A multigrid algorithm for the lowest-order Raviart-Thomas mixed triangular finite element method*, SIAM J. Numer. Anal., **29** (1992), pp. 647–678.

[6] F. Brezzi and M. Fortin, *Mixed and Hybrid Finite Element Methods*, Springer-Verlag, New York, Berlin, Heidelberg, 1991.

[7] Z. Chen, *Analysis of mixed methods using conforming and nonconforming finite element methods*, RAIRO, Math. Model. Numer. Anal., **27** (1993), pp. 9–34.

[8] L.C. Cowsar, *Domain decomposition methods for nonconforming finite elements spaces of Lagrange-type*, Preprint TR93-11, March 1993.

[9] T. Rusten and R. Winther, *A preconditioned iterative method for saddle-point problems*, SIAM J. Matrix Anal. Appl., **13** (1992), pp. 887–904.

[10] M. Sarkis, *Two-level Schwarz methods for nonconforming finite elements and discontinuous coefficients*, Preprint, 1993.

[11] P. Vassilevski and R. Lazarov, *Preconditioning saddle-point problems arising from mixed finite element discretization of elliptic equations*, UCLA CAM Report 92-46, 1992.

[12] P.S. Vassilevski and J. Wang, *Multilevel iterative methods for mixed finite element discretizations of elliptic problems*, Numer. Math., **63** (1992), pp. 503–520.

[13] A. Weiser and M. Wheeler, *On convergence of block-centered finite-differences for elliptic problems*, SIAM J. Numer. Anal., **25** (1988), pp. 351–375.

R. Ewing, R. Lazarov, S. Maliassov:
Institute for Scientific Computation
Department of Mathematics
Texas A&M University
College Station
TX 77843, U.S.A.

Yu. Kuznetsov:
Institute of Numerical Mathematics
Russian Academy of Sciences
32-a Leninski Prospect, 117334 Moscow
Russia

R.D. Lazarov:
Institute of Mathematics
Bulgarian Academy of Sciences
1113 Sofia,
Bulgaria

Gradient algorithms for linear systems arising from the air pollution model on MPS 38

Krassimir Georgiev *

Abstract

The BiCGstab, CGS and LSCG methods for iterative solving of systems of linear equations with nonsymmetric coefficient matrices arising from the large scale air pollution problems are discussed. Numerical experiments on a minisupercomputer MPS 38 for pure advection and advection dominated advection-diffusion problems are presented.

1 Introduction

The iterative methods are in many situations quite attractive for use on vector or parallel computers [2]. Unfortunately, no single iterative method is robust enough to solve all sparse linear systems accurately and efficiently. Generally, an iterative method is suitable only for a class of problems since the rate of convergence depends on spectral properties of the corresponding matrix.

This study is oriented to efficient iterative solvers for systems of linear equations with nonsymmetric coefficient matrices arising from the discretization of the differential equations in the 2D-model for a transport of air pollutants.

The discretization in the large-scale air pollution model, even in two dimensions, leads to huge systems of algebraic equations (up to $O(10^5)$) equations per time-step [8]. Therefore, a real-time solution one can get only using efficient numerical methods and run them on vector processors and/or on parallel computers. The development of robust and reliable software for handling the mathematical model under consideration is an important step in the solution of the problem.

The content of the paper can be outlined as follows. In Section 2 the conjugate gradients methods for systems with nonsymmetric matrices are discussed. A short description and comments of the iterative algorithms which are implemented in the software are given. In Section 3 we will stop at some main characteristics of the minisupercomputer MPS 38. In Section 4 some numerical results for the systems of linear

*This work was supported in part by the Bulgarian Ministry of Education and Science under Grants MM/212-92 and MM/75-91.

equations arising from pure advection and advection dominated advection diffusion problems are presented. A brief discussion of the results and some conclusions are given in Section 5.

2 Conjugate gradients for nonsymmetric linear systems of equations.

For solving linear systems with a real, symmetric and positive definite (SPD) coefficient matrices, the Conjugate Gradient (CG) algorithm seems to be one of the most efficient among the simple iterative techniques. There are some generalizations (see: Axelsson, Mantenffel, Van der Vorst, etc.) of CG in the case when the matrix is not SPD as in the case in the discretization of nonselfadjoint elliptic differential equations.

2.1 Polynomial equivalent of the CG method.

Let us consider the system of linear equations

$$A\mathbf{x} = \mathbf{b},$$

where A is $N \times N$ matrix.

If A is SPD and $\mathbf{x}_0$ is an arbitrary initial value for the solution $\mathbf{x}$ then the CG algorithm for the above system is [6]:

$\mathbf{r}_0 = A\mathbf{x}_0 - \mathbf{b}, \quad \mathbf{p}_{-1} = 0, \quad \rho_{-1} = 1;$
$n = 0;$
while *residual* > *tolerance* **do**
begin

$$\rho_n = \mathbf{r}_n^T\mathbf{r}_n; \qquad \beta_n = \frac{\rho_n}{\rho_{n-1}}; \quad \mathbf{p}_n = \mathbf{r}_n + \beta_n\mathbf{p}_{n-1};$$
$$\sigma_n = \mathbf{p}_n^T A\mathbf{p}_n; \qquad \alpha_n = \frac{\rho_n}{\sigma_n}; \quad \mathbf{r}_{n+1} = \mathbf{r}_n - \alpha_n A\mathbf{p}_n;$$
$$\mathbf{x}_{n+1} = \mathbf{x}_n - \alpha_n\mathbf{p}_n; \quad n = n+1;$$

end

The residuals $\mathbf{r}_n$ satisfy $\mathbf{r}_n = A\mathbf{x}_n - \mathbf{b}$ and the vectors $\mathbf{p}_n$ are search directions. The CG algorithm gives the exact solution after at most N steps. At each step of the CG algorithm

$$E_n = \mathbf{r}_n^T A^{-1}\mathbf{r}_n$$

is minimized. E_n can be regarded as the square of the norm of the residual $\mathbf{r}_n$. The residuals are mutually orthogonal i.e. $\mathbf{r}_n^T\mathbf{r}_m = \rho_n\delta_{nm}$, and the search directions are mutually A orthogonal i.e. $\mathbf{p}_n^T A\mathbf{p}_m = \sigma_n\delta_{nm}$. It is proved [6] that for the residuals and search directions the next two equalities are true

$$\mathbf{r}_n = P_n(A)\mathbf{r}_0 \qquad \mathbf{p}_n = T_n(A)\mathbf{r}_0, \tag{1}$$

where P_n and T_n are polynomials of degree not higher than n. Denote by Π^n the set of real polynomials of degree up to n and define the symmetric bilinear form $(.,.)$ on Π^n by

$$(P,T) = [P(A)\mathbf{r}_0]^T T(A)\mathbf{r}_0.$$

As long as A is real and symmetric matrix this bilinear form is semidefinite. Now the polynomial variant of CG method can be written

$$
\begin{array}{llll}
P_0 \equiv 1, \quad T_{-1} \equiv 0, \quad \rho_{-1} = 1; & & & \\
n = 0; & & & \\
\textbf{while} \ \ \textit{residual} > \textit{tolerance} \ \textbf{do} & & & \\
\textbf{begin} & & & \\
\quad \rho_n = (P_n, P_n); & \beta_n = \dfrac{\rho_n}{\rho_{n-1}}; & T_n(\tau) = P_n(\tau) + \beta_n T_{n-1}(\tau); & (2) \\
\quad \sigma_n = (T_n, V(\tau)T_n); & \alpha_n = \dfrac{\rho_n}{\sigma_n}; & P_{n+1}(\tau) = P_n(\tau) - \alpha_n \tau T_n(\tau); & \\
\quad n = n + 1; & & & \\
\textbf{end.} & & &
\end{array}
$$

Here $V(\tau) \equiv \tau$ and the residuals and search direction has to be computed by formulae (1).

2.2 BiCG algorithm.

It has been proved [6] that if $[.,.]$ is any symmetric bilinear form defined on Π^n which satisfies

$$[PQ, T] = [P, QT] \tag{3}$$

for any $P, Q, T \in \Pi^n$ and P_n and T_n are constructed according to (2) but using $[.,.]$ instead of $(.,.)$ then

$$[P_n, T_m] = \rho_n \delta_{nm}; \qquad [P_n, V\,T_m] = \rho_n \delta_{nm}, \quad \text{for any} \quad n, m.$$

Now defining $[.,.]$ in an appropriate way it is possible to generalize the CG algorithm for nonsymmetric systems of linear equations and to get BiCG algorithm as it has been given in [3]. Let, in addition to the assumptions in Section 2.1, $\tilde{\mathbf{r}}_0$ be a suitable chosen vector (in the practice very often $\tilde{\mathbf{r}}_0 \equiv \mathbf{r}_0$). If

$$[P,T] = \tilde{\mathbf{r}}_0^T P(A)\, T(A)\, \mathbf{r}_0 = (P(A^T)\tilde{\mathbf{r}}_0)^T\, T(A)\, \mathbf{r}_0$$

and

$$\mathbf{p}_{-1} = \tilde{\mathbf{p}}_{-1} = 0, \quad \mathbf{r}_n = P_n(A)\mathbf{r}_0, \quad \tilde{\mathbf{r}}_n = P_n(A^T)\tilde{\mathbf{r}}_0, \quad \mathbf{p}_n = T_n(A)\mathbf{r}_0, \quad \tilde{\mathbf{p}}_n = T_n(A^T)\tilde{\mathbf{r}}_0,$$

then the vectors $\mathbf{r}_n$, $\tilde{\mathbf{r}}_n$, $\mathbf{p}_n$ and $\tilde{\mathbf{p}}_n$ are the same as in the standard Fletcher's BiCG algorithm [3].

2.3 CGS algorithm.

The computational work of the BiCG algorithm is twice the computational work of the CG algorithm. Another disadvantage of the BiCG method is the necessity of programming of the operation $A^T\mathbf{v}$, especially when A is stored with a general data structure. For improving the situation one can use that the bilinear form $[.,.]$ has the property (3) and to compute scalars ρ_n and σ_n from (2) by formulae

$$\rho_n = [P_0, P_n^2], \quad \sigma_n = [P_0, VT_n^2]. \tag{4}$$

It follows from (4) that it is possible to construct an algorithm which generates P_n^2 and T_n^2 rather than P_n and T_n. Now, defining

$$\Phi_n = P_n^2, \quad \theta_n = P_n T_{n-1} \quad \text{and} \quad \xi_n = T_{n-1}^2,$$

one can get the algorithm

$\Phi_0 = 1, \quad \theta_0 = \xi_{-1} = 0, \quad \rho_{-1} = 1; \quad n = 0;$
while *residual* > *tolerance* **do**
begin

$$\rho_n = [1, \Phi_n]; \qquad \beta_n = \frac{\rho_n}{\rho_{n-1}}; \qquad \gamma_n = \Phi_n + \beta_n\theta_n;$$

$$\alpha_n = \gamma_n + \beta_n(\theta_n + \beta_n\xi_{n-1}); \quad \sigma_n = [1, V\xi_n]; \qquad \alpha_n = \frac{\rho_n}{\sigma_n};$$

$$\theta_{n+1} = \gamma_n - \alpha_n V\xi_n; \qquad \Phi_{n+1} = \Phi_n - \alpha_n V(\gamma_n + \theta_{n+1}); \quad n = n + 1;$$

end.

If after that we define $\mathbf{r}_n = \Phi_n(A)\mathbf{r}_0$, $\mathbf{q}_n = \theta_n(A)\mathbf{r}_0$ and $\mathbf{p}_n = \xi_n(A)\mathbf{r}_0$ then the standard CGS algorithm [1] is obtained. Detailed description of this procedure can be found in [6]. This algorithm requires computational work not more then BiCG but does not need A^T.

3 BiCGstab algorithm.

The residual in the CGS algorithm has the property

$$\mathbf{r}_i = P_i^2(A)\mathbf{r}_0,$$

where $P_i(A)\mathbf{r}_0$ is the BiCG residual, i.e.

$$(P_i(A)\mathbf{r}_0)^T P_j(A^T)\tilde{\mathbf{r}}_0 = 0, \quad j < i.$$

In order to improve the BiCG algorithm one can construct the algorithm where the residual $\mathbf{r}_i$ is obtained as follows:

$$\mathbf{r}_i = \tilde{P}_i(A)P_i(A)\mathbf{r}_0,$$

where

$$\tilde{P}_i(A) = (1 - \omega_1 A)(1 - \omega_2 A)\dots(1 - \omega_i A),$$

with appropriate parameters ω_k, $k = 1, \dots, i$ and obtain the BiCGstab algorithm [7].

$\mathbf{r}_0 = \mathbf{b} - A\mathbf{x}_0, \quad \mathbf{v}_0 = \mathbf{p}_0 = 0;$
Choose $\tilde{\mathbf{r}}_0$ *so that* $\mathbf{r}_0^T\tilde{\mathbf{r}}_0 \neq 0;$
$\rho_0 = \alpha = \omega_0 = 1, \quad n = 0;$
while *residual* > *tolerance* **do**
begin

$$\rho_n = \tilde{\mathbf{r}}_n^T\mathbf{r}_n; \quad \beta_n = \frac{\rho_n}{\rho_{n-1}}\frac{\alpha}{\omega_{n-1}}; \quad \mathbf{p}_n = \mathbf{r}_{n-1} + \beta(\mathbf{p}_{n-1} - \omega_{n-1}\mathbf{v}_{n-1});$$

$$\mathbf{v}_n = A\mathbf{p}_n; \quad \alpha = \frac{\rho_n}{(\tilde{\mathbf{r}}_0^T, \mathbf{v}_n)}; \quad \mathbf{s} = \mathbf{r}_{n-1} - \alpha\mathbf{v}_n; \quad \mathbf{t} = A\mathbf{s};$$

$$\omega_n = \frac{(\mathbf{t}^T\mathbf{s})}{(\mathbf{t}^T, \mathbf{t})}; \quad \mathbf{x}_n = \mathbf{x}_{n-1} + \alpha\mathbf{p}_n + \omega_n\mathbf{s}; \quad \mathbf{r}_n = \mathbf{s} - \omega_n\mathbf{t}; \quad n = n + 1;$$

end.

4 Least squares conjugate gradients algorithm.

A robust but rather slowly converging iterative method for linear systems is obtained by using CG method to the normal equation of the system

$$A^T A\mathbf{x} = A^T\mathbf{b}.$$

For linear systems with square nonsingular matrices, this approach has some disadvantages because of computing two matrix-vector products per iteration step (with A and A^T) or only one but with A^TA which has less sparse structure and its condition number is square of the condition number of A. Nevertheless, there are some situations (indefinite systems e.g.) where this approach is useful. If $\mathbf{s}_0 = \mathbf{b} - A\mathbf{x}_0$ and $\mathbf{r}_0 = A^T\mathbf{b} - A^TA\mathbf{x}_0$, the least squares conjugate gradient (LSCG) algorithms reads

$\mathbf{s}_0 = \mathbf{b} - A\mathbf{x}_0, \quad \mathbf{r}_0 = A^T\mathbf{s}_0;$
$\rho_{-1} = 0, \quad \mathbf{p}_{-1} = 0, \quad i = 0;$
while *residual* > *tolerance* **do**
begin

$$\rho_i = \mathbf{r}_i^T\mathbf{r}_i; \quad \beta_i = \frac{\rho_i}{\rho_{i-1}}; \quad \mathbf{p}_i = \mathbf{r}_i + \beta_i\mathbf{p}_{i-1};$$

$$\sigma_i = \mathbf{p}_i^T(A^TA)\mathbf{p}_i = (A\mathbf{p}_i)^TA\mathbf{p}_i; \quad \alpha_i = \frac{\rho_i}{\sigma_i}; \quad \mathbf{x}_{i+1} = \mathbf{x}_i + \alpha_i\mathbf{p}_i;$$

$$\mathbf{s}_{i+1} = \mathbf{s}_i - \alpha_iA\mathbf{p}_i; \quad \mathbf{r}_{i+1} = A^T\mathbf{s}_{i+1}; \quad i = i + 1;$$

end.

5 Minisupercomputer MPS 38.

MPS 38 is a middle class pipeline supercomputer. It is a peripheral device that operates independently but under the direction of a host processor. MPS 38 has a multiunits parallel structure and performs high-speed arithmetic calculations by distributing them among separate and independent functional units. It has separate memories for program instructions and data; separate arithmetic units for floating-point multiplications and additions which are designed as pipelines; a separate integer arithmetic unit for addresses and indexes calculations; separate busses to move data between the different functional units. Six types of optimizations can be performed on five optimization levels. The code overlapping allows unrelated operations to be started before previous operations have finished. The local optimizations are performed on separate basic blocks. The global optimizations effect multiple basic blocks in the program. The software pipelining is a code-generating technique that MPS 38 pipeline architecture uses to produce code which overlaps operations performed during one iteration of a loop with operations performed during other iterations of the same loop. The peak performance of the minisupercomputer MPS 38 is 12 Mflops but the performance which can be reached for solving real problems is between 40% and 70%.

6 Numerical examples.

The numerical experiments in this section are intended to show the behaviour of the CGS, BiCGstab and LSCG algorithms. The iterations were stopped when $||\mathbf{r}_k||_2/||\mathbf{r}_0||_2 < 10^{-10}$. The nonsymmetric matrices in all examples are five diagonal matrices and they are expressed by five vectors. The gradient methods under consideration are simply vectorized because of the created algorithms which use vector operations (sum of two vectors, SAXPY operations ($\alpha x + y$), dot products and matrix-vector products, which may be coded as a sequence of n SAXPY operations). The codes were written on Fortran 77.

6.1 Example 1 - pure advection in two dimensions.

A sample test-example for pure advection in two dimensions has been proposed by Molenkamp [5] and after that used in numerous studies. The mathematical description of this example can be given as follows

$$\begin{aligned} \mathcal{L}u \equiv & \frac{\partial c}{\partial t} + a(x,y)\frac{\partial c}{\partial x} + b(x,y)\frac{\partial c}{\partial y} = 0, \quad x,y \in \Omega \\ & c(x,y,0) = c_0(x,y) \\ & c(x,y,t)|_{\Gamma=\partial\Omega} = 0, \end{aligned} \tag{5}$$

where $\Omega = [0,2] \times [0,2]$, $t \in [0, 2\pi]$, $a(x,y) = y - 1$, $b(x,y) = x - 1$. The initial value condition u_0 is given by

$$c_0(x,y) = \begin{cases} 100(1 - \frac{r^*}{r}), & r^* < r \\ 0, & r^* < r, \end{cases} \quad (6)$$

where $r^* = \sqrt{(x-0.5)^2 + (y-1)^2}$ and $r = 0.25$. The exact solution at the end of the time interval is $c(x,y,2\pi) = c(x,y,0)$, i.e. the cone defined by the non-trivial part of (6) is rotated about a line orthogonal to the horizontal plane and cutting the plane at the point $(1,1)$. To discritize the differential equations in (5) the second order "up-wind" scheme with approximation accuracy $O(h^2 + \tau^2)$ is used [4]. The results obtained by 64×64 uniform grid in the horizontal plane and 320 time-steps are given in the Table 1.

Method	Number of iterations	Execution time (in sec.)	Performance in Mflops
LSCG	2761	140	3.2 (26.7%)
BiCGstab	1537	75	3.4 (28.3%)
CGS	2368	130	3.0 (25.0%)

Table 1. *Unknowns* - 3969, *Time steps* - 320.

It can be seen that the best results with respect to the number of iterations and the execution time is reached by the BiCGstab algorithm. There is no big difference between the performance of the three algorithms.

6.2 Example 2 - advection dominated second order problem.

In this subsection we consider an advection dominated second order partial differential equation with Dirichlet boundary conditions on the unit square Ω

$$u_{xx} + u_{yy} + 100\,u_x = F(x,y), \quad (x,y) \in \Omega.$$

The function F is defined by the exact solution $u^*(x,y) = \exp(xy)\sin(\pi x)\sin(\pi y)$. This equation was discretized using central finite difference scheme. The results obtained by (80×80) uniform grid are presented in the Table 2.

Method	Number of iterations	Execution time (in sec.)	Performance in Mflops
LSCG	1002	70	4.3 (35.8%)
BiCGstab	136	5	6.8 (56.7%)

Table 2. *Unknowns* - 6241.

It can be seen the good results obtained by the BiCGstab algorithm according all parameters especially the performance has been reached.

6.3 Example 3 - advection-diffusion problem.

Consider the boundary value problem

$$-\varepsilon(u_{xx} + u_{yy}) + a(x,y)u_x + b(x,y)u_y = 0,$$

on the unit square Ω, where

$$a(x,y) = 4x(x-1)(1-2y) \qquad \text{and} \qquad b(x,y) = 4y(1-y)(1-2x).$$

The boundary condition are given by formula

$$u(x,y) = \sin \pi x + \sin 13\pi x + \sin \pi y + \sin 13\pi y.$$

The parameter ε was taken to be equal to 0.1. The results obtained using finite difference method on (80 $\times$ 80 uniform grid are presented in the table 4.

Method	Number of iterations	Execution time (in sec.)	Performance in Mflops
LSCG	1688	140	3.7 (30.8%)
BiCGstab	97	9	2.7 (22.5%)
CGS	107	10	2.6 (21.7%)

Table 4. *Unknowns* - 6241.

7 Conclusions.

From our experiments we have learned that the BiCGstab may be an attractive method for iterative solving of the linear systems of equation with nonsymmetric matrices arising from the discretization of the convection and convection-diffusion problems. The results show that this three iterative algorithms can be successfully and efficiently implemented on the pipeline minisupercomputer MPS 38 for solving the problems under consideration.

8 Acknowledgement.

The author is grateful to S.Palikarev and A.Radeva from Multiprocessor system Ltd., Bulgaria for the possibility to use MPS 38 and his graduate student V.Peykova for the codes of the algorithms.

References

[1] O.Axelsson, *A generalized conjugate direction method and its application on a singular perturbation problem*, Lecture Notes in Mathematics, Vol.773, 1980, pp.1-11.

[2] J.Dongara, I.Duff, D.Sorensen, H.A.van der Vorst, *Solving linear systems on vector and shared memory computers*, SIAM, 1991.

[3] R.Fletcher, *Conjugate gradient methods for indefinite systems*, Lecture Notes in Mathematics, Vol.506, 1976.

[4] I.Lirkov, S.Margenov, *Circulant preconditioning of second order up-wind approximation of advection problems*, in Scientific Computation and Mathematical Modeling, S.Markov (Ed.), DATECS Publ., 1993.

[5] C.Molenkamp, *Accuracy of finite difference methods applied to the advection equation*, J. Appl. Meteor., Vol.7, pp.160-167.

[6] P.Sonnenveld, *CGS: a fast Lanczos-type solver for nonsymmetric linear systems*, SIAM J. Sci. Statist. Comput., Vol.10, 1989, pp.36-52.

[7] H.A.van der Vorst, *BiCGstab: A fast and smoothly converging variant of BiCG for the solution of nonsymmetric linear systems*, SIAM J. Sci. Statist. Comput., Vol.13, 1992, pp.631-644.

[8] Z.Zlatev, R.Berkowicz, *Numerical treatment of large-scale air pollution models*, Comput. Math. Appl., Vol.16, 1988, pp.93-109.

Address of Author:
Bulgarian Academy of Sciences
Acad.G.Bonchev str. Bl.25-A, 1113 Sofia
Bulgaria

Monte-Carlo evaluation of multivariate normal integrals and sensitivity to variate ordering

G.J.Gibson C.A.Glasbey D.A.Elston

1 Introduction.

This paper considers a stochastic simulation method for the evaluation of multivariate normal integrals of the form

$$p = \frac{1}{\sqrt{|\mathbf{V}|(2\pi)^n}} \int_{a_1}^{b_1} \int_{a_2}^{b_2} \cdots \int_{a_n}^{b_n} e^{-\frac{1}{2}\mathbf{x}^T\mathbf{V}^{-1}\mathbf{x}} dx_n \ldots dx_1,$$

where $\mathbf{x} = (x_1, \ldots, x_n)$, $E[\mathbf{x}] = \mathbf{0}$, $\mathbf{V} = E[\mathbf{x}\mathbf{x}^\mathbf{T}]$, and $-\infty \leq a_i, b_i \leq \infty, 1 \leq i \leq n$. The need to compute integrals of this form arises in a number of applications such as order-restricted statistical inference (Robertson *et al*, 1984), signal detection, and sensitivity analysis. The method we consider in the paper was developed independently by Genz (1992) and involves simulating each element x_i in turn, conditional on realized values of the preceding terms $(x_1, \ldots, x_{i-1})$. A key feature of the method is the use of a "swindle" in the estimation of p as a product of conditional probabilities. Genz studied the efficiency of the simulation scheme in comparison with that of quadrature methods. In particular he demonstrated that the former is often superior to the latter, in terms of the computer time required to give a specified accuracy, when the dimension of the integrals is high.

Our chief concern in this paper is to investigate the sensitivity of performance of the Monte-Carlo method to the ordering of the variates $x_1, \ldots, x_n$. In particular we show that the standard error of the estimated integral (for a fixed number of simulations) can vary by on order of magnitude depending on this ordering. A heuristic method for identifying a suitable ordering of the variates prior to simulation of the integral developed. The performance of the method when this ordering is applied to variates is investigated through simulation. It is shown that use of the ordering results in estimates whose standard errors are generally greatly reduced in comparison to random orderings of variates, or simpler ordering heuristics.

2 Calculation of normal integrals by stochastic simulation.

We retain the notation introduced in Section 1. Let $S \subset R^n$ be defined by

$$S = [a_1, b_1] \times [a_2, b_2] \times \ldots \times [a_n, b_n].$$

It is clear that an estimate of $p = P(\mathbf{x} \in S)$ could be obtained simply by simulating values of $\mathbf{x}$ from the appropriate multivariate normal distribution and estimating the proportion of instances for which $\mathbf{x} \in S$. However, it is well known that such a crude strategy necessitates many simulations to yield an accurate estimate. Instead, we consider a method which operates recursively employing a "swindle" to give an altogether superior scheme. For $1 \leq i \leq n$, let $S_i \subset R^n$ be the set defined by

$$S_i = \{\mathbf{x} \in R^n \mid a_j \leq x_j \leq b_j, 1 \leq j \leq i\}, \text{ and let } p_i = P(\mathbf{x} \in S_i).$$

By definition $p_0 = 1$. The simulation operates by generating recursively a sequence of probabilities $\hat{p}_i, 1 \leq i \leq n$, along with a sequence of variates $\xi_i, 1 \leq i \leq n-1$, as follows.
(i) Set $i = 1, \hat{p}_l = q_l = P(x_l \in [a_1, b_1])$.
(ii) Simulate ξ_1 from $x_1 \sim N(0, V_{11})$ conditional on $x_1 \in [a_1, b_1]$.
(iii) Increment i by 1.
(iv) Calculate $\hat{p}_i = \hat{p}_{i-l} q_i$ where $q_i = P(\mathbf{x} \in [a_i, b_i] \mid x_j = \xi_j, 1 \leq j \leq i-1)$.
(v) If $i = n$ then stop.
(vi) Simulate ξ_i from $x_i \in [a_i, b_i] \mid x_j = \xi_j, 1 \leq j \leq i-1)$.
(vii) Go to (iii).
An inductive argument demonstrates that $E(\hat{p}_i) = p_i, 1 \leq i \leq n$.

Hence, the value of p generated by the method is an unbiased estimate of the corresponding normal integral. An efficient implementation of the above procedure can be coded in FORTRAN by making use of some standard library routines (NAG, 1991). The main complexity arises in step (iv) above where the mean, μ_i, and variance, σ^2, of x_i conditional on the values of $x_1, \ldots, x_{i-1}$ is required. This calculation is done using the Cholesky decomposition of the full covariance matrix $\mathbf{V}$. That is, we factorise $\mathbf{V} = \mathbf{L}\mathbf{L}^T$ where $\mathbf{L}$ is a lower triangular matrix. Using standard results on the multivariate normal distribution (see e.g. Tong, (1990)), it can be shown that

$$\sigma_i^2 = L_{ii}^2,$$

and

$$\mu_i = \sum_{j=1}^{i-1} L_{ij}(\xi_j - \mu_j)/L_{jj},$$

where $\mu_1 = 0$. The Cholesky factorisation is carried out using the NAG routine FO3AEF before any simulations are generated. Having specified the conditional distribution of x_i, the probability q_i is calculated by using NAG routines S15ACF and S15ABF. These routines calculate for the standard normal the excluded probabilities , α and β, respectively above and below the bounds $(b_i - \mu_i)/\sigma_i$ and $(a_i - \mu_i)/\sigma_i$, so that $q_i = 1 - \alpha - \beta$.

The simulation of ξ_i (steps (ii) and (vi) above) is effected, after appropriate rescaling, by simulating a variate z from a standard normal conditional on z lying between the above bounds. First we generate a variate $y \in [0, 1]$ with a uniform distribution using NAG routine GO5CAF, and consider the probability $\alpha + (1 - \alpha - \beta)y$. Using NAG routine GO1CDF we invert the normal integral to calculate the value z such that

$$\alpha + (1 - \alpha - \beta)y = (2\pi)^{-\frac{1}{2}} \int_{-\infty}^{z} e^{-\frac{x^2}{2}} dx.$$

Then z has the required distribution, and $\xi_i = \mu_i + z\sigma_i$.

The above method has also been developed independently by Genz (1992), who compared it the MULNOR algorithm (Shervish, 1984) and the SCHURE algorithm (Berntsen *et al*, 1991). In particular, he demonstrated that the Monte-Carlo method out-performed the other methods when $n \geq 6$ and the covariance matrix was selected randomly using the method of Marsaglia and Olkin (1984).

3 Effect of variate ordering on algorithm performance.

This section discusses how the performance of the simulation method is influenced by the choice of variate ordering. We first specify how the relative efficiency of different orderings is measured for the purpose of this paper. Suppose that we estimate a normal integral, p, by taking the mean, $\bar{p}$, of m independent estimates $p(k), 1 \leq k \leq m$. Now $\bar{p}$ is itself a random variable (normally distributed for large m), with expectation p, and standard error which can be estimated as

$$\hat{\sigma}_p = \left\{ \frac{1}{m(m-1)} \sum_{i=1}^{m} (p(i) - \bar{p})^2 \right\}^{\frac{1}{2}}.$$

The efficiency of different orderings is compared by examining the values of $\hat{\sigma}_p$ which they produce for a fixed value of m.

In Genz's investigation of the method the lower limits of integration a_i were set to $-\infty$, and the upper limits b_i were chosen randomly, Genz utilised the ordering suggested by Shervish (1984), ordering variates so that the largest b_i are dealt with first. We tested the rule-of-thumb used by Genz for $n = 2$ by extensive simulation in the equivalent case where $b_i = \infty$ and a_i is finite, $1 \leq i \leq n$. The variates x_1 and x_2 had marginal distribution $N(0, 1)$, while their covariance was varied from -0.9 to 0.9 in steps of 0.1. For each covariance structure, the lower bounds a_1 and a_2 were varied independently from -2.0 to 2.0 in steps of 0.4. For each of

Order	$\hat{\sigma}_p(\times 10^{-6})$
(1,2,3)	6.55
(1,3,2)	10.15
(2,1,3)	11.82
(2,3,1)	11.83
(3,1,2)	7.68
(3,2,1)	7.23

Table 1: **Variation of $\hat{\sigma}_p$ with variate ordering for simulation of normal integral in Example 1.**

the $19 \times 11 \times 11$ combinations of covariance and lower bounds, the corresponding integral was estimated from 1000 simulations.In every case a simulation with $a_1 > a_2$ produced a smaller value of $\hat{\sigma}_p$ than the simulation with the order of the bounds reversed. This suggests that the rule is valid when $n = 2$, though this has not been proved analytically. However, we can easily show that it breaks down for higher dimensional integrals.

Example 1. Let $n = 3, b_i = \infty, i = 1,2,3$, and let $a_1 = 1.0, a_2 = 0$, and $a_3 = 0.5$. Suppose that the covariance matrix $\mathbf{V}$ is defined by $V_{ij} = (-0.8)^{|i-j|}, 1 \le i,j \le 3$. When the integral is estimated by simulation using the ordering $1,2,3$, with $m = 10000$, the values $\overline{p} = 7.27 \times 10^{-4}$ and $\hat{\sigma}_p = 7 \times 10^{-6}$ are observed. When the variates are taken in the order $1,3,2$ (in accordance with the scheme used by Genz) we find that the value of $\hat{\sigma}_p$ increases to 1.015×10^{-5}, which represents a significant fall-off in accuracy. Table 1 summarises the value of $\hat{\sigma}_p$ for the six distinct orderings of the three variables, showing that the original ordering is not bettered.

The above result is not surprising when the calculation is examined in more detail. Let $\mathbf{x}'$ denote the permutation (x_1, x_3, x_2) of the original vector (x_1, x_2, x_3) so that the corresponding covariance matrix is given by

$$\mathbf{V}' = \begin{bmatrix} 1.0 & 0.64 & -0.8 \\ 0.64 & 1.0 & -0.8 \\ -0.8 & 0.64 & 1.0 \end{bmatrix}.$$

If a value $x'_1 = \xi$ is simulated conditional on $x'_1 > a'_1 = 1.0$, then the mean and covariance of (x'_2, x'_3) conditional on $x'_1 = \xi$, are given respectively by

$$(0.64\xi, -0.8\xi)^T$$

and the matrix

$$\mathbf{V}'' = \begin{bmatrix} 0.5004 & -0.288 \\ -0.288 & 0.36 \end{bmatrix}.$$

If we transform the variates x'_2 and x'_3 so that they have zero mean and unit variance, then the corresponding lower bounds become

$$a''_2 = (0.5 - 0.64\xi)/(0.5004)^{1/2}$$

and

$$a_3'' = 0.8\xi/(0.36)^{1/2}.$$

Since $\xi > 1$, we have $a_2'' < a_3''$. In this case our experience for the 2-dimensional case suggests that x_3' should be selected before x_2' when simulation proceeds to the subsequent variate, i.e. the original ordering should be retained.

Example 1 demonstrates why a simple reordering of variates by lower bound is not an effective strategy even when all upper bounds are infinite, and it illustrates the important interaction between bounds and conditional covariances which takes place as the simulation proceeds. It also suggests an alternative approach to ordering of variates. Given that a "swindle" operates by simulating outcomes conditional on the occurrence of a desired event, it seems plausible that its efficiency may be maximised if, at each stage of the simulation, the event selected is that with the smallest probability, conditional on previous outcomes. This philosophy leads to the following strategy. The first variate to be simulated, x_1', should be chosen to be x_i, where i minimises $P(x_i \in [a_i, b_i])$. Thereafter, x_j' should be chosen to be x_i where i minimises $P(x \in [a_i, b_i] \mid x_1' = \xi_1', \ldots, x_{j-1}' = \xi_{j-1}'$. As formulated, this strategy requires the variates to be ordered separately for each of the m simulations as it proceeds, and the calculation and comparison of the above probabilities would be computationally costly. Therefore we propose the following approximation to the strategy which fixes the order of variates once and for all, prior to the first simulation. The selection x_1' is unchanged from the above. Thereafter, x_j' should be chosen to be x_i', where i minimises

$$P(x_i' \in [a_i, b_i] \mid x_1' = \overline{\xi}_1', \ldots, x_{j-1}' = \overline{\xi}_{j-1}'),$$

and

$$\overline{\xi}_i' = E(x_i \in [a_i, b_i] \mid \overline{\xi}_1', \ldots, x_{j-1}' = \overline{\xi}_{j-1}').$$

When the strategy is applied in the situation of Table 1, the ordering 1, 2, 3 is selected, and this is indeed the ordering which achieves the smallest value of $\hat{\sigma}_p$ was produced by an alternative ordering. The following is one such example.

Example 2. Let $a_1 = -1.0, a_2 = -0.5, a_3 = 0.0, a_4 = 0.5$ and $a_5 = 1.0$, let $b_i = \infty, 1 \le i \le 5$, and let $V_{ij} = 0.9^{|i-j|}$. Evaluating the 5-dimensional integral from 5000 simulations, using the original variate ordering, $(1, 2, 3, 4, 5)$, produces a value of $p = 0.1487$, and $\hat{\sigma}_p = 3.1 \times 10^{-3}$. The order selected by the above strategy is $(5, 4, 2, 3, 1)$, which results in $p = 0.1463$ and $\hat{\sigma}^p = 1.77 \times 10^{-4}$. However, the optimal ordering, as found by exhaustive simulation, is $(5, 4, 3, 2, 1)$ which gives an estimate for which $\hat{\sigma}_p = 1.68 \times 10^{-4}$. Nevertheless, the ordering chosen by the selection strategy is close to optimal, and considerably better than a randomly chosen ordering of the variates. This behaviour appears to occur in general, with the strategy selecting an ordering which, if not actually optimal, is close to optimal.

Figure 1 plots the value of $\hat{\sigma}_p$ for each of 120 different ordering of the variates for this example. There is a clear tendency for the values of $\hat{\sigma}_p$ to cluster in distinct

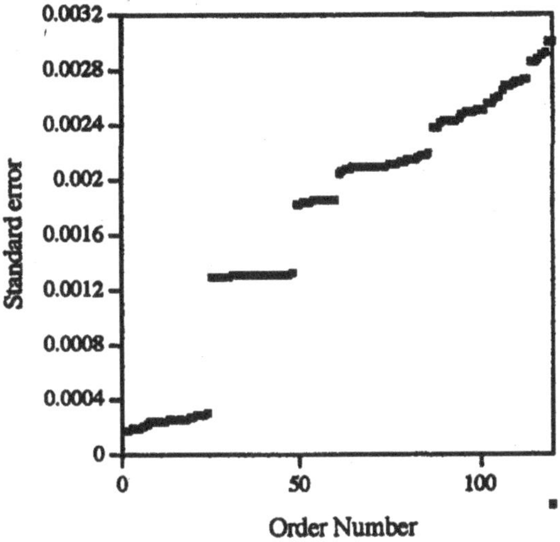

Figure 1: Variation of $\hat{\sigma}_p$ with ordering of variates for integral of Example 2.

bands. In this example, the 24 best orderings are those for which variable 5 are chosen first, while the orderings in the "second-best" group are those in which variable 4 is selected first. The other groupings apparent in Figure 1 cannot be characterised so easily. At least so far as this example is concerned, it appears that a near-optimal value of $\hat{\sigma}_p$ can be achieved by correctly choosing the first variable. At the time of writing we have not investigateds the extent to which this is true in general but it is our ordering method is significantly superior to simpler heuristics.

4 Conclusions.

In this paper we have consider how the efficiency of a Monte-Carlo method for calculating multivariate normal integrals is influenced by the order in which variates are simulated. The examples given demonstrate that the standard errors in the estimated integrals can vary by an order of magnitude depending on the choice of variate ordering. A simple mathod of ordering variates prior to simulation has been proposed and the illustratory examples suggest that the method may enable the selection of an ordering for which the standard error is close to optimal. By the use of this ordering strategy, it may be possible to improve the general afficiency of the Monte-Carlo method and to extend the range of scenarios in which it is superior to alternative methods of estimating multivariate normal integrals.

There are a number of areas related to the paper where further research would be worthwhile. We intend to carry out a more extensive investigation of ordering variates using a range of dimensions and correlation structures for the integral. Monte-Carlo method itself might be generalised to allow the estimation of integrals over domains which are non-rectangular.

Acknowledgement

This workwas supported by funding from the Scottish Office Agriculture and Fisheries Department.

References

[1] Bernsten, J., Espelid, T. O., and Genz, A. (1991). "Algorithm 698:SCUHRE - an adaptive multidimensional integration routine for a vector of integrals," *Association for Computing Machinery Transactions on Mathematical Sftware*, 17, 452-456.

[2] Genz, A. (1992). "Numerical computation of multivariate normal probabilities," *J. Comput. Graph. Stat.* 1, 141-150.

[3] Marsaglia, G. and Olkin, I. (1984). "Generating correlation matrices," *SIAM Journal of Scientific and Statistical Computing,* 5, 470-475.

[4] NAG (1991). *NAG FORTRAN Library Manual, Mark 15,* NAG, Oxford.

[5] Robertson, T., Wright, F. T. and Dykstra, R. L. (1988). *Order Restricted Statistical Inference,* John Wiley and Sons Ltd., Chichester.

[6] Schervish, M. J. (1984). "Multivariate Normal Probabilities with Error Bound: Algorithm AS195." *Applied Statistics, 33*, 81-87.

[7] Tong, Y. L. (1990). *The Multivariate Normal Distribution,* Springer-Verlag, New York.

G.J.Gibson, C.A.Glasbey
Scottish Agricultural Statistics Service,
James Clerk Maxwell Building,
University of Edinburgh,
Mayfield Rd., Edinburgh, EH9 3JZ,UK

D.A.Elston
SASS, Macaulay Land Use Research Institute,
Craigiebuckler, Aberdeen, AB9 2QJ, UK

Monte Carlo Methods for Nonlinear Equations

Todor V. Gurov*

Abstract

The paper deals with the analysis of the probable error of the Monte Carlo methods for solving integral equation with polynomial nonlinearity and systems of nonlinear algebraic equations of special type.

1 Introduction

Monte Carlo methods for estimation of the inner product

$$(g, \phi) = \int_G g(x)\phi(x)dx \tag{1}$$

are considered.

Hereafter integration means integration in the domain $G \in \mathbb{R}^n$. The function $g(x)$ belongs to the space $L_2(G)$ and $x = (x^{(1)}, x^{(2)}, \dots, x^{(n)}) \in G$ is a point in the Euclidean space $\mathbb{R}^n$. The function $\phi(x)$ is a solution of the integral equation with polynomial nonlinearity, i.e.

$$\phi(x) = \int \dots \int k(x, y_1, y_2, \dots, y_m) \prod_{i=1}^{m} \phi(y_i) \prod_{i=1}^{m} dy_i + f(x) \tag{2}$$

or the operator type is $\phi = \mathbb{K}\phi^{(m)} + f$, where $m \geq 2$, is natural number. The kernel $k(x, y_1, y_2, \dots, y_m) \in L_2\underbrace{(G \times \dots \times G)}_{m+1}$ and $f(x) \in L_2(G)$ are given functions.

It is assumed that this equation has an iteration solution corresponding to the iteration process in the metric of Banah space $\mathbb{F}$, i.e.

$$\phi_{s+1} = \mathbb{K}\phi_s^{(m)} + f, \quad \phi_s^{(m)} = \prod_{i=1}^{m} \phi_s(y_i), \quad \phi_0 = f. \tag{3}$$

If the following equation

*This work was supported in part by the Ministry of Education and Science of Bulgaria under grants I-210/92 and MU MSA-3/94.

$$\phi(x) = \int \cdots \int | k(x, y_1, y_2, \ldots, y_m) \prod_{i=1}^{m} \phi(y_i) | \prod_{i=1}^{m} dy_i + | f(x) | \qquad (4)$$

has an iteration solution, then the use of a branching stochastic process [2] enables us to construct random variables whose mathematical expectations are equal to functional (1).

We note that branching stochastic processes, which describe phenomena such as generation and death of particles in nature, are closely connected with the integral equations with polynomial nonlinearity.

2 Monte Carlo method

In this section we describe the Monte Carlo method, which is applied to the integral equation (2). We refer to Ermakov/Mikhailov [2] for more details.

Consider the following branching stochastic process: Suppose that any particle with initial frequency function $p_0(x) \geq 0$, $(\int p_0(x)dx = 1)$ is born in the domain $G \in \mathbb{R}^n$ in a random point x_0. In the next moment this particle either dies out with probability $h(x)$, $(0 \leq h(x) < 1)$ or generates a posterity of m analogical particles in the next random points $x_{00}, x_{01}, \ldots, x_{0m-1}$ with probability $p_m(x) = 1 - h(x)$ and transition frequency function

$$\frac{p(x_0, x_{00}, \ldots, x_{0m-1})}{p_m(x_0)} \geq 0,$$

where

$$\int \cdots \int p(x_0, x_{00}, x_{01}, \ldots, x_{0m-1}) \prod_{i=0}^{m-1} dx_{0i} = 1 - h(x_0).$$

The generated particles behave in the next moment as the initial one and etc. The traces of such a process is a tree from the type sketched in Fig.1.

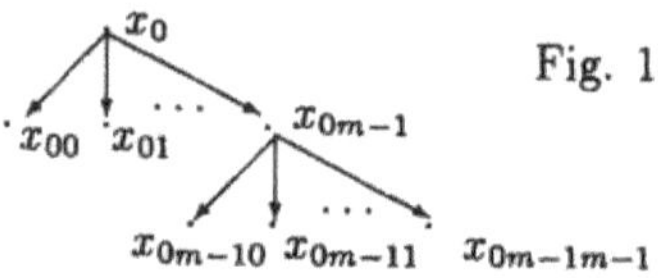

Fig. 1.

The used index numbers in Fig.1 are called multi-indices.The particle from the zero generation is enumerated with zero index, i.e. x_0. Its direct inheritors are enumerated with the indices $00, 01, \ldots, 0m-1$, i.e. next points $x_{00}, x_{01}, \ldots, x_{0m-1}$ are first generation. If a particle (a point) from the q-th generation of the tree γ has the multi-index $\gamma[q]$, then the multi-index of the l-th inheritor of this particle has the following form $\gamma[q+1] = \gamma[q]l$, where $l = 0, 1, ..., m-1$. In this way the multi-index is a number written in m-th numerical system.

Denote by **A** the set of points, which generate new points and denote by $\mathbb{B}$ the set of points, which die out [7].

There is a connection between the branching stochastic process and the iteration process (3). This connection allows to construct a procedure of a random choice of tree and to calculate the values of some random variable (r.v.). This r.v. corresponds to the random choice tree (see [2], pp.254). Thus we receive arbitrary trees γ when modulating the branching process and we correspond to every one of them the r.v. $\Theta_{[g]}(\gamma)$ in the following way:

$$\Theta_{[g]}(\gamma) = \frac{g(x_0)}{p_0(x_0)} \quad \text{if the tree consists of the initial point only.} \tag{5}$$

If the tree consists of other points, then $\Theta_{[g]}(\gamma)$ is constructed simultaneously with the construction of γ. When we have a transition from the random point $x_{\gamma[q]}$ to points $x_{\gamma[q]0}, x_{\gamma[q]1}, \ldots, x_{\gamma[q]m-1}$ then we multiply by,

$$\frac{k(x_{\gamma[q]}, x_{\gamma[q]0}, \ldots, x_{\gamma[q]m-1})}{p(x_{\gamma[q]}, x_{\gamma[q]0}, \ldots, x_{\gamma[q]m-1})}$$

and when the point $x_{\gamma[r]}$ dies we multiply by $f(x_{\gamma[r]}) \,/\, h(x_{\gamma[r]})$.

Then the r.v. $\Theta_{[g]}(\gamma)$ which corresponds to γ is:

$$\Theta_{[g]}(\gamma) = \frac{g(x_0)}{p_0(x_0)} \prod_{x_{\gamma[q]} \in \mathbf{A}} \frac{K(x_{\gamma[q]})}{P(x_{\gamma[q]})} \prod_{x_{\gamma[r]} \in \mathbb{B}} \frac{f(x_{\gamma[r]})}{h(x_{\gamma[r]})} \tag{6}$$

with the frequency function

$$p_\gamma = p_0(x_0) \prod_{x_{\gamma[q]} \in \mathbf{A}} P(x_{\gamma[q]}) \prod_{x_{\gamma[r]} \in \mathbb{B}} h(x_{\gamma[r]}), \tag{7}$$

$1 \le q \le k-1, \quad 1 < r \le k-1,$

where $K(x_{\gamma[q]}) = k(x_{\gamma[q]}, x_{\gamma[q]0}, \ldots, x_{\gamma[q]m-1})$; $P(x_{\gamma[q]}) = p(x_{\gamma[q]}, x_{\gamma[q]0}, \ldots, x_{\gamma[q]m-1})$.

Thus the mathematical expectation is:

$$E\Theta_{[g]}(\gamma) = (g, \phi) = \int g(x)\phi(x)dx, \tag{8}$$

where $\phi(x)$ is the solution of (2) (see [1],[2]).

Suppose the initial frequency function and the transition frequency function are admissible with respect to the functions $g(x)$ and $k(x_{\gamma[q]}, x_{\gamma[q]0}, \ldots, x_{\gamma[q]m-1})$ correspondingly (see [2],[4],[7]).

Besides, we propose with probability one that all the trees have a finite number of generations and the arithmetic mean over particles which are born in any generation is also finite. In this way the procedure of estimation exists.

Then, the series

$$\frac{1}{N}(\sum_{j=1}^{N}\Theta_{[g]}(\gamma)_j) \approx (g,\phi)$$

is an estimate of the functional (1). $(\Theta_{[g]}(\gamma)_j)$ is the value of $\Theta_{[g]}(\gamma)$ for the "j" trial of the tree γ; N is the number of the trials. In this case the probable error is

$$r = 0.6745\sigma(\Theta_{[g]}(\gamma))N^{-0.5},$$

where $\sigma^2(\Theta_{[g]}(\gamma))$ is the dispersion.

3 Almost optimal algorithm

The problem of optimization of the Monte Carlo algorithms consists in the minimization of the dispersion, i.e. minimization of the second moment $E\Theta^2_{[g]}(\gamma)$ of the r.v. $\Theta_{[g]}(\gamma)$. This is possible by means of a suitable chosen frequency function p_γ.

Introduce the functions

$$\Phi(x) = \frac{\mid f(x) \mid}{(h(x))^{\frac{1}{2}}};$$

$$\Phi(x) = \left(\int \dots \int \frac{k^2(x,y_1,y_2,\dots,y_m)}{p(x,y_1,y_2,\dots,y_m)} \prod_{i=1}^{m} \Phi^2(y_i) \prod_{i=1}^{m} dy_i\right)^{\frac{1}{2}}; \quad (9)$$

$$\hat{\Phi}(x) = \int \dots \int \mid k(x,y_1,y_2,\dots,y_m) \prod_{i=1}^{m} \Phi(y_i) \mid \prod_{i=1}^{m} dy_i. \quad (10)$$

Suppose that r.v. has finite dispersion [2]. The following statements hold [7]:

Lemma 1. The transition frequency function

$$p(x,y_1,y_2,\dots,y_m) = \frac{\mid k(x,y_1,y_2,\dots,y_m)\prod_{i=1}^{m}\Phi(y_i) \mid}{\int \dots \int \mid k(x,y_1,y_2,\dots,y_m)\prod_{i=1}^{m}\Phi(y_i) \mid \prod_{i=1}^{m} dy_i}$$

minimizes $\Phi(x)$ for any $x \in G$ and

$$\min_p \Phi(x) = \hat{\Phi}(x).$$

Lemma 2. The initial frequency function

$$p_0(x_0) = \mid g(x_0)\Phi(x_0) \mid / \left(\int \mid g(x_0)\Phi(x_0) \mid dx_0\right)$$

minimizes the functional

$$\int g^2(x_0)\Phi^2(x_0)p_0^{-1}(x_0)dx_0.$$

The minimum of this functional is equal to

$$\left(\int \mid g(x_0)\Phi(x_0) \mid dx_0\right)^2.$$

Theorem 1. The frequency function

$$\hat{p}_\gamma = c \mid g(x_0) \mid \prod_{x_{\gamma[q]}\in \mathbf{A}} \mid K(x_{\gamma[q]}) \mid \prod_{x_{\gamma[r]}\in \mathbb{B}} \Phi(x_{\gamma[r]})h(x_{\gamma[r]}) \tag{11}$$

minimizes the second moment $E\Theta^2_{[g]}(\gamma)$ of the r.v. $\Theta_{[g]}(\gamma)$ for any tree γ, i.e.

$$\min_{p_\gamma} E\Theta^2_{[g]}(\gamma) = E\hat{\Theta}^2_{[g]}(\gamma),$$

where $p(\gamma) = \hat{p}(\gamma)$ and $c = \left(\int \mid g(x_0)\hat{\Phi}(x_0) \mid dx_0\right)^{-1}$.

It is possible to show that

$$\overline{p}_\gamma = \overline{c} \mid g(x_0) \mid \prod_{x_{\gamma[q]}\in \mathbf{A}} \mid K(x_{\gamma[q]}) \mid \prod_{x_{\gamma[r]}\in \mathbb{B}} h(x_{\gamma[r]}) \tag{12}$$

is "almost" optimal. It is easy to show this in the case when the branching stochastic process has only one generation, only one tree γ_0 is constructed, i.e.

$$\phi(x_0) = \int \ldots \int k(x_0, y_1, y_2, \ldots, y_m) \prod_{i=1}^m f(y_i) \prod_{i=1}^m dy_i.$$

The next equalities hold:

$$\Phi(x_0) = \left(\int \ldots \int \frac{k^2(x_0, y_1, y_2, \ldots, y_m)}{p(x_0, y_1, y_2, \ldots, y_m)} \prod_{i=1}^m \Phi^2(y_i) \prod_{i=1}^m dy_i\right)^{\frac{1}{2}}; \tag{13}$$

$$\hat{\Phi}(x_0) = \int \ldots \int \mid k(x_0, y_1, y_2, \ldots, y_m) \prod_{i=1}^m \Phi(y_i) \mid \prod_{i=1}^m dy_i = \parallel K(x_0)\Phi^{(m)} \parallel; \tag{14}$$

$$\hat{p}_{\gamma_0} = \hat{c}_0 \mid g(x_0) \mid \ \mid k(x_0, y_1, \ldots, y_m) \mid \prod_{i=1}^m \Phi(y_i)h(y_i); \tag{15}$$

$$\overline{p}_{\gamma_0} = \overline{c}_0 \mid g(x_0) \mid \ \mid k(x_0, y_1, \ldots, y_m) \mid \prod_{i=1}^m h(y_i), \tag{16}$$

where $(\hat{c}_0)^{-1} = \parallel g \parallel \ \parallel K(x_0)\Phi^{(m)} \parallel$; $(\overline{c}_0)^{-1} = \parallel g \parallel \ \parallel K(x_0) \parallel$.

Present the frequency function $\overline{p}_{\gamma_0}$ in the following type:

$$\overline{p}_{\gamma_0} = \overline{c}_0 \mid g(x_0) \mid \; \mid k(x_0, y_1, \ldots, y_m) \mid \prod_{i=1}^{m} h(y_i) =$$

$$= \frac{\mid g(x_0) \mid}{\parallel g \parallel} \frac{\mid k(x_0, y_1, \ldots, y_m) \mid}{\parallel K(x_0) \parallel} \prod_{i=1}^{m} h(y_i) = \overline{p}_0(x_0)\overline{p}(x_0, y_1, \ldots, y_m) \prod_{i=1}^{m} h(y_i).$$

In fact by substituting $\overline{p}(x_0, y_1, \ldots, y_m)$ into the expression for $\Phi(x_0)$ we get:

$$\Phi^2(x_0) = \left(\int \ldots \int \frac{k^2(x_0, y_1, y_2, \ldots, y_m)}{\mid k(x_0, y_1, y_2, \ldots, y_m) \mid} \parallel K(x_0) \parallel \prod_{i=1}^{m} \Phi^2(y_i) \prod_{i=1}^{m} dy_i\right) =$$

$$= \parallel K(x_0) \parallel \; \parallel K(x_0)(\Phi^{(m)})^2 \parallel .$$

A necessary and sufficient condition for validity of the equality

$$\parallel K(x_0) \parallel \; \parallel K(x_0)(\Phi^{(m)})^2 \parallel = \parallel K(x_0)(\Phi^{(m)}) \parallel^2$$

is the following condition:

$$\parallel K(x_0)(\Phi^{(m)})^2 \parallel = c_0^2 \parallel K(x_0) \parallel, \tag{17}$$

where c_0 is constant. It follows that the $\overline{p}(x_0, y_1, \ldots, y_m)$ minimizes $\Phi(x_0)$ for any $x_0 \in G$ in the case of $\Phi = (c_0)^{\frac{2}{m}} = const$. The value of this minimum is $\hat{\Phi}(x_0)$. In this case the desired functional is

$$E\Theta^2_{[g]}(\gamma_0) = \int g^2(x_0)\Phi^2(x_0)p_0^{-1}(x_0)dx_0 = c_0^2 \int \mid g(x_0) \mid \; \parallel K(x_0) \parallel^2 \; \parallel g \parallel dx_0 =$$

$$= c_0^2 \parallel g \parallel \int \mid g(x_0) \mid \; \parallel K(x_0) \parallel^2 dx_0.$$

Here we use $\overline{p}_0(x_0) = \mid g(x_0) \mid \; \parallel g \parallel^{-1}$ instead of $p_0(x_0)$. It see that $\overline{p}_0(x_0)$ minimizes $E\Theta^2_{[g]}(\gamma_0)$ if $\parallel K(x_0) \parallel = const$.

These arguments show that the frequency function $\overline{p}_{\gamma_0}$ is "almost optimal" and the concept "almost optimal" (i.e. concept of nearness of $\overline{p}_{\gamma_0}$ to $\hat{p}_{\gamma_0}$) is connected with the concept of collineation of norms $\parallel K(x_0)(\Phi^{(m)})^2 \parallel$ and $\parallel K(x_0) \parallel$.

4 Solving nonlinear algebraic systems of special type

We shall consider the inner product of given vector $g = (g_1, g_2, \ldots, g_n)$ with a vector-solution $\phi = (\phi_1, \phi_2, \ldots, \phi_n)$ of a nonlinear algebraic system of the following special type:

$$\phi_j = \sum_{\beta_1+\beta_2+\cdots+\beta_n=m} k^{(j)}_{\beta_1\beta_2\ldots\beta_n} \phi_1^{\beta_1}\phi_2^{\beta_2}\ldots\phi_n^{\beta_n} + f_j, \quad j = 1, 2, \ldots, n \tag{18}$$

and

$$(g, \phi) = \sum_{j=1}^{n} g_j \phi_j. \tag{19}$$

Consider the integral equation with polynomial nonlinearity (2), where $G \equiv [0, n) \subset \mathbb{R}^1$. Let us cut G into intervals G_α, so that $x \in G_\alpha$ if $\alpha - 1 \leq x < \alpha$, $\alpha = 1, 2, \ldots, n$.

Suppose that functions $\phi(x)$, $f(x)$ and the kernel $k(x, y_1, y_2, \ldots, y_m)$ are constants in any intervals G_α. Thus we obtain:

$$\phi(x) = \sum_{\alpha_1=1}^{n} \int_{G_{\alpha_1}} \cdots \sum_{\alpha_m=1}^{n} \int_{G_{\alpha_m}} k(x, y_1, y_2, \ldots, y_m) \prod_{i=1}^{m} \phi(y_i) \prod_{i=1}^{m} dy_i + f(x) \tag{20}$$

Introduce the notations:

$$\phi_\alpha = \phi(x), \; f_\alpha = f(x) \text{ for any } x \in G_\alpha,$$

$$k^{(\alpha)}_{\alpha_1\alpha_2\ldots\alpha_m} = k(x, y_1, y_2, \ldots, y_m) \text{ for any } x \in G_\alpha \text{ and for any } y_i \in G_{\alpha_i},$$

$$\alpha, \alpha_i = 1, 2, \ldots, n \text{ and } i = 1, 2, \ldots, m.$$

Having in mind that functions $\phi(x)$, $f(x)$ and the kernel $k(x, y_1, y_2, \ldots, y_m)$ are constants in any intervals G_α we obtain the following nonlinear algebraic system

$$\phi_\alpha = \sum_{\alpha_1=1}^{n} \cdots \sum_{\alpha_m=1}^{n} k^{(\alpha)}_{\alpha_1\alpha_2\ldots\alpha_m} \phi_{\alpha_1} \phi_{\alpha_2} \ldots \phi_{\alpha_m} + f_\alpha. \tag{21}$$

In this system the number of coefficients is n^m. If we combine similar terms we get the system (18), where the number of coefficients is $\binom{n+m-1}{n}$. Thus these systems correspond to the integral equation with polynomial nonlinearity (2).

The r.v. $\Theta_{[g]}(\alpha)$ for which $E\Theta_{[g]}(\alpha) = (g, \phi) = \sum_{j=1}^{n} g_j \phi_j$ is

$$\Theta_{[g]}(\alpha) = \frac{g_\alpha}{p^0_\alpha} \prod_{x_{\alpha_i} \in \mathbb{A}} \frac{k^{(\alpha_i)}_{\alpha_1\alpha_2\ldots\alpha_m}}{p^{(\alpha_i)}_{\alpha_1\alpha_2\ldots\alpha_m}} \prod_{x_\beta \in \mathbb{B}} \frac{f_\beta}{h_\beta} \tag{22}$$

with the frequency function

$$p_\alpha = p^0_\alpha \prod_{x_{\alpha_i} \in \mathbb{A}} p^{(\alpha_i)}_{\alpha_1\alpha_2\ldots\alpha_m} \prod_{x_\beta \in \mathbb{B}} h_\beta, \tag{23}$$

where $p^0_\alpha = p_0(x) \geq 0$ for any $x \in G_\alpha$, $h_\beta = h(x)$ for any $x \in G_\beta$ and $1 \geq h_\beta > 0$, $p^{(\alpha)}_{\alpha_1\alpha_2\ldots\alpha_m} = p(x, y_{\alpha_1}, y_{\alpha_2}, \ldots, y_{\alpha_m}) \geq 0$ for any $x \in G_\alpha$, $\alpha = 1, 2, \ldots, n$ and for any $y_{\alpha_i} \in G_{\alpha_i}$, $\alpha_i = 1, 2, \ldots, n$, $i = 1, 2, \ldots, m$.

The following equalities hold

$$\sum_{\alpha_1=1}^{n} \cdots \sum_{\alpha_m=1}^{n} p^{(\alpha)}_{\alpha_1\alpha_2\ldots\alpha_m} = 1 - h_\alpha \quad \text{and} \quad \sum_{\alpha=1}^{n} p^0_\alpha = 1.$$

Then, the series

$$\frac{1}{N}(\sum_{j=1}^{N}\Theta_{[g]}(\alpha)_j) \approx (g,\phi)$$

is an estimate of the functional (19). $(\Theta_{[g]}(\alpha)_j)$ is the value of $\Theta_{[g]}(\alpha)$ for the "j" trial of a tree which we obtain if it is started by a random point $x_\alpha \in G_\alpha$; N is the number of the trials.

The vectors Φ and $\hat{\Phi}$ with components Φ_α and $\hat{\Phi}_\alpha$ are analogous to the functions $\Phi(x_0)$ and $\hat{\Phi}(x_0)$, where for any $\alpha = 1,2,\ldots,n$

$$\Phi_\alpha = \left(\sum_{\alpha_1=1}^{n}\cdots\sum_{\alpha_m=1}^{n}\frac{(k^{(\alpha)}_{\alpha_1\alpha_2\ldots\alpha_m})^2\Phi^2_{\alpha_1}\cdots\Phi^2_{\alpha_m}}{p^{(\alpha)}_{\alpha_1\alpha_2\ldots\alpha_m}}\right)^{\frac{1}{2}},$$

$$\hat{\Phi}_\alpha = \sum_{\alpha_1=1}^{n}\cdots\sum_{\alpha_m=1}^{n} | k^{(\alpha)}_{\alpha_1\alpha_2\ldots\alpha_m}\Phi_{\alpha_1}\cdots\Phi_{\alpha_m} | = \| K^{(\alpha)}\Phi^{(m)} \| \quad \text{and} \quad \Phi_\alpha = \frac{| f_\alpha |}{(h_\alpha)^{\frac{1}{2}}}.$$

The following statements hold (they are analogous of lemmas 1, 2 and theorem 1):
Lemma 3. The transition probability

$$\hat{p}^{(\alpha)}_{\alpha_1\alpha_2\ldots\alpha_m} = \frac{| k^{(\alpha)}_{\alpha_1\alpha_2\ldots\alpha_m}\Phi_{\alpha_1}\cdots\Phi_{\alpha_m} |}{\| K^{(\alpha)}\Phi^{(m)} \|}, \quad \text{minimizes } \Phi_\alpha$$

$$\text{and} \qquad \min_p \Phi_\alpha = \hat{\Phi}_\alpha \quad \text{for any } \alpha = 1,2,\ldots,n.$$

Lemma 4. The probability $\hat{p}^0_\alpha = | g_\alpha\hat{\Phi}_\alpha | \,/\, \| g\hat{\Phi} \|$ minimizes the sum

$$\sum_{\alpha=1}^{n}(\frac{g^2_\alpha\Phi^2_\alpha}{p^0_\alpha}) \qquad \text{and} \qquad \min_{p^0}\sum_{\alpha=1}^{n}(\frac{g^2_\alpha\Phi^2_\alpha}{p^0_\alpha}) = \| g\hat{\Phi} \|^2 .$$

Theorem 2. Introduce the next constant $\hat{c}^{-1} = \| g\hat{\Phi} \|$.
Then the frequency function

$$\hat{p}_\alpha = \hat{c} \mid g_\alpha \mid \prod_{x_{\alpha_i}\in\mathbf{A}} | k^{(\alpha_i)}_{\alpha_1\alpha_2\ldots\alpha_m} | \prod_{x_\beta\in\mathbb{B}} \Phi_\beta h_\beta,$$

minimizes the second moment $E\Theta^2_{[g]}(\alpha)$ of the r.v. $\Theta_{[g]}(\alpha)$, i.e.

$$\min_{p_\alpha} E\Theta^2_{[g]}(\alpha) = E\hat{\Theta}^2_{[g]}(\alpha), \quad \text{when } p_\alpha = \hat{p}_\alpha.$$

If we choose vector $e^{(j)} = \underbrace{(0,\ldots,0,\overbrace{1}^{j},0,\ldots,0)}_{n}$ instead of vector $g = (g_1,g_2,\ldots,g_n)$ then it is estimated the j-th component of the solution of the algebraic system (21).

References

[1] Sobol I.M., Monte Carlo numerical methods. Moskow, "Nauka", 1968 (in Russian)

[2] Ermakov S.M., Mikhailov G.A., Statistical simulation. Moskow, "Nauka", 1982 (in Russian)

[3] Shreider Yu.A., Method of Statistical Testing. Monte Carlo method. Elsevier Publishing company, 335 Jan Van Galenstraat, P.O. Box 211, 1964. Amsterdam (Netherlands).

[4] Dimov I.T., Minimization of the probable error for Monte Carlo methods. Application of Mathematics in Technology. Differential equations and applications. Varna 1986, Sofia 1987, pp.161-164. (Bulgaria).

[5] Ermakov S.M., On summation of series connected with integral equation. Vestnik Leningrad Univ. Math. Vol.16 (1984) pp.57-63. (Russian).

[6] Ermakov S.M., Nekrutkin V.V., Sipin A.S., Random processes for solving classical equations of the mathematical physics, Moskow, "Nauka", 1984, (in Russian)

[7] Gjurov T., Minimization of the probable error of the Monte Carlo method for solving of nonlinear integral equation, Mathematica Balkanica, Vol.6, 1992, pp.237–249.

Center of Informatics and Computer Technology
Bulgarian Academy of Sciences
Acad.G.Bontchev St., Bl.25A
1113 Sofia, BULGARIA

On circulant preconditioning of elliptic problems in L-shaped domain

Ivan D. Lirkov   Svetozar D. Margenov

Abstract

An application of circulant block-factorization preconditioners for Dirichlet boundary value problems in L-shaped domain is studied. New estimates of the convergence rate and of the total computational complexity are obtained. The proofs are sketched only. The accent of this article is towards the algorithmic aspects of the iterative scheme. Various numerical tests illustrate the scope of efficient applications of the proposed preconditioners.

1 Introduction.

In this paper we are concerned with the numerical solution of second order elliptic boundary value problems. Using finite differences or finite elements, such problems reduce to linear systems of the form $A\mathbf{u} = \mathbf{f}$, where A is a sparse matrix. The Preconditioned Conjugate Gradient (PCG) method is ideally suited to exploit the sparsity of A. In the case when A is symmetric and positive definite, the convergence rate of the PCG algorithm is a function of the relative condition number $\kappa(C^{-1}A)$ (namely $N_{it} = O(\sqrt{\kappa})$). Let C be a preconditioner for the original matrix A. The preconditioning strategy can be specified by the following two criteria: a) the number of iterations N_{it} to be as small as possible; b) the computational cost of the solution of the preconditioning system $C\mathbf{v} = \mathbf{w}$ to be significantly less than of the original system $A\mathbf{u} = \mathbf{f}$. The simultaneous satisfaction of the above criteria is one of the main objectives of the recent studies in the field of constructing efficient iterative methods.

The incomplete LU factorization is one of the most wide-spread preconditioning technique, see e.g., [1, 6, 12, 4]. For some of the pointwise incomplete factorization preconditioners it is proved (e.g., Gustafsson [7]) that the relative condition number $\kappa(M^{-1}A) = O(\sqrt{N})$, where N is the size of the discrete problem. The pointwise incomplete factorization algorithms are known as very robust with respect to problems with variable coefficient, but a possible disadvantage is their inherently sequential nature.

Another class of preconditioners based on a diagonal by diagonal averaging of the entries of a given matrix A to form a circulant approximation C was proposed in [2] (see also [13] and [8]). The use of the circulant approximations is motivated by their fast inversion based on the FFT. It is shown for the model problem, that the block circulant preconditioners can be constructed such that $\kappa(C^{-1}A) = O(\sqrt{N})$, which is asymptotically the same result as for certain (modified) ILU type preconditioners. The circulant preconditioners are highly parallelizable, see e.g., [10] and [11], but they are substantially more sensitive to a possible high variation of the coefficients of the elliptic operator.

The circulant block-factorization preconditioners, recently introduced in [9], relax the sensitivity of the circulant approximation with respect to possible high variation of the problem coefficients. The computational complexity and the parallel efficiency of this new algorithm is as high as of the block circulant one ([2, 11]).

The goal of this study is to expand the range of applicability of the circulant block-factorization preconditioners to elliptic problems in nonregular domains (different than rectangular ones). As a first step in this direction we have considered an application of these preconditioners to iterative solution of elliptic problems in L-shaped domain.

The outline of the paper is as follows. The studied problem is formulated in Section 2. The circulant block-factorization preconditioner is defined in Section 3. The relative condition number for the model problem is estimated. Section 4 contains a description of the preconditioning algorithm. An estimate of the computational complexity is given at the end. The numerical tests are presented in Section 5. The convergence rate of the proposed algorithm is computationally analyzed. One of the aims of this section is to specify the stability of the PCG algorithm for some problems with variable coefficients.

2 Formulation of the problem.

The following 2D elliptic problem is considered

$$\begin{aligned} -(a(x,y)u_x)_x-(b(x,y)u_y)_y &= f(x,y), \qquad \forall (x,y)\in\Omega, \\ u(x,y)=0, \qquad \forall (x,y)\in\Gamma &= \partial\Omega, \\ 0<c_{\min}\le a(x,y),\, b(x,y) &\le c_{\max}, \end{aligned} \tag{1}$$

where $\Omega=\Omega_1\cup\Omega_2=[0,1]\times[0,\frac{1}{2}]\cup[0,\frac{1}{2}]\times[\frac{1}{2},1]$ (see the L-shaped domain, Fig. 1(a)).

The week formulation of (1) has the following form: find $u\in H_0^1(\Omega)$, such that:

$$a_\Omega(u,v)=(f,v) \qquad \forall v\in H_0^1(\Omega), \tag{2}$$

where

$$a_\Omega(u,v)=\int_\Omega [a(x,y)u_xv_x+b(x,y)u_yv_y]\,dxdy.$$

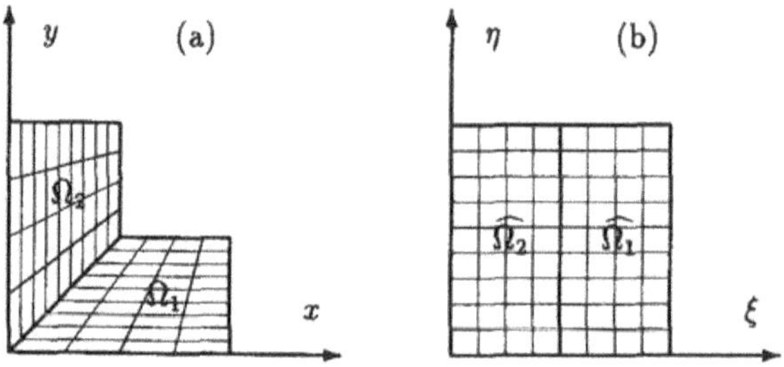

Figure 1: Domain transform into the unit square and corresponding grids

The circulant block-factorization is introduced and analyzed in [9], where elliptic problems in rectangular domain are considered. In order to expand the applicability of this approach to the problem (2), we transform the L-shaped domain Ω into the unit square (Fig. 1(b)) by peace-wise bilinear transform as follows

$$\Omega_1 \to \widehat{\Omega_1} \left| \begin{array}{lcl} x & = & 1-(2-\eta)(1-\xi) \\ y & = & \frac{\eta}{2} \end{array} \right. \qquad \Omega_2 \to \widehat{\Omega_2} \left| \begin{array}{lcl} x & = & \frac{\eta}{2} \\ y & = & 1-(2-\eta)\xi \end{array} \right. . \tag{3}$$

In such a way we rewrite the bilinear form a_Ω in the equivalent form

$$\mathcal{A}_{\widehat{\Omega}}(\hat{u},\hat{v}) = \mathcal{A}_{\widehat{\Omega_1}}(\hat{u},\hat{v}) + \mathcal{A}_{\widehat{\Omega_2}}(\hat{u},\hat{v}) \tag{4}$$

$$\mathcal{A}_{\widehat{\Omega_1}}(\hat{u},\hat{v}) = \int_{\widehat{\Omega_1}} \left\{ \left[\frac{a(\xi,\eta)+4(1-\xi)^2 b(\xi,\eta)}{2(2-\eta)} \right] \hat{u}_\xi \hat{v}_\xi + \right.$$
$$\left. 2(1-\xi)b(\xi,\eta)(\hat{u}_\xi\hat{v}_\eta + \hat{u}_\eta\hat{v}_\xi) + 2(2-\eta)b(\xi,\eta)\hat{u}_\eta\hat{v}_\eta \right\} d\xi d\eta$$

$$\mathcal{A}_{\widehat{\Omega_2}}(\hat{u},\hat{v}) = \int_{\widehat{\Omega_2}} \left\{ \left[\frac{b(\xi,\eta)+4\xi^2 a(\xi,\eta)}{2(2-\eta)} \right] \hat{u}_\xi \hat{v}_\xi + \right.$$
$$\left. 2\xi a(\xi,\eta)(\hat{u}_\xi\hat{v}_\eta + \hat{u}_\eta\hat{v}_\xi) + 2(2-\eta)a(\xi,\eta)\hat{u}_\eta\hat{v}_\eta \right\} d\xi d\eta.$$

Now we consider the numerical solution of the transformed week formulation of the studied problem, i.e.: find $\hat{u} \in H_0^1(\widehat{\Omega})$, such that

$$\mathcal{A}_{\widehat{\Omega}}(\hat{u},\hat{v}) = (\hat{f},\hat{v}) \qquad \forall \hat{v} \in H_0^1(\widehat{\Omega}), \tag{5}$$

The Finite Element Method (FEM) with bilinear finite elements is applied to problem (5). A uniform rectangular mesh on the unit square $\widehat{\Omega}$ is used with mesh-sizes $h_\xi = \frac{1}{m}$, $h_\eta = \frac{1}{n}$. The meshparameter m is chosen to be odd number from a algorithmic reason. As a result the boundary value problem is reduced to the linear algebraic system

$$A\mathbf{u} = \mathbf{f}. \tag{6}$$

The stiffness matrix A admits a block tridiagonal structure. A can be written in the form $A = tridiag(A_{i,i-1}, A_{i,i}, A_{i,i+1}), \quad i = 1, 2, \dots n$, where A_{ij} are tridiagonal matrices.

In what follows we consider as a model problem the Poisson problem

$$-\Delta u = f, \qquad (x, y) \in \Omega \tag{7}$$

In this case, the stiffness matrix A can be written as a sum of tensor products in the form

$$A = P_1 \otimes Q_1 + P_2 \otimes Q_2 + P_3 \otimes Q_3 \tag{8}$$

where

$$\begin{aligned}
P_1 &= \frac{1}{h_\xi} tridiag \begin{pmatrix} -1-\left((2i-1)h_\xi\right)^2, & 2+8(ih_\xi)^2+2h_\xi^2, & -1-\left((2i+1)h_\xi\right)^2 \\ -1-\left((2i-1)h_\xi-2\right)^2, & 2+8(ih_\xi-1)^2+2h_\xi^2, & -1-\left((2i+1)h_\xi-2\right)^2 \end{pmatrix} \begin{matrix} \text{for } i < \frac{m}{2} \\ \text{for } i > \frac{m}{2} \end{matrix} \\
Q_1 &= \frac{h_\eta}{6} tridiag \left(\frac{1}{4-(2j-1)h_\eta}, \quad \frac{4}{4-(2j-1)h_\eta} + \frac{4}{4-(2j+1)h_\eta}, \quad \frac{1}{4-(2j+1)h_\eta} \right) \\
P_2 &= tridiag \begin{pmatrix} (i-\frac{1}{2})h_\xi, & 0, & -(i+\frac{1}{2})h_\xi \\ (i-\frac{1}{2})h_\xi - 1, & 0, & 1-(i+\frac{1}{2})h_\xi \end{pmatrix} \begin{matrix} \text{for } i < \frac{m}{2} \\ \text{for } i > \frac{m}{2} \end{matrix} \\
Q_2 &= tridiag \left(-1, \quad 0, \quad 1 \right) \\
P_3 &= \frac{h_\xi}{6} tridiag \left(1, \quad 4, \quad 1 \right) \\
Q_3 &= \frac{1}{h_\eta} tridiag \left(-4 + (2j-1)h_\eta, \quad 8 - 4jh_\eta, \quad -4 + (2j+1)h_\eta \right).
\end{aligned}$$

3 Circulant block-factorization. Model analysis.

Using the standard LU factorization procedure, we first split $A = D + L + U$ into its block-diagonal and strictly block-triangular parts respectively. Then the exact block-factorization can be written in the form, $A = (X + L)(I + X^{-1}U)$, where $X = diag(X_1, X_2, \dots X_n)$. We have $A = X + L + U + LX^{-1}U$. Therefore $X = D - LX^{-1}U$, which rewritten component-wise gives

$$X_1 = A_{1,1}, \text{ and } X_i = A_{i,i} - A_{i,i-1} X_{i-1}^{-1} A_{i-1,i}, \qquad i = 2, \dots m \tag{9}$$

It is known that the above factorization exists if A is, for example, positive definite. This factorization can be used to solve system (6), which requires solution of linear systems involving the blocks X_i. Note that $\{X_i\}$ are in general full matrices and the resulting (direct) Gaussian elimination is too expensive. The common idea of the block-ILU factorization methods is to approximate X_i (or X_i^{-1}) by sparse (band) matrices. The idea introduced in [9] is to modify the original matrix A in such a way, that the resulting blocks from the exact factorization of the modified matrix (in place of X_i) to be circulant.

Let us recall that a circulant matrix C has the form $C_{k,j} = c_{(j-k)\mathrm{mod}n}$, where n is the size of C (for more details see [5]). Any circulant matrix can be factored as

$$C = F\Lambda F^*, \tag{10}$$

where Λ is a diagonal matrix containing the eigenvalues of C, and F is the Fourier matrix $F = \frac{1}{\sqrt{n}}\left\{e^{2\pi\frac{jk}{n}\mathrm{i}}\right\}_{0\leq j,k\leq n-1}$. Here i stands for the imaginary unit.

We use a circulant block-factorization approximation C as a preconditioner for the matrix A.

$$C = tridiag(-C_{i,i-1}, C_{i,i}, -C_{i,i+1}) \qquad i = 1, 2, \ldots m,$$

where $C_{i,j} = Circulant(A_{i,j})$ is some circulant approximation of the corresponding block $A_{i,j}$. The preconditioner C is factored exactly. Note that the recursion (9), performed for C, is closed in the class of circulant matrices. That is, the corresponding blocks X_i are circulant and therefore the solution of the preconditioned system involving the matrix C can be performed efficiently based on the FFT.

Following the averaging approach from [3] to the blocks Q_i, $i = 1, 2, 3$ from (8), we construct the circulant block-factorization preconditioner for the stiffness matrix of the model problem (7) in the form

$$C = P_1 \otimes C_1 + P_2 \otimes C_2 + P_3 \otimes C_3, \tag{11}$$

where

$$\begin{aligned} C_1 &= \frac{h_\eta \ln 2}{12}(4, 1, 0, \ldots 0, 1) \\ C_2 &= (0, 1, 0, \ldots 0, -1) \\ C_3 &= \frac{3}{h_\eta}(2, -1, 0, \ldots 0, -1). \end{aligned}$$

The following theorem holds.

Theorem 1 *The preconditioning matrix C defined by (11) satisfies the relative condition number estimate*

$$\kappa(C^{-1}A) \leq 12n \tag{12}$$

where A is the stiffness matrix (8).

The proof follows the technique from [2, 9].

Note that, in difference to the block circulant preconditioners from [2], the construction (11) is based on an averaging procedure along the η coordinate direction only. This provides some attractive advantages of the presented method, that are demonstrated by the numerical test in the last section.

4 Preconditioning algorithm.

In order to calculate C^{-1} we rewrite the circulant blocks of the preconditioner C in factored form

$$C_{i,j} = F\Lambda_{i,j}F^*$$
$$X_i = FD_i^{-1}F^*$$

and that yields

$$\begin{aligned} D_1^{-1} &= \Lambda_{1,1} \\ D_i^{-1} &= \Lambda_{i,i} - \Lambda_{i,i-1}D_{i-1}\Lambda_{i-1,i}. \end{aligned}$$

Let us denote $\Lambda = tridiag(\Lambda_{i,i-1}, \Lambda_{i,i}, \Lambda_{i,i+1})$. Then

$$Cw = u \Longleftrightarrow (I \otimes F)\Lambda(I \otimes F^*)w = u.$$

The solution of the preconditioned system $Cw = u$ is split in three stages

$$\begin{aligned} \bar{u} &= (I \otimes F^*)u \\ \Lambda\bar{w} &= \bar{u} \\ w &= (I \otimes F)\bar{w} \end{aligned}$$

First and third stages can be performed by block-FFT (see [10]), and second stage consist solving of two recurrence equations

$$\left| \begin{array}{l} v_1 = D_1 u_1 \\ v_i = D_i(u_i - \Lambda_{i,i-1}v_{i-1}) \end{array} \right. \qquad \left| \begin{array}{l} w_m = v_m \\ w_i = v_i - D_i\Lambda_{i,i+1}w_{i+1} \end{array} \right. \tag{13}$$
$$i = 2, 3, \ldots m \qquad\qquad i = m-1, m-2, \ldots 1$$

The total computational cost of the preconditioner is

$$\begin{aligned} \aleph_{C^{-1}u} &= 2\aleph_{FFT} + 2\aleph_{rec} \qquad (14) \\ \aleph_{FFT} &= 5mn\log n \\ \aleph_{rec} &= (3m-2)n \end{aligned}$$

Theorem 2 *The computational complexity of the PCG iterative solution of the problem (7) by the circulant preconditioner is*

$$\aleph_{PCG} \le 2(\sqrt{3n}\ln\frac{2}{\varepsilon} + 1)(5m\log n + 3m - 2)n$$

where ε is the relative error.

The proof follows from the estimate of number of iterations (see [6]), Theorem 1 and (14).

Remark 1 *Block-FFT is well parallelizable and (13) consist n independent systems which can be calculated in parallel.*

Table 1: Number of iterations, maximal error and L_2 norm of the error for the Poisson problem.

m	n	N_{it}	$\|u-u^h\|_C$	$\|u-u^h\|_{L_2}$
7	8	13	1.67×10^{-1}	2.333×10^{-2}
15	8	14	4.67×10^{-2}	6.309×10^{-3}
15	16	18	4.53×10^{-2}	5.377×10^{-3}
31	16	17	1.28×10^{-2}	1.620×10^{-3}
31	32	22	1.16×10^{-2}	1.311×10^{-3}
63	32	23	3.29×10^{-3}	4.135×10^{-4}
63	64	30	2.93×10^{-3}	3.258×10^{-4}
127	64	33	8.28×10^{-4}	1.047×10^{-4}
127	128	41	7.35×10^{-4}	8.135×10^{-5}
255	128	49	2.08×10^{-4}	2.635×10^{-5}
255	256	64	1.84×10^{-4}	2.034×10^{-5}

5 Numerical tests.

The computations are done with double precision on IBM 4381 mainframe. The iteration stopping criterion is $\|r^{N_{it}}\|/\|r^0\| < 10^{-6}$, where r^j stands for the residual at the jth iteration step of the preconditioned conjugate gradient method. We denote by u the exact solution, and by u^h the obtained approximate solution. The number of iterations is presented as a measure of the convergence rate of the preconditioners.

Table 1 shows the number of iterations, the maximal error and the L_2 norm of the error $(u - u^h)$ for problem (7). The presented data demonstrate a slightly slower growth of the iterations than $O(\sqrt{n})$. However, the numerical results are in a good agreement with the theoretical estimate (12).

Remark 2 *Let us note that the same convergence rate of the PCG iterative method is obtained when the problem coefficients are*

$$a(x,y) = b(x,y) = \begin{cases} 1 & (x,y) \in \Omega_1 \\ \sigma & (x,y) \in \Omega_2 \end{cases} \quad \textit{for } \sigma = 10, 10^2, 10^3.$$

The explanation of this fact easy follows from the independence of our algorithm of the coefficient's jump through the interface between Ω_1 and Ω_2.

It was already mentioned, that the circulant preconditioners are known as sensitive with respect to high variation of the coefficients of the elliptic operator. Since the circulant block-factorization preconditioners are constructed by averaging of the coefficients only along one of the coordinate directions (along the η direction in the case of L-shaped domain), our method is less sensitive to the varying coefficients. The next experiments illustrate the behavior of the block-factorization circulant preconditioner for problems with variable coefficients. The test problem we used is taken from [2], and is (1) where the coefficients are $a(x,y) = 1 + \epsilon e^{x+y}$ and $b(x,y) = 1 + \frac{\epsilon}{2}\sin(2\pi(x+y))$. Here $\epsilon \in [0,1]$ is a parameter. Tables 2–3 demonstrate a minor sensitivity when ϵ

Table 2: Number of iterations, maximal error and L_2 norm of the error.

		$\epsilon = 0.01$			$\epsilon = 0.1$		
m	n	N_{it}	$\|u - u^h\|_C$	$\|u - u^h\|_{L_2}$	N_{it}	$\|u - u^h\|_C$	$\|u - u^h\|_{L_2}$
7	8	13	1.68×10^{-1}	2.337×10^{-2}	13	1.72×10^{-1}	2.378×10^{-2}
15	8	14	4.73×10^{-2}	6.341×10^{-3}	14	5.29×10^{-2}	6.785×10^{-3}
15	16	18	4.58×10^{-2}	5.441×10^{-3}	16	5.00×10^{-2}	6.091×10^{-3}
31	16	18	1.31×10^{-2}	1.649×10^{-3}	18	1.76×10^{-2}	2.272×10^{-3}
31	32	22	1.22×10^{-2}	1.384×10^{-3}	22	1.75×10^{-2}	2.251×10^{-3}
63	32	23	3.72×10^{-3}	4.552×10^{-4}	24	9.32×10^{-3}	1.430×10^{-3}
63	64	30	3.55×10^{-3}	4.141×10^{-4}	30	9.40×10^{-3}	1.490×10^{-3}
127	64	33	1.40×10^{-3}	1.884×10^{-4}	34	7.41×10^{-3}	1.323×10^{-3}
127	128	42	1.42×10^{-3}	1.970×10^{-4}	45	7.46×10^{-3}	1.346×10^{-3}
255	128	49	9.05×10^{-4}	1.504×10^{-4}	49	6.98×10^{-3}	1.308×10^{-3}
255	256	65	9.20×10^{-4}	1.559×10^{-4}	66	6.99×10^{-3}	1.314×10^{-3}

increases. Such kind of sensitivity is typical, e.g., for the modified ILU methods (see in [2]).

Table 4 shows results obtained when $a = 1 + \frac{1-\max(x,y)}{1-\min(x,y)}$ and $b = 1$. One can see that the number of iterations is almost the same as for the Poisson problem. The explanation is, that in such a case, the coefficients are constants along the η-direction, and the averaging procedure provides the same relative conditional number as for the Poisson problem.

Acknowledgment

This paper was partially supported by the National Science Foundation under grant INT–92–20287 and also in part by the Bulgarian Ministry of Education and Science under grants MM/75-91 and MU MSA-3/94.

References

[1] O. Axelsson and V.A. Barker, *Finite Element Solution of Boundary Value Problems: Theory and Computations*, Academic Press, Orlando, Fl.(1983).

[2] R.H. Chan and T.F. Chan, Circulant preconditioners for elliptic problems, *J. Numerical Lin.Alg.Appl.*, **1** (Mar. 1992), 77-101.

Table 3: Number of iterations, maximal error and L_2 norm of the error, $\epsilon = 1$.

m	n	N_{it}	$\|u-u^h\|_C$	$\|u-u^h\|_{L_2}$
7	8	13	2.02×10^{-1}	2.860×10^{-2}
15	8	15	9.19×10^{-2}	1.304×10^{-2}
15	16	19	8.39×10^{-2}	1.187×10^{-2}
31	16	20	5.41×10^{-2}	8.532×10^{-3}
31	32	25	5.18×10^{-2}	8.342×10^{-3}
63	32	28	4.45×10^{-2}	7.615×10^{-3}
63	64	36	4.38×10^{-2}	7.589×10^{-3}
127	64	38	4.21×10^{-2}	7.419×10^{-3}
127	128	52	4.20×10^{-2}	7.415×10^{-3}
255	128	60	4.15×10^{-2}	7.373×10^{-3}
255	256	82	4.15×10^{-2}	7.373×10^{-3}

Table 4: Number of iterations, maximal error and L_2 norm of the error, for $a = 1 + \frac{1-\max(x,y)}{1-\min(x,y)}$ and $b = 1$.

m	n	N_{it}	$\|u-u^h\|_C$	$\|u-u^h\|_{L_2}$
7	8	13	1.74×10^{-1}	2.343×10^{-2}
15	8	14	5.02×10^{-2}	6.760×10^{-3}
15	16	18	4.77×10^{-2}	5.474×10^{-3}
31	16	18	1.40×10^{-2}	1.764×10^{-3}
31	32	22	1.23×10^{-2}	1.345×10^{-3}
63	32	24	3.61×10^{-3}	4.531×10^{-4}
63	64	31	3.11×10^{-3}	3.353×10^{-4}
127	64	33	9.11×10^{-4}	1.150×10^{-4}
127	128	44	7.81×10^{-4}	8.384×10^{-5}
255	128	51	2.29×10^{-4}	2.896×10^{-5}
255	256	67	1.95×10^{-4}	2.097×10^{-5}

[3] T.F. Chan, An Optimal Circulant Preconditioner for Toeplitz Systems, *SIAM J.Sci. Stat.Comp.* **9** (Jul. 1988) 766-771.

[4] T.F. Chan and P.S. Vassilevski, A framework for block-ILU factorizations using block-size reduction, *Math.Comp*, to appear.

[5] P.J. Davis, *Circulant matrices*, John Wiley, New York (1979).

[6] G.H. Golub and C.F. van Loan, *Matrix Computations*, 2nd edition, Johns Hopkins Univ.Press, Baltimore (1989).

[7] I. Gustafsson, A class of first-order factorization methods, *BIT*, **18** (1978), 142–156.

[8] S. Holmgren and K. Otto, Iterative solution methods for block-tridiagonal systems of equations, *SIAM J.Matr.Anal.Appl.*, **13** (1992), 863–886.

[9] I.D. Lirkov, S.D. Margenov and P.S. Vassilevski, Circulant block - factorization preconditioners for elliptic problems, *UCLA CAM Report 93–29*, (1993) (to appear in Computing).

[10] C. van Loan, *Computational frameworks for the fast Fourier transform*, SIAM, Philadelphia (1992).

[11] S.D. Margenov and I.T. Lirkov, Preconditioned conjugate gradient iterative algorithms for transputer based systems, *in Parallel and distributed processing*, K. Boyanov ed., Sofia (1993), 406–415.

[12] G. Meurant, A review on the inverse of symmetric tridiagonal and block tridiagonal matrices, *SIAM J.Matr.Anal.Appl.*, **13** (1992), 707–728.

[13] G. Strang, A proposal for Toeplitz matrix calculations, *Stud. Appl.Math.*, **74** (1986), 171–176.

Center for Informatics and Computer Technology,
Bulgarian Academy of Sciences,
Acad.G.Bontchev Str.Bl.25A,
1113 Sofia, Bulgaria

DIFFERENCE SCHEMES ON THE GRIDS LOCALLY REFINING IN SPACE AS WELL AS IN TIME

P.P. Matus[1] and P.N. Vabishchevich[2]
[1]Institute of Mathematics BAS
Minsk, Byelorussia
[2]Institute for Mathematical Modeling RAS
Moscow, Russia

June 19, 1994

1 Introduction

At present, more attention is focused on the development of adaptive methods for solving problems with peculiarities [1], [2], [3]. The simplest classes of adaptive methods are connected with using the grids locally refining in space as well as in time. This work is concerned with the construction and examination of such schemes on the basis of the general theory of difference schemes [4].

Concerning to the approximate solution of the elliptic boundary value problems on the grids with local refinement [5], the main problem is connected with constructing schemes on the boundary of an adaptation region. New schemes, heritaging the fundamental properties of the initial differential problem (selfadjoint and conservation properties, the maximum principle, *etc.*), are suggested in [6].

Using the domain decomposition technique, there are developed in the work [7] three general classes of schemes adapting in time. They are constructed via additional fractional time-levels in an adaptation region. The stability and convergence properties of such schemes are investigated in the work.

2 Elliptic problems on grids with local refinement

Let us consider the Dirichlet problem for the Poisson equation in a square

$$\Omega = \{x \mid x = (x_1, x_2), \quad -1 < x_\alpha < 1, \quad \alpha = 1, 2\}.$$

We shall find the solution of equation

$$-\Delta u = f(x), \quad x \in \Omega, \tag{1}$$

with boundary conditions

$$u(x) = g(x), \quad x \in \partial\Omega. \tag{2}$$

In equation (1)

$$\Delta u \equiv \sum_{\alpha=1}^{2} \frac{\partial^2 u}{\partial x_\alpha^2}. \tag{3}$$

To demonstrate usage of the grids with local refinement for problem (1)–(3), we consider only the case when the grid step is divided in two in an adaptation region. We recognize two types of such a grid. In the first case the boundary of an adaptation region passes through the points of the basic coarse grid, in the second – lies between them.

The main difficulties in using the grids with local refinement are connected with difference approximations in the points located at and in the vicinity of the adaptation subdomain boundary. It is necessary to construct the schemes that inherit the fundamental properties of the initial differential problem. For example, for problem (1)–(3) the selfadjoint and positive definite grid operator is desired. Besides, we may demand the maximum principle fulfillment for the difference equation.

Using integro-interpolation (control volume) method [4] for the scheme construction we can satisfy the conservation and selfadjoint properties. But the maximum principle for the difference solution does not hold in this case. On the other hand, it is easy to construct the schemes which satisfy the maximum principle using linear interpolation for the additional nodes at the boundary of an adaptation region (the first-type grid). But in this case the selfadjoint and conservation properties are not inherited. Difference schemes that meet the selfadjoint and conservation properties as well as the maximum principle are constructed here for elliptic second-order boundary value problems on the basis of the integro-interpolation method where the control volumes are nothing but the cells of a Voronoi mesh.

We shall use locally refining grids with the boundary of an adaptation region passing between the points of the basic grid (the second-type grid). Standard five-point approximations (for 2D case) are used in the internal points of the basic grid. The difference Laplace operator for these points can be represented as follows

$$\Lambda y = \frac{4}{h^2}\left\{\begin{matrix} & 1 & \\ 1 & -4 & 1 \\ & 1 & \end{matrix}\right\}.$$

Special approximations are needed in the points of an adaptation subregion boundary as well as in the points of the basic grid in the vicinity of these boundary. It is necessary to determine the characteristic points and the corresponding control volumes (the cells a Voronoi mesh for these points)

The following six-point stencil is employed to formulate the difference equation in the point of an adaptation subregion boundary (this point is located at the center of a basic grid cell):

$$(\Lambda y)_1 = \frac{16}{5h^2}\left\{\begin{matrix} 1/2 & & 1/2 \\ 1 & -4 & 1 \\ & 1 & \end{matrix}\right\}.$$

Such an approximation results from the standard five-point approximation of the Laplace operator and linear interpolation of the difference solution between the points of the basic grid (within a factor).

For the point of the basic grid in the vicinity of the boundary we obtain

$$(\Lambda y)_2 = \frac{16}{15h^2}\left\{\begin{matrix} & 1 & \\ 1/2 & -4 & 1/2 \\ 1/2 & 1 & 1/2 \end{matrix}\right\}.$$

Next, for the corner point of the fine grid we get

$$(\Lambda y)_3 = \frac{8}{3h^2}\left\{\begin{matrix} 1/2 & & 1 \\ 1/2 & -4 & \\ & 1 & 1/2 \end{matrix}\right\}.$$

This approximation can also be treated as the result of linear interpolation between the points of the basic grid. And finally, for the point of the basic grid near the corner of the adaptation subregion we have the following formula

$$(\Lambda y)_4 = \frac{8}{7h^2}\left\{\begin{matrix} 1/2 & 1 & \\ & -4 & 1 \\ 1 & & 1/2 \end{matrix}\right\}.$$

It is easy to check that the selfadjoint property and the maximum principle hold for such approximations.

3 Schemes with varying time-steps

Refinement of the temporal grid in some part of calculation domain may be useful for the solution of time-dependent problems. Difference schemes adapting in time have been constructed in [8] on the basis of a special implementation of computations out of the adaptation region. Here we consider three general classes of time-adaptive schemes that contain a number of suggested early schemes as a particular case.

The problems of the development and investigation of the schemes with locally varying time-step will be presented for the simplest one-dimensional problem. There is no problem to extend this approach to more general problems. In $\Omega = (0, l)$ we shall find the solution of the parabolic equation

$$\frac{\partial u}{\partial t} - \frac{\partial}{\partial x}(k(x)\,\frac{\partial u}{\partial x}) = f(x,t), \quad x \in \Omega,\ t > 0, \tag{4}$$

supplemented with the homogeneous boundary conditions (the Dirichlet problem)

$$u(0,t) = 0, \quad u(l,t) = 0, \quad t > 0 \tag{5}$$

and the initial condition

$$u(x,0) = u_o(x), \quad x \in \Omega. \tag{6}$$

For the simplicity, let us suppose that in region Ω a uniform grid with spacing h is introduced and ω is the set of internal points. Here we will use the standard index-free notation of the theory of the difference schemes theory [4]. Let us define the grid operator

$$Ay = -(a(x)y_{\bar{x}})_x, \quad x \in \omega, \tag{7}$$

on the set of grid functions $y \in H$ such as $y(x) \equiv 0, x \notin \omega$ where, for example, $a(x) = 0.5(k(x) + k(x-h))$. In a Hilbert space H we shall introduce the scalar product and the norm as follows

$$(y,v) = \sum_{x\in\omega} y(x)v(x)h, \quad \|y\| = (y,y)^{1/2}.$$

For operator A, defined as (7), in space H we have $A = A^* > 0$ (see e.g. [4]).

Let us write for problem (4)–(6) the following semidiscrete problem (discrete in space but continuous in time):

$$\frac{dy}{dt} + Ay = \varphi, \quad x \in \omega \tag{8}$$

for the given $y(x,0), x \in \omega$. Difference schemes with locally varying time-steps will be constructed for the problem (8). Let us suppose that in some subregion $\Omega^* \subset \Omega$ (the region of adaptation) it is necessary to use essentially smaller time-step than out of this zone. We will consider that for the solution of the Cauchy problem for equation (8) out of the adaptation region we use the time-step $\tau > 0$. In the adaptation region we introduce fractional time-levels $\tau_p = \tau/p, p > 1$. In the transition region (between the first and second regions) time-steps are monotonically changed.

Difference schemes will be constructed via formal implementation of p fractional steps in order to achieve new time-level in the whole calculation region. Out of the adaptation region real computations are carried out only one time, other p steps are fictitious (time-step for them is equal to zero). For the formal presentation of this approach we shall use the following unit splitting of a computational region.

Let us define functions χ_α, $\alpha = 1, 2, ..., p$ in such a way that for all x the following equality holds

$$\sum_{\alpha=1}^{p} \chi_\alpha(x,t) = 1. \tag{9}$$

Unit splitting (9) is employed for transition from one time-level to another. When we solve unsteady problems, an adaptation region may be changed (dynamic adaptation) – for different time levels we can use it's own splitting. That is why a dependence of function χ_α, $\alpha = 1, 2, ..., p$ on time is reflected in (9).

Out of the adaptation region $\chi_1(x,t) = 1$, whereas other $\chi_\alpha(x,t) = 0$, $\alpha = 2, 3, ..., p$. In the adaptation region all $\chi_\alpha(x,t)$ are equal each other and therefore from (9) it follows that $\chi_\alpha(x,t) = 1/p$, $\alpha = 1, 2, ..., p$. Let us suppose that all $\chi_\alpha(x,t)$, $\alpha = 1, 2, ..., p$ are continuous and in the transitional region they are linear.

Let us connect with (9) the following additive presentation of operator A:

$$A = \sum_{\alpha=1}^{p} A_\alpha(t). \tag{10}$$

Each operator A_α is connected with function $\chi_\alpha(x,t)$, $\alpha = 1, 2, ..., p$ and so it is connected with the decomposition of the computational region onto separate subregions. We will point out main opportunities on operator $A'_\alpha, \alpha = 1, 2, ..., p$ choice:

$$A_\alpha = \chi_\alpha(x,t)A, \quad \alpha = 1, 2, ..., p. \tag{11}$$

$$A_\alpha = A\chi_\alpha(x,t), \quad \alpha = 1, 2, ..., p. \tag{12}$$

Taking into account that function $\chi_\alpha(x,t)$ is variable for $\alpha = 1, 2, ..., p$ we get $A_\alpha \neq A_\alpha^*$, $\alpha = 1, 2, ..., p$ for (11) and (12).

Choice of selfadjoint operators of the decomposition seems to be more traditional. For one-dimensional problem (4)–(6) one can obtain operators $A_\alpha, \alpha = 1, 2, ..., p$ by analogy with (7):

$$A_\alpha y = -(a_\alpha(x) y_{\bar{x}})_x, \quad x \in \omega, \tag{13}$$

where for example

$$a_\alpha(x) = 0,5(\chi_\alpha(x)k(x) + \chi_\alpha(x-h)k(x-h)), \quad \alpha = 1, 2, ..., p.$$

Such operators of decomposition correspond to the problem with degenerated coefficients $\chi_\alpha(x,t)k(x,t)$ instead of $k(x,t)$. In this case $A_\alpha = A_\alpha^*, \alpha = 1, 2, ..., p$.

Construction of the difference scheme for equation (8) is performed using splitting (10) on the basis of general results of the difference schemes theory [4]. Let us denote by y^n the difference solution at the time moment $t^n = n\tau$, were $\tau > 0$ – time-step. If we use schemes of component-splitting than the transition at a new time-level for the approximate solution of the Cauchy problem (7), (10) is carried out in correspondence with

$$\frac{y^{n+\alpha/p} - y^{n+(\alpha-1)/p}}{\tau} + A_\alpha^n y^{n+\alpha/p} = \varphi_\alpha^n, \quad \alpha = 1, 2, ..., p, \tag{14}$$

where $A_\alpha^n = A_\alpha(t^n)$ – operators of decomposition for the transition from $n-$ to $(n+1)-$time level .

It was found, that for the approximation error of difference scheme (10), (14) with operators $A_\alpha, \alpha = 1, 2, ..., p$ from equations (11), (12) and (13) the following estimate take place

$$\left\| z^{n+1} \right\|_D \leq M \sum_{k=1}^{n} \tau \sum_{\alpha=1}^{p} \left\| \chi_\alpha^k \right\|_A (\tau + h^2),$$

where $D = A, A^{-1}$ and E respectively. Thus the convergence rate of the difference scheme with the grid locally refining in time depends on the width of the transition region. In the limit case (the width of the transition region is $O(h)$) the convergence rate is $O(h^{-1/2}(\tau + h^2))$. It is easy to extend the results obtained here in different directions.

4 Generalization

Difference schemes with local grid refinement were constructed for basic boundary value problems for elliptic second-order equations. Non-selfadjoint problems of convection/diffusion are studied separately.

Schemes with local refinement in time are investigated as the operator-difference schemes. So there is no problem to use the basic results obtained here for more general unsteady problems. More accurate in time schemes were studied too.

On the basis of a model one-dimensional parabolic equation the schemes with dynamic local spatial refinement were constructed and studied [9]. The problems of constructing such schemes were examined along with the derivation of the corresponding apriori estimates for the solution error. The particular emphasis was on the problem of scheme construction on new additional points of spatial grids.

References

[1] *Adaptive Computational Methods for Partial Differential Equations* (Eds. I. Babuška, J. Chandra and J.F.Flaherty). SIAM, Philadelphia, 1983.

[2] *Accuracy Estimates and Adaptive Refinements in Finite Element Computations* (Eds. I. Babuška, O.C. Zienkievicz, J.R. Gago and E.R. de Arantes e Oliveira) . Wiley, London, 1986.

[3] *Adaptive Methods for Partial Differential Equations* (Eds. J.F. Flaherty, P.J. Paslow, M.S. Shephard and J.D. Vasilakis). SIAM, Philadelphia, 1989.

[4] A.A. Samarskii. *Theory of Difference Schemes.* Nauka, Moscow, 1989 (in Russian).

[5] S.F. McCormick. *Multilevel Adaptive Methods for Partial Differential Equations.* SIAM, Philadelphia, 1989.

[6] P.N. Vabishchevich and G.I. Shishkin. Difference schemes on grids with local refinement (accepted in for publication).

[7] P.N. Vabishchevich and P.P. Matus. Difference schemes with varying time-steps (accepted in for publication).

[8] P.P. Matus. *Differential Equations*, 1990, 26, N^{0}7, 1241–1254 (in Russian).

[9] P.N. Vabishchevich, P.P. Matus and V.G. Rychagov. Difference schemes with dynamically refining grids (accepted in for publication).

Summation of Slowly Convergent Series via Quadratures*

Gradimir V. Milovanović

Abstract

In this paper we consider a few methods for summation of slowly convergent series using quadrature formulas of Gaussian type. Such summation/integration procedures are based (i) on the Laplace transform method, and (ii) on an integration over some contours in the complex plane. Especially, we investigate some series with irrational terms. Numerical results are included to illustrate these methods.

1 Introduction

We consider convergent series of the type

$$T = \sum_{k=1}^{+\infty} a_k \qquad \text{and} \qquad S = \sum_{k=1}^{+\infty} (-1)^k a_k \tag{1.1}$$

and introduce the notation: $T = T^{(m-1)} + T_m^{(\infty)}$, $S = S^{(m-1)} + S_m^{(\infty)}$,

$$T_m^{(n)} = \sum_{k=m}^{n} a_k, \qquad S_m^{(n)} = \sum_{k=m}^{n} (-1)^k a_k,$$

where $T^{(m-1)}$ and $S^{(m-1)}$ are the corresponding partial sums of (1.1).

Some methods of summation these series can be found, for example, in the books of Henrici [5], Lindelöf [6], and Mitrinović and Kečkić [8].

Recently, a few new summation/integration procedures for slowly convergent series are developed (see [3], [1], [2], [7]). In this paper we will give a short account of these methods as well as some new approaches to this subject, including numerical examples in order to illustrate and compare these methods. Especially, we investigate some series with irrational terms.

*This work was supported in part by the Science Fund of Serbia, grant number 0401F.

2 Laplace Transform Method

Suppose that the general term of T (and S) is expressible in terms of the derivative of a Laplace transform, or in terms of the Laplace transform itself. Namely, let $a_k = F'(k)$, where

$$F(p) = \int_0^{+\infty} e^{-pt} f(t)\,dt, \qquad \operatorname{Re} p \geq 1.$$

Then

$$\sum_{k=1}^{+\infty} F'(k) = -\sum_{k=1}^{+\infty} \int_0^{+\infty} te^{-kt} f(t)\,dt = -\int_0^{+\infty} \frac{t}{e^t - 1} f(t)\,dt.$$

Similarly, for "alternating" series, one obtains

$$\sum_{k=1}^{+\infty} (-1)^k F'(k) = \int_0^{+\infty} \frac{t}{e^t + 1} f(t)\,dt$$

and

$$\sum_{k=1}^{+\infty} (-1)^k F(k) = -\int_0^{+\infty} \frac{1}{e^t + 1} f(t)\,dt.$$

In a joint paper with Gautschi [3] we considered the construction of Gaussian quadrature formulas on $(0, +\infty)$,

$$\int_0^{+\infty} g(t)w(t)\,dt = \sum_{\nu=1}^{n} \lambda_\nu g(\tau_\nu) + R_n(g), \tag{2.1}$$

with respect to the weight functions

$$w(t) = \varepsilon(t) = \frac{t}{e^t - 1} \qquad \text{(Einstein's function)}$$

and

$$w(t) = \varphi(t) = \frac{1}{e^t + 1} \qquad \text{(Fermi's function).}$$

We obtained the first $n = 40$ coefficients in the three-term recurrence relation for the corresponding orthogonal polynomials accurately to 25 decimal digits (see [3, Tables 1 and 3]). If the series T and S are slowly convergent and the respective function f on the right of the equalities above is smooth, then low-order Gaussian quadrature (2.1) applied to the integrals on the right, provides a possible summation procedure. Numerical examples show fast convergence of this procedure (see [3, §4]). In the sequel we refer to this procedure as the *Laplace transform method.* A problem which arises with this procedure is the determination of the original function f for a given series. For some other applications see [1] and [2].

3 Contour Integration Over a Rectangle

Suppose that $a_k = f(k)$, where $z \mapsto f(z)$ is a holomorphic function in the region

$$G_m = \{z \in \mathbb{C} \mid \operatorname{Re} z \geq \alpha,\ m-1 < \alpha < m\}, \quad m \in \mathbb{N}.$$

In [7] we derived an alternative summation/integration method for the series (1.1) which requires the indefinite integral F of f chosen so as to satisfy the following decay conditions:

(C1) *F is a holomorphic function in the region G_m;*

(C2) $\lim_{|t|\to+\infty} e^{-c|t|} F(x + it/\pi) = 0$, *uniformly for $x \geq \alpha$;*

(C3) $\lim_{x\to+\infty} \int_{-\infty}^{+\infty} e^{-c|t|}\, |F(x+it/\pi)|\, dt = 0,$

where $c = 2$ or $c = 1$, when we consider $T_m^{(n)}$ or $S_m^{(n)}$, respectively.

Namely, taking $\Gamma = \partial G$ and

$$G = \Big\{ z \in \mathbb{C} \mid \alpha \leq \operatorname{Re} z \leq \beta,\ |\operatorname{Im} z| \leq \frac{\delta}{\pi} \Big\},$$

where $m - 1 < \alpha < m$, $n < \beta < n+1$ $(m, n \in \mathbb{Z}, m \leq n)$, we obtain that

$$T_m^{(n)} = \frac{1}{2\pi i} \oint_\Gamma f(z) \frac{\pi}{\tan \pi z}\, dz \quad \text{and} \quad S_m^{(n)} = \frac{1}{2\pi i} \oint_\Gamma f(z) \frac{\pi}{\sin \pi z}\, dz.$$

After integration by parts, these formulas reduce to

$$T_m^{(n)} = \frac{1}{2\pi i} \oint_\Gamma \Big(\frac{\pi}{\sin \pi z}\Big)^2 F(z)\, dz \tag{3.1}$$

and

$$S_m^{(n)} = \frac{1}{2\pi i} \oint_\Gamma \Big(\frac{\pi}{\sin \pi z}\Big)^2 \cos \pi z\, F(z)\, dz, \tag{3.2}$$

where $z \mapsto F(z)$ is an integral of $z \mapsto f(z)$.

Taking $\alpha = \alpha_m = m - 1/2$, $\beta = \beta_n = n + 1/2$, and letting $\delta \to +\infty$ and $n \to +\infty$, under conditions (C1) – (C3), the integrals in (3.1) and (3.2) over Γ reduce to integrals along the line $z = \alpha_m + iy$ $(-\infty < y < +\infty)$.

After some calculations, we reduce T and S to a problem of quadrature on $(0, +\infty)$ with respect to the weight functions

$$w_1(t) = \frac{1}{\cosh^2 t} \quad \text{and} \quad w_2(t) = \frac{\sinh t}{\cosh^2 t}, \tag{3.3}$$

respectively. Thus,

$$T = T^{(m-1)} + \int_0^{+\infty} \Phi(\alpha_m, t/\pi)\, w_1(t)\, dt$$

and

$$S = S^{(m-1)} + \int_0^{+\infty} \Psi\left(\alpha_m, t/\pi\right) w_2(t)\, dt,$$

where w_1 and w_2 are defined in (3.3) and

$$\Phi(x,y) = -\frac{1}{2}\left[F(x+iy) + F(x-iy)\right],$$

$$\Psi(x,y) = \frac{(-1)^m}{2i}\left[F(x+iy) - F(x-iy)\right].$$

Numerical quadratures of Gaussian type with respect to the weights w_1 and w_2 were constructed in [7]. The first $n = 40$ coefficients in the three-term recurrence relation for the corresponding orthogonal polynomials were obtained accurately to 30 decimal digits.

Numerical experiments shows that is enough to use only the quadrature with respect to the first weight $w_1(t) = 1/\cosh^2 t$. Namely, in the series S we can include the hyperbolic sine as a factor in the corresponding integrand so that

$$S = S^{(m-1)} + \int_0^{+\infty} \Psi\left(\alpha_m, t/\pi\right) \sinh(t) w_1(t)\, dt. \tag{3.4}$$

4 Series With Irrational Terms

In this section we consider some series of the form

$$U_\pm(a,\nu) = \sum_{k=1}^{+\infty} \frac{(\pm 1)^{k-1}}{(k^2+a^2)^{\nu+1/2}}.$$

In 1916 Kapteyn (see [9, p. 386]) proved the formula

$$U_+(a,\nu) = \sum_{k=1}^{+\infty} \frac{1}{(k^2+a^2)^{\nu+1/2}} = \frac{\sqrt{\pi}}{(2a)^\nu \Gamma(\nu+1/2)} \int_0^{+\infty} \frac{t^\nu}{e^t-1} J_\nu(at)\, dt$$

which is valid when $\operatorname{Re}\nu > 0$ and $|\operatorname{Im} a| < 1$. Here, J_ν is the Bessel function of the order ν. Since for $F(p) = 1/(p^2+a^2)^{\nu+1/2}$ ($\operatorname{Re}\nu > -1/2$, $\operatorname{Re} p > |\operatorname{Im} a|$) the original function is

$$f(t) = \frac{\sqrt{\pi}}{(2a)^\nu \Gamma(\nu+1/2)} t^\nu J_\nu(at),$$

using the Laplace transform method we find

$$U_-(a,\nu) = \sum_{k=1}^{+\infty} \frac{(-1)^{k-1}}{(k^2+a^2)^{\nu+1/2}} = \frac{\sqrt{\pi}}{(2a)^\nu \Gamma(\nu+1/2)} \int_0^{+\infty} \frac{t^\nu}{e^t+1} J_\nu(at)\, dt.$$

Thus, this method leads to an integration of the Bessel function $t \mapsto J_\nu(at)$ with Einstein's weight $\varepsilon(t)$ or Fermi's weight $\varphi(t)$. For some special values of ν, we can use also quadratures with respect to the weights $t^{\pm 1/2}\varepsilon(t)$ and $t^{\pm 1/2}\varphi(t)$ (see [3] and [1]).

In order to sum the series $U_-(a,0)$, $a > 0$, we can integrate the function $z \mapsto F(z) = \sqrt{z^2+a^2}g(z)$, with $g(z) = \pi/\sin \pi z$, over the circle

$$C_n = \Big\{ z \in \mathbb{C} \ \Big| \ |z| = n + \frac{1}{2} \Big\}, \qquad n > a,$$

with cuts along the imaginary axis, so that the critical singularities ia and $-ia$ are eliminated (cf. [8, p. 217]). Precisely, the contour of integration Γ is given by $\Gamma = C_n^1 \cup l_1 \cup \gamma_1 \cup l_2 \cup C_n^2 \cup l_3 \cup \gamma_2 \cup l_4$, where C_n^1 and C_n^2 are parts of the circle C_n, γ_1 and γ_2 are small circular parts of radius ε and centres at $\pm ia$, and l_k $(k = 1,2,3,4)$ are the corresponding line segments.

Let $F^*(z)$ be the branch of $F(z)$ which corresponds to the value of the square root which is positive for $z = 1$. Since

$$\oint_\Gamma F^*(z)\,dz = 2\pi i \sum_{k=-n}^{n} \frac{(-1)^k}{\sqrt{k^2+a^2}},$$

and $\int_{\gamma_1} \to 0$, $\int_{\gamma_1} \to 0$, when $\varepsilon \to +0$, and $\int_{C_n^1 \cup C_n^2} \to 0$, when $n \to +\infty$, we obtain

$$\sum_{k=1}^{+\infty} \frac{(-1)^k}{\sqrt{k^2+a^2}} = -\frac{1}{2a} + \int_a^{+\infty} \frac{du}{\sinh \pi u \sqrt{u^2-a^2}},$$

i.e.,

$$\sum_{k=1}^{+\infty} \frac{(-1)^{k-1}}{\sqrt{k^2+a^2}} = \frac{1}{2a} - \frac{1}{2}\int_{-1}^{+1} \left(t \sinh \frac{\pi a}{t} \right)^{-1} \frac{dt}{\sqrt{1-t^2}}.$$

Thus, we reduced $U_-(a,0)$ to a problem of Gauss-Chebyshev quadrature. Since $t \mapsto (t\sinh(\pi a/t))^{-1}$ is an even function we can apply the $(2n)$-point Gaussian approximations with only n functional evaluations, so that we have

$$U_-(a,0) \approx GC(n) = \frac{1}{2a} - \frac{\pi}{2n} \sum_{k=1}^{n} \left(\tau_k \sinh \frac{\pi a}{\tau_k} \right)^{-1}, \tag{4.1}$$

where $\tau_k = \cos((2k-1)\pi/(4n))$, $k = 1, \dots, n$.

Remark 4.1. The same method can be applied to the summation of the series

$$\sum_{k=-\infty}^{+\infty} f(k, \sqrt{k^2+a^2}) \quad \text{and} \quad \sum_{k=-\infty}^{+\infty} (-1)^k f(k, \sqrt{k^2+a^2}) \quad (a > 0),$$

where f is a rational function. Then we integrate the function $z \mapsto F(z) = f(z, \sqrt{z^2+a^2})g(z)$, with $g(z) = \pi/\tan \pi z$ and $g(z) = \pi/\sin \pi z$, respectively, over the circle C_n with the cuts.

5 Numerical Results

In this section we illustrate the previous methods taking the series $U_-(a,0)$, with $a =$ 0.25, 0.5, 1, 2, 4. All computations were done in Q-arithmetic on the MICROVAX 3400 computer (machine precision $\approx 1.93 \times 10^{-34}$).

Table 5.1 shows the relative errors in Gaussian approximations $GC(n)$ (cf. (4.1)) and $GF(n)$ (the Laplace transform method with Fermi's weight), as well as

$$S_{n,m} = S^{(m-1)} + \sum_{\nu=1}^{n} \lambda_\nu \Psi(\alpha_m, \tau_\nu/\pi) \sinh(\tau_\nu), \quad m = 1, 2, 3, 4,$$

(cf. (3.4)) for $n = 5(5)40$ and $a = 0.25$. (Numbers in parentheses indicate decimal exponents.)

TABLE 5.1

Relative errors in Gaussian approximation of the sum $U_-(a,0)$ for $a = 0.25$

n	$GC(n)$	$GF(n)$	$S_{n,1}$	$S_{n,2}$	$S_{n,3}$	$S_{n,4}$
5	1.8(−2)	2.7(−10)	2.4(−4)	1.6(−4)	1.4(−4)	1.2(−4)
10	7.2(−4)	8.8(−19)	9.2(−7)	4.2(−9)	3.9(−9)	3.6(−9)
15	1.9(−4)	1.8(−24)	1.8(−8)	9.2(−14)	8.6(−14)	8.2(−14)
20	2.9(−5)		5.6(−10)	1.3(−16)	1.7(−18)	1.7(−18)
25	1.2(−5)		1.2(−10)	1.5(−18)	2.6(−23)	3.2(−23)
30	9.0(−8)		1.1(−11)	3.0(−20)	8.7(−27)	6.1(−28)
35	1.0(−6)		6.1(−13)	2.5(−23)	3.1(−28)	1.1(−32)
40	3.5(−7)		7.0(−13)	1.8(−23)	2.0(−30)	1.6(−33)

The Bessel function J_0 was evaluated by means of the rational approximations in [4] indexed 5852, 6553 and 6953, with precision 23.22, 23.37 and 23.46, respectively. Because of that, some entries in third column are empty.

TABLE 5.2

Relative errors in Gaussian approximation of the sum $U_-(a,0)$ for $a = 1$

n	$GC(n)$	$GF(n)$	$S_{n,1}$	$S_{n,2}$	$S_{n,3}$	$S_{n,4}$
5	3.1(−5)	2.0(−4)	2.0(−4)	2.4(−4)	2.1(−4)	1.8(−4)
10	5.1(−7)	2.1(−7)	6.2(−6)	6.3(−9)	5.9(−9)	5.4(−9)
15	9.4(−9)	9.0(−12)	2.2(−6)	1.6(−12)	1.3(−13)	1.2(−13)
20	5.0(−10)	2.3(−14)	3.8(−7)	5.9(−15)	2.4(−18)	2.5(−18)
25	4.5(−11)	2.5(−19)	3.9(−8)	2.1(−16)	1.1(−22)	4.9(−23)
30	3.6(−12)	2.3(−21)	1.8(−8)	2.5(−17)	2.0(−24)	1.0(−27)
35	6.6(−14)	7.0(−26)	3.3(−9)	1.4(−18)	4.8(−26)	6.4(−31)
40	4.4(−14)		2.3(−9)	3.4(−20)	3.1(−27)	6.3(−33)

TABLE 5.3
Relative errors in Gaussian approximation of the sum $U_-(a,0)$ for $a = 4$

n	$GC(n)$	$GF(n)$	$S_{n,1}$	$S_{n,2}$	$S_{n,3}$	$S_{n,4}$
5	8.2(−10)	1.2	9.4(−4)	7.6(−4)	6.3(−4)	5.4(−4)
10	9.0(−14)	6.5(−2)	7.0(−8)	1.3(−8)	2.0(−8)	1.9(−8)
15	4.9(−17)	1.5(−1)	3.6(−8)	1.6(−10)	6.5(−13)	4.3(−13)
20	1.5(−19)	3.3(−3)	2.5(−9)	7.0(−12)	1.5(−15)	8.9(−18)
25	9.2(−22)	1.7(−2)	2.4(−8)	3.1(−13)	2.3(−17)	4.2(−21)
30	1.1(−24)	2.8(−4)	8.6(−9)	6.4(−14)	1.1(−18)	1.4(−24)
35	8.9(−26)	1.9(−3)	4.8(−9)	7.9(−15)	5.0(−20)	1.2(−25)
40	1.4(−27)	1.1(−4)	1.9(−9)	3.4(−17)	2.2(−21)	5.7(−27)

The corresponding relative errors for $a = 1$ and $a = 4$ are presented in Tables 5.2 and 5.3, respectively. Also, we mention here the exact sums $U_-(a,0)$ (determined as $S_{40,5}$ to 30 significant digits):

$$
\begin{aligned}
U_-(0.25,0) &= 0.666326189064665806052832629421,\\
U_-(0.50,0) &= 0.599262331208773941764013859373,\\
U_-(1.00,0) &= 0.440917473865185397183787033140,\\
U_-(2.00,0) &= 0.248166827542466167909044168541,\\
U_-(4.00,0) &= 0.124997557589011259481281086361,
\end{aligned}
$$

As we can see, the Laplace transform method ($GF(n)$) is very efficient for a small parameter a. However, when a increases, the integrand $J_0(at)$ becomes a highly oscillatory function and the convergence of the process slows down considerably. On the other hand, the convergence of Gauss-Chebyshev approximations $GC(n)$ is slightly faster if the parameter a is larger. Also, we can see a rapidly increasing of convergence of the summation process $S_{n,m}$ as m increases.

References

[1] W. Gautschi, *A class of slowly convergent series and their summation by Gaussian quadrature*, Math. Comp. **57** (1991), 309–324.

[2] W. Gautschi, *On certain slowly convergent series occurring in plate contact problems*, Math. Comp. **57** (1991), 325–338.

[3] W. Gautschi and G.V. Milovanović, *Gaussian quadrature involving Einstein and Fermi functions with an application to summation of series*, Math. Comp. **44** (1985), 177–190.

[4] J.F. Hart et al., *Computer Approximations*, Wiley, New York-London-Sydney, 1968.

[5] P. Henrici, *Applied and Computational Complex Analysis*, Vol. 1, Wiley, New York, 1984.

[6] E. Lindelöf, *Le calcul des résidus*, Gauthier–Villars, Paris, 1905.

[7] G.V. Milovanović, *Summation of series and Gaussian quadratures*, In: Approximation and Computation (R. V. M. Zahar, ed.), ISNM, Birkhäuser, Boston, 1994 (to appear).

[8] D.S. Mitrinović and J.D. Kečkić, *The Cauchy Method of Residues - Theory and Applications*, Reidel, Dordrecht, 1984.

[9] G.N. Watson, *Theory of Bessel Functions*, Cambridge, 1944.

Address of Author:
University of Niš, Faculty of Electronic Engineering
Department of Mathematics, P.O. Box 73
18000 Niš, Serbia
Yugoslavia

The Short Length AMLI Method - II. Computational and Communication complexity *

Maya Neytcheva

Abstract

The classical way of measuring the efficiency of a preconditioner implies estimations of the spectral equivalence relations with the original matrix and of the costs of an action if its inverse on a vector. Nowadays, the increasing use of parallel computers impose an extra efficiency criteria also of communication complexity and the ability to implement it on parallel computers. Many recent results indicate that there exists a conflict between the usage of easily parallelizable but slow converging methods, like the diagonally preconditioned conjugate gradient method on one hand, and the usage of fast converging (of optimal order) preconditioning methods, such as full multigrid or multilevel iteration methods, which are more difficult to parallelize on the other hand. It is already shown in the framework of hierarchical basis function methods that short-length multilevel methods parallelize well, preserve their optimality with respect to computational complexity and have an optimal order of parallel efficiency. This was done for block diagonally preconditioned matrices and for the full block factorized preconditioner. This presentation deals with the computational and communication complexity of the full block factorized preconditioning method.

Keywords: multilevel iteration, optimal order, computational complexity, communication complexity, hierarchical basis functions

1 Introduction

To solve any problem on a $n \times n$ (or $n \times n \times n$) mesh when the condition number is $O(n^2)$ or larger we need in general at least $O(n)$ iterations using an unpreconditioned (or diagonally preconditioned) conjugate gradient method. During $O(n)$ iterations

*The research of the author was supported by The Netherlands Organization for Scientific Research N.W.O. under grant no. 611-302-021.

data from any point in the mesh has had a chance to reach any other point, because the longest distance between any pair of points is at most $2n$ (or $3n$). Clearly, such an information exchange is required for any solution method. Hence, it seems that any method would require at least an $O(n)$ computer cycles irrespective of the number of processors, at least on a mesh array computer. Thus, there seems to be a parallelization conflict: either one must accept more computation (iterations) to get a high megaflop rate or one must accept much communications to get a fast rate of convergence. A way to overcome this is to let the global exchange of information occur on a coarse mesh and then use local operators to distribute the globally spread data from each coarse mesh point to its surrounding neighbors in the fine mesh. As such local operators, the operators arising in the hierarchical basis function method will be used here.

The full multilevel method uses of all levels from the finest to the very coarsest, consisting of a fixed small number of meshpoints. It suffers, therefore, from much recursion and communication overhead, together with idleness of processors while performing on the coarser levels. The above can be easily overcome if a short-length version of the method is used, where the coarsest level for the algorithm corresponds to a fairly fine discretization level. In [2], it has been shown that it can be done without loosing the optimality, neither of the computational complexity nor of the rate of convergence. The arithmetic complexity per iteration step may increase slightly but there is a gain in a reduction of the condition number of the preconditioned matrix. It turns out that this reduction is particularly significant for the V-cycle version of the method so that the resulting V-cycle variant is almost optimally conditioned also. Therefore, the presented method is an improvement of the full-length AMLI both for serial and parallel computers. Like the full-length version, the improved method can be treated completely in a finite element context or by purely algebraic means (see the references in [2], for example). In [2] and [3] a short-length multilevel version of the hierarchical basis finite element method was presented using a block diagonal preconditioner based on hierarchical basis functions and for the corresponding full block-matrix factorization method.

The remainder of the paper is as follows. Section 2 contains some basic results for the full block-matrix factorization for the multilevel hierarchical basis function (h.b.f.) method. The algebraic and communication complexity of the method is analyzed in Section 3. An upper bound of the level number for the coarsest mesh is also derived, for which there is no loss of optimal order computational complexity even if one uses a short-length variant of the AMLI method. Because of the ease of implementing the V-cycle version, it is analyzed in particular detail. Numerical results are presented in Section 4. Conclusions can be found in the final section.

2 Full block-matrix factorization preconditioning for the multilevel hierarchical basis function method

Let an elliptic and bounded bilinear form $a(\cdot,\cdot)$ be given, corresponding to the weak variational formulation of some model second order boundary value problem $\mathcal{L}u = f$ in $\Omega \subset \mathbf{R}^n$. Let $\mathcal{L}$ be elliptic and selfadjoint and Ω be bounded. Consider $\ell - k_0$ successive refinements of Ω, described by corresponding triangulations $\{\tau_s\}, s = k_0, \cdots, \ell$, where τ_ℓ is the finest and τ_{k_0} is the coarsest. Let $\{\mathcal{V}^{(s)}\}_{s=k_0}^{\ell}$ be a sequence of nested finite element spaces, associated with τ_s, $\{N_s\}$ be the corresponding sets of nodes in τ_s with degrees of freedom n_s, $\{\widehat{\phi}_i^{(s)}\}_{i=1}^{n_s}$ be the standard nodal basis functions in $\mathcal{V}^{(s)}$ and $\{\phi_i^{(s)}\}_{i=1}^{n_s}$ be the hierarchical basis functions.
To keep this presentation self-consistant we include some basic notations and results from [3]. As in [3], the h-version of the h.b.f. method is mainly considered.
The h.b. functions in the multilevel case are defined as follows: $\{\phi_i^{(k_0)}\} = \{\widehat{\phi}_i^{(k_0)}\}$, $\{\phi_i^{(s)}\} = \left\{ \{\widehat{\phi}_i^{(s)}\}, i \in N_s \backslash N_{s-1};\ \{\phi_i^{(s-1)}\}, i \in N_{s-1} \right\}, s = k_0+1, \cdots \ell$.
After a recursive two-by-two block partitioning of each two-level h.b.f. matrix $\widetilde{A}^{(s)} = \begin{bmatrix} A_{11}^{(s)} & A_{12}^{(s)} \\ A_{21}^{(s)} & \widetilde{A}^{(s-1)} \end{bmatrix} \begin{matrix} n_s - n_{s-1} \\ n_{s-1} \end{matrix}$, $s = \ell, \ell-1, \cdots, k_0+1$, the corresponding h.b.f. matrix takes the form:

$$A^{(\ell)} = \begin{bmatrix} A_{11}^{(\ell)} & & \widetilde{A}_{12}^{(\ell)} & & \\ & A_{11}^{(\ell-1)} & & \widetilde{A}_{12}^{(\ell-1)} & \\ \widetilde{A}_{21}^{(\ell)} & & \ddots & & \\ & \widetilde{A}_{21}^{(\ell-1)} & & \ddots & \\ & & & & \widetilde{A}^{(k_0)} \end{bmatrix} \begin{matrix} \} n_\ell - n_{\ell-1} \\ \} n_{\ell-1} - n_{\ell-2} \\ \\ \\ \} n_{k_0} \end{matrix} \tag{2.1}$$

where $\widetilde{A}^{(k_0)} = \widehat{A}^{(k_0)}$ is the standard nodal basis functions matrix and the row-, respectively, column-matrices $\widetilde{A}_{12}^{(s)}$ and $\widetilde{A}_{21}^{(s)}$ contain the corresponding couplings between the nodes at different levels.

We define recursively a preconditioner $C^{(\ell)}$ to $A^{(\ell)}$ as follows:

$$C^{(k_0)} = \widehat{A}^{(k_0)},\ C^{(s)} = \begin{bmatrix} I_1^{(s)} & 0 \\ \widetilde{A}_{21}^{(s)} B_{11}^{(s)^{-1}} & I_2^{(s)} \end{bmatrix} \begin{bmatrix} B_{11}^{(s)} & \widetilde{A}_{12}^{(s)} \\ 0 & \widetilde{C}^{(s-1)} \end{bmatrix}, s = k_0+1, \cdots, \ell, \tag{2.2}$$

$$\widetilde{C}^{(s-1)^{-1}} = \left[I - P_{\nu_s}\left(C^{(s-1)^{-1}} A^{(s-1)} \right) \right] A^{(s-1)^{-1}}. \tag{2.3}$$

Here it is assumed that $A_{11}^{(s)} \le B_{11}^{(s)} \le \alpha A_{11}^{(s)}$, $\widetilde{A}^{(s-1)} \le \widetilde{B}_{22} \le \beta \widetilde{A}^{(s-1)}$ hold for some $\alpha \ge 1$ and $\beta \ge 1$, where $\widetilde{B}_{22} = C^{(s-1)} + A_{21}^{(s)} B_{11}^{(s)^{-1}} A_{12}^{(s)}$ is a sparse approximation of $A^{(s-1)}$. $P_{\nu_s}(t)$ is a polynomial of degree ν_s, normalized so that $P_{\nu_s}(0) = 1$. The polynomials P_{ν_s} are chosen so that the value of P_{ν_s} is minimized in the interval, defined by the eigenvalues of the matrices $C^{(s-1)^{-1}}A^{(s-1)}$. In general,

$$P_{\nu_s}(t) = \left[T_{\nu_s}\left(\frac{\overline{\lambda}_{s-1}+\underline{\lambda}_{s-1}-2t}{\overline{\lambda}_{s-1}-\underline{\lambda}_{s-1}}\right)+1\right] \Big/ \left[T_{\nu_s}\left(\frac{\overline{\lambda}_{s-1}+\underline{\lambda}_{s-1}}{\overline{\lambda}_{s-1}-\underline{\lambda}_{s-1}}\right)+1\right], \tag{2.4}$$

where $T_{\nu_s}(t)$ are the Chebyshev polynomials of degree ν_s, $\overline{\lambda}_{s-1} = 1$, $\underline{\lambda}_{s-1}$ are the largest, respectively, the smallest eigenvalue of $C^{(s-1)^{-1}}A^{(s-1)}$ and $A^{(s-1)} = \begin{bmatrix} A_{s-1,s-1} & \tilde{A}_{s-1,*} \\ \tilde{A}_{*,s-1} & A^{(s-2)} \end{bmatrix}$. As shown in [3], there exists a relation between the value of the constant γ in the strengthened Cauchy-Bunyakowski-Schwarz (C.-B.-S.) and the polynomial degree ν_s, for which the condition number $\varkappa_s = \varkappa(C^{(s)^{-1}}A^{(s)})$ is bounded above independently of s.

Theorem 2.1 *Let $A^{(\ell)}$ be the hierarchical basis functions matrix, permuted and partitioned as in (2.1). Let $C^{(\ell)}$ be a preconditioner to $A^{(\ell)}$, defined by (2.2) - (2.4). Denote by $\varkappa_s$ the relative condition number of $C^{(s)^{-1}}A^{(s)}$ for $s = k_0, \cdots, \ell$. Then, if $B_{11}^{(s)} = A_{11}^{(s)}$ for $s = k_0, \cdots, \ell$, the following estimates for $\varkappa_s$ hold:*

(i) if $\nu_s = 1, s = k_0, \cdots, \ell$, then $\varkappa_\ell \leq (1-\gamma^2)^{-(\ell-k_0)}\varkappa_{k_0}$.

(ii) if $\nu_s > \sqrt{1-\gamma^2}^{-1}$, then $\varkappa_s \leq \sqrt{1-\gamma^2}\Psi(\hat{\varkappa}^{-1/2})$, where $\hat{\varkappa}$ is a fixed point of the recursion $t_s = \sqrt{1-\gamma^2}\Psi_{\nu_s}(t_{s-1})$, where $\Psi_{\nu_s}(t) = \left[\frac{(1+\sqrt{t})^{\nu_s}-(1-\sqrt{t})^{\nu_s}}{(1+\sqrt{t})^{\nu_s}+(1-\sqrt{t})^{\nu_s}}\right]$ and $t_{s-1} = \varkappa_{s-1}^{-1/2}$. If $\gamma^2 < 3/4$, then ν can be chosen equal to two and the upper bound for $\varkappa_s$ is found to be $\varkappa_s \leq \frac{1}{2\theta-1}$. It $\gamma^2 < 8/9$, then for $\nu = 3$, $\varkappa_s \leq \frac{3-\theta}{3\theta-1}$. Here $\theta = \sqrt{1-\gamma^2}$.

3 Computational complexity

Following [3], we analyze now the dependence of the computational complexity of the multilevel preconditioner ((2.2)–(2.4)) on the polynomial degree ν, for a fixed ν. The computational complexity of one action of the preconditioner is

$$w_\ell \leq C(n_\ell + \nu(n_{\ell-1} + \nu(n_{\ell-2} + \cdots + n_{k_0+1}))) + \nu^{\ell-k_0}C_{k_0}n_{k_0}, \tag{3.1}$$

where n_s is the degree of freedoms at level s and the constant C depends on the sparsity of the matrices involved in the preconditioner but is independent of ℓ and k_0. C_{k_0} depends on the solution method used when solving systems on the coarse mesh k_0.

Let for simplicity $n_s = 4^s = n_{s+1}/4$, which holds for a regular mesh. Relation (3.1) shows then that if $\nu < 4$, $w_\ell \leq n_\ell\left[C\left(1-\left(\frac{\nu}{4}\right)^{\ell-k_0}\right)\Big/\left(1-\left(\frac{\nu}{4}\right)+C_{k_0}\left(\frac{\nu}{4}\right)^{\ell-k_0}\right)\right]$, or, for the work per meshpoint,

$$w_\ell/n_\ell \leq 4C/(4-\nu) + (C_{k_0} - 4C/(4-\nu))\left(\nu/4\right)^{\ell-k_0}. \tag{3.2}$$

As is seen, the condition number of $C^{(\ell)^{-1}}A^{(\ell)}$ increases when k_0 decreases. Also, the communication overhead increases. The aim is, therefore, to choose k_0 as the largest number for which the cost for each action of the preconditioner is bounded, independent of ℓ. Since C_{k_0} depends on the method used for the coarse mesh system we consider two such methods.

Assume first the diagonally preconditioned conjugate gradient method is used to solve systems with $\widehat{A}_{k_0k_0}$. As the condition number of the system is $O(n_{k_0})$, the number of iterations required is $C_0 n_{k_0}^{\frac{1}{2}}$ for some constant C_0, which does not depend on k_0. Since $n_{k_0} = 4^{k_0}$, to balance the two terms in (3.2) best for any value of ℓ, we must then let $n_{k_0}^{\frac{1}{2}} = (4/\nu)^{\ell-k_0}$, that is $2^{k_0} = (4/\nu)^{\ell-k_0}$ or

$$k_0 = \widehat{k}_0 \equiv \lceil (2-\log_2\nu)/(3-\log_2\nu)\ell \rceil \,. \tag{3.3}$$

From (3.3) we find then the values of $\widehat{k}_0$ as a function of ν, given in Table 1.

ν	3	2	1
$\widehat{k}_0$	$\lceil\frac{2}{7}\ell\rceil$	$\lceil\frac{1}{2}\ell\rceil$	$\lceil\frac{2}{3}\ell\rceil$

Table 1: Values of the coarse level number for a diagonally preconditioned method

ν	3	2	1
$\widehat{k}_0$	$\lceil\frac{4}{9}\ell\rceil$	$\lceil\frac{2}{3}\ell\rceil$	$\lceil\frac{4}{5}\ell\rceil$

Table 2: Values of the coarse level number for the modified incomplete factorization method

(The exact value of $\widehat{k}_0$ for $\nu = 3$ is $\approx 0.2933\ell$.) The work per meshpoint is then bounded by $w_\ell/n_\ell \le 4C/(4-\nu) + C_0$.

Consider now the use of a faster convergent method for the coarse mesh, namely, the conjugate gradient method, preconditioned with a modified incomplete factorization or a block diagonal domain decomposition method. For these methods the condition number of the preconditioned system grows as $C_0\, 2^{k_0}, k_0 \to \infty$ for some constant C_0. (For a recent parallelizable incomplete factorization method, see [6].) As for (3.3) we find now that the optimal value to minimize the work per meshpoint of one action of the preconditioner satisfies $2^{\frac{k_0}{2}} = (4/\nu)^{\ell-k_0}$ or $k_0 = 2(\ell-k_0)(2-\log_2\nu)$, i.e., $\widehat{k}_0 = \left\lceil \frac{4-2\log_2\nu}{5-2\log_2\nu}\,\ell \right\rceil$, which leads to the values for $\widehat{k}_0$ as a function of ν, given in Table 2.

Theorem 3.1 *If $\gamma^2 < \frac{8}{9}$ then for $\nu = 3$, or if $\gamma^2 < \frac{3}{4}$ then for $\nu = 2$ (or 3), the preconditioned system $C^{(\ell)^{-1}}A^{(\ell)}$ has an optimal order of condition number, bounded above by a fixed number on all levels [i.e., $C^{(\ell)}$ is spectrally equivalent to $A^{(\ell)}$]. In addition, if (a) the diagonally preconditioned conjugate gradient method is used on the coarsest mesh level k_0, where $k_0 \le \frac{2-\log_2\nu}{3-\log_2\nu}\ell$, or (b) the modified incomplete factorization preconditioned conjugate gradient method is used on the coarsest mesh level k_0, where $k_0 \le \frac{4-\log_2\nu}{5-\log_2\nu}\ell$, then the arithmetic cost per iteration step is $O(n_\ell)$. Hence the preconditioner has an optimal order of computational complexity, i.e., the cost to solve systems with $A^{(\ell)}$ is $O(n_\ell)\cdot O(\log\frac{1}{\varepsilon})$ where ε is the ratio of the final and initial residuals.*

Consider the parallel efficiency of the method. Denote by $E = T_1^*/pT_p$ the parallel efficiency of the short length AMLI method, where T_1^* is the computing time using the best algorithm on a single processor and T_p is the computing time for the short-length AMLI method with an optimal choice of k_0 on a p-processor machine. Assume that $p \le n_{k_0}$, i.e., the nodepoints on the coarsest mesh are mapped with at least one meshpoint per processor. Then we find

$$E = \frac{O(n_\ell)}{p\left[\frac{n_{k_0}^{3/2}}{p} + \frac{n_\ell}{n_{k_0}}\right]s} \ge \frac{O(n_\ell)}{n_{k_0}\left(\frac{n_\ell}{n_{k_0}}\right)s} \ge C > 0, \ell \to \infty. \tag{3.4}$$

Here, n_ℓ/n_{k_0} is the size of the subgrid on each processor and the amount of work during one iteration on each processor is proportional to this plus the work $n_{k_0}^{3/2}/p$ to solve the coarse mesh problem. Since $p \le n_{k_0}$, there will occur no (or minor amount of) idleness of processors and all communications required will be between nearest neighbor processors, assuming that a natural mapping of subgrids to processors is used. In (2.4) s is the number of iterations, which is bounded, independent of ℓ, if $\nu \ge 2$ and γ^2 is sufficiently small, as is shown in Theorem 2.1. Hence, under these assumptions, the parallel efficiency is bounded below by a positive number C, independent of ℓ. Because even on a serial computer there occurs much recursion overhead in the full-length AMLI algorithm, the optimal algorithm on an one-processor machine will also be a short length one, although perhaps of a somewhat longer length than on a p-processor machine. Note that the cost per iteration step, associated with the coarse mesh solver can be made arbitrarily small by decreasing the level number of the coarsest grid. Hence, assuming still $p \le n_{k_0}$, we find that the total amount of all arithmetic operations on the parallel computer is about equal to that on the uniprocessor. Since all operations are fully parallelizable, the parallel efficiency can, therefore, be made very close to its optimal value 1, if $p \le n_{k_0}$ and k_0 is somewhat smaller than the values given in Tables 3 and 4. It is readily seen that the parallel efficiency of the diagonally preconditioned conjugate gradient method is bounded above by $O(n_\ell^{-1})$ fo a $n_\ell \times n_\ell$ mesh. The parallel efficiency of the full multigrid or multilevel iteration method is bounded above by $O(\ell^{-2})$, $\ell \to \infty$ due to increasing distances over which data must be communicated when the meshes get coarser. Hence, in both cases the efficiency approaches zero as ℓ increases. The above results for k_0 as a function of ℓ can be further improved using polynomials of cyclically varying degrees as in [7] and [1].

Remark 3.1 A similar computational complexity analysis can be done for three space dimensional problems. One finds that $\frac{w_\ell}{n_\ell} \le const$ if $k_0 = \left\lceil \frac{3-\log_2 \nu}{4-\log_2 \nu}\ell \right\rceil$ and, for $\nu = 1$, the condition number grows as $\varkappa_{\ell,k_0} \le const\,(1/(1-\gamma^2))^{\frac{\ell}{4}}$. For $\gamma^2 = 5/7$ we find $\varkappa_{\ell,k_0} \le (1.35)^\ell$, which increases considerably slower than 2^ℓ (for the classical hierarchical basis function matrix the condition number is $O(2^\ell)$).

Grid	Diag. Precond.			full AMLI			short AMLI		
	Iter.	Time (sec)	Mflops	Iter.	Time (sec)	Mflops	Iter.	Time (sec)	Mflops
256^2	1027	14.05	115	31	16.53	7.25	17	2.62	16.64
512^2	2090	81.15	162	40	85.10	7.45	17	9.20	20.87

Table 3: Coefficient jump 1/0.001. Performance of the PCG on CM-200

4 Numerical experiments

The following test problem was implemented on CM-200 computer with 256 processing elements and clock frequency 10 MHz: $-\nabla(a\nabla u) = f$, in $\Omega \equiv (0,1)^2$ with homogeneous Dirichlet boundary conditions, where $a = \varepsilon \ll 1$ in $\tilde{\Omega} = [\frac{1}{3}, \frac{2}{3}]^2 \subset \Omega$ and $a = 1$ elsewhere. The problem was discretized using regular isosceles triangles and piece-wise linear basis functions. The meshsize parameter h was taken to be equal to $1/2^n$ for some integer n and has been chosen always such that there will be meshlines along the edges of $\tilde{\Omega}$. The stopping criterion used for the preconditioned conjugate gradient method was made dependent on the coefficient jumps, namely: $\frac{g_i^T M^{-1} g_i}{g_1^T M^{-1} g_1} < 2\pi^2 h^2 \min(a_{|\tilde{\Omega}}, a_{|\Omega/\tilde{\Omega}})\, 10^{-12}$. The AMLI preconditioner was constructed as described in [1]. A diagonally preconditioned conjugate gradient method was used as a coarsest level solver. The results of solving the discretized problem by different iterative methods and the corresponding Mflops are given in Table 3. Note that an additional advantage of using a few level AMLI method is that the number of (outer) iterations decreases due to a smaller condition number of ${C^{(s)}}^{-1} A^{(s)}$. In the short AMLI implementation here $\nu = 1$, but the distance $\ell - k_0$ was kept constant (=3). For more details see [5].

5 Conclusions

The estimates in Section 3 of the work per meshpoint show that there is no need to continue multilevel or multigrid methods to a very coarse level to achieve an optimal order of computational complexity. It suffices and is even recommendable to use only few levels because the recursion and communication overhead is then also small. In addition, the corresponding preconditioning has a smaller condition number, which is most significant for the V-cycle case ($\nu = 1$). This condition number is more robust also with respect to other problem parameters than the meshsize (see [3]).

Acknowledgements

The author hereby gratefully acknowledges the help she was given by Prof. O. Axelsson while deriving the these results and writing the text of the paper.

References

[1] O. Axelsson and M. Neytcheva, *Algebraic multilevel iteration method for Stieltjes matrices*, Num. Lin. Alg. Appl., to appear.

[2] O. Axelsson and M. Neytcheva, *The Short Length AMLI Method. I*, submitted.

[3] O. Axelsson and M. Neytcheva, *The Short Length AMLI Method - II. Rate of convergence*, Faculty of Mathematics and Informatics, University of Nujmegen, Toernooiveld, 6525 ED Nijmegen, The Netherlands. Internal report in progress.

[4] J.F. Maitre and F. Musy, *The contraction number of a class of two-level Methods; An Exact Evaluation for some Finite Element Subspaces and Model Problems*, Multigrid Methods (W. Hackbusch and U. Trottenberg, eds.), Proceedings, Köln-Porz 1981, Lecture Notes in Math., vol 960, Springer, 1982, pp. 535-544.

[5] M. Neytcheva, *Implementing the AMLI algorithm on the Connection Machine*, Report UNIC-93-15.

[6] Yvan Notay, *On Parallelization of Approximate Factorization Preconditioning*, Université Libre de Bruselles, Report IT/IF/14-12, December 1993.

[7] P.S. Vassilevski, *Hybrid V-cycle algebraic multilevel preconditioners*, Math. Comp., 58 (1992), pp. 489–512.

Faculty of Mathematics and Informatics
University of Nijmegen
Toernooiveld, 6525 ED Nijmegen
The Netherlands

Arnoldi's procedure and angles between Krylov subspaces

B. Philippe and M. Sadkane
IRISA-INRIA
Campus Universitaire de Beaulieu
35042 Rennes Cedex, France.

Abstract

When a Krylov subspace method is applied to a non-Hermitian matrix $A \in \mathbb{C}^{n\times n}$ it produces, at step m, a matrix V_m, whose columns are orthonormal and candidate for spanning an invariant subspace for A. The purpose of this note is to propose a way to evaluate the evolution of the canonical angles between $span\{V_m\}$ and $span\{AV_m\}$ and between $span\{V_m\}$ and $span\{A^m v_1\}$ where $V_m = [v_1, v_2, \ldots, v_m]$ is generated by a block-Arnoldi method.

1 Outline of the block-Arnoldi method

We begin this note with a brief review of the block-Arnoldi method for computing estimates of a few outermost eigenpairs of a non-Hermitian matrix $A \in \mathbb{C}^{n\times n}$. For more details on block Krylov subspace methods see for example [1, 2, 4]. Throughout this note, $\| x \|$ denotes the Euclidean norm of the vector x. $\| A \|$ and A^H denote the spectral norm and the conjugate transpose of the matrix A respectively. The non-zero singular values: $\sigma_1(B) \geq \sigma_2(B) \geq \ldots$, of a matrix B are always labelled in decreasing order.

Let $v_1 \in \mathbb{C}^{n\times l}$ be a rectangular matrix with l orthonormal columns. An orthonormal basis for the block Krylov subspace $K_m(A, v_1) = span\{v_1, Av_1, \ldots, A^{m-1}v_1\}$ may be obtained via the following Modified Gram-Schmidt process where $m > l$ is the size of the considered basis ($ml \leq n$)

Algorithm 1

for $j = 1, \ldots, m$ **do**
  $w_j = Av_j$;
  for $i = 1, \ldots, j$ **do**
    $h_{i,j} = v_i^H w_j$;
    $w_j = w_j - v_i h_{i,j}$;

end for
$w_j = v_{j+1}h_{j+1,j}$; *QR factorization*;
end for

When, in the last step of Algorithm 1, the QR factorization of w_j reveals a rank deficient matrix, the new v_{j+1} is limited to a basis of $span\{w_j\}$. We assume throughout this note, that w_j is of full rank and therefore $h_{j+1,j}$ is regular.

With $V_m = [v_1, v_2, \ldots, v_m] \in \mathbb{C}^{n \times ml}$ and $H_{m,m} = (h_{i,j})_{1 \le i,j \le m} \in \mathbb{C}^{ml \times ml}$, we have from the above algorithm

$$AV_m = V_m H_{m,m} + v_{m+1} h_{m+1,m} E_m^H. \quad (1)$$

The columns of V_m are orthonormal, the matrix $H_{m,m}$ is a block upper Hessenberg matrix and $E_m \in \mathbb{C}^{ml \times l}$ stands for the matrix $\begin{pmatrix} 0 \\ I_l \end{pmatrix}$ where I_l is the identity matrix of order l. The relation (1) can be written as

$$AV_m = V_{m+1} H_{m+1,m} \quad (2)$$

with

$$H_{m+1,m} = \begin{pmatrix} H_{m,m} \\ h_{m+1,m} E_m^H \end{pmatrix}. \quad (3)$$

Now, by successive applications of (2), we obtain

$$A^m v_1 = A^{m-1}(Av_1) = (A^{m-1}V_2)H_{2,1} = \ldots = V_{m+1} H_{m+1,m} H_{m,m-1} \cdots H_{3,2} H_{2,1}. \quad (4)$$

Noticing that (3) holds for each $k = 1, \ldots, m$, we see that (4) can be written in the form

$$A^m v_1 = V_m H_{m,m} g_m + v_{m+1} g_{m,m} = V_{m+1} g_{m+1} \quad (5)$$

with $g_m = H_{m,m-1} \cdots H_{3,2} H_{2,1}$ and $g_{m,m} = h_{m+1,m} h_{m,m-1} \cdots h_{3,2} h_{2,1}$. From (2) and (4) we see in particular that $span\{A^m v_1\} \subset span\{AV_m\} \subset span\{V_{m+1}\}$.

2 Computation of canonical angles

We would like to characterize the evolution of $span\{V_m\}$ during the iterations. Note that if the columns of V_m span an approximate invariant subspace of A, then the matrix $v_{m+1} h_{m+1,m} E_m^H$ from (1) will have a small norm. However $span\{V_m\}$ may contain an invariant subspace even if the norm of $v_{m+1} h_{m+1,m} E_m^H$ is not small.

One way to monitor the degree of invariance of $span\{V_m\}$ is to estimate the canonical angles between $span\{V_m\}$ and $span\{AV_m\}$. We see from (2), that the intersection between $span\{V_m\}$ and $span\{AV_m\}$ is a subspace of dimension $(m-1)l$, and therefore the $(m-1)l$ smallest canonical angles between $span\{V_m\}$ and $span\{AV_m\}$ are equal to zero. The l remaining canonical angles give some ideas about the invariance. In other words, the smallness of the largest canonical angle signals that a subspace of $span\{V_m\}$ is about to become invariant. The sine of the largest of these canonical angles is the distance between $span\{V_m\}$ and $span\{AV_m\}$. The smallest of them measures the distance to singularity of $h_{m+1,m}$. We see from (4) that the intersection between $span\{A^m v_1\}$ and $span\{V_m\}$ is the null subspace as long as our assumption on the regularity of the matrices $h_{j+1,j}$ $j = 1, \ldots, m$ holds.

The purpose of this section is to estimate the non-zero canonical angles $span\{V_m\}$ and $span\{AV_m\}$. between $span\{A^m v_1\}$ and $span\{V_m\}$. Let

$$H_{m+1,m} = U \begin{pmatrix} \Sigma \\ 0 \end{pmatrix} W^H \tag{6}$$

be the singular value decomposition of $H_{m+1,m}$ where $U \in \mathbf{C}^{(m+1)l \times (m+1)l}$ and $W \in \mathbf{C}^{ml \times ml}$ are unitary matrices and $\Sigma = diag(\sigma_1, \ldots, \sigma_{ml})$ with $\sigma_1 \geq \ldots \geq \sigma_{ml} > 0$ the singular values of $H_{m+1,m}$. To see that $\sigma_{ml} > 0$, assume that there exists a vector $t \in \mathbf{C}^{ml}$ such that

$$H_{m+1,m} t = 0. \tag{7}$$

If we write $t = \begin{pmatrix} t_1 \\ t_2 \end{pmatrix}$, where $t_1 \in \mathbf{C}^{(m-1)l}$ and $t_2 \in \mathbf{C}^l$, we see that (3) implies that

$$h_{m+1,m} t_2 = 0 \tag{8}$$

and since $h_{m+1,m}$ is regular, we conclude that $t_2 = 0$ and that (7) reduces to

$$H_{m,m-1} t_1 = 0. \tag{9}$$

We can repeat this argument and prove that all components of t vanish.

Now if we partition U as follows

$$U = \begin{pmatrix} U_{11} & U_{12} \\ U_{21} & U_{22} \end{pmatrix} \tag{10}$$

with $U_{11} \in \mathbf{C}^{ml \times ml}$, $U_{12} \in \mathbf{C}^{ml \times l}$, $U_{21} \in \mathbf{C}^{l \times ml}$, and $U_{22} \in \mathbf{C}^{l \times l}$, then we have the following theorem

Theorem 2.1 *Let $0 < \theta_1 \leq \theta_2 \leq \ldots \leq \theta_l \leq \frac{\pi}{2}$ be the non-zero canonical angles between $span\{V_m\}$ and $span\{AV_m\}$. Then*

$$\sin \theta_k = \sigma_{l-k+1}(U_{21}) = \sqrt{1 - \sigma_k^2(U_{22})}, \quad k = 1, 2, \ldots, l \tag{11}$$

where $\sigma_{l-k+1}(U_{21})$ and $\sigma_k(U_{22})$ are the $(l - k + 1) - th$ and the $k - th$ largest singular values of U_{21} and U_{22} respectively.

Proof.

From (2), (3) and (4) we have

$$\begin{aligned} AV_m &= V_{m+1} U \begin{pmatrix} \Sigma W^H \\ 0 \end{pmatrix} \\ &= V_{m+1} \begin{pmatrix} U_{11} \\ U_{21} \end{pmatrix} \Sigma W^H. \end{aligned}$$

So $span\{AV_m\} \equiv span\{V_{m+1} \begin{pmatrix} U_{11} \\ U_{21} \end{pmatrix}\}$, and since the ml columns of $V_{m+1} \begin{pmatrix} U_{11} \\ U_{21} \end{pmatrix}$ are orthonormal, they form an orthonormal basis of the subspace $span\{AV_m\}$. The non-zero canonical angles θ_k, $k = 1, 2, \ldots, l$ are then defined by:

$$\begin{aligned} \cos\theta_k &= \sigma_k \left(V_m^H V_{m+1} \begin{pmatrix} U_{11} \\ U_{21} \end{pmatrix} \right) \\ &= \sigma_k(U_{11}) \\ &= \sqrt{1 - \sigma_{l-k+1}^2(U_{21})} \quad \text{since } U_{11}^H U_{11} + U_{21}^H U_{21} = I_{ml} \\ &= \sigma_k(U_{22}). \end{aligned}$$

□

The above expression (11) of $\sin\theta_k$ may be computed iteratively during the block-Arnoldi process. Indeed, let us denote by $U^{(m)} \in \mathbb{C}^{(m+1)l \times l}$ the matrix whose columns are the singular vectors corresponding to the null singular values of $H_{m+1,m}$, using the notation of (10),

$$U^{(m)} = \begin{pmatrix} U_{12} \\ U_{22} \end{pmatrix}. \tag{12}$$

Note that

$$H_{m+2,m+1} = \begin{pmatrix} H_{m+1,m} & h_{m+1} \\ 0 & h_{m+2,m+1} \end{pmatrix} \tag{13}$$

where

$$h_{m+1} = \begin{pmatrix} h_{1,m+1} \\ h_{2,m+1} \\ \vdots \\ h_{m+1,m+1} \end{pmatrix} \tag{14}$$

is the last block column of the matrix $H_{m+1,m+1}$. If we denote by $S^{(m)}$ the matrix $S^{(m)} = -h_{m+2,m+1}^{-H} h_{m+1}^H U^{(m)}$, then

$$H_{m+2,m+1}^H \begin{pmatrix} U^{(m)} \\ S^{(m)} \end{pmatrix} = 0. \tag{15}$$

It suffices then to orthonormalize the columns of $\begin{pmatrix} U^{(m)} \\ S^{(m)} \end{pmatrix}$ to obtain $U^{(m+1)}$, and the canonical angles between $span\{AV_{m+1}\}$ and $span\{V_{m+1}\}$ are given by the singular values of the last l–block of $U^{(m+1)}$.

3 A convergence criterion

The computation of canonical angles we discussed in the previous section can be useful in the case where one is interested in deciding whether the whole subspace $K_m(A, v_1)$ is invariant by A. This is important in some applications such as controllability. If however one is interested in computing an invariant subspace of $K_m(A, v_1)$, then a criterion based only on the computation of these canonical angles is too severe to be recommended. In this section we discuss a more flexible convergence criterion of Algorithm 1. *We restrict the study to the case of the standard version of Arnoldi's method* ($l = 1$). The reason for this restriction is due to the fact that we want to use some optimality property that are known only in the unblocked version of Algorithm 1.

In order to express the relation between the two basis V_m and $K_m = [v_1 \, ; \, Av_1 \, ; \, \cdots \, ; \, A^{m-1}v_m]$ of the the Krylov subspace $K_m(A, v_1)$, we consider the non singular square upper-triangular matrix $G_m = [e_1 \, ; \, g_2 \, ; \, \cdots \, ; \, g_m]$ where g_i ($i = 1, \cdots, m$) are the vectors previously introduced but eventually filled with zero entries in their last positions to obtain the correct length. From (5) we obtain

$$K_m = V_m G_m. \tag{16}$$

Theorem 3.1 *When expressed in the basis K_m , the Arnoldi relation (5) becomes*

$$AK_m = K_m C_m + v_{m+1} g_{m+1,m+1} e_m^H \tag{17}$$

where $C_m = \begin{bmatrix} 0 & . & . & 0 & \beta_0 \\ 1 & . & . & . & \beta_1 \\ . & 1 & . & . & . \\ . & . & . & . & . \\ 0 & . & . & 1 & \beta_{m-1} \end{bmatrix}$ *is the companion matrix of $H_{m,m}$ and satisfies* $C_m = G_m{}^{-1} H_{m,m} G_m$.

Proof. For $i = 1, \cdots, m$, we have the relation $g_i = H_{m,m}{}^{i-1} e_1$. Let $p(z) = z^m - \sum_{i=0}^{m-1} \beta_i z^i$ the characteristic polynomial of $H_{m,m}$. From the Cayley-Hamilton theorem, the following expression holds:

$$H_{m,m} g_m = (H_{m,m})^m e_1 = \sum_{i=0}^{m-1} \beta_i (H_{m,m})^i e_1 = \sum_{i=0}^{m-1} \beta_i g_{i+1}.$$

Since for $i = 1, \cdots, m-1$, we obviously have the relation $H_{m,m} g_i = g_{i+1}$, we obtain that $V_m H_{m,m} G_m = K_m C_m$ which proves that $G_m{}^{-1} H_{m,m} G_m = C_m$.

The conclusion easily follows by multiplying the Arnoldi relation (1) by the matrix G_m on the right side and from the relation $g_{m+1,m+1} = h_{m+1,m} g_{m,m}$. □

Let us consider now a Ritz pair (λ, x) obtained at the step m and its residual $r = Ax - \lambda x$, i.e. $x = V_m y$ where $H_{m,m} y = \lambda y$. In the basis K_m , the vector x is expressed by $x = K_m u$ with $u = G_m{}^{-1} y$. From the previous theorem u is an eigenvector of C_m

corresponding to λ . By denoting ν_m its last component, it can be expressed by $u = \nu_m t$ where

$$t = \begin{pmatrix} \beta_0/\lambda \\ \vdots \\ p_{i-1}(\lambda)/\lambda^i \\ \vdots \\ p_{m-2}(\lambda)/\lambda^{m-1} \\ 1 \end{pmatrix} \quad \text{where } p_i(\lambda) = \sum_{j=0}^{i} \beta_j \lambda^j. \tag{18}$$

We obtain therefore the following estimations.

Proposition 3.1 *The residual* $r = Ax - \lambda x$ *can be expressed by*

$$r = v_{m+1} h_{m+1,m} (e_m{}^H y), \quad \|r\| = h_{m+1,m} |e_m{}^H y| \tag{19}$$

$$r = v_{m+1} g_{m+1,m+1} (e_m{}^H u), \quad \|r\| = g_{m+1,m+1} |\nu_m| \tag{20}$$

$$\frac{\|r\|}{\|x\|} = \frac{g_{m+1,m+1}}{\|G_m t\|} = \tan\phi_m \frac{\|H_{m,m} g_m\|}{\|G_m t\|}, \tag{21}$$

where ϕ_m *is the angle* $\angle(A^m v_1, K_m(A, v_1))$.

Remark 3.1 *The angle* ϕ_m *depends on* $g_{m+1,m+1}$ *which contains information from the beginning of the iteration whereas* $h_{m+1,m}$ *is related to the invariance of* $K_m(A, v_1)$ *by* A *. We already mentioned that the subspace* $K_m(A, v_1)$ *may be far from invariant even when it contains an almost invariant subspace of smaller dimension.*

The quantity $\frac{\|H_{m,m} g_m\|}{\|G_m t\|}$ *, which is smaller than the very crude bound* $\|A\|\ \chi_2(G_m)$ *, (*χ_2 *stands for the spectral condition number) is expected to be often of moderate size.*

The quantity $g_{m+1,m+1}$ can also be expressed from the characteristic polynomial of $H_{m,m}$. More precisely, let $\mathcal{P}_m$ be the set of all polynomials of degree m, and let $\hat{p}_m$ be the characteristic polynomial of the Hessenberg matrix $H_{m,m}$ of order m, then

Theorem 3.2 *We have*

$$h_{m+1,m} v_{m+1} = \frac{\hat{p}_m(A) v_1}{\|\hat{p}_{m-1}(A) v_1\|}. \tag{22}$$

$$\|\hat{p}_m(A) v_1\| = \min_{p_m \in \mathcal{P}_m} \|\hat{p}_m(A) v_1\|. \tag{23}$$

Let $\theta_1, \theta_2, \ldots, \theta_m$ *be the eigenvalues of* $H_{m,m}$*, then*

$$\begin{aligned} (I - V_m V_m^H) A^m v_1 &= g_{m+1,m+1} v_{m+1} && (24) \\ &= \hat{p}_m(A) v_1 && (25) \\ &\equiv (A - \theta_1 I)(A - \theta_2 I) \ldots (A - \theta_m I)\ v_1 && (26) \end{aligned}$$

Proof. The proof of the two first statements can be found in [3, 5]. The third statement comes from (5) and the first statement. □

Let us consider now examples which illustrate the situations. The first matrix is a random matrix of order 50, with one real eigenvalue at about 25 and the others in the rectangle $[-2, 2.5] + i * [-2, +2]$. The initial vector is randomly chosen. The residual which are displayed in Table 1 are the absolute residuals which correspond to the largest eigenvalue ($\lambda_1 = 25.238$).

iteration	$\tan \phi_m$	residual norm
2	9.6388e-01	5.0147e+00
3	2.1908e-01	1.6500e+00
4	1.8736e-02	4.6767e-01
5	1.7013e-03	4.2821e-02
6	1.1654e-04	3.0407e-03
7	7.0238e-06	1.8365e-04
8	5.3667e-07	1.4247e-05
9	4.0028e-08	1.0255e-06
10	2.9416e-09	7.5943e-08

Table 1: Behaviour of $\tan \phi_k$ with matrix A and random initial vector

In the second experiment we consider the matrix B which is matrix $B = A - \tilde{\lambda_1} I$ where $|\lambda_1 - \tilde{\lambda_1}| < 10^{-5}$. The largest eigenvalue then becomes the smallest eigenvalue in modulus. The initial vector is not changed. The residuals displayed in Table 2 are the absolute residuals corresponding to the smallest eigenvalue ($\lambda = 7.0519 \times 10^{-5}$).

iteration	$\tan \phi_m$	residual norm
2	2.8335e-02	5.0147e+00
3	6.1077e-03	1.6500e+00
4	5.0267e-04	4.6767e-01
5	4.3672e-05	4.2821e-02
6	2.8475e-06	3.0407e-03
7	1.6268e-07	1.8365e-04
8	1.1745e-08	1.4247e-05
9	8.2586e-10	1.0255e-06
10	5.7115e-11	7.5943e-08

Table 2: Behaviour of $\tan \phi_k$ with matrix B and random initial vector

In the third experiment, we turn back to matrix A but with a new initial vector which is obtained by perturbing an eigenvector of A corresponding to an internal real eigenvalue ($\lambda_2 = -7.4860 \times 10^{-1}$). In Table 3 are reported the absolute residuals corresponding to the approximations of λ_1 and λ_2 which are the smallest residuals of the Ritz pairs.

iteration	$\tan\phi_m$	λ_2-residual	λ_1-residual
2	3.1815e-06	8.8122e-08	1.0409e+01
3	1.0349e-05	1.0451e-07	1.1878e+00
4	2.7298e-05	2.7294e-07	9.4513e-02
5	6.2317e-05	2.1904e-06	6.6599e-03
6	1.6243e-05	2.6726e-07	4.2136e-04
7	1.2391e-06	1.4301e-07	3.2334e-05
8	8.6570e-08	2.5911e-07	2.2636e-06
9	7.0698e-09	3.0883e-07	1.8714e-07
10	4.5753e-10	2.7545e-07	1.2139e-08

Table 3: Behaviour of $\tan\phi_k$ with matrix A and an approximate eigenvector as initial vector

From the three experiments, the $\tan\phi_k$ appears to give a correct evaluation of the residuals evolution for the best Ritz-pair. We consider that, when building the Arnoldi matrix, it is useful to check $\tan\phi_k$ and compute the eigendecomposition of $H_{k,k}$ as soon as $\tan\phi_k$ becomes smaller than the tolerance threshold. This often saves computation since it can stop the process before reaching the maximum size m.

All the study of this part has been done under the assumption $l = 1$, however we noticed that the tests we have run with other sizes of blocks had a similar behaviour.

References

[1] G. H. Golub and C. F. Van Loan. Matrix Computations. The Johns Hopkins University Press, 1989.

[2] A. Ruhe, Implementation aspects of band Lanczos algorithms for computation of eigenvalues of large sparse symmetric matrices. Math. Comp. 33:680-687(1979).

[3] Y. Saad, Projection method for solving large sparse eigenvalue problems. Lect. Notes Math, Vol. 973, p. 121-144, 1982.

[4] M. Sadkane. A block Arnoldi-Chebyshev method for computing the leading eigenpairs of large sparse unsymmetric matrices. Numer. Math. 64:181-193(1993).

[5] D. C. Sorensen, Implicit application of polynomial filters in a k-step Arnoldi method. SIAM J. Matrix Anal. Appl. Vol. 13, No. 1, pp.357-385 (1992)

ITERATIVE APPLICATION OF FFT METHODS FOR THE SOLUTION OF LINEAR ELLIPTIC PARTIAL DIFFERENTIAL EQUATIONS

W. M. Pickering & P. J. Harley

Abstract

A method for the numerical solution of linear elliptic partial differential equations is described which involves the iterative application of FFT techniques. The general approach is to choose an equation 'near' (in some sense) to the equation to be solved and perform an iterative refinement on the solution of this nearby equation. This equation is appropriately chosen from the general class of equations soluble directly by FFT methods and the iterative technique employed is of D'Yakonov-Gunn form which is equivalent to a block-Jacobi iteration for the relevant Fourier harmonic amplitudes. Shifting and scaling are employed to improve convergence rates and the method allows the use of a 'variable' shift which can significantly improve the rate of convergence. The overall operations count is proportional to $\mu N^2 \log_2 N + \nu N^2$, where μ and ν are constants, for an $N \times N$ grid.

Numerical results are presented for illustrative examples in the unit square with Dirichlet boundary conditions and it is demonstrated that, with suitable shifting and scaling, convergence rates can be achieved which vary little with grid-size.

1 INTRODUCTION

We consider solving a linear elliptic partial differential equation of the form

$$\mathcal{L}\phi(x,y) = q(x,y), \tag{1}$$

where

$$\mathcal{L}\phi \equiv a(x,y)\phi_{xx} + b(x,y)\phi_x + c(x,y)\phi_{yy} + d(x,y)\phi_y - f(x,y)\phi \tag{2}$$

and $a, c > 0, f \geq 0$, over a rectangular region with (for simplicity) Dirichlet boundary conditions. It is assumed that the coefficient functions a, b, c, d and f are well-behaved throughout the solution domain. The basis of the method is to solve (1) by repeatedly solving an equation which is in some sense 'near' to (1) and which can be solved directly by the fast Fourier transform method. Similar ideas have been proposed, for example, by Concus and Golub [4] for the Helmholtz equation using Bunemann's cyclic reduction Poisson solver [3] and by Bank [2] for certain non-separable equations of self-adjoint form using marching methods. An advantage of the current formulation compared with other similar techniques is that it permits the solution of any equation of the general (elliptic) form (1) to be obtained by the same method.

For our purposes the most suitable 'nearby' equation which may be solved directly by FFT (with, for illustrative reasons, the Fourier transform in the x-direction) is of the form

$$\overline{\mathcal{L}}\phi \equiv \overline{a}\phi_{xx} + \overline{b}\phi_x + \overline{c}(y)\phi_{yy} + \overline{d}(y)\phi_y - \overline{f}(y)\phi = q(x,y), \tag{3}$$

where $\overline{a}$ and $\overline{b}$ are constants and $\overline{c}, \overline{d}$ and $\overline{f}$ may be functions of y. The general form of iteration is defined as

$$\overline{\mathcal{L}}_h\phi^{(m+1)} = q - \left(\mathcal{L}_h - \overline{\mathcal{L}}_h\right)\phi^{(m)},\ (m = 0, 1, 2, ...) \tag{4}$$

where $\mathcal{L}_h$ and $\overline{\mathcal{L}}_h$ respectively denote suitably discretized versions of $\mathcal{L}$ and $\overline{\mathcal{L}}$ and m denotes iteration count. Iteration (4) is a D'Yakonov-Gunn [5],[6] iteration of the form

$$\overline{\mathcal{L}}_h\phi^{(m+1)} = \overline{\mathcal{L}}_h\phi^{(m)} - \tau\left(\mathcal{L}_h\phi^{(m)} - q\right), \tag{5}$$

with iteration parameter $\tau = 1$ and it is well known that, for self-adjoint problems with smooth coefficients, this form of iteration has a convergence rate which is essentially independent of grid-size. Both Bank's [2] iteration and the Concus & Golub [4] iteration are of this type and it is worth noting here that actual convergence rates depend on the derivatives of the coefficient functions in the equation, as discussed in some detail by Bank [2]. Thus, for such problems, scaling the original equation can have a significant effect on convergence rates.

The above theoretical results do not, of course, apply to our equation (2) since it is not in self-adjoint form. However, (4) is easily shown to be equivalent to a block-Jacobi iteration for the relevant Fourier harmonic amplitudes and a convergence analysis of this iteration together with the practical computational behaviour of the method suggest that convergence rates can be achieved which vary little with grid-size. In some examples such behaviour was found to be significantly dependent on the choice of suitable scaling; the choice of 'nearby' equation (3) is discussed in Section 3.

2 THE METHOD

We make use of a second-order difference scheme proposed by Roscoe [10] which is based on a local exponential approximation. For elliptic equations of the general form (2) this scheme always leads to a diagonally dominant system of linear equations and, for equations of the form (3), the resulting linear system may be solved by FFT techniques (Pickering [8]). Similar schemes were originally developed for use with methods such as successive over-relaxation (see, for example, Allen [1]).

Assuming that (2) is satisfied in a rectangular region and using a square mesh of side h, at a mesh point (x_i, y_j) $(i = 1, 2, \ldots, M;\ j = 1, 2, \ldots, N)$, Roscoe's scheme may be written as

$$\begin{aligned}\alpha_{i,j}\phi_{i+1,j} - \beta_{i,j}\phi_{i,j} \ &+\ \gamma_{i,j}\phi_{i-1,j} + \\ \delta_{i,j}\phi_{i,j+1} \ &-\ \xi_{i,j}\phi_{i,j} + \eta_{i,j}\phi_{i,j-1} - h^2 f_{i,j}\phi_{i,j} = h^2 q_{i,j},\end{aligned} \tag{6}$$

where, with the notation $e_1 = \exp(-b_{i,j}h/a_{i,j}), e_2 = \exp(-d_{i,j}h/c_{i,j})$, the coefficients in (6) may be written as $\alpha_{i,j} = hb_{i,j}(1 - e_1)^{-1}$, $\delta_{i,j} = hd_{i,j}(1 - e_2)^{-1}$, $\beta_{i,j} = \alpha_{i,j}(1 + e_1)$, $\gamma_{i,j} = \alpha_{i,j}e_1$, $\xi_{i,j} = \delta_{i,j}(1 + e_2)$ and $\eta_{i,j} = \delta_{i,j}e_2$. Applying the same scheme to equation (3) leads to a relation of the form (6) with coefficients $\overline{\alpha}, \overline{\beta}, \overline{\gamma}, \overline{\delta}_j, \overline{\xi}_j, \overline{\eta}_j$ and $\overline{f}_j$, where, for example, $\overline{\alpha} = h\overline{b}(1 - \exp(-\overline{b}h/\overline{a}))^{-1}, \overline{\delta}_j = h\overline{d}_j(1 - \exp(-\overline{d}_j h/\overline{c}_j))^{-1}$.

By defining $\overline{\alpha}_{i,j} = \alpha_{i,j} - \overline{\alpha}$, $\overline{\delta}_{i,j} = \delta_{i,j} - \overline{\delta}_j$ together with similar relations for $\overline{\beta}_{i,j}, \overline{\gamma}_{i,j}, \overline{\xi}_{i,j}, \overline{\eta}_{i,j}$ and $\overline{f}_{i,j}$, the relevant discretized iteration (4) may be written in point-wise form as

$$\begin{aligned}\overline{\gamma}\phi^{(m+1)}_{i-1,j} - (\overline{\beta} + \overline{\xi}_j + h^2\overline{f}_j)\phi^{(m+1)}_{i,j} + \overline{\alpha}\phi^{(m+1)}_{i+1,j} + \overline{\delta}_j\phi^{(m+1)}_{i,j+1} + \overline{\eta}_j\phi^{(m+1)}_{i,j-1} \ &=\ h^2 q_{i,j}\ - \\ \overline{\gamma}_{i,j}\phi^{(m)}_{i-1,j} + (\overline{\beta}_{i,j} + \overline{\xi}_{i,j} + h^2\overline{f}_{i,j})\phi^{(m)}_{i,j} - \overline{\alpha}_{i,j}\phi^{(m)}_{i+1,j} - \overline{\delta}_{i,j}\phi^{(m)}_{i,j+1} \ &-\ \overline{\eta}_{i,j}\phi^{(m)}_{i,j-1}.\end{aligned} \tag{7}$$

Using the above relation and making the discrete change of variable $\phi_{i,j} = (\overline{\gamma}/\overline{\alpha})^{i/2}\psi_{i,j}$, we may express the linear equations on one row (j = constant) of our grid in the matrix form

$$A_j\mathbf{\Psi}^{(m+1)}_j + \overline{\delta}_j I\mathbf{\Psi}^{(m+1)}_{j+1} + \overline{\eta}_j I\mathbf{\Psi}^{(m+1)}_{j-1} = \mathbf{Q}^*_j - F_j\mathbf{\Psi}^{(m)}_j - D_j\mathbf{\Psi}^{(m)}_{j+1} - E_j\mathbf{\Psi}^{(m)}_{j-1}, \tag{8}$$

where $\mathbf{\Psi}_j = (\psi_{1,j}, \psi_{2,j}, \ldots, \psi_{M,j})^T$, A_j denotes the Mth order (symmetric) tridiagonal matrix given by

$$A_j = (\overline{\alpha}\overline{\gamma})^{1/2}J - k_j I \tag{9}$$

where J is the Mth order (constant) tridiagonal matrix with diagonal elements -2 and sub- and super-diagonal elements unity and I is the unit matrix of order M, $k_j = \overline{\beta} + \overline{\xi}_j + h^2\overline{f}_j - 2(\overline{\alpha}\overline{\gamma})^{1/2}$, $\mathbf{Q}^*_j = \mathbf{Q}_j - \mathbf{\Psi}_B$, where $\mathbf{Q}_j$ is a vector with components $h^2(\overline{\alpha}/\overline{\gamma})^{i/2}q_{i,j}$ $(i = 1, 2, \ldots, M)$ and $\mathbf{\Psi}_B$ is a vector with zero components except the first and last, which are, respectively, $\gamma_{1,j}(\overline{\alpha}/\overline{\gamma})^{1/2}\psi_{0,j}$ and $\alpha_{M,j}(\overline{\gamma}/\overline{\alpha})^{1/2}\psi_{M+1,j}$, where $\psi_{0,j}$ and $\psi_{M+1,j}$ denote boundary values of $\psi_{i,j}$.

The matrices D_j and E_j in relation (8) are given by $D_j = diag(\overline{\delta}_{1,j}, \overline{\delta}_{2,j}, \ldots, \overline{\delta}_{M,j})$, $E_j = diag(\overline{\eta}_{1,j}, \overline{\eta}_{2,j}, \ldots, \overline{\eta}_{M,j})$ and the matrix F_j is a tridiagonal matrix with (for $i = 1, 2, \ldots, M$) diagonal elements $\overline{\beta}_{i,j} + \overline{\xi}_{i,j} + h^2\overline{f}_{i,j}$, sub-diagonal elements $-\overline{\gamma}_{i,j}(\overline{\alpha}/\overline{\gamma})^{1/2}$ and super-diagonal elements $-\overline{\alpha}_{i,j}(\overline{\gamma}/\overline{\alpha})^{1/2}$.

For $j = 1$ and $j = N$, boundary values of $\psi_{i,0}$ and $\psi_{i,N+1}$ may be absorbed into $\mathbf{Q}^*_1$ and $\mathbf{Q}^*_N$, in the usual fashion.

It is clear that the full system of linear equations derived from (8) for $\Psi_j^{(m+1)}, j = 1,2,...,N$ has a tridiagonal block structure in which every sub-matrix has the same eigenvectors, namely the eigenvectors $\mathbf{x}_s, s = 1,2,...,M$ of the matrix J. It is well-known that these eigenvectors are given by $\mathbf{x}_s = (\sin\theta_s, \sin 2\theta_s, \ldots, \sin M\theta_s)$, where $\theta_s = s\pi/(M+1)$, and satisfy the orthogonality condition $\mathbf{x}_r^T\mathbf{x}_s = (M+1)\Delta_{rs}/2$, where here Δ_{rs} denotes the Kronecker delta function. The corresponding eigenvalues, λ_s of J are given by $\lambda_s = -4\sin^2(\theta_s/2)$. Thus, assuming that $\Psi_j^{(m)}, j = 1,2,\ldots,N$, are known, the right-hand sides of (8) may be evaluated and Fourier analyzed using the usual FFT techniques (see, for example, van Loan [7], Pickering [9]). Analyzing each $\Psi_j^{(m+1)}$ similarly, in the form

$$\Psi_j^{(m+1)} = \sum_{s=1}^{M} w_{j,s}^{(m+1)}\mathbf{x}_s, \tag{10}$$

we may derive, in the usual fashion, the M (independent) tridiagonal systems of equations (each of order N) satisfied by the Fourier harmonic amplitudes $w_{j,s}^{(m+1)}$ and these equations may be solved by the usual Thomas algorithm. It should be noted that using approximation (6) ensures that each tridiagonal system is diagonally dominant. Each $\Psi_j^{(m+1)}$ may be synthesized as usual and the overall procedure repeated until convergence. Thus, for $M = N$, the overall operations count $T(N)$ is of the form

$$T(N) = m_0(\mu N^2 \log_2 N + \nu N^2), \tag{11}$$

where m_0 denotes the number of iterations required for convergence and μ and ν are constants.

An alternative implementation of the above computational method is to perform the iterations entirely in the transformed plane. The relevant equations for the Fourier harmonic amplitudes are readily derived from the iterative relation (8) and have a structure which is essentially of block-Jacobi form. A typical block of equations may be written as

$$Y_r\mathbf{w}_r^{(m+1)} = \mathbf{v}_r - \frac{2}{M+1}\sum_{s=1}^{M} Z_{rs}\mathbf{w}_s^{(m)}, \ (r = 1,2,\ldots,M) \tag{12}$$

where Y_r is a tridiagonal matrix with diagonal elements which are the eigenvalues of (9), that is $(\overline{\alpha\gamma})^{1/2}\lambda_r - k_j$, sub-diagonal elements $\overline{\eta}_j$ and super-diagonal elements $\overline{\delta}_j$. Z_{rs} is a tridiagonal matrix with diagonal elements $\mathbf{x}_r^T F_j \mathbf{x}_s$ and sub- and super-diagonal elements respectively $\mathbf{x}_r^T E_j \mathbf{x}_s$ and $\mathbf{x}_r^T D_j \mathbf{x}_s$, $\mathbf{v}_r$ denotes a vector with components $2\mathbf{x}_r^T\mathbf{Q}_j^*/(M+1)$ and $\mathbf{w}_s = (w_{1,s}, w_{2,s}, \ldots, w_{N,s})^T$. It is a straightforward matter to show that, at each iteration, the right-hand side of (12) may be evaluated by Fourier synthesis and that this approach requires approximately three times the work (since Z_{rs} is tridiagonal) of the Fourier analysis of the right-hand side of (8). Consequently, the practical implementation of the iteration is based on relation (8). An analysis of the convergence properties of (12) does, however, give some useful insight into the choice of 'nearby' equation (3), as discussed in more detail in the following Section.

3 CONVERGENCE OF THE ITERATIONS

In general the matrices Y_r and Z_{rs} do not commute and an analysis of the convergence properties of iteration (12) is not straightforward. For this reason, we concentrate here on a particular case in which these matrices can be arranged to be of a relatively simple form. If the diagonal matrices D_j and E_j are constant matrices and the tridiagonal matrix F_j is chosen so that $\mathbf{x}_s$ is one of its eigenvectors then $Z_{rs} = 0,\ r \neq s$. This can be achieved by choosing F_j to be a constant matrix with equal sub- and super-diagonal elements. If Y_r is also chosen to be a constant matrix then Y_r and Z_{rr} will have a common set of eigenvectors if the ratio of the sub- and super- diagonal elements of Y_r is the same as the corresponding ratio for Z_{rr} . After some algebra, the above requirements lead to the relations

$$\overline{a}/a = \overline{b}/b = k_1, \qquad \overline{c}/c = \overline{d}/d = k_2, \tag{13}$$

where k_1 and k_2 are constants ($k_1,\ k_2 \neq 0$) and $a,\ b,\ c,\ d,\ \overline{a},\ \overline{b},\ \overline{c}$, and $\overline{d}$ are coefficients in equations (2) and (3) (here assumed constant). Hence it may be shown that the eigenvalues of $Y_r^{-1} Z_{rr}$ are given by

$$\zeta_{r,t} = \frac{a(1 - k_1)U(r) + c(1 - k_2)V(t) + h^2(1 - k_3)f}{ak_1U(r) + ck_2V(t) + h^2k_3f}, \quad \begin{pmatrix} r = 1,2,...,M \\ t = 1,2,...,N \end{pmatrix} \tag{14}$$

where $k_3 = \overline{f}/f$ and $U(r)$ and $V(t)$ are positive functions given by

$$\begin{aligned} U(r) &= 4\left[\sinh^2\left(\frac{bh}{4a}\right) + \sin^2\left(\frac{r\pi}{2(M+1)}\right)\right]\frac{bh/2a}{\sinh(bh/2a)} \\ V(t) &= 4\left[\sinh^2\left(\frac{dh}{4c}\right) + \sin^2\left(\frac{t\pi}{2(N+1)}\right)\right]\frac{dh/2c}{\sinh(dh/2c)}. \end{aligned} \tag{15}$$

The corresponding eigenvectors of $Y_r^{-1} Z_{rr}$ are σ_t with components $e^{\kappa dh/2c} \sin \kappa\pi t/(N+1)$, $\kappa = 1, 2, \ldots, N$. It follows from (14) and (15) that $|\zeta_{r,t}| < 1$ for $k_1, k_2, k_3 > 1/2$. If $k_1 = k_2 = k_3 = k$, $\zeta_{r,t} = (1-k)/k$ which is clearly independent of h and, furthermore, with $k = 1$ for constant coefficient equations (2) and (3), convergence is expected in one iteration. Both these results were verified by practical computation.

For variable coefficient equations, conditions (13) cannot be satisfied exactly and we have attempted to satisfy these conditions in an average sense. Thus $\overline{a}$ and $\overline{b}$ were determined from (13) by replacing a and b in (13) by averages of, respectively, grid-point values of $a_{i,j}$ and $b_{i,j}$ over the solution domain. The coefficients $\overline{c}_j$, $\overline{d}_j$ and $\overline{f}_j$ may be similarly determined or, alternatively, N values for each coefficient may be determined using averages of, respectively, grid-point values of $c_{i,j}$, $d_{i,j}$ and $f_{i,j}$ on each row (j = constant) of the grid. These two options will be called 'fixed' or 'variable' shifts and correspond to choosing either a constant coefficient 'nearby' equation of the form (3) or a 'nearby' equation (3) with $\overline{c}, \overline{d}$ and $\overline{f}$ which are functions of y. Strategies using other than simple average values may be tried for the above calculations and these may or may not result in faster convergence depending on the behaviour of the coefficients in equation (2).

4 PRACTICAL EXAMPLES

The method has been applied successfully to a large number of test problems, two of which we present here. In all cases we set $M = N = 2^{\Gamma} - 1$, where $\Gamma \geq 2$ is an integer and convergence was assumed when successive iterates agreed to 7 significant figures. As far as the authors are aware, the problem of finding an optimal scaling for iterations of this type has not been solved, a point also noted by Bank [2].

4.1 Example 1

The first example is an equation used by Bank [2], (his test problem 1) as a test case for his generalized marching method. As used by Bank this equation was in self-adjoint form but, for our purposes, we write this equation in the form (2) as

$$\left(1 + (x + y)^2\right)^2 \frac{\partial^2 \phi}{\partial x^2} + 4\left(1 + (x + y)^2\right)(x + y)\frac{\partial \phi}{\partial x} +$$
$$\left(1 + \sin^2(x + y)\right)^2 \frac{\partial^2 \phi}{\partial y^2} + 2\left(1 + \sin^2(x + y)\right)\sin 2(x + y)\frac{\partial \phi}{\partial y} = q(x,y), \quad (16)$$

where the right-hand side and (Dirichlet) boundary conditions are constructed to be consistent with the analytic solution $\phi(x,y) = \sin x \sin y$ over the unit square. Tables 1(a) and 1(b) show the required number of iterations, m_0, as a function of N for various scalings of (16) and using the FFT in either the x- or y-directions. In Table 1(a) case (i) is unscaled (scaling by a constant does not affect convergence rates) and cases (ii) and (iii) are, respectively, scaled by (in the notation of (2)) $1/c$ and $(ac)^{-1/4}$ and use a variable shift. Case (iii) is the scaling found by Bank to give fast convergence, which is clearly not the case for the present method. For all these cases convergence was not achieved in ≤ 100 iterations for $N \geq 7$. Case (iv) is scaled by $(ac)^{-1/2}$ and uses a variable shift and cases (v) and (vi) are both scaled by $1/a$, the difference between these two cases is that (v) is for a fixed shift (in the y-direction) whereas (vi) is for a corresponding variable shift. The advantage of using a variable shift is clear and (vi) is the most rapidly convergent of all the cases chosen in Table 1(a). (Other scalings were tried but showed no improvement over this case). All the results in Table 1(a) were obtained using $k_1 = k_2 = 1$ and the choice of coefficients in the relevant nearby equation was made using average values, as described in the previous Section. Using values of k_1 and k_2 other than unity produced no improvement in convergence rates, and, as suggested by the analysis of Section 3, when both k_1 and k_2 were reduced towards the value 0.5, convergence became very difficult to achieve. Table 1(b) shows a similar set of results using the FFT in the y-direction with $k_1 = k_2 = 1$. Case (i) is unscaled and cases (ii), (iv) and (v) are scaled by, respectively, $1/c, 1/a$, and $(a^2 + c^2)^{-1/2}$ and use a variable shift in the x-direction. Cases (iii) and (vi) are scaled by $(ac)^{-1/2}$ and (iii) is for a fixed shift whereas (vi) is for a corresponding variable shift. The advantage of using a variable shift is more marked than for the relevant results presented in Table 1(a) and case (vi) of Table 1(b) is clearly the most rapidly convergent iteration of all the cases considered. Furthermore, for this case the number of iterations required for convergence increases only slowly with N .

	(a) m_0 for various scalings of equation (16) with FFT in x-direction						(b) m_0 for various scalings of equation (16) with FFT in y-direction					
N	(i)	(ii)	(iii)	(iv)	(v)	(vi)	(i)	(ii)	(iii)	(iv)	(v)	(vi)
3	72	23	23	10	12	10	24	13	9	13	13	8
7	*	*	*	20	19	13	53	28	17	21	16	12
15	*	*	*	40	28	17	*	76	33	31	24	16
31	*	*	*	90	37	23	*	*	69	40	33	20
63	*	*	*	*	46	30	*	*	*	49	39	21
127	*	*	*	*	53	39	*	*	*	57	45	23

Table 1: **(a) & (b)** *Required iterations, m_0, for $N = 2^\Gamma - 1$, $(2 \leq \Gamma \leq 7)$, for various scalings (see text) with FFT in (a) x-direction and in (b) y-direction (* denotes > 100 iterations).*

A possible reason why the method can produce quicker convergence using the FFT in the y-direction rather than the x-direction for equation (16) may be found by examining the coefficients in (16). In this equation it is clear that the coefficients of the x-derivatives vary more rapidly and have larger magnitude than the coefficients of the corresponding y-derivatives. As the iteration is based on a nearby equation whose derivatives in either x or y have to have constant coefficients in order to use the FFT, it thus seems preferable to choose this direction to be the y-direction for equation (16). This point of view is supported by the unscaled results (a)(i) and (b)(i) of Table 1, but of course, such arguments may be modified by other choices of scaling. However, for the various scalings employed, it is clear that the convergence rates observed in Table 1(b) are generally better than those in Table 1(a).

4.2 Example 2

The second example we consider is the equation

$$(1 + 2x - x^2)\left(\frac{\partial^2\phi}{\partial x^2} + \frac{1}{1-x}\frac{\partial\phi}{\partial x}\right) + 4\frac{\partial^2\phi}{\partial y^2} + 4\left(2e^{2y} - 3\right)\frac{\partial\phi}{\partial y} - 8\phi = q(x,y), \quad (17)$$

over the unit square with right-hand side and Dirichlet boundary conditions consistent with the analytic solution $\phi(x,y) = \exp(xy)$. An equation similar to (17) arises in the study of the fluid dynamics of tornadoes (Sozou & Pickering [11]) and is essentially the equation for the vorticity. (The coefficient of ϕ in (17) has opposite sign from Sozou *et al* and the region and boundary conditions are also different from their paper).

Tables 2(a) and 2(b) show m_0 against N for various scalings and using the FFT in the x- and y-directions. In Table 2(a), with the FFT in the x-direction all the cases use a variable shift in the y-direction and (i) is unscaled, (ii) is scaled by $(a^2+c^2)^{-1/2}$, (iii) is scaled by $1/a$

	(a) m_0 for various scalings of equation (17) with FFT in x-direction				(b) m_0 for various scalings of equation (17) with FFT in y-direction			
N	(i)	(ii)	(iii)	(iv)	(i)	(ii)	(iii)	(iv)
3	9	8	9	9	16	16	13	13
7	13	12	13	13	19	19	13	11
15	17	16	16	16	21	21	13	12
31	25	22	18	17	25	25	13	13
63	35	32	21	19	28	35	13	11
127	*	44	24	23	30	44	11	11

Table 2: **(a) & (b)** *Required iterations, m_0, for $N = 2^\Gamma - 1$, $(2 \le \Gamma \le 7)$, for various scalings (see text) with FFT in (a) x-direction and in (b) y-direction (* denotes > 50 iterations).*

and for these cases $k_1 = k_2 = k_3 = 1$. Case (iv) is unscaled but $k_1 = 1.2$ (and $k_2 = k_3 = 1$). For $N \geq 31$ this case shows a considerable improvement in convergence rates compared with (i) and this improvement is brought about solely by the change in k_1 . The value of k_1 used in (iv) was found by computer experiment to be close to the optimal for this case and varying (additionally) k_2 and k_3 was not found to further improve rates of convergence. In Table 2(b) cases (ii), (iii) and (iv) use a variable shift in the x-direction and (ii) is scaled by $1/a$, (iii) is scaled by $(a^2+c^2)^{-1/2}$ and (iv) is unscaled. Case (i) is also unscaled but uses a fixed shift in the x-direction. It is clear that scalings (iii) and (iv) give very similar results and comparison of (i) and (iv) shows that use of a variable rather than constant shift can considerably improve convergence rates. This comparison also shows that, for this example, using a variable shift can also affect the variation in overall iteration count with grid-size. For (iii) and (iv) the number of iterations required is in the range 11-13 and hence is effectively independent of grid-size. In all cases shown in Table 2(b) the values $k_1 = k_2 = k_3 = 1$ were used (other values of these parameters were tried but no significant improvements were found compared with the results shown in Figure 2(b)). Comparing Tables 2(a) and 2(b) it is clear that the best of these results is 2(b) cases (iii) or (iv) (scaled by $(a^2+c^2)^{-1/2}$ and unscaled, respectively) and the general reasons why FFT in the y-direction is the preferred choice for equation (17) are essentially the same as those given in connection with equation (16).

For both examples 1 and 2 root mean square errors (obtained by comparison of the numerical solution with the grid-point values of the analytic solution) were determined and these errors were found to decrease by a factor of approximately 4 for each successive halving of the grid, as is expected in view of the second-order nature of approximation (6). Furthermore, the execution time per iteration was found to increase by a factor of approximately 4 for each successive halving of the grid, as is indicated by the form of operations count (11).

The program was written in Fortran 77 and run on a 486 DX (33 Mhz) PC, and, for $N = 63$ the time per iteration was approximately 0.41s. The overall execution times for Ex. 1(b)(vi) and Ex. 2(b)(iv) (for $N = 63$) were, respectively, 8.6s and 4.5s with corresponding RMS errors 2.8,-6 and 1.8,-5.

5 CONCLUSIONS

It has been shown that linear elliptic equations of general form (2) may be solved by iterative application of FFT techniques using an iteration of D'Yakonov-Gunn form. The examples considered demonstrate that, with an appropriate choice of FFT in the x-direction or y-direction (see Section 4) and with suitable shifting and scaling, convergence rates can be achieved which vary little with grid-size. In such cases execution times are relatively small. The use of a variable rather than a constant shift was found to improve convergence rates in all cases (as illustrated by 1(b)(iii) and (vi), 2(b)(i) and (iv)) and, for many problems, the choice of parameters $k_1 = k_2 = k_3 = 1$ was found to be close to the optimum. The problem of finding an optimum scaling for this form of iteration has not been investigated but, if scaling is required, scalings of the form $(a^2 + c^2)^{-1/2}$ or $(ac)^{-1/2}$ were found to give good or reasonably good convergence properties for the examples considered. These examples assumed Dirichlet boundary conditions, but the method can easily be extended to other suitable boundary conditions, such as Neumann or periodic conditions on a rectangular region.

References

[1] Allen, D. N. de G. *A suggested approach to finite-difference representations of differential equations, with an application to determine temperature distributions near a sliding contact*, Q. J. Mech. Appl. Math. 15, 1962, pp 11-33.

[2] Bank, R. E. *Marching algorithms for elliptic boundary value problems II: the variable coefficient case*, SIAM J. Numer. Anal. 14, 1977, pp 950-970.

[3] Bunemann, O. *A compact non-iterative Poisson solver*, Report 294, Stanford University Institute for Plasma research, Stanford, California, USA, 1969.

[4] Concus, P. and Golub, G. H. *Use of fast direct methods for the efficient numerical solution of nonseparable elliptic equations*, SIAM J. Numer. Anal. 10, 1973, pp 1103-1120.

[5] D'Yakonov, E. G. *On an iterative method for the solution of finite-difference equations*, Dokl. Aka. Nauk SSSR, 138, 1961, pp 522-525.

[6] Gunn, J. E. *The solution of difference equations by a semi-explicit iterative technique*, SIAM J. Numer. Anal. 2, 1965, pp 24-25.

[7] van Loan, C. "Computational Frameworks for the Fast Fourier Transform", Ch. 4, SIAM, Philadelphia, USA, 1992.

[8] Pickering, W. M. *Nonpolynomial finite-difference schemes and the use of the fast Fourier transform*, J. Comput. Phys. 51, 1983, pp 357-363.

[9] Pickering, W. M. "An Introduction to Fast Fourier Transform Methods for Partial Differential Equations, with Applications", Ch 2 & 3, Research Studies Press, Taunton, England, 1986.

[10] Roscoe, D. F. *New methods for the derivation of stable difference representations for differential equations*, J. Inst. Maths. Applics. 16, 1975, pp 291-301.

[11] Sozou, C. and Pickering, W. M. *Development of the Stokes flow field associated with a line vortex perpendicular to a plane wall*, Phys. Fluid. A5, 1993, pp 1105-1112.

School of Mathematics & Statistics,
Applied Mathematics Section,
P O Box 597,
The University,
Sheffield S10 2UN,
England

Parallel Approximation with Self-Similar Functions

Bl. Sendov

Abstract

A representation of the Hausdorff continuous function is introduced, which is useful for parallel approximation. This representation employs a special type of self-similar functions, called here binary odd functions.

1 Introduction

An usual situation in the numerical analysis is to work with functions f from a Hilbert space H with orthonormal basis $\{\phi_i\}_{i=0}^{\infty}$. Every function $f \in H$ is represented in the form $f(x) = \sum_{i=0}^{\infty} \phi_i(f)\phi_i(x)$, where $\phi_i(f)$ are the generalized Fourier coefficients of f, and an approximation of f with error $\epsilon = \|f - \sum_{i=0}^{n} \phi_i(f)\phi_i\| = \left(\sum_{i=n+1}^{\infty} \phi_i(f)^2\right)^{1/2}$ is given with the final set of numbers $\{\phi_i(f)\}_{i=0}^{n}$.

If the Hilbert space H is divided in orthogonal subspaces $H = \cup_{p=1}^{\infty} H_p$, then every function $f \in H$ will have a uniquely defined representation

$$f(x) = \sum_{p=1}^{\infty} f_p(x), \quad \text{where} \quad f_p \in H_p. \tag{1.1}$$

If the division of H is chosen appropriately, a partial sum from the representation (1.1) may be used for a parallel approximation of f, calculating in parallel the different members of this sum.

Our purpose in this paper is to present a division of the class of Hausdorff continuous functions [9] of one variable with L_2 norm. The proposed method may be generalized for functions defined in the m-dimensional cube. We shall explain here only the basic idea using the functions defined on the unit interval $[0, 1]$.

2 Self-similar functions

The notion of self-similarity is studied extensively [1] - [7], [10] in connection with fractal structures. G. B. Lewellen [7] describe the self-similarity of a set in short: "The small-scale geometry of a self-similar set reproduces after a fashion the large-scale geometry of the set, such as seen in the Cantor space." In this paper we study self-similar functions and our "fashion" shall be the oddness of a function defined on a self-similar set.

Most studies of self-similarity have dealt with compact sets [5], [7], [4]. A function may be self-similar if it is defined on a self-similar set. And if this set is for example the unit interval, every self-similar function have to be continuous.

In this paper we consider not only continuous functions and to have self-similarity these functions are defined in a dense self-similar subset of the unit interval.

A self-similar object is produced by a collection $\mathcal{J}$ of contraction maps, which may be finite [5], countable [4] or uncountable [7]. We shall use the self-similar set $O \subset [0,1]$, defined by the collection $\mathcal{J} = \{w_0, w_1\}$ of contractive mappings $w_0(x) = x/2$, $w_1(x) = (x+1)/2$. This set $O = (0;0)$ has the properties $O = w_0(O) \cup w_1(O)$ and $w_0(O) \cap w_1(O) = \emptyset$. These properties remind us the *open set condition* [5, p. 735] in a stronger form. In fact $O = (0;0)$ is the set of all binary irrational points in the unit interval $[0,1]$.

Every number $x \in (0,1)$ may be represented in the binary form

$$x = a_1(x)2^{-1} + a_2(x)2^{-2} + a_3(x)2^{-3} + \dots, \tag{2.2}$$

where $a_n(x)$ is 0 or 1. The number x is binary irrational, if $a_n(x) = 0$ for infinitely many indices n and $a_n(x) = 1$ for infinitely many indices n.

Let $(k;p)$, where k, p are non negative integers and $p < 2^k$, be the set of all binary irrational numbers from the interval $(p2^{-k}, (p+1)2^{-k})$. The set $(k;p)$ is called a *binary interval* of rank k. The binary interval $(0;0) \subset [0,1]$ is of rank 0. Obviously $(k;p) = (k+1;2p) \cup (k+1;2p+1)$ and $(k;p) \subset (k-1;[p/2])$, where $[x]$ is the integer part of x.

The functions $a_n(x)$ from (2.2) are uniquely defined in $(0;0)$, but this is not true if x is binary rational. Every binary rational x has two different representations of the form (2.2), one with finite number of $a_n(x) = 1$ and the other with finite number of $a_n(x) = 0$.

We shall consider the set $\mathcal{BC}$ of all functions defined, bounded and continuous on $(0;0)$.

Every function f from $\mathcal{BC}$ is Lebesgue integrable in the interval $[0,1]$ as f is not defined only in the set with measure zero in this interval. In the functional space $\mathcal{BC}$ we shall use the uniform norm $\|f\| = \sup\{|f(x)| : x \in (0;0)$ and the L_2 norm $\|f\|_2 = \left(\int_0^1 f^2(x)\,dx\right)^{\frac{1}{2}}$.

Definition 2.1 *We call the function $f \in \mathcal{BC}$* **binary odd** *if*

$$f(x) + f(1-x) = 0, \quad \text{for} \quad x \in (0;0),$$

and for every binary interval $(k;p)$

$$f(2^{-k}p + x) + f(2^{-k}(p+1) - x) = f_{k,p} = \text{constant} \quad \text{for} \quad x \in (k;0).$$

The set of all binary odd functions is denoted by $\mathcal{BO}_0$.

An example of a binary odd function is the linear function $l(x) = 2x - 1$. More generally, the fixed point of the iterated equation

$$\phi(x) = \begin{cases} \lambda\phi(2x) - 1 + \lambda & \text{for} \quad x \in (1;0) \\ \lambda g(2x-1) + 1 - \lambda & \text{for} \quad x \in (1;1) \end{cases},$$

where $|\lambda| < 1$, is a binary odd function. For $\lambda = 1/2$, $\phi(x) = 2x - 1$.

Definition 2.2 *We call the function $f \in \mathcal{B}C$* **binary even** *if*

$$f(x) = \begin{cases} g(2x) & \text{for} \quad x \in (1;0) \\ g(2-2x) & \text{for} \quad x \in (1;1) \end{cases},$$

where the function g is binary odd. The set of all binary even functions is denoted by $\mathcal{BE}_0$.

Definition 2.3 *If the function $f \in \mathcal{B}C$ is such that for a natural number k and $p = 0, 1, 2, \dots, 2^k - 1$ all the functions*

$$h_{k,p}(x) = f(2^{-k}(x+p)); \quad p = 0, 1, 2, 3, \dots, 2^k - 1, \quad x \in (0;0)$$

are binary even, then f is called **binary even of rank** k. *The set of all binary even functions of rank k is denoted by $\mathcal{BE}_k$.*

The binary even functions (Definition 2.2) are binary even of rank 0.

3 Self-similar vectors

Now we fix the natural number s and shall consider the Euclidean space E^{2^s} of 2^s dimensional vectors $\{b_p\}_{p=0}^{2^s-1}$.

Definition 3.1 *A vector $B = \{b_p\}_{p=0}^{2^s-1} \in E^{2^s}$ is called* **binary odd** *if*

$$b_{2^{s-1}-p} + b_{2^{s-1}+p-1} = 0, \quad \text{for} \quad p = 1, 2, 3, \dots, 2^{s-1} - 1,$$

and for every two nonnegative integers l, q; $l < s, q < 2^l$

$$b_{(q+1)2^l-p} + b_{q2^l+p-1} = b_{l,q}, \quad \text{for} \quad p = 1, 2, 3, \dots, 2^{l-1} - 1.$$

The vector $(-5, 3, -2, 6, -6, 2, -3, 5)$ is binary odd in E_3.

It is easy to see that a binary odd vector in E^{2^s} is determined by s numbers and that all binary odd vectors form a linear subspace $H_0 \subset E^{2^s}$ of dimension s.

In an obvious way we define a **binary even** vector, and a **binary even vector of rank** k . The vector $(-5, 3, -3, 5, 5, -3, 3, -5)$ is binary even, or binary even of rank 0, and $(-5, 5, 5, -5, 2, -3, -3, 2)$ is binary even of rank 1.

All binary even (of rank 0) vectors in E^{2^s} form a subspace H_1 of dimension $s-1$. All binary even of rank k vectors in E^{2^s} form a subspace H_{k+1} of dimension $2^k s - k - 1$.

A binary even vector in E^{2^s} may have the highest rank $s-2$. On the other hand

$$\sum_{k=0}^{s-2} 2^k(s-k-1) = 2^s - s - 1. \tag{3.3}$$

Let I be the one dimensional subspace of E^{2^s} consisting of all vectors collinear with the unit vector. Then the subspaces $I, H_0, H_1, \ldots, H_{s-1}$ are orthogonal and

$$E^{2^s} = I \cup H_0 \cup H_1 \cup \ldots \cup H_{s-1}. \tag{3.4}$$

3.1 Pixel functions

Definition 3.2 *A function $f \in \mathcal{BC}$ is called a* **pixel** *function with* **resolution** *s if f is constant in every binary interval of rank s.*

The set of all pixel functions with resolution s is denoted by $\mathcal{P}_s$.

It is obvious that $\mathcal{P}_s$ is a 2^s dimensional linear subspace of $\mathcal{BC}$.

Let $\mathcal{BOP}_s$ be the set of all binary odd pixel functions with resolution $\leq s$.

Lemma 3.1 *Let $b_k(x) = 2a_k(x) - 1;\quad x \in (0;0), k = 1,2,3,\ldots$, (see (2.2)). The functions*

$$b_1(x),\ b_2(x),\ b_3(x),\ \ldots,\ b_s(x) \tag{3.5}$$

form an orthonormal basis in $\mathcal{BOP}_s$ with the L_2 norm.

Proof. From the definition of b_k it follows that

$$\int_0^1 b_i(x)b_j(x)\,dx = \begin{cases} 0 & \text{for } i \neq j, \\ 1 & \text{for } i = j, \end{cases}$$

hence the system (3.5) is orthonormal with s members.

To complete the proof it remains to see that $\mathcal{BOP}_s$ is a s-dimensional space. This follows from the fact that the values of a function from $\mathcal{BOP}_s$ form a binary odd vector of dimension 2^s, which completes the proof.

Corollary 3.1 *Every function $f \in \mathcal{BOP}_s$ has the representation $f(x) = \sum_{i=1}^{s} b_i(f)b_i(x)$, where $b_i(f) = \int_0^1 f(x)b_i(x)\,dx$.*

Let

$$c_n(x) = \begin{cases} b_n(x) & \text{for } x \in (1;0), \\ b_n(1-x) & \text{for } x \in (1;1), \\ 0 & \text{for } x \notin [0,1] \end{cases},$$

and

$$c_{n,k,p}(x) = 2^{k/2}c_n(2^k x - p); k = 0,1,2,\dots,s-2,$$
$$n = k+2, k+3, \dots, s, \quad p = 0,1,2,\dots,2^k - 1.$$

Let $\mathcal{BEP}_{k,s} = \mathcal{BE}_k \cup \mathcal{P}_s$ be the set of all binary even of rank k pixel functions with resolution s. It is easy to prove:

Lemma 3.2 *The functions*

$$c_{n,k,p}(x); \quad k = 0,1,2,\dots,s-2, \quad n = k+2, k+3, \dots, s, \tag{3.6}$$
$$p = 0,1,2,\dots,2^k - 1$$

form an orthonormal basis in $\mathcal{BEP}_{k,s}$ with the L_2 norm.

Corollary 3.2 *Every function $f \in \mathcal{BEP}_{k,s}$ has the representation*

$$\sum_{i=1}^{s} \sum_{p=0}^{2^k-1} c_{i,k,p}(f) c_{i,k,p}(x), \tag{3.7}$$

where $c_{i,k,p}(f) = \int_0^1 f(2^{-k}(x+p))c_i(x)\,dx$.

We shall use the linear operator

$$\Phi_s(f;x) = f^{<s>}(x) = 2^s \int_{(s;p)} f(t)\,dt \quad \text{for} \quad x \in (s;p),$$
$$p = 0,1,2,\dots,2^s - 1.$$

It is easy to see that Φ_s is a projection of $\mathcal{BC}$ in $\mathcal{P}_s$.

Definition 3.3 *Let $f \in \mathcal{BC}$ and*

$$v((k;p),f) = \sup\{|f(x') - f(x'')| : \ x', x'' \in (k;p)\}.$$

We call

$$\tau(f;2^{-k}) = \max\left\{v((k;p),f) : \ p = 0,1,2,\dots,2^k - 1\right\} \tag{3.8}$$

the **fractal modulus** *of continuity.*

Definition 3.4 *We call a function $f \in \mathcal{BC}$* **Hausdorff continuous** *[8], [9], if*

$$\lim_{s\to\infty} \tau(f;2^{-s}) = 0.$$

The set of all Hausdorff continuous fractal functions is denoted by $\mathcal{HC}$.

Lemma 3.3 *For every $f \in \mathcal{BC}$ the inequality*

$$\|f - \Phi_s(f))\|_2 \leq \tau(f; 2^{-s}) \tag{3.9}$$

holds.

Proof. Let $x \in (s; p) \subset (0; 0)$. Then

$$(f(x) - \Phi_s(f; x))^2 = \left(2^s \int_{(s;p)} f(x)\, dt - 2^s \int_{s;p)} f(t)\, dt\right)^2$$

$$\leq 2^s \int_{(s;p)} (f(x) - f(t))^2\, dt \leq \left(\tau(f; 2^{-s})\right)^2.$$

From this (3.9) follows and the Lemma is proved.

4 Representation by self-similar functions

From Corollary 3.1, Corollary 3.2 and Lemma 3.9 follows:

Theorem 4.1 *Every Hausdorff continuous $f \in \mathcal{BC}$, has the representation*

$$f(x) = b_0(f) + \sum_{n=1}^{\infty} b_n(f) b_n(x) + \sum_{k=0}^{\infty} \sum_{n=k+2}^{\infty} \sum_{p=0}^{2^k-1} c_{n,k,p}(f) c_n(2^k x - p),$$

where

$$b_n(f) = \int_0^1 f(x) b_n(x)\, dx, \quad c_{n,k,p}(f) = \int_0^1 f(2^{-k}(x+p)) c_n(x)\, dx,$$

and the representation

$$f(x) = b_0(f) + \sum_{k=0}^{\infty} f_k(x), \tag{4.10}$$

where

$$f_0(x) = \sum_{n=1}^{\infty} b_n(f) b_n(x),$$

is binary odd and

$$f_{k+1}(x) = \sum_{n=k+2}^{\infty} c_{n,k,p}(f) c_n(2^k x - p); \quad \text{for } x \in (k; p), \;\; p = 0, 1, 2, \ldots, 2^k - 1$$

is binary even of rank k.

The representation (4.10) is of type (1.1) and may be used for parallel approximation of functions.

After the constant $b_0(f)$, the first and the second functions $f_0(x)$ and $f_1(x)$ in (4.10) get the global *oddness* end the global *evenness* of the function f. The function $f_k(x)$ in (4.10) gets the *evenness* locally, separately for every binary interval of rank k. This property of the representation (4.10) is important for approximation of functions with local singularities.

The approximation with self-similar functions is appropriate for not smooth functions as signals and real world images.

In a forthcoming paper we use the representation (4.10) to obtain another representation through self-similar fractal functions with given fractal dimension.

References

[1] BANDT, C. AND R. TEKLEHAIMANOT (1992), *Self-Similar Sets as Inverse Limit of Finite Topological Spaces*, Topology, Measures and Fractals (Warnemünde, 1991) Math. Res., **66**, Academie-Verlag, Berlin, 41 - 46.

[2] DELIU, A., J. S. GERONIMO, R. SHONKWILER AND D. HARDIN (1991), *Dimensions Associated with Recurrent Self-Similar Sets*, Math. Proc. Cambridge Philos. Soc. **110**, 327 - 336.

[3] FALCONER, K. J. (1988), *The Hausdorff Dimension of Self-Affine Fractals*, Math. Proc. Cambridge Philos. Soc., **103**, no. 2, 339 - 350.

[4] HATA, M. (1985), *On the Structure of the Self-Similar sets*, Japan J. Appl. Math.,**2**, 381 - 414.

[5] HUTCHINSON, J. (1981), *Fractals and Self-Similarity* , Indiana Univ. J. of Math., **30**, 713 - 747.

[6] LALLEY, S. P. (1988), *The Packing and Covering Functions of Some Self-Similar Fractals*, Indiana Univ. Math. J. **37**, no. 3, 699 - 710.

[7] LEWELLEN, G. B. (1993), *Self-Similarity*, Rocky Mountain J. of Math., 3, **23**, 1023 - 1040..

[8] SENDOV, BL.(1969), *Certain questions in the theory of approximation of functions and sets in the Hausdorff metric*, Uspehi Mat. Nauk **24**, # 5, 141 - 178 (Russian Math. Surveys, **24**, 1969, # 5, 143 - 183).

[9] SENDOV, BL. (1990), *Hausdorff Approximations*, Kluwer Academic Publishers, Dordrecht / Boston / London.

[10] URBAŃSKI, M. (1990), *The Probability Distribution and Hausdorff Dimension of Self-Affine Functions*, Probab. Theory Related Fields, **84**, 377 - 391.

Center of Informatics and Computer Technology
Bulgarian Academy of Sciences
"Acad. G. Bonchev" street, Block 25A,
1113 Sofia, BULGARIA

A TWO-LEVEL STABILIZING FRAMEWORK FOR INTERFACE DOMAIN DECOMPOSITION PRECONDITIONERS

PANAYOT S. VASSILEVSKI AND OWE AXELSSON

ABSTRACT. We study the classical interface domain decomposition method in the case of inexact subdomain solvers. We propose a two-level scheme based on approximate "harmonic" extension of the functions defined on the interface into the interior of the subdomains. Then by introducing a corresponding two-level hierarchical decomposition (or basis) the original problem is transformed to a more stable one, which allows inexact subdomain solvers. The additive (block-diagonal) version of the studied DD algorithms is illustrated with numerical examples.

1. INTRODUCTION

In this paper we study the classical interface domain decomposition (or DD) method in the case of inexact subdomain solvers. Such a situation was studied by several authors, e.g., Börgers [6], Haase, Langer and Meyer [13], Y. Vasilevskij [16], Axelsson and Vassilevski [4], Ewing and Vassilevski [12] and others. All of these results show good resulting condition number estimates if the subdomain solvers are accurate enough; in particular, they should be viewed as methods, which give sufficiently accurate approximate inverses to the subdomain stiffness matrices. In this paper we propose to first stabilize the original stiffness matrix and then apply the DD technique and the inexact subdomain solvers. By stabilization here we mean that we change the standard nodal basis by a special two-level hierarchical one in which basis the corresponding stiffness matrix is more stable with respect to perturbations due to replacing (approximating) some of its blocks that arise using DD ordering. The two-level hierarchical basis is obtained by choosing functions that form a basis on the interface and their energy norm in the whole domain (i.e., the norm defined by the given bilinear form) is bounded by a meshindependent constant. These functions we call approximate "harmonic". In the subdomains we keep the standard nodal basis functions (vanishing on the interface). Note that the approximate harmonic functions have large support inside the subdomains, hence they have essential overlap with the standard nodal basis functions in the subdomains and between the approximate harmonic functions themselves. This is a situation similar to the classical hierarchical basis (cf. Yserentant [18]). Hence, the corresponding stiffness matrix computed with respect to this two-level hierarchical basis is not as sparse as in the nodal basis case. It is a standard tool then to make a transformation of the coefficient vectors corresponding to the two-level hierarchical basis functions and to avoid the explicit computation of the two-level stiffness matrix. Because of the meshindependent bounded norm of the approximate harmonic functions, one can prove that the cosine of the abstract angle (denoted further by γ) between the space spanned by the approximate harmonic functions and its hierarchical complement space (i.e., the space spanned by the nodal basis functions vanishing on the interface) is bounded away from one, uniformly in the mesh size $h \to 0$. It depends only on the energy norm of the approximate harmonic extension operator (i.e., the operator that extends a given function on the interface to an approximate harmonic function in the interior of the subdomains). Applying then the classical results for two-level iterative preconditioners (cf. [1], see also [5]) one easily gets optimal convergence results for the DD method with inexact (but of optimal order) subdomain solvers.

1991 *Mathematics Subject Classification.* Primary 65N30, Secondary 65F10.

Key words and phrases. domain decomposition, two-level hierarchical method, approximate block-factorization, preconditioning, finite element elliptic problems.

This research of the first author was partially supported by the US National Science Foundation under grant NSF INT-9220287 and also in parts by Bulgarian Ministry for Science and Education under grant MM-91 # 75 and grant MM-91 # 78.

We can also generalize the results of the two-domain case to the multi domain case (to be presented elsewhere) by a recursive application of the two-domain decomposition preconditioner with inexact subdomain solvers. As a result we get an alternative convergence analysis of certain product (and non-overlapping) DD methods with inexact subdomain solvers. It is an alternative to the already existing ones, cf., e.g., Bramble, Pasciak, Wang, and Xu [7], or the survey papers of Xu [17] and Dryja, Smith and Widlund [11]. The results presented in the present paper can be viewed as an application of the convergence analysis technique for the algebraic multilevel methods as used in Vassilevski [14], Axelsson and Vassilevski [3], Vassilevski [15] (see also the recent survey of this topic by Axelsson and Neytcheva [2]), but now applied for the DD methods with inexact subdomain solvers. A major tool in the algebraic multilevel theory is a strengthened Cauchy–Schwarz inequality (similar to (8), see below) which turned out to be equivalent to an estimate of certain interpolation (extension or prolongation) operator (e.g., an estimate of the form (A) or (7), see below) and these are also the main tools used in the present analysis. In this way, besides presenting a new algorithm, we also provide a new alternative framework for analyzing certain DD algorithms.

The remainder of the paper is organized as follows. In Section 2 the basic formulation of the interface DD method and its transformation using the approximate harmonic extension is given. In Section 3 the condition number estimates are presented. Finally in Section 4 a numerical illustration of the studied additive (block–diagonal) two-domain DD methods with inexact subdomain solvers is presented.

2. Interface DD preconditioning and its stabilization

In this section we briefly recall the interface DD method and then introduce the two-level hierarchical transformation based on approximate harmonic functions.

Consider the model second order elliptic bilinear form

$$a(u,\phi) = \int_\Omega \sum_{i,j=1}^{2} a_{i,j}(x)\frac{\partial u}{\partial x_i}\frac{\partial \phi}{\partial x_j},$$

where the coefficient matrix $\{a_{i,j}(x)\}$ is assumed symmetric and positive definite uniformly in $x \in \Omega$ and the entries $a_{i,j}(x)$ are measurable and bounded functions in Ω. The given domain Ω is assumed polygonal. Given a finite element discretization space $V = V_h$ associated with a triangulation $\mathcal{T}_h$ one is interested in the efficient solution of the linear algebraic system arising from the discrete problem

$$a(u_h,\phi) = (f,\phi) \quad \text{for all } \phi \in V,$$

where $f \in L^2(\Omega)$ is given and $u_h \in V$ is the sought approximate solution. Let $\{\phi_i \in V\}_{i=1}^{n}$ be a given computational basis of V. Normally in the application $\{\phi_i\}$ is the standard nodal (or Lagrangian) basis of V associated with a set of nodes on the elements of $\mathcal{T}_h$. The above system, with respect to a given computational basis can be represented as a linear system of algebraic equations

$$Ax = \mathbf{b}. \tag{1}$$

In the interface domain decomposition method one chooses a separator (interface boundary) Γ that is a meshline, and partitions the set of unknowns in three groups with respect to the following partitioning of Ω,

$$\Omega = \Omega_1 \cup \Gamma \cup \Omega_2.$$

Ordering the unknowns corresponding to the interface Γ last we get the following block-form of (1),

$$\begin{bmatrix} A_2 & 0 & A_{20} \\ 0 & A_1 & A_{10} \\ A_{02} & A_{01} & A_0 \end{bmatrix} \begin{bmatrix} \mathbf{x}_2 \\ \mathbf{x}_1 \\ \mathbf{x}_0 \end{bmatrix} \begin{matrix} \}\ \Omega_2 \\ \}\ \Omega_1 \\ \}\ \Gamma \end{matrix} = \begin{bmatrix} \mathbf{b}_2 \\ \mathbf{b}_1 \\ \mathbf{b}_0 \end{bmatrix} \begin{matrix} \}\ \Omega_2 \\ \}\ \Omega_1 \\ \}\ \Gamma \end{matrix}. \tag{2}$$

In the interface DD methods the following block-factorization plays a major role,

$$A = \begin{bmatrix} A_2 & 0 & 0 \\ 0 & A_1 & 0 \\ A_{02} & A_{01} & S \end{bmatrix} \begin{bmatrix} I & 0 & A_2^{-1}A_{20} \\ 0 & I & A_1^{-1}A_{10} \\ 0 & 0 & I \end{bmatrix}.$$

First, replacing here the dense matrix

$$S = A_0 - A_{01}A_1^{-1}A_{10} - A_{02}A_2^{-1}A_{20}$$

(called DD Schur complement) by an easier to invert preconditioner Σ, one gets a DD preconditioner with exact subdomain solvers. Note, that to implement one action of S or to solve systems with A_1 and A_2 one must actually perform the subdomain solutions. Since these can be costly a natural step is to further replace A_1 and A_2 by spectrally equivalent preconditioners M_1 and M_2, respectively. However, as shown, e.g., in Börgers [6], this straightforward approach requires that M_1^{-1} and M_2^{-1} are not only spectrally equivalent to A_1^{-1} and A_2^{-1}, but sufficiently close to them. This can be achieved by performing sufficiently many (of order $\log h^{-1}$) inner iterations (with an optimal iterative method) in the subdomains and as a result the procedure can become too costly as well. In this paper we propose a way to deal with this difficulty.

We first develop the algorithm based on a change of the basis and then make the assumptions under which the resulting method will be analyzed in the following section.

Given a vector $\mathbf{v}_0$ on the interface we suppose that there are given two extension operators R_{10} and R_{20} that map $\mathbf{v}_0$ into the subdomains Ω_1 and Ω_2 respectively. Examples of such operators will be given later. Then we consider the following transformation of the vectors

$$\hat{\mathbf{v}} \equiv \begin{bmatrix} I & 0 & -R_{20} \\ 0 & I & -R_{10} \\ 0 & 0 & I \end{bmatrix} \mathbf{v}, \quad \text{where } \mathbf{v} = \begin{bmatrix} \mathbf{v}_2 \\ \mathbf{v}_1 \\ \mathbf{v}_0 \end{bmatrix},$$

and also the transformation of the original problem (1) to

$$J^T A J (J^{-1}\mathbf{x}) = J^T \mathbf{b}.$$

Here

$$J = \begin{bmatrix} I & 0 & R_{20} \\ 0 & I & R_{10} \\ 0 & 0 & I \end{bmatrix}. \tag{3}$$

The blocks of the transformed matrix $\hat{A} \equiv J^T A J$, denoted by $\hat{A}_{01}, \hat{A}_{02}$, etc. (similarly to the blocks of A in (2)), are readily computed (from the equality $\hat{A} = J^T A J$ using (3)) as follows

$$\begin{aligned} \hat{A}_1 &= A_1, \\ \hat{A}_2 &= A_2, \\ \hat{A}_{12} &= 0, \\ \hat{A}_{21} &= 0, \\ \hat{A}_{10} &= A_{10} + A_1 R_{10}, \\ \hat{A}_{20} &= A_{20} + A_2 R_{20}, \\ \hat{A}_{01} &= A_{01} + R_{10}^T A_1, \\ \hat{A}_{02} &= A_{02} + R_{20}^T A_2, \\ \hat{A}_0 &= A_0 + R_{10}^T A_{10} + R_{20}^T A_{20} + R_{10}^T A_1 R_{10} + R_{20}^T A_2 R_{20} \\ &\quad + A_{01} R_{10} + A_{02} R_{20}. \end{aligned} \tag{4}$$

Using (3) we also see that $\hat{A}$ admits the following block-factorization

$$\hat{A} = \begin{bmatrix} A_2 & 0 & 0 \\ 0 & A_1 & 0 \\ A_{02} + R_{20}^T A_2 & A_{01} + R_{10}^T A_1 & S \end{bmatrix} \begin{bmatrix} I & 0 & A_2^{-1}(A_{20} + A_2 R_{20}) \\ 0 & I & A_1^{-1}(A_{10} + A_1 R_{10}) \\ 0 & 0 & I \end{bmatrix}.$$

The important fact is that after the transformation of the matrix the resulting factorization still involves the blocks A_1, A_2 and the same Schur complement S (of the original matrix). The changes only occur in the offdiagonal blocks. Under some assumptions on R_{10} and R_{20} we will show in the next section that the above factorization is stable under perturbations, i.e., we will be able to replace A_1 and A_2 as well as S with any available spectrally equivalent preconditioners and get as a result a spectrally equivalent preconditioner to $\hat{A}$. Examples of preconditioners for A_1, A_2 and S will be given later.

The main assumption we need reads: there is a constant $\eta \geq 1$ such that for any $\mathbf{v}_0$ defined on Γ and any $\mathbf{v}_1$ and $\mathbf{v}_2$ supported in Ω_1 and Ω_2, respectively, the following stability estimate holds

$$\text{(A)} \qquad \begin{bmatrix} R_{20}\mathbf{v}_0 \\ R_{10}\mathbf{v}_0 \\ \mathbf{v}_0 \end{bmatrix}^T A \begin{bmatrix} R_{20}\mathbf{v}_0 \\ R_{10}\mathbf{v}_0 \\ \mathbf{v}_0 \end{bmatrix} \leq \eta \inf_{\mathbf{v}_2,\mathbf{v}_1} \begin{bmatrix} \mathbf{v}_2 \\ \mathbf{v}_1 \\ \mathbf{v}_0 \end{bmatrix}^T A \begin{bmatrix} \mathbf{v}_2 \\ \mathbf{v}_1 \\ \mathbf{v}_0 \end{bmatrix}.$$

Note that, as is well known, the infimum of the right-hand side for any $\mathbf{v}_2$, $\mathbf{v}_1$, equals $\eta \mathbf{v}_0^T S \mathbf{v}_0$. If we use the function-bilinear form notation the above assumption takes a more compact form,

$$a(v_0^H, v_0^H) \leq \eta \inf_{v,\; v|_\Gamma = v_0} a(v, v),$$

where v_0^H is the approximate harmonic extension in Ω of a given v_0 on Γ and v is any extension of v_0 in the interior of Ω_1 and Ω_2 with $v|_\Gamma = v_0$. In other words, v_0^H is a quasioptimal extension. The optimal extension corresponds to the case $\eta = 1$ and in that case we have to solve two subdomain problems with given Dirichlet data v_0 on Γ. In the latter case we have $R_{10} = -A_1^{-1}A_{10}$ and $R_{20} = -A_2^{-1}A_{20}$. As is readily seen, then $\hat{A}$ becomes block-diagonal. However, as already mentioned we have then to perform exact subdomain solvers (when computing the actions of S) which we wanted to avoid.

3. Convergence of the stabilized DD preconditioners with inexact subdomain solvers

In this section we analyze the condition number of $M^{-1}\hat{A}$, where M is defined as follows

$$(5) \qquad M = \begin{bmatrix} M_2 & 0 & 0 \\ 0 & M_1 & 0 \\ \hat{A}_{02} & \hat{A}_{01} & \Sigma \end{bmatrix} \begin{bmatrix} I & 0 & M_2^{-1}\hat{A}_{20} \\ 0 & I & M_1^{-1}\hat{A}_{10} \\ 0 & 0 & I \end{bmatrix}.$$

We also consider the block-diagonal preconditioner

$$(6) \qquad M_D = \begin{bmatrix} M_2 & 0 & 0 \\ 0 & M_1 & 0 \\ 0 & 0 & \Sigma \end{bmatrix}.$$

Here M_1, M_2 and Σ are given spectrally equivalent preconditioners to A_1, A_2 and S, respectively. Such are found, for M_1 and M_2 based on multigrid/multilevel methods (cf. e.g., [8], [3], [15]) and for Σ in the existing DD literature, cf., e.g., Dryja, Smith and Widlund [11] and many others.

Now we look at assumption (A). It can be rewritten in terms of the block entries of $\hat{A}$ as

$$(7) \qquad \mathbf{v}_0^T \hat{A}_0 \mathbf{v}_0 \leq \eta \mathbf{v}^T \hat{A} \mathbf{v}, \quad \text{where } \mathbf{v} = \begin{bmatrix} \mathbf{v}_2 \\ \mathbf{v}_1 \\ \mathbf{v}_0 \end{bmatrix},$$

for any $\mathbf{v}_0$, $\mathbf{v}_1$ and $\mathbf{v}_2$. Note that $\hat{A}_0$ is a principal submatrix of $\hat{A}$. This inequality implies the following strengthened Cauchy–Schwarz inequality (cf. Lemma 2 in Vassilevski [15]),

$$(8) \qquad \mathbf{v}_0^T \left[\hat{A}_{02}, \hat{A}_{01}\right] \begin{bmatrix} \mathbf{v}_2 \\ \mathbf{v}_1 \end{bmatrix} \leq \gamma \left[\mathbf{v}_0^T \hat{A}_0 \mathbf{v}_0\right]^{\frac{1}{2}} \left[\mathbf{v}_1^T A_1 \mathbf{v}_1 + \mathbf{v}_2^T A_2 \mathbf{v}_2\right]^{\frac{1}{2}},$$

where $\gamma = \sqrt{1 - \frac{1}{\eta}}$. Then the condition numbers of $M^{-1}\hat{A}$ and $M_D^{-1}\hat{A}$ are estimated, e.g., as in Axelsson and Gustafsson [1], since M (and M_D) is nothing but a two-level hierarchical preconditioner with inexact blocks for the transformed matrix $\hat{A}$. The condition number estimates depend only on γ (or equivalently on η) and on the constants in the spectral equivalence relations between M_1 and A_1, M_2 and A_2, and Σ and S. In particular, if these constants are meshindependent, the estimates of the condition number of $M^{-1}\hat{A}$ (and $M_D^{-1}\hat{A}$) will be meshindependent as well. More precisely we can state the following result.

Theorem 1. *Assume that the following spectral equivalence relations hold:*

$$\begin{array}{ccccc} \mathbf{v}_2^T A_2 \mathbf{v}_2 & \leq & \mathbf{v}_2^T M_2 \mathbf{v}_2 & \leq & (1+\delta_2)\mathbf{v}_2^T A_2 \mathbf{v}_2, \\ \mathbf{v}_1^T A_1 \mathbf{v}_1 & \leq & \mathbf{v}_1^T M_1 \mathbf{v}_1 & \leq & (1+\delta_1)\mathbf{v}_1^T A_1 \mathbf{v}_1, \\ \mathbf{v}_0^T \hat{A}_0 \mathbf{v}_0 & \leq & \mathbf{v}_0^T \Sigma \mathbf{v}_0 & \leq & (1+\delta_0)\mathbf{v}_0^T \hat{A}_0 \mathbf{v}_0, \end{array}$$

for some positive constants δ_2, δ_1 and δ_0. Then the following estimates hold

$$\mathbf{v}^T \hat{A} \mathbf{v} \leq \mathbf{v}^T M \mathbf{v} \leq \frac{1}{1-\gamma^2}\left\{\frac{1}{2}\left[\delta+\delta_0+\sqrt{(\delta-\delta_0)^2+4\delta\delta_0\gamma^2}\right]+1\right\}\mathbf{v}^T \hat{A}\mathbf{v} \quad \textit{for all } \mathbf{v},$$

and

$$(1-\gamma)\Delta_0 \mathbf{v}^T M_D \mathbf{v} \leq \mathbf{v}^T \hat{A} \mathbf{v} \leq (1+\gamma)\mathbf{v}^T M_D \mathbf{v} \quad \textit{for all } \mathbf{v}.$$

Here $\delta = \max\{\delta_1, \delta_2\}$, ($\gamma = \sqrt{1-\frac{1}{\eta}}$) and

$$\Delta_0 = \frac{2(1+\gamma)}{1+\delta_0}\left[1+\Delta+\sqrt{(\Delta-1)^2+4\Delta\gamma^2}\right]^{-1}, \quad \Delta = \frac{1+\delta}{1+\delta_0}.$$

Proof. A detailed proof can be found in Axelsson and Gustafsson [1]. □

Remark 1. Commonly in the DD literature the following estimate is established

$$c_1 \mathbf{v}_0^T S \mathbf{v}_0 \leq \mathbf{v}_0^T \Sigma \mathbf{v}_0 \leq c_2 \mathbf{v}_0^T S \mathbf{v}_0$$

for some general positive constants c_1 and c_2. We assumed in Theorem 1 that

$$\mathbf{v}_0^T \hat{A}_0 \mathbf{v}_0 \leq \mathbf{v}_0^T \Sigma \mathbf{v}_0 \leq (1+\delta_0)\mathbf{v}_0^T \hat{A}_0 \mathbf{v}_0 \tag{9}$$

holds instead. Noting however that (A) implies (see Vassilevski [15]) the spectral equivalence

$$(1-\gamma^2)\mathbf{v}_0^T \hat{A}_0 \mathbf{v}_0 \leq \mathbf{v}_0^T S \mathbf{v}_0 \leq \mathbf{v}_0^T \hat{A}_0 \mathbf{v}_0,$$

we can see that the above alternative assumption is not actually a restriction. Also, for practical use one can scale Σ appropriately (as in (9), important for multi-level generalization of M, (5), to the multidomain case). The scaling can be based on estimates obtained by the Lanczos method. Note that this is possible since the actions of $\hat{A}_0$ are available through the actions of A, R_{10}, R_{20}, R_{10}^T and R_{20}^T (see (4)).

4. Numerical illustration

In this section we show some numerical illustration of the two-domain interface DD preconditioners with inexact subdomain solvers. We consider a model L-shaped domain as shown in Figure 1, with interface boundary $\Gamma = \{(x,y) : x = 0,\ 0 < y < 1\}$. The bilinear form is defined by

$$a(u,\phi) = \int_\Omega c(x,y)\nabla u \cdot \nabla \phi,$$

where $c(x,y) = 1+x^2+y^2$. The approximate harmonic extension operators are defined by extending a given Dirichlet data on Γ in the interior of the domains $\Omega_1^{(1)} \subset \Omega_1 = (-2,0)\times(0,2)$ and $\Omega_2^{(1)} \subset \Omega_2 = (0,1)\times(0,1)$ by solving discretized Poisson equations with given Dirichlet data on Γ (and zero on the rest of the boundary). We varied in the test the size of the subdomains $\Omega_1^{(1)}$ and $\Omega_2^{(1)}$. The meshsize is $h = \frac{1}{m+1}$ and the number of interior nodes in Ω_1 is $n^2 = (2m+1)^2$ and in Ω_2 is m^2. On Γ the number of interior nodes is m. In the test we vary $m = 15, 31, 63$ and $n = 2m+1 = 31, 63, 127$ respectively. The domain $\Omega_1^{(1)}$ is a rectangle with sides of length $(n_1+1)h \times (m_1+1)h$ and $\Omega_2^{(1)}$ is also a rectangle with sides of length $(n_2+1)h \times (m+1)h$ for various given integers n_1, m_1 and n_2. We have $1 \leq n_1 \leq n$, $m \leq m_1 \leq n$ and $1 \leq n_2 \leq m$. In the subdomains we use preconditioners M_1 and M_2 for A_1 and A_2, respectively that are based on the approximate block-factorization method from Chan and Vassilevski [9]. These approximations can have guaranteed convergence properties. More precisely, the condition numbers of $M_1^{-1}A_1$ and of $M_2^{-1}A_2$ depend quadratically on a parameter (denoted further by m_1^f and m_2^f); namely, they are proportional to the quotients $\left(\frac{n+1}{m_1^f+1}\right)^2$ $(1 \leq m_1^f \leq n)$ and $\left(\frac{m+1}{m_2^f+1}\right)^2$ $(1 \leq m_2^f \leq m)$, respectively. In particular, the case $m_1^f = n$ and $m_2^f = m$ corresponds to exact subdomain solvers (or exact factorization of the subdomain stiffness matrices). The interface preconditioner Σ corresponds to the square root of the discrete negative second derivative along Γ as proposed in Dryja [10]. We show next in Tables 1–2, the convergence properties of the block–diagonal interface DD preconditioner M_D, (6), as introduced in Section 3, corresponding to the already specified method parameters R_{10}, R_{20}, Σ, and M_1 and M_2. Note that this is not as widely used preconditioner in the DD literature. In the test we show the sensitivity of the method with respect to the choice of the accurate subdomain preconditioners M_1 and M_2 (respectively, parameters m_1^f and m_2^f) and accurate harmonic extension operators R_{10} and R_{20}, (respectively, parameters n_1, m_1 and n_2). As one can expect, the results are monotonic with respect to these parameters; namely, by improving the quality of the subdomain preconditioners (or choosing m_1^f close to n and $m_2^f = \frac{1}{2}m_1^f$) the method requires fewer iterations. Also, by getting better harmonic prolongations, i.e., having n_1 and m_1 closer to n and also n_2 closer to m (which gives rise to a smaller constant η in the estimate (A) or (7)) the method improves in quality (or equivalently, in number of iterations). We used the preconditioned CG method for the transformed matrix $\hat{A}$ with stopping criterion of getting the norm of the preconditioned residual less than 10^{-9} and the initial iterate was $M_D^{-1}J^T$(r.h.s.), where r.h.s. was the right–hand side of the discrete problem corresponding to the original problem with $u = e^{-x}(x^2+y^2)$ as an exact solution. J is the transformation matrix defined by (3) in Section 2. In the Tables 1–2 we show the number of iterations, *iter* and the average reduction factor ρ defined by

$$\rho = \left(\frac{\|\mathbf{r}_{iter}\|_{M_D^{-1}}}{\|\mathbf{r}_0\|_{M_D^{-1}}}\right)^{\frac{1}{iter}},$$

where $\mathbf{r}_{iter}$ is the last residual and $\mathbf{r}_0$ is the initial one.

FIGURE 1. L-shaped domain Ω partitioned by interface $\Gamma = \{(0,y) : \; 0 < y < 1\}$; $\Omega_1^{(1)} \subset \Omega_1 = (-2,0) \times (0,2)$ and $\Omega_2^{(1)} \subset \Omega_2 = (0,1) \times (0,1)$ are used for harmonic estension of data given on Γ

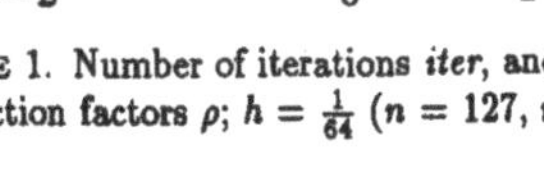

TABLE 1. Number of iterations *iter*, and average reduction factors ρ; $h = \frac{1}{64}$ $(n = 127, m = 63)$

	m_1=75 n_1=75 n_2=45	m_1=63 n_1=63 n_2=31	m_1=63 n_1=10 n_2=10	m_1=63 n_1=1 n_2=1
m_1^f=32 m_2^f=16	*iter*=31 ρ=0.440	*iter*=31 ρ=0.440	*iter*=32 ρ=0.450	*iter*=58 ρ=0.637
m_1^f=16 m_2^f=8	*iter*=50 ρ=0.593	*iter*=50 ρ=0.590	*iter*=49 ρ=0.588	*iter*=69 ρ=0.681

TABLE 2. Number of iterations *iter*, and average reduction factors ρ; $h = \frac{1}{32}$ ($n = 63$, $m = 31$)

	$m_1=45$ $n_1=45$ $n_2=24$	$m_1=31$ $n_1=31$ $n_2=15$	$m_1=31$ $n_1=5$ $n_2=5$	$m_1=31$ $n_1=1$ $n_2=1$
$m_1^f=32$ $m_2^f=16$	$iter=22$ $\rho=0.327$	$iter=22$ $\rho=0.336$	$iter=24$ $\rho=0.361$	$iter=33$ $\rho=0.489$
$m_1^f=16$ $m_2^f=8$	$iter=29$ $\rho=0.410$	$iter=28$ $\rho=0.409$	$iter=29$ $\rho=0.416$	$iter=39$ $\rho=0.510$
$m_1^f=8$ $m_2^f=4$	$iter=44$ $\rho=0.552$	$iter=44$ $\rho=0.551$	$iter=44$ $\rho=0.554$	$iter=50$ $\rho=0.600$

REFERENCES

1. O. Axelsson and I. Gustafsson, *Preconditioning and two-level multigrid methods of arbitrary degree of approximations*, Math. Comp. **40** (1983), 219–242.
2. O. Axelsson and Maya Neytcheva, *The algebraic multilevel iteration methods - theory and applications*, preprint (1993).
3. O. Axelsson and P.S. Vassilevski, *Algebraic multilevel preconditioning methods, II*, SIAM J. Numer. Anal. **27** (1990), 1569–1590.
4. O. Axelsson and P.S. Vassilevski, *Construction of variable-step preconditioners for inner–outer iterative methods*, Iterative Methods in Linear Algebra, Proceedings of the IMACS Int. Symposium. on iterative methods in linear algebra (R. Beauwens and P. de Groen, eds.), North Holland, Amsterdam, 1992, pp. 1–14.
5. R.E. Bank and T. Dupont, *Analysis of a two-level scheme for solving finite element equations*, Report CNA 159, Center for Numerical Analysis, The University of Texas at Austin (1980).
6. C. Börgers, *The Neumann–Dirichlet domain decomposition method with inexact solvers on the subdomains*, Numer. Math. **55** (1989), 123–136.
7. J.H. Bramble, J.E. Pasciak, J. Wang, and J. Xu, *Convergence estimates for product iterative methods with applications to domain decompositions*, Math. Comp. **57** (1991), 23–46.
8. J.H. Bramble, *Multigrid Methods*, Cornell Mathematics Department Lecture Notes (1992).
9. T.F. Chan and P.S. Vassilevski, *A framework for block-ILU factorization using block size reduction*, Math. Comp., to appear.
10. M. Dryja, *Capacitance matrix method for Dirichlet problem on polygon region*, Numer. Math. **39** (1982), 51–64.
11. M. Dryja, B.F. Smith and O.B. Widlund, *Schwarz analysis of iterative substructuring algorithms for elliptic problems in three dimensions*, SIAM J. Numer. Anal., submitted.
12. R.E. Ewing and P.S. Vassilevski, *Two-level iterative refinement preconditioners*, Domain decomposition methods for partial differential equations, V, SIAM, Philadelphia, 1992, pp. 262–270.
13. G. Haase, U. Langer and A. Meyer, *Domain decomposition preconditioners with inexact subdomain solvers*, J. Numer. Lin. Alg. Appl. **1** (1991), 27–41.
14. P.S. Vassilevski, *Nearly optimal iterative methods for solving finite element equations based on multilevel splitting of the matrix*, Preprint # 1989–09, EORI, University of Wyoming, Laramie, Wyoming, USA, 1989.
15. P.S. Vassilevski, *Hybrid V-cycle algebraic multilevel preconditioners*, Math. Comp. **58** (1992), 489–512.
16. Y. Vasilevskij, *Domain decomposition methods with inexact subregion solvers*, Numerical methods and mathematical software, Y.A. Kuznetzov, ed., Dept. Numer. Math., Russian Acad. Sciences, Moscow, 1990, pp. 23–29, (Russian).
17. J. Xu, *Iterative methods by space decomposition and subspace correction*, SIAM Review **34** (1992), 581–613.
18. H. Yserentant, *On the multilevel splitting of finite element spaces*, Numer. Math. **49** (1986), 379–412.

CENTER OF INFORMATICS AND COMPUTER TECHNOLOGY, BULGARIAN ACADEMY OF SCIENCES, "ACAD. G. BONTCHEV" STREET, BLOCK 25A, 1113 SOFIA, BULGARIA

E-mail address: panayot@bgearn.bitnet

FACULTY OF MATHEMATICS AND INFORMATICS, UNIVERSITY OF NIJMEGEN, TOERNOOIVELD, 6525 ED NIJMEGEN, THE NETHERLANDS

E-mail address: u641007@hnykun11.bitnet

MIXED FINITE ELEMENT METHODS OVER QUADRILATERALS *

Junping Wang Tarek Mathew

Abstract

We extend the rectangular mixed finite elements of Raviart and Thomas [14], Brezzi, Douglas, Fortin, and Marini [5], and Douglas and Wang [8] to arbitrary convex quadrilaterals. The stability and error estimates are established for the quadrilateral mixed finite element solutions.

1. Introduction

We are concerned with the approximation of the second-order elliptic equation:

$$-\nabla \cdot (a\nabla u) = f, \qquad \text{in } \Omega \tag{1.1}$$

with the Dirichlet boundary condition $u = -g$ on $\Gamma = \partial\Omega$. Here Ω is an open bounded domain in $\mathbf{R}^2$ and $a = (a_{ij}(x_1, x_2))$ is a symmetric matrix with entries $a_{ij} \in L^\infty(\Omega)$ that satisfies for some $\mu_i > 0$

$$\mu_1 \sum_{i=1}^{2} \eta_i^2 \le \sum_{i,j=1}^{2} a_{ij}\eta_i\eta_j \le \mu_2 \sum_{i=1}^{2} \eta_i^2, \quad \forall (x_1, x_2) \in \Omega, \quad \eta \in \mathbf{R}^2.$$

The mixed finite element method [14, 6] is employed to discretize (1.1) which provides a direct approximation for the flux variable

$$\mathbf{q} = -a\nabla u. \tag{1.2}$$

Of interest in this paper, we consider finite element partitions of Ω into convex quadrilaterals. Our objective is to extend the rectangular elements of Raviart and Thomas (RT) [14], Brezzi, Douglas, Fortin, and Marini ($BDFM$) [5], and Douglas and Wang (DW) [8] to arbitrary convex quadrilaterals.

Raviart and Thomas discussed some quadrilateral mixed finite elements only on parallelograms [14]. Arbogast, Wheeler, and Yotov [1] considered a mixed method

*This research was supported in part by NSF Grant No. EHR-910-8774 and INT-9309286.

which assumes that the physical domain can be mapped onto a rectangular computational domain by an isomorphism. To the author's knowledge, the mixed finite element method on arbitrary quadrilateral partition is an open problem. Needless to say, quadrilateral elements are important in theory and applications (especially for problems in $\mathbf{R}^3$) and deserve a careful study.

The paper is organized as follows. In Section 2, we review the standard mixed finite element method for (1.1) and outline a general framework for stability and convergence. In Section 3, we present some quadrilateral elements, while their stability is established in Section 4. Finally, we derive some error estimates for the quadrilateral approximations in Section 5.

2. Preliminaries

Let $m \geq 0$ be any integer and denote by

$$H^m(\Omega) = \{\phi;\ D^\alpha \phi \in L^2(\Omega), |\alpha| \leq m\}$$

the usual Sobolev space with norm

$$\|\phi\|_{m,\Omega} = \left(\sum_{|\alpha|\leq m} \int_\Omega |D^\alpha \phi|^2 dx \right)^{1/2}.$$

To describe a mixed variational form for (1.1), set

$$H(div;\Omega) = \{\mathbf{v};\ \mathbf{v} \in [L^2(\Omega)]^2, \nabla\cdot\mathbf{v} \in L^2(\Omega)\}$$

provided with the norm

$$\|\mathbf{v}\|_H = \left(\|\mathbf{v}\|_{0,\Omega}^2 + \|\nabla\cdot\mathbf{v}\|_{0,\Omega}^2\right)^{1/2}.$$

A mixed variational formulation for (1.1) seeks $(\mathbf{q}, u) \in H(div;\Omega) \times L^2(\Omega)$ such that

$$(2.1)\qquad \begin{aligned} (\tau\mathbf{q},\mathbf{v}) - (\nabla\cdot\mathbf{v}, u) &= \langle g, \mathbf{v}\cdot\nu\rangle, && \mathbf{v} \in H(div;\Omega),\\ (\nabla\cdot\mathbf{q}, w) &= (f, w), && w \in L^2(\Omega), \end{aligned}$$

where $\tau(x) = a(x_1,x_2)^{-1}$ is the reciprocal of the coefficient function $a(x_1,x_2)$. Here $(\cdot,\cdot)$ denotes the L^2-inner product in either $L^2(\Omega)$ or $[L^2(\Omega)]^2$ as appropriate, and $\langle\cdot,\cdot\rangle$ denotes the inner product on $L^2(\Gamma)$ with ν as the outward normal direction.

To approximate (2.1), let V_h and W_h be finite dimensional subspaces of $H(div;\Omega)$ and $L^2(\Omega)$, respectively. An approximate solution of $(\mathbf{q},u)$ is given by $(\mathbf{q}_h, u_h) \in V_h \times W_h$ satisfying

$$(2.2)\qquad \begin{aligned} (\tau\mathbf{q}_h,\mathbf{v}) - (\nabla\cdot\mathbf{v}, u_h) &= g(\mathbf{v}), && \mathbf{v} \in V_h,\\ (\nabla\cdot\mathbf{q}_h, w) &= f(w), && w \in W_h, \end{aligned}$$

where $g \in (H(div;\Omega))'$ and $f \in L^2(\Omega)$ are given by $g(\mathbf{v}) = \langle g, \mathbf{v}\cdot\nu\rangle$ and $f(w) = (f, w)$.

The mixed finite element method is a method of type (2.2) where V_h and W_h are chosen as finite element spaces associated with a prescribed partition $\mathcal{T}_h$ with mesh parameter h. The finite element spaces V_h and W_h should be constructed to satisfy the *inf-sup* condition of Brezzi [3] and Babuška [2]. More precisely, there should be a constant $\beta^* > 0$, independent of the mesh parameter h such that

$$\inf_{w\in W_h}\sup_{\mathbf{v}\in V_h}\frac{(\nabla\cdot\mathbf{v}, w)}{\|\mathbf{v}\|_H\|w\|_{0,\Omega}} \geq \beta^*. \tag{2.3}$$

Set

$$Z_h = \{\mathbf{v}\in V_h;\ (\nabla\cdot\mathbf{v}, w) = 0,\ \forall w\in W_h\}.$$

One has the following stability result.

Theorem 2.1. *Let V_h and W_h be closed subspaces of $H(div;\Omega)$ and $L^2(\Omega)$, respectively. In addition to (2.3), assume that there exists a constant $\alpha_0 > 0$ such that*

$$(\tau\mathbf{v}, \mathbf{v}) \geq \alpha_0\|\mathbf{v}\|_H^2, \qquad \mathbf{v}\in Z_h. \tag{2.4}$$

Then for any $(g, f)\in V_h'\times W_h'$ there exists a unique solution $(\mathbf{q}_h, u_h)$ satisfying (2.2). Moreover, there exists a constant $C = C(\mu_1, \mu_2, \alpha_0, \beta^*)$ such that

$$\|\mathbf{q}_h\|_H + \|u_h\|_{0,\Omega} \leq C\left(\sup_{w\in W_h}\frac{(f,w)}{\|w\|_{0,\Omega}} + \sup_{\mathbf{v}\in V_h}\frac{g(\mathbf{v})}{\|\mathbf{v}\|_H}\right). \tag{2.5}$$

Let $\mathcal{T}_h$ be a finite element partition of Ω into ordinary rectangles. There are several finite elements associated with $\mathcal{T}_h$ that are stable. For any element K and non-negative integers i and j, denote by $Q_{i,j}(K)$ the space of polynomials of degree no more than i and j in x_1 and x_2, respectively. Also let $P_j(K)$ be the space of polynomials on K of total degree no more than j. Raviart and Thomas [14] introduced a family of elements by setting for each $K\in\mathcal{T}_h$ and $j\geq 0$

$$RT_j(K) = V(j,K)\times W(j,K),$$

where

$$V(j,K) = Q_{j+1,j}(K)\times Q_{j,j+1}(K), \quad W(j,K) = Q_{j,j}(K).$$

The corresponding finite element spaces are defined by

$$V_h = V(j,\mathcal{T}_h) = \{\mathbf{v}\in H(div;\Omega); \mathbf{v}|_K\in V(j,K),\quad \forall K\in\mathcal{T}_h\}, \tag{2.6}$$

$$W_h = W(j,\mathcal{T}_h) = \{w\in L^2(\Omega); w|_K\in W(j,K),\quad \forall K\in\mathcal{T}_h\}. \tag{2.7}$$

The family of mixed elements introduced by Brezzi, Douglas, Fortin, and Marini [5] is defined on K by

$$V(j,K) = P_{j+1}(K)\backslash\{y^{j+1}\}\times P_{j+1}(K)\backslash\{x^{j+1}\}, \quad W(j,K) = P_j(K).$$

The elements discovered by Douglas and Wang [8] is given by

$$V(j,K) = \left[P_j(K) \oplus \mathrm{span}\{x^{j+1}, xy^j\}\right] \times \left[P_j(K) \oplus \mathrm{span}\{y^{j+1}, yx^j\}\right]$$

and

$$W(j,K) = P_{j-1}(K) \oplus \mathrm{span}\{x^j, y^j\}.$$

Notice that the mixed finite element spaces corresponding to them should be defined by (2.6) and (2.7).

The condition (2.3) is usually verified by using a method proposed by Fortin [12]. The method assumes the existence of an operator $\Pi_h : [H^1(\Omega)]^2 \to V_h$ such that for any $\mathbf{q} \in [H^1(\Omega)]^2$,

$$(\nabla \cdot (\mathbf{q} - \Pi_h \mathbf{q}), w) = 0, \qquad \forall w \in W_h \tag{2.8}$$

and

$$\|\Pi_h \mathbf{q}\|_H \le C \|\mathbf{q}\|_{1,\Omega}, \tag{2.9}$$

where C is a constant independent of $\mathbf{q}$ and the mesh parameter h. For details, we refer to [12, 6].

3. Quadrilateral mixed finite elements

Let $\hat{K}$ denote the unit square $[0,1] \times [0,1]$ in the $\hat{x} = (\hat{x}_1, \hat{x}_2)$ reference space, with vertices denoted by $\hat{p}_j, 1 \le j \le 4$ (see Figure 1). With the notations of Figure 1, for each convex quadrilateral K with vertices p_j, there exists exactly one invertible mapping $\hat{F}_K \in [Q_{1,1}(\hat{K})]^2$ that maps $\hat{K}$ onto K and is such that

$$\hat{F}_K(\hat{p}_j) = p_j, \qquad 1 \le j \le 4.$$

For each $\hat{x}$, let $B = B(\hat{x}) = \{\partial \hat{F}_i / \partial \hat{x}_j\}_{i,j}$ denote the derivative of $\hat{F}_K$ at the point $\hat{x}$.

For any $\hat{\phi}$ defined on $\hat{K}$, set $\phi = \hat{\phi} \circ F_K^{-1}$ and by the chain rule

$$\hat{\nabla}\hat{\phi} = B^T \nabla \phi, \tag{3.1}$$

where $\hat{\nabla}$ and ∇ are the gradient operators with respect to the variables $\hat{x}$ and x, respectively. If $\hat{\phi} = \hat{x}_1$ then the vector-valued function $\hat{\nabla}\hat{\phi}$ coincides with the outward normal direction of $\hat{K}$ on the edge $\hat{e}_2$. Moreover, the vector-valued function $\nabla\phi$ is proportional to the normal direction of K on e_2. Thus the relation (3.1) indicates that the matrix B^{-T} maps the outward normal direction of $\hat{K}$ to a direction that is outward normal to K on its boundary. Let $J_K = \det(B)$ be the Jacobian of the bilinear transformation $\hat{F}_K$. Set

$$M = J_K^{-1} B \tag{3.2}$$

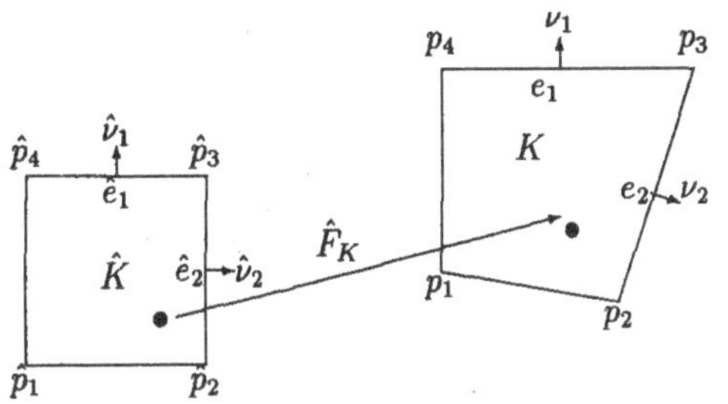

Figure 1. Quadrilateral K and the reference element $\hat{K}$

Lemma 3.1. Let the linear transformation M be defined by (3.2). With the notation of Figure 1, the linear transformation M^{-T} maps $\hat{\nu}_i$ to $|e_i|\nu_i$ for $1 \le i \le 4$.

Proof. Let $A = \{a_{ij}\}_{2\times 2}$ be a matrix whose first and second columns are vectors $p_2 - p_1$ and $p_4 - p_1$, respectively. Set $\mathbf{g} = p_1 - p_2 + p_3 - p_4$ and $\mathbf{p} = p_1$, It is not hard to see that

$$x = \hat{F}_K(\hat{x}) \equiv A\hat{x} + \mathbf{g}\hat{x}_1\hat{x}_2 + \mathbf{p} \tag{3.3}$$

and

$$B = A + R, \qquad R = \begin{bmatrix} g_1\hat{x}_2 & g_1\hat{x}_1 \\ g_2\hat{x}_2 & g_2\hat{x}_1 \end{bmatrix}.$$

Furthermore, a simple computation yields

$$M^{-1} = \begin{bmatrix} a_{22} & -a_{12} \\ -a_{21} & a_{11} \end{bmatrix} + \begin{bmatrix} g_2\hat{x}_1 & -g_1\hat{x}_1 \\ -g_2\hat{x}_2 & g_1\hat{x}_2 \end{bmatrix}. \tag{3.4}$$

The lemma is then verified by (3.4) and some trivial calculation.

Let $\hat{V}_j(\hat{K}) \times \hat{W}_j(\hat{K})$ be any family of rectangular elements of index j that is stable. Let $\mathcal{T}_h$ be a finite element partition of Ω into quadrilaterals. For each element K of $\mathcal{T}_h$, we introduce the finite dimensional spaces:

$$V_j(K) = \{\mathbf{q} = M\tilde{\mathbf{q}} \circ \hat{F}_K^{-1};\ \tilde{\mathbf{q}} \in \hat{V}_j(\hat{K})\}$$
$$W_j(K) = \{w = \hat{w} \circ \hat{F}_K^{-1};\ \hat{w} \in \hat{W}_j(\hat{K})\}.$$

Observe that $V_j(K) \subset [C^\infty(K)]^2$ and $W_j(K) \subset C^\infty(K)$ are no longer spaces of polynomials on K (unless K is a parallelogram). Our quadrilateral mixed finite elements are defined as follows:

$$V_h = \{\mathbf{v} \in H(div;\Omega);\quad \mathbf{v}|_K \in V_j(K),\quad \forall K \in \mathcal{T}_h\}$$
$$W_h = \{w \in L^2(\Omega);\quad w|_K \in W_j(K),\quad K \in \mathcal{T}_h\}.$$

The corresponding mixed finite element problem seeks $(\mathbf{q}_h, u_h) \in V_h \times W_h$ such that (2.2) is satisfied.

4. Stability

In this section we verify the conditions of Theorem 2.1 for the quadrilateral elements. The result can be stated as follows.

Theorem 4.1. *The inequalities (2.3) and (2.4) are satisfied for the quadrilateral mixed finite elements introduced in Section 3. Thus, the quadrilateral mixed method is stable.*

Proof. The argument relies on some results to be derived in this section. By assuming them at this moment, we have from Lemma 4.4 that (2.4) is trivially satisfied. Next, the projection operator Π_h satisfying (2.8) is defined by (4.8). The equation (2.8) is verified by Lemma 4.3 and the estimate (2.9) is concluded by the local nature of Π_h (see Section 5 for details in the estimate). It follows that (2.3) holds true.

In the rest of this section, we introduce a locally-defined operator Π_h and establish some properties for the quadrilateral elements. For simplicity, we consider the Raviart-Thomas element of index j. For each quadrilateral K of $\mathcal{T}_h$, let $\mathbf{v}$ be a vector-valued function on K. Introduce the following notation:

$$\hat{\mathbf{v}} = \mathbf{v} \circ \hat{F}_K \qquad \tilde{\mathbf{v}} = M^{-1}\hat{\mathbf{v}}.$$

Lemma 4.1. *Let K be a convex quadrilateral of $\mathcal{T}_h$ and $\mathbf{v} \in [H^1(K)]^2$ be any vector-valued function. Then*

$$\int_K \nabla \cdot \mathbf{v}\ w dx = \int_{\hat{K}} \hat{\nabla} \cdot \tilde{\mathbf{v}}\ \hat{w} d\hat{x}, \qquad \forall w \in L^2(K). \tag{4.1}$$

Proof. Integrating by parts yields

$$\int_K \nabla \cdot \mathbf{v}\ w dx = -\int_K \mathbf{v} \cdot \nabla w dx + \int_{\partial K} \mathbf{v} \cdot \nu w ds. \tag{4.2}$$

To deal with the boundary integral, we transform ∂K to $\partial \hat{K}$ by using the transformation $\hat{F}_K$, yielding

$$\int_{e_i} \mathbf{v} \cdot \nu_i w ds = |e_i| \int_{\hat{e}_i} \hat{\mathbf{v}} \cdot \nu_i \hat{w} d\ hats. \tag{4.3}$$

Notice that $M^{-T}\hat{\nu}_i = |e_i|\nu_i$. Thus,

$$|e_i|\hat{\mathbf{v}} \cdot \nu_i = M^{-1}\hat{\mathbf{v}} \cdot \hat{\nu}_i = \tilde{\mathbf{v}} \cdot \hat{\nu}_i.$$

Substituting the above into (4.3) we obtain

$$\int_{e_i} \mathbf{v} \cdot \nu_i w ds = \int_{\hat{e}_i} \tilde{\mathbf{v}} \cdot \hat{\nu}_i \hat{w} d\hat{s}. \tag{4.4}$$

The area integral in (4.2) can be handled as follows. First notice that by (3.1)

$$\nabla w = B^{-T}\hat{\nabla}\hat{w}.$$

Using the transformation $\hat{F}_K$ we obtain

$$\int_K \mathbf{v}\cdot\nabla w dx = \int_{\hat{K}} \hat{\mathbf{v}}\cdot B^{-T}\hat{\nabla}\hat{w}J_K d\hat{x} = \int_{\hat{K}} M^{-1}\hat{\mathbf{v}}\cdot\hat{\nabla}\hat{w}d\hat{x} = \int_{\hat{K}} \tilde{\mathbf{v}}\cdot\hat{\nabla}\hat{w}d\hat{x}. \tag{4.5}$$

Now substituting (4.4) and (4.5) into (4.2) and using the Green's identity we obtain (4.1).

For any $\mathbf{v}\in[H^1(K)]^2$ define $\Pi_K\mathbf{v}\in V_j(K)$ by setting

$$\Pi_K\mathbf{v} = M\hat{\Pi}_{\hat{K}}\tilde{\mathbf{v}}, \tag{4.6}$$

where $\hat{\Pi}_{\hat{K}}$ is the standard projection operator defined on the reference element $\hat{K}$ by the following linear equations:

$$\begin{aligned} \int_{\hat{e}_i} \hat{\Pi}_{\hat{K}}\tilde{\mathbf{v}}\cdot\hat{\nu}_i\hat{p}d\hat{s} &= \int_{\hat{e}_i} \tilde{\mathbf{v}}\cdot\hat{\nu}_i\hat{p}d\hat{s}, \qquad \forall\hat{p}\in P_j(\hat{e}_i) \\ \int_{\hat{K}} \hat{\Pi}_{\hat{K}}\tilde{\mathbf{v}}\cdot\mathbf{z}d\hat{x} &= \int_{\hat{K}} \tilde{\mathbf{v}}\cdot\mathbf{z}d\hat{x}, \qquad \forall\mathbf{z}\in Q_{j-1,j}(\hat{K})\times Q_{j,j-1}(\hat{K}). \end{aligned} \tag{4.7}$$

Now for any vector-valued function $\mathbf{v}\in[H^1(\Omega)]^2$, one can define $\Pi_h\mathbf{v}$ by setting

$$\Pi_h\mathbf{v} = \Pi_K\mathbf{v}, \qquad \text{on } K. \tag{4.8}$$

The following lemma shows that Π_h is a linear operator from $[H^1(\Omega)]^2$ to V_h.

Lemma 4.2. *The function $\Pi_h\mathbf{v}$ defined by (4.8) belongs to the quadrilateral space V_h.*

Proof. It suffices to show that $\Pi_h\mathbf{v}$ is continuous across each of the internal edges in the normal direction. For any element $K\in\mathcal{T}_h$, the normal component $\Pi_K\mathbf{v}\cdot\nu_i$ is given by

$$\Pi_K\mathbf{v}\cdot\nu_i = M\hat{\Pi}_{\hat{K}}\tilde{\mathbf{v}}\cdot\nu_i = \hat{\Pi}_{\hat{K}}\tilde{\mathbf{v}}\cdot\hat{\nu}_i/|e_i|. \tag{4.9}$$

The right-hand side of (4.9) is determined by the following set of linear equations:

$$\frac{1}{|e_i|}\int_{\hat{e}_i}\hat{\Pi}_{\hat{K}}\tilde{\mathbf{v}}\cdot\hat{\nu}_i\hat{p}ds = \frac{1}{|e_i|}\int_{\hat{e}_i}\tilde{\mathbf{v}}\cdot\hat{\nu}_i\hat{p}d\hat{s} = \int_{\hat{e}_i}\hat{\mathbf{v}}\cdot\nu_i\hat{p}d\hat{s}, \quad \forall\hat{p}\in P_j(\hat{e}_i).$$

Thus, the polynomial $\hat{\Pi}_{\hat{K}}\tilde{\mathbf{v}}\cdot\hat{\nu}_i/|e_i|$ is uniquely determined by the function $\mathbf{v}\cdot\nu_i$ on the edge e_i. It follows that the vector-valued function $\Pi_h\mathbf{v}$ is continuous across each of the internal edges in the normal direction.

Lemma 4.3. *Let $\Pi_h\mathbf{v}$ be given by (4.8). Then we have*

$$(\nabla\cdot(\mathbf{v}-\Pi_h\mathbf{v}), w) = 0, \qquad \forall w\in W_h. \tag{4.10}$$

Proof. Since

$$(\nabla\cdot(\mathbf{v}-\Pi_h\mathbf{v}),w)=\sum_{K\in\mathcal{T}_h}\int_K\nabla\cdot(\mathbf{v}-\Pi_h\mathbf{v})w\,dx$$

and by (4.1) and (4.7)

$$\int_K\nabla\cdot(\mathbf{v}-\Pi_h\mathbf{v})w\,dx=\int_{\hat K}\hat\nabla\cdot(\tilde{\mathbf{v}}-\hat\Pi_K\tilde{\mathbf{v}})\hat w\,d\hat x=0,$$

it follows that (4.10) is valid.

Next we show that the subspace Z_h contains only divergence-free vector-valued functions in V_h.

Lemma 4.4. *Let* $\mathbf{v}\in V_h$ *be such that*

$$(\nabla\cdot\mathbf{v},w)=0,\qquad\forall w\in W_h. \tag{4.11}$$

Then, $\nabla\cdot\mathbf{v}=0$ *in* Ω.

Proof. Let K be a quadrilateral of $\mathcal{T}_h$. By (4.1) we have

$$\int_K\nabla\cdot\mathbf{v}w\,dx=\int_{\hat K}\hat\nabla\cdot\tilde{\mathbf{v}}\hat w\,d\hat x,\qquad\forall w\in L^2(K). \tag{4.12}$$

Thus, by choosing $\hat w=\hat\nabla\cdot\tilde{\mathbf{v}}$ and then using (4.11) we obtain

$$\hat\nabla\cdot\tilde{\mathbf{v}}=0.$$

The lemma is then verified by taking $w=\nabla\cdot\mathbf{v}$ in (4.12).

5. Error estimates

Let $\hat Q_j$ be the standard L^2 projection operator onto the reference pressure space $\hat W_j(\hat K)$. For any $w\in L^2(\Omega)$ set

$$Q_hw=(\hat Q_j\hat w)\circ\hat F_K^{-1},\qquad\text{on }K.$$

The stability of the quadrilateral elements implies the following error estimate for the mixed method [13, 11, 7, 6]

$$\|\mathbf{q}-\mathbf{q}_h\|_H+\|u-u_h\|_{0,\Omega}\le C\left(\inf_{\mathbf{v}\in V_h}\|\mathbf{q}-\mathbf{v}\|_H+\inf_{\chi\in W_h}\|u-\chi\|_{0,\Omega}\right). \tag{5.1}$$

Furthermore, using the operator Π_h and Q_h it is not hard to see that

$$\|\mathbf{q}_h-\Pi_h\mathbf{q}\|_H+\|u_h-Q_hu\|_{0,\Omega}\le C\|\mathbf{q}-\Pi_h\mathbf{q}\|_{0,\Omega} \tag{5.2}$$

for some constant C. Thus, it suffices to estimate the right-hand side of (5.1) and (5.2) in order to derive error estimates for the quadrilateral mixed finite element solution.

For each integer $s \geq 0$ and $i = 1, 2$ we set

$$[\phi]_{s,\Omega,x_i} = \|\partial^s \phi / \partial x_i^s\|_{0,\Omega}.$$

For simplicity, we consider only the Raviart-Thomas element in this section.

Lemma 5.1. *Let* $\mathbf{q} \in [H^1(\Omega)]^2$ *be a vector-valued function and* Π_h *be given by (4.8). Then there exists a constant* C *such that*

$$\|\mathbf{q} - \Pi_h \mathbf{q}\|_{0,\Omega} \leq C h^{j+1} \|\mathbf{q}\|_{j+1,\Omega}, \tag{5.3}$$

provided that $\mathbf{q} \in [H^{j+1}(\Omega)]^2$. *Moreover, if* $\mathbf{q} \in [H^{j+2}(\Omega)]^2$, *then we have*

$$\|\nabla \cdot (\mathbf{q} - \Pi_h \mathbf{q})\|_{0,\Omega} \leq C h^{j+1} \|\mathbf{q}\|_{j+2,\Omega}. \tag{5.4}$$

Proof. Let K be any quadrilateral element of $\mathcal{T}_h$. Since on K, $\mathbf{q} - \Pi_h \mathbf{q} = \mathbf{q} - M \hat{\Pi}_{\hat{K}} \tilde{\mathbf{q}}$, then we have by passing to $\hat{K}$

$$\int_K |\mathbf{q} - \Pi_h \mathbf{q}|^2 dx \leq C \int_{\hat{K}} |\tilde{\mathbf{q}} - \hat{\Pi}_{\hat{K}} \tilde{\mathbf{q}}|^2 d\hat{x}.$$

If $\tilde{\mathbf{q}} = (\tilde{q}_1, \tilde{q}_2)^T$, then by the construction of $\hat{V}_j(\hat{K})$ we obtain

$$\int_{\hat{K}} |\tilde{\mathbf{q}} - \hat{\Pi}_{\hat{K}} \tilde{\mathbf{q}}|^2 d\hat{x} \leq C \left([\tilde{q}_1]^2_{j+1,\hat{K},\hat{x}_2} + [\tilde{q}_2]^2_{j+1,\hat{K},\hat{x}_1} \right)$$

for some constant C. Observe that $\tilde{\mathbf{q}} = M^{-1} \hat{\mathbf{q}}$ and M^{-1} is a 2×2 matrix with linear entries in the reference variable $\hat{x}$, and the first row of M^{-1} involves only $\hat{x}_1$ and the second row involves only $\hat{x}_2$. Thus,

$$\int_{\hat{K}} |\tilde{\mathbf{q}} - \hat{\Pi}_{\hat{K}} \tilde{\mathbf{q}}|^2 d\hat{x} \leq C h^2 \left([\hat{q}_1]^2_{j+1,\hat{K},\hat{x}_2} + [\hat{q}_2]^2_{j+1,\hat{K},\hat{x}_1} \right). \tag{5.5}$$

By transforming $\hat{K}$ back to K with $\hat{F}_K$,

$$\int_K |\mathbf{q} - \Pi_h \mathbf{q}|^2 dx \leq C h^{2(j+1)} \|\mathbf{q}\|^2_{j+1,K},$$

which verifies (5.3).

To derive (5.4), by setting $w = \nabla \cdot (\mathbf{q} - \Pi_h \mathbf{q})$ and then using Lemma 4.1

$$\begin{aligned} \int_K |\nabla \cdot (\mathbf{q} - \Pi_h \mathbf{q})|^2 dx &= \int_K \nabla \cdot (\mathbf{q} - \Pi_h \mathbf{q}) w dx \\ &= \int_{\hat{K}} \hat{\nabla} \cdot (\tilde{\mathbf{q}} - \hat{\Pi}_{\hat{K}} \tilde{\mathbf{q}}) \hat{w} d\hat{x} \leq \|\hat{\nabla} \cdot (\tilde{\mathbf{q}} - \hat{\Pi}_{\hat{K}} \tilde{\mathbf{q}})\|_{0,\hat{K}} \|\hat{w}\|_{0,\hat{K}}. \end{aligned} \tag{5.6}$$

Substituting the obvious inequality $\|\hat{w}\|_{0,\hat{K}} \le Ch^{-1}\|w\|_{0,K}$ to (5.6) and then eliminating $\|w\|_{0,K}$ we obtain

$$\begin{aligned}\int_K |\nabla\cdot(\mathbf{q}-\Pi_h\mathbf{q})|^2 dx &\le Ch^{-2}\int_{\hat{K}} |\hat{\nabla}\cdot(\tilde{\mathbf{q}}-\hat{\Pi}_{\hat{K}}\tilde{\mathbf{q}})|^2 d\hat{x}\\ &\le Ch^{-2}\int_{\hat{K}} |(I-\hat{Q}_j)\hat{\nabla}\cdot\tilde{\mathbf{q}}|^2 d\hat{x}.\end{aligned}$$

It follows that

$$\int_K |\nabla\cdot(\mathbf{q}-\Pi_h\mathbf{q})|^2 dx \le Ch^{-2}\left([\hat{\nabla}\cdot\tilde{\mathbf{q}}]^2_{j+1,\hat{K},\hat{x}_1} + [\hat{\nabla}\cdot\tilde{\mathbf{q}}]^2_{j+1,\hat{K},\hat{x}_2}\right). \tag{5.7}$$

Using the representation (3.4) for the matrix M^{-1} we arrive at

$$[\hat{\nabla}\cdot\tilde{\mathbf{q}}]^2_{j+1,\hat{K},\hat{x}_i} \le Ch^2\left([\hat{\mathbf{q}}]^2_{j+1,\hat{K},\hat{x}_2} + [\hat{\mathbf{q}}]^2_{j+1,\hat{K},\hat{x}_1} + |\hat{\mathbf{q}}|^2_{j+2,\hat{K}}\right), \quad i=1,2,$$

where $|\hat{\mathbf{q}}|_{j+2,\hat{K}}$ denotes the full seminorm of $\mathbf{q}$ of order $j+2$. Substituting the above estimate into (5.7) and then transforming $\hat{K}$ back to K we obtain

$$\int_K |\nabla\cdot(\mathbf{q}-\Pi_h\mathbf{q})|^2 dx \le Ch^{2(j+1)}\|\mathbf{q}\|^2_{j+2,K},$$

which completes the proof of the lemma.

Combining the estimates (5.3) and (5.4) with (5.1) and (5.2) we obtain the following result.

Theorem 5.1. *Let the true flux* $\mathbf{q}$ *of (1.2) be in* $[H^{j+1}(\Omega)]^2$ *for* $j \ge 0$ *and* $(\mathbf{q}_h, u_h)$ *be the quadrilateral mixed finite element solution of* $(\mathbf{q}, u)$ *obtained by solving (2.2). Then there exists a constant* C *such that*

$$\|\mathbf{q}-\mathbf{q}_h\|_{0,\Omega} + \|u_h - Q_h u\|_{0,\Omega} \le Ch^{j+1}\|\mathbf{q}\|_{j+1,\Omega}. \tag{5.8}$$

Moreover, if $\mathbf{q} \in [H^{j+2}(\Omega)]^2$ *then we have*

$$\|\mathbf{q}-\mathbf{q}_h\|_H + \|u - u_h\|_{0,\Omega} \le Ch^{j+1}\|\mathbf{q}\|_{j+2,\Omega}. \tag{5.9}$$

Error estimates in negative norms can be derived by using a standard duality argument and some regularity assumptions for the augmented problems. We refer to [7] for a treatment of the standard mixed method. Finally, flux superconvergence [9, 10] can be derived if each quadrilateral of $\mathcal{T}_h$ is almost a parallelogram (i.e., the vector $\mathbf{g}$ in (3.3) is of size $\mathcal{O}(h^2)$).

Remark 5.1. For the $BDFM$ element the estimate (5.8) holds true for the corresponding quadrilateral mixed finite element approximation $(\mathbf{q}_h, u_h)$. But the estimate (5.9) has to be replaced by

$$\|\mathbf{q}-\mathbf{q}_h\|_H + \|u - u_h\|_{0,\Omega} \le Ch^{j+\sigma_j}\|\mathbf{q}\|_{j+1+\sigma_j,\Omega},$$

where $\sigma_j = 1$ for $j = 0$ and $\sigma_j = 0$ for $j \geq 1$. The lack of accuracy in the above estimate is caused by the pressure space which consists of polynomials of *total* degree no more than j. Similarly, the approximate solution by using the DW element has the following error estimates:

$$\|\mathbf{q} - \mathbf{q}_h\|_{0,\Omega} + \|u_h - Q_h u\|_{0,\Omega} \leq Ch^{j+\sigma_j}\|\mathbf{q}\|_{j+\sigma_j,\Omega},$$

where $\sigma_j = 1$ for $j = 0, 1$ and $\sigma_j = 0$ for $j \geq 2$.

References

[1] T. Arbogast, M. F. Wheeler, and I. Yotov, *Logically rectangular mixed methods for groundwater flow and transport on general geometry*, Tech. Report, TR94-03(1994), Dept. Compt. Appl. Math., Rice University.

[2] I. Babuška, *The finite element method with Lagrangian multipliers*, Numer. Math., **20**(1973), 179–192.

[3] F. Brezzi, *On the existence, uniqueness, and approximation of saddle point problems arising from Lagrangian multipliers*, R.A.I.R.O., **8**(1974), 129–151.

[4] F. Brezzi, J. Douglas, Jr., R. Duran, and M. Fortin, *Mixed finite elements for second order elliptic problems in three variables*, Numer. Math., **51**(1987), 237–250.

[5] F. Brezzi, J. Douglas, Jr., M. Fortin, and L. D. Marini, *Efficient rectangular mixed finite elements in two and three space variables*, M^2AN, **21**(1987), 581–604.

[6] F. Brezzi and M. Fortin, "Mixed and Hybrid Finite Element Methods", Springer-Verlag, New York, 1991.

[7] J. Douglas, Jr. and J. E. Roberts, *Global estimates for mixed methods for second order elliptic equations*, Math. Comp., **44**(1985), 39–52.

[8] J. Douglas, Jr. and J. Wang, *A new family of mixed finite element spaces over rectangles*, Comp. Appl. Math., **12**(1993), 183–197.

[9] J. Douglas, Jr. and J. Wang, *Superconvergence of mixed finite element methods on rectangular domains*, Calcolo, **26**(1989), 121–133.

[10] R. E. Ewing, R. Lazarov, and J. Wang, *Superconvergence of the velocity along the Gauss lines in mixed finite element methods*, SIAM J. Numer. Anal., **28**(1991), 1015–1029.

[11] R. S. Falk and J. E. Osborn, *Error estimates for mixed methods*, R.A.I.R.O. Anal. Numer., **14**(1980), 309–324.

[12] M. Fortin, *An analysis of the convergence of mixed finite element methods*, R.A.I.R.O. Anal. Numer., **11**(1977), 341–354.

[13] V. Girault and P.-A. Raviart, "Finite Element Methods for Navier-Stokes Equations", Springer-Verlag, Berlin Heidelberg, 1986.

[14] P.-A. Raviart and J.-M. Thomas, *A mixed finite element method for second order elliptic problems*, Mathematical Aspects of the Finite Element Method, Lecture Notes in Mathematics 606, Springer-Verlag, Berlin and New York, 1977, 292–315.

Department of Mathematics
University of Wyoming
Laramie
WY 82071

Part II

Parallel Algorithms

NON UNIFORM RECURRENCE EQUATIONS ON 2D REGULAR ARRAYS

V.N. Aleksandrov* S.Fidanova

Abstract

A special class of non uniform recurrent equations producing data dependency graphs which involve non-constant data dependencies on the input data $f(i)$ is considered in this paper. This class includes for example Dynamic Programming recurrence equations for Knapsack Type Problems. $2D$ regular arrays for such equations are derived. It is proved that the mathematical expectation of the execution time on the fixed sized $(q \times p)$ regular arrays is $ET = O(mc/pq + m + q)$, i.e. optimal, for the case when $f(i) \leq t << c$ and it is measured by $ET = O(mc^2/p^2q + mc/pq + m + q)$ in the case when $t = c$ under the assumption that $f(i)$ are independently drawn from the uniform distribution over $\{1, 2, ..., t\}$.

1 Introduction

Many problems in computer science can be solved by the use of dynamic programming techniques. These include the Unbounded Knapsack Problem (UKP) [12], 0/1 Knapsack Problem (0/1KP) [12], Subset Sum Problem (SSP) [12], Change Making Problem (CMP) [12], Longest Common Subsequence (LCS) [8], etc. Consider the following restricted class of non-uniform recurrent equations:

$$\begin{array}{lll} v[i,j] = v[i-1,j] & 0 \leq j < f(i), & 1 \leq i \leq m \\ v[i,j] = F(v[i-1,j], v[i-l, j-f(i)] + g(i)) & f(i) \leq j \leq c, & 1 \leq i \leq m \end{array} \quad (1)$$

where the initial values $v[0,j]$ for $j = 0, ..., c$ are given, the integer functions $f(i)$ and $g(i)$, where $1 \leq f(i) \leq t$, $1 \leq g(i)$ are known in advance for $i \in [1, m]$ and finally the values $v[m,j]$ for $j = 0, ..., c$ must be computed. Equations (1) are run time dependent [13] and express the generic instance of Dynamic Programming (DP) forward part for such problems like:

- (i1). UKP If F is max, $l = 0, f(i) = w_i, g(i) = p_i, v[0,j] = 0$ for $0 \leq j \leq c$;

*Visiting Member of Staff at the Department of Computing Science, University of Newcastle. Lecturer, Department of Statistics and Computational Mathematics, University of Liverpool.

- (i2). CMP If F is min, $l = 0, f(i) = w_i, g(i) = 1, v[0, j] = 0$ for $0 \leq j \leq c$;
- (i3). SSP If F is max, $l = 1, f(i) = w_i, g(i) = w_i, v[0, j] = 0$ for $0 \leq j \leq c$;
- (i4). 0/1KP If F is max, $l = 1, f(i) = w_i, g(i) = p_i, v[0, j] = 0$ for $0 \leq j \leq c$;

The problems $(i1) - (i4)$ are NP-hard [9, 12] have a wide range of applications [12] and can be solved sequentially in $O(mc)$ time.
The $2D$ systolic arrays for UKP are treated by Andonov and Gruau in [4, 5] The systolic algorithm for UKP with time complexity $O(\frac{mc^2}{q\alpha} + c + m)$ on the ring of q cells with memory size α is presented by Andonov, Aleksandrov, Benaini [2]. More efficient algorithm on the similar architecture as in [2] is considered by Andonov and Quinton [3]. Its execution time T is: $mw_{\min}(c+q)/q\alpha + m \leq T \leq mw_{\max}(c+q)/q\alpha$, where $\alpha \leq w_{\max}, q \leq mw_{\max}/\alpha$ and $w_{\max}$ and $w_{\min}$ are bounded when c grows. The values $w_{\min}, w_{\max}$ denote the minimum and maximum of weights w_i. The algorithm on similar architecture but with an improved backtracking part is presented in [6]. A linear array where non-linear allocation function is used and probabilistic bound on execution time is found is presented in [1].
In our paper 2D regular arrays for recurrences (1) are investigated. A heuristic mapping method applied in [1] for deriving linear arrays for such recurrences which produces non-linear allocation functions is used to map recurrences (1) onto 2D fixed sized regular arrays. The mathematical expectation of the execution time in case when $f(i)$ $1 \leq i \leq m$ are uniformly distributed is found.
The terminology and problem formulation are in Section 2. The execution time and mathematical expectation of time for the fixed sized arrays are obtained in section 3.

2 Problem Formulation

We make the following assumptions:

- The system of equations is presented in a normal form. It is obtained by series of transformations such as variable normalization , input output separation , fully indexing, pipelining. (see for details [15]). We will concentrate on the computation equations and will ignore input/output equations.
- The algorithm is presented by Data Dependence Graph (DDG) , which is a direct acyclic graph $G = (N, E)$ with finite set of nodes N and finite set of edges E. There is an 1 to 1 mapping of points z in D and nodes in N. The DDG edges present dependence vectors, embedded into Z^2.
- The design produce modular arrays, i.e. the arrays consist of identical cells with fixed memory size α independent of the problem size.

The normal form of computation equations (1) is exhibited on FIG1.

$$\begin{array}{ll} 1 \le j \le c,\ 1 \le i \le m & v1[i,j] = v1[i,j-1] \\ 0 \le j < f(i),\ 1 \le i \le m & v[i,j] = v[i-1,j] \\ f(i) \le j \le c,\ 1 \le i \le m & v[i,j] = F(v[i-1,j], v[i-l, j-f(i)] + v1[i,j]) \end{array} \quad (2)$$

FIG 1. Normal form of equations

where the initial values $v[0,j]$ for $0 \le j \le c$ and $v1[i,0] = g(i)$ for $1 \le i \le m$ are given and $D = \{(i,j) : \ 0 \le j \le c, 1 \le i \le m\}$ is the domain of computations. The following example will illustrate our approach.

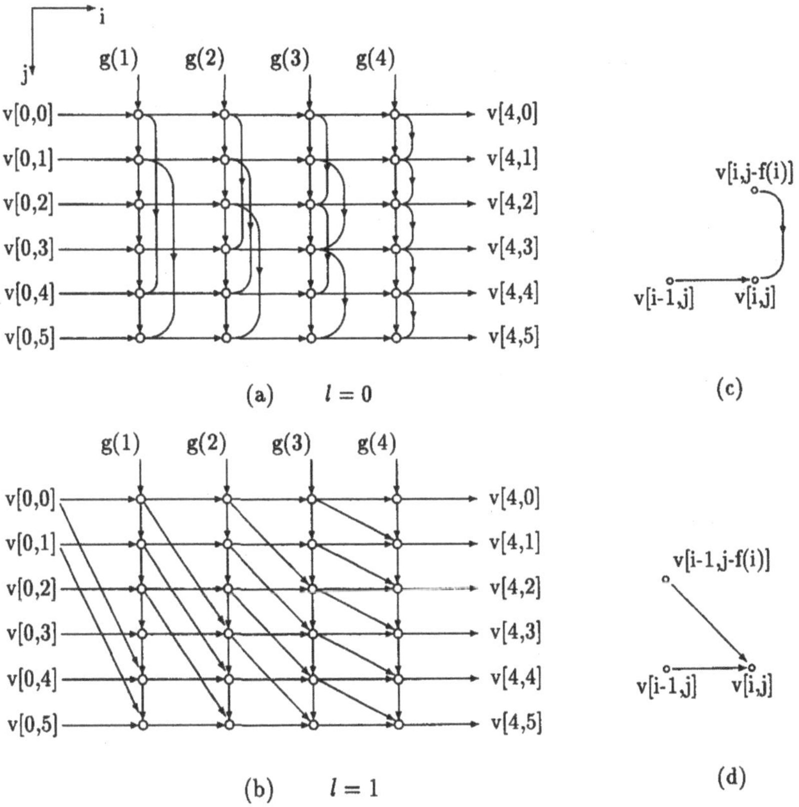

FIG. 2. The *DDGs* for the given example

Example: Let $l = 0$ or $1, m = 4, c = 5, g(1) = 2, g(2) = 3, g(3) = 4, g(4) = 5, f(1) = 4, f(2) = 3, f(3) = 2$ and $f(4) = 1$. The corresponding DDG's for $l = 0$ and 1 are

presented on FIG.2 (a) and (b). The dependences in any point $(i,j) \in D$ for variable v are depicted in details on FIG.2 (c) and (d).
Consider the DDGs. They have the following properties, as noted in [7] for $0/1KP$ and UKP, which make them suitable for pipeline computation:

- (ii1) In each stage the same operations are performed. (F or identity.).
- (ii2) Data dependencies occur in two cases only: between adjacent columns of DDGs and within the same column.
- (ii3) If $v[i,j] j = 0,\ldots,c$ are computed in an increasing order of j, then $v[i,j]$ can be computed whenever values $v[i-1,j]$ $j = 0,\ldots,c$ are available.

In addition the following observation is valid [1]:

- (ii4) Only f(i) memory registers are required to store the values from i-th column of DDGs during the computations.

Now based on the previous properties and observations our task is:
First determine the corresponding timing and allocation functions pair $(t(z), alloc(z))$, which maps the DDGs computations onto modular 2D regular array. The allocation function alloc(z): $Z^2 \to \bar{E}(\bar{E} \subset Z^2)$ maps many nodes onto a single processor and the non negative timing function $time(z) : Z^2 \to Z$ schedules the computations of DDG nodes in such a way that $t(x) \neq t(y)$ if alloc(x)=alloc(y) and $t(x) > t(y)$ if x depends on y and $x, y \in$D.
Second find the mathematical expectation of time (ET) under the assumption that $f(i)$ are independent random variables uniformly distributed on $\{1, 2, ..., t\}$.

3 Fixed sized array

Now we will show how to find the probabilistic time bound in case when $f(i)$ are independent random variables uniformly distributed on $\{1, 2, ..., t\}$ $t \leq c$.
The array is depicted on FIG.3, where **S MFIFO** is a Small Multichannel FIFO memory, **SL** is a Selector which can switch the data stream to pass through the busses or to be input into the **L MFIFO** Large Multichannel FIFO memory.

The following allocation function is used here:

$$
\begin{array}{lll}
a(i,j) = (k,r) & i = 1,2,...,m & \text{where} \\
r = (j \bmod f_{max}) \bmod p + 1 & k = \left\lfloor \frac{j \bmod f_{max}+1}{p} \right\rfloor \bmod q + 1 & \text{if } f(i) \geq p^2 \\
 = (j \bmod f_{max}) \bmod p + 1 & k = \left(\left\lfloor \frac{j}{f(i)} \right\rfloor p + \left\lceil \frac{j \bmod f_{max}+1}{p} \right\rceil \right) \bmod q + 1 & \text{if } f(i) < p^2
\end{array}
\tag{3}
$$

where $k \quad (1 \leq k \leq q)$ - is the number of a column of cells and $r \quad (1 \leq r \leq p)$ is the number of the cell in the column of cells in the array.
The idea of algorithm is to compute the values $v[i,j]$ in cell (r,k) according to the allocation function (3) , and after that to propagate two copies of the so computed data forward into the array. The first copy is required for value $v[i+1,j]$ computation, the second copy is required for value $v[i,j+f(i)]$ computation. The second copies must be properly shifted in order to meet allocation function (3).

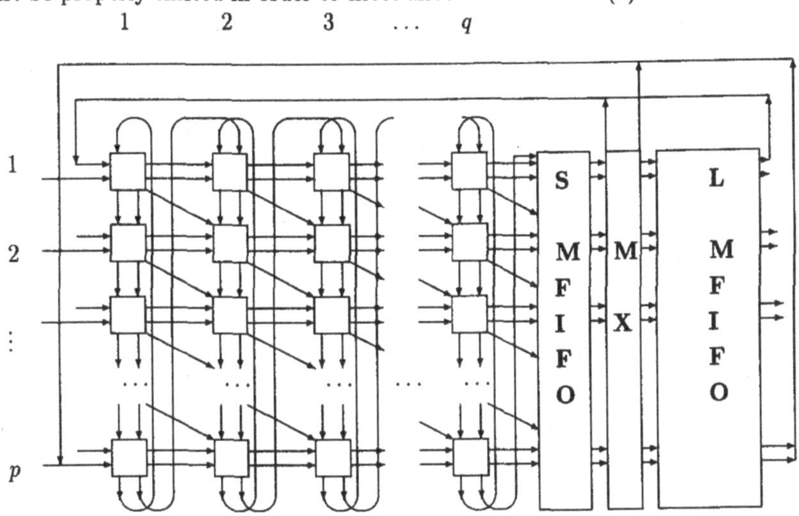

FIG.3 2D Fixed Sized Regular Array for recurrence (1)

We partition values $v[i,j] \quad j=0,1,...,c$ in $\lceil (c+1)/f_{max} \rceil \lceil f_{max}/p \rceil$ data sets which are input from the left into the array in portions of p data each time step. (The initial values are $v[0,j] \quad j=0,1,...,c$). The computations are performed in m bands one after another i.e. one column after another. First f_{max} data are propagated $\lceil f_{max}/pq \rceil$ times through the array using S MFIFO memory of size $7 \times \lceil f_{max}/p \rceil$, since f_{max} values must be computed and two copies of values with corresponding allocation addresses and control data are needed during the computation. At the last $\lceil f_{max}/pq \rceil$-th propagation of this data set the first copy of these f_{max} values is switched by Selectors to L MFIFO memory in order to wait the computation of the next $(\lceil (c+1)/f_{max} \rceil - 1) f_{max}$ data sets and to be input for the next band computation. The second copy of the data set properly shifted is transferred to the first column of cells in order to be input together with the next set of f_{max} values. Further we will suppose that $p^3 q \leq (t+1)/2, \quad pq \leq f_{max} \leq t$ and $t \leq c$. There are two possibilities for $f(i) \quad (1 \leq i \leq m)$: $\quad f(i) < p^2$ or $f(i) \geq p^2$.
Suppose at first that for all $f(i) \quad (1 \leq i \leq m)$ we have $f(i) \geq p^2$ and that they are sorted in a decreasing order. Now we will determine more precisely how the array

operates. Consider the following case $p^2 \leq f(i) \leq f_{max}$.
The computation is following: The values $v[i-1,j]$ of previous column are propagated in portions of $\lceil f_{max}/p \rceil$ values from left to the right in the array. The new values $v[i,j]$ are computed according to the allocation function (3), i.e. whenever the value $v[i-1,j]$ meets the cell where the value $v[i,j]$ must be computed. The first copy of $v[i,j]$ is carried without shifting since it is required for the value $v[i+1,j]$ computation and according to (3) it is allocated in the same cell. The second copy of $v[i,j]$ must be delivered according to (3) in cell (r',k'), where $r' = ((j+f(i)) \bmod f_{max}) \bmod p + 1$ and $k' = \left\lfloor \frac{(j+f(i)) \bmod f_{max}+1}{p} \right\rfloor \bmod q+1$ which means that these p values in each column must be shifted no more than p times to reach the required allocation. The shifting is performed propagating forward these data from column of cells to column of cells using diagonal and spiral links. Note that no more that p columns of cells are required for the full shifting in one column. During the next propagation of current $\lceil f_{max}/p \rceil$ data sets through the array these values are stored in corresponding cells according to (3) when necessary and at least memory of size $\alpha = 1$ in each cell is required to store them. Note that due to the data dependencies the last $f(i)$ values of each column of $c+1$ values are not required for further computations and the second copy of these values is not necessary to be carried.

Based on the previous discussion observe that to compute one column of $c+1$ values we need $\lceil (c+1)/f_{max} \rceil \lceil f_{max}/pq \rceil \lceil f_{max}/p \rceil$ time steps. Since we have m such columns then the time of algorithm is bounded by:

$$T \leq m \lceil (c+1)/f_{max} \rceil \lceil f_{max}/pq \rceil \lceil f_{max}/p \rceil \qquad (4)$$

From (4) it follows that :

Theorem 1 *The execution time of algorithm for the problems described by (1) is* $T = O(\frac{mc}{pq} + m + q)$ *when* $f_{max} \leq t << c$ *and* $T = O(\frac{mc^2}{p^2q} + \frac{mc}{pq} + m + q)$ *when* $f_{max} = c$ *on* $q \times p$ *cells.*

Must be noted that for some $f(i)$ the following condition $f(i) < p^2$ might be valid. Here we apply again (3) when $f(i) < p^2$, computing $f(i)$ values in $\lceil f(i)/p \rceil$ consecutive columns of cells. We proceed in the same manner as in previous cases. The second copy is shifted again to meet (3). Since $f(i) < p^2$ then there are columns through which the values are propagated without computation. Thus at least p columns of cells are required to compute each $f(i)$ values.

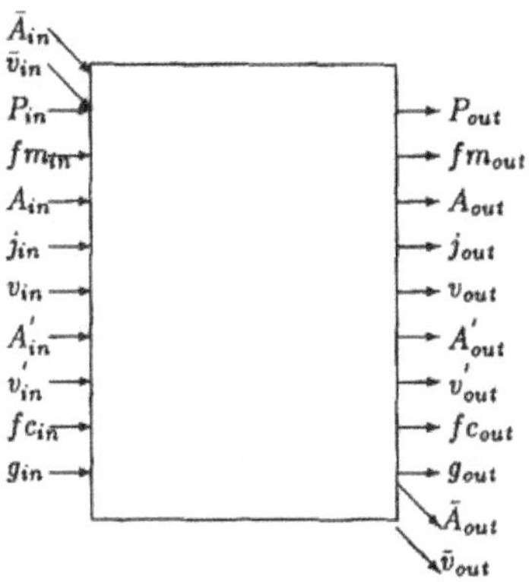

if $(A_{in} = 1)$ **then**
{
  $Fc = F(v_{in}, M + g)$; $j_{out} = j_{in}$; $v_{out} = Fc$;
  $A_{out} = \lfloor (j_{in} \bmod fm + 1)/P \rfloor$;
  $K2 = ((j_{in} + fc) \bmod fm) \bmod P + 1$;
  $K = (j_{in} \bmod fm) \bmod P + 1$;
  if $(K2 = K)$ **then**
    {$A'_{out} = \lceil (j_{in} + fc)/P \rceil$; $v'_{out} = Fc$;
    $\bar{A}_{out} = 0$; $\bar{v}_{out} = nil$; }
  if $(K2 > K)$ **then**
    {$A'_{out} = \lceil (j_{in} + fc)/P \rceil$; $v'_{out} = nil$;
    $\bar{A}_{out} = K2 - K$; $\bar{v}_{out} = Fc$; }
  if $(K2 < K)$ **then**
    {$A'_{out} = \lceil (j_{in} + fc)/P \rceil$; $v'_{out} = nil$;
    $\bar{A}_{out} = P - K + K2$; $\bar{v}_{out} = Fc$; }
}
else
{
  $j_{out} = j_{in}$; $v_{out} = v_{in}$; $A_{out} = A_{in} - 1$;
  if $(\bar{A}_{in} > 1)$ **then**
    { $\bar{A}_{out} = \bar{A}_{in} - 1$; $\bar{v}_{out} = \bar{v}_{in}$;
    $A'_{out} = A'_{in} - 1$; $v'_{out} = v'_{in}$; }
  if $(\bar{A}_{in} = 1)$ **then** { $\bar{A}_{out} = 0$; $\bar{v}_{out} = nil$;
    if $(A'_{in} = 1)$
    then { $M = v'_{in}$; $A'_{out} = 0$; $v'_{out} = nil$; }
    else { $A'_{out} = A'_{in} - 1$; $v'_{out} = \bar{v}_{in}$; }
  }
  if $(\bar{A}_{in} = 0)$ **then** { $\bar{A}_{out} = \bar{A}_{in}$; $\bar{v}_{out} = \bar{v}_{in}$;
    if $(A'_{in} = 1)$ **then**
    { $M = v'_{in}$; $A'_{out} = 0$; $v'_{out} = nil$; }
    else { $A'_{out} = A'_{in} - 1$; $v'_{out} = \bar{v}_{in}$; }
  }
}

FIG.4 (a) The Cell (b) Cell operation in case when $l = 0$ and all $f(i) \geq p^2$

Now we will find the mathematical expectation of time assuming that $f(i)$ are discrete random variables independently drawn from the uniform distribution over $\Omega = \{1, 2, ..., t\}$, $t \leq c$. Denote by η the number of $f(i)$'s such that $f(i) < p^2$. The probability that $\eta = k$, where $k = 0, 1, ..., m$ is given by:

$$P(\eta = k) = C_m^k((p^2 - 1)/t)^k(1 - (p^2 - 1)/t)^{m-k} \tag{5}$$

since $P(f(i) < p^2) = P(f(i) \leq p^2 - 1) = (p^2 - 1)/t$ and $P(p^2 \leq f(i) \leq m) = 1 - (p^2 - 1)/t$.

The mathematical expectation of η is :

$$E\eta = \sum_{k=0}^{m} kC_m^k((p^2 - 1)/t)^k(1 - (p^2 - 1)/t)^{m-k} = m(p^2 - 1)/t \tag{6}$$

In case $f(i) < p^2$ we have to compute $\lceil f_{max}/f(i) \rceil p \lceil (c+1)/f_{max} \rceil$ such columns of values over $q \times p$ processors for $m(p^2-1)/pqt$ cases in average. Thus the mathematical expectation of time is $\{m(p^2-1)/pqt\} \lceil f_{max}/f(i) \rceil \lceil (c+1)/f_{max} \rceil pq$. Note that choosing $p^3 q \leq (t+1)/2$ the mathematical expectation of time reaches $O(mc/pq+m)$. Therefore the mathematical expectation of the execution time of the algorithm is:

$$\begin{aligned} ET &= E\{(m-\eta) \lceil (c+1)/f_{max} \rceil \lceil f_{max}/pq \rceil \lceil f_{max}/p \rceil \\ &\quad + \eta \lceil f_{max}/f(i) \rceil \lceil (c+1)/f_{max} \rceil pq\} \\ &= m(1-(p^2-1)/t) \lceil (c+1)/f_{max} \rceil \lceil f_{max}/pq \rceil \lceil f_{max}/p \rceil \\ &\quad + \{m(p^2-1)/pqt\} \lceil f_{max}/f(i) \rceil \lceil (c+1)/f_{max} \rceil pq \end{aligned} \tag{7}$$

and the following Theorem 2 is valid:

Theorem 2 *Under the assumption that $f(i)$ are uniformly drawn from $\{1, 2, ..., t\}$ $t \leq c$ the mathematical expectation of the execution time of algorithm is $ET = O(\frac{mc}{pq} + m + q)$ when $f_{max} \leq t << c$, and $ET = O(\frac{mc^2}{p^2 q} + \frac{mc}{pq} + m + q)$ when $f_{max} = c$ on $q \times p$ cells*

Based on previous explanation we will define the cell and cell operation. Assume the values $f(i), g(i), f_{max}, P$ are propagated with the first p data when the computation of each column started and are stored in corresponding registers fc, g, fm, p in each cell. Thus the cell is presented in FIG. 4 .(a) and the cell operation in case when $l = 0$ and all $f(i) \geq p^2$ is outlined in FIG. 4 .(b).

Remark 1 In case $l = 1$ the array operates in the same way. Only the minor changes of cell operation concerning the processing of the second copies of data are required.

4 Conclusion

A special class of non uniform recurrent equations producing data dependency graphs which involve non-constant data dependencies on the input data $f(i)$ is considered in this paper. This class includes for example Knapsack Type Problems. $2D$ regular arrays for such equations are discussed.

Here we apply allocation function for mapping of computations onto fixed sized modular array, which enables us to obtain the probabilistic bound on execution time and to minimize cell memory. It is proved that the mathematical expectation of the execution time is $ET = O(mc/pq + m + q)$, i.e. optimal, for the case when $f(i) \leq t << c$ and it is measured by $ET = O(mc^2/p^2q + mc/pq + m + q)$ in the case when $t = c$ under the assumption that $f(i)$ are independently drawn from the uniform distribution over $\{1, 2, ..., t\}$, $p^3 q \leq (t+1)/2$ and memory size of each cell $\alpha = 1$.

Acknowledgements: Thanks to Dr.G.M.Megson and Dr. I.Dimov for the fruitful discussions and comments. The first author is grateful to British Council, Open Society Fund and National Academic Foundation for the financial support.

The authors are also financially supported by the BG MSHE YS2, YS20 and NATO research grant HTECH.CRG 931298.

References

[1] Aleksandrov V. N. and Fidanova S., *On the Expected Execution Time for a Class of non-Uniform Recurrence Equations Mapped onto 1D Regular Arrays*, to appear in J. Parallel Algorithms and Applications, Vol.1, No 4.

[2] Andonov, R. A., Aleksandrov V. N. and Benaini A.,*A Linear Systolic Array for the Knapsack Problem*, Technical Report No12, CICT , Sofia May, 1991

[3] Andonov R. and P.Quinton, "Efficient linear systolic array for the knapsack problem, Publ. Interne *No*639, March 1992.

[4] Andonov, R. A., Gruau, *A 2D Toroidal Systolic Array for the Knapsack Problem*, in Algorithms and Parallel VLSI Architectures II, Bonas, France, June 1991.

[5] Andonov, R. A., Gruau, *A 2D Modular Toroidal Systolic Array for the Knapsack Problem*, in ASAP'91, Barcelona, Spain, September 1991.

[6] Andonov, R. A., Raimbault F. and Quinton P., *Dynamic Programming Parallel Implementations for the Knapsack Problem*, TR No 740, IRISA, France, June 1993.

[7] Chen, G.H., Chern, M.S., and Jang, J.H. Pipeline architectures for dynamic programming algorithms. *Parallel Computing* ,13 (1990), pp. 111-117.

[8] Cormen T.H., C.E. Leiserson, R.L.Rivest, *Introduction to Algorithms*, The MIT Press, 1990

[9] Garey, M., and Johnson, D. *Computer and Intractability: A guide to the theory of NP-completeness.* Freeman, San Francisco, 1979.

[10] Lee, J., Shragowitz, E., and Sahni, S. A hypercube algorithm for the 0/1 knapsack problem. *J. of Parallel and Distributed Computing* **5**, (1988), 438-456.

[11] Lin J. and J.Storer "Processor efficient hypercube algorithm for the knapsack problem", *J. of Parallel and Distributed Computing*, vol.13, pp. 332-337,1991.

[12] Martello S. and P.Toth, *Knapsack Problems: Algorithms and Computer Implementation.* John Wiley and Sons, 1990

[13] G.M.Megson, *Towards a Formal Model of Algorithmic Engineering* To appear in Proc. of Workshop on Parallel Algorithms, August 18-22, 1992, Sofia, Bulgaria.

[14] G.M.Megson, The derivation of Uniform Recurrence Equations for the Knapsack problem, *Journal of Parallel Algorithms and Applications.* , Vol.1, No1, 1993.

[15] P. Quinton, V. Van Dongen, *The Mapping of Linear Recurrence Equations on regular Arrays*, Journal of VLSI Signal Processing, 1, pp. 95-113 (1989)

Center of Informatics and Computer Technology,
Bulgarian Academy of Sciences,
Acad. G.Bonchev str., block 25-A, 1113 Sofia,
BULGARIA

On the Implementation of Numerical Methods on Parallel Computers

G. De Pietro

1 Introduction

The implementation of numerical methods on parallel computers is one of the most popular parallel processing working area. Computer scientists, mathematicians, physicists spend a lot of time in this activity, following two different (but not necessarily disjoined) approaches. The first consists in the porting of existing numerical techniques, mainly looking at how an existing method can be efficiently implemented on a particular class (i.e. MIMD or SIMD) of parallel architectures. Moreover, great efforts are being done in porting mathematic packages (e.g. BLAS) on several parallel computers. The second consists in designing a new parallel algorithm for the same problem solved by the sequential algorithm; this approach usually requires more efforts than the previous ones but in this way the initial problem can be often solved in very efficiently on parallel computers. The choice between these two different approaches depends on many considerations. On one hand, there is the demand for the re-use of a lot of existing sequential programs which are well-tested both in terms of the numerical methods they adopt, and in terms of program bugs; in other words, there is a great demand for porting activities. On the other hand, the efficient use of parallel architecture, the presence of new parallel languages, lead the programmers to the development of new parallel algorithms, many times highly tailored for a particular machine or class of machines.

The main goal of this paper is to give a brief overview of some techniques concerning the implementation of some numerical methods on both MIMD and SIMD parallel computers. Moreover two practical applications, an implementation of an iterative solver for linear equations system on a MIMD machine, and an implementation of a Monte Carlo method on a SIMD machine will be presented.

2 Parallelization techniques

As we said in the previous section, there are two main strategies for developing parallel programs. Of course, there are no methodologies for designing new parallel algorithms, rather there are some general criteria to follow (e.g. to minimize the data exchange, task granularity suitable for the target processing nodes, etc). Instead, many parallelization strategies have been proposed for the parallel implementation of existing algorithms[1][2]:

- domain decomposition
- loop transformation
- recursive doubling
- odd-even reduction
- coloring schemes

2.1 Domain decomposition

Domain Decomposition is one the most used technique; the computational domain D is subdivided into a certain number of subdomains D_i on which the calculation is performed in parallel until some data exchange between subdomains is required. Depending on the particular numerical method on which domain decomposition is applied, this technique can lead to the realization of new methods (as it will be shown in section 3) or simply to a parallel version of the original algorithm. It should be pointed out that the meaning computer scientists give to the word "domain decomposition" is not the same given by mathematicians; many times it is more convenient to name this technique "data distribution".

2.2 Loop Transformation

This technique is based on the use of operations which change both the structure and the execution order by maintaining the equivalence to the original program. There are several types of transformations; the simplest uses the input-output equivalence criteria, in which a set of input variables is mapped into the same set of output variables for all input-output equivalent programs[2]. The widespread loop transformations are:

- Cycle Shrinking, which is applicable when the dependence cycle of a loop involves dependencies between loops that are far apart;
- Loop Interchanging, which consists in performing a valid interchange of the execution order of the loops in order to increase parallelism;
- Loop Fusion, which transform two consecutive loops into a single loop;
- Loop collapsing, which transform two nested loops into a single one in order to increase the vector length for vector processing nodes.

2.3 Recursive Doubling

Let O be an associative operation on pairs of mathematical objects

$$(a \; O \; b) \; O \; C = a \; O \; (b \; O \; c)$$

then the expression

$$a_1 O a_2 O \ldots\ldots\ldots a_n$$

can be rewritten as

$$(a_1 O a_2) O (a_3 O a_4) O \ldots\ldots\ldots\ldots O (a_{n-1} O a_n)$$

the operation in brackets can be carried out in parallel; if the size of the set of the object is n, the result is produced in $log_2 n$ steps

2.4 ODD-EVEN Reduction

Let AX=Y be a triangular system

1 $A_{i-1}X_{i-2} + B_{i-1}X_{i-1} + C_{i-1}X_i = Y_{i-1}$

2 $A_i X_{i-1} + B_i X_i + C_i X_{i+1} = Y_i$

3 $A_{i+1}X_i + B_{i+1}X_{i+1} + C_{i+1}X_{i+2} = Y_{i+1}$

by subtracting eqn 1 and 3 (properly multiplied) from the 2, we obtain

$$A_i^{(1)} X_{i-2} + B_i^{(1)} X_i + C_i^{(1)} X_{i+2} = Y_i^{(1)}$$

after k steps with k=$log_2 n$ we have

$$B_i^{(k)} X_i = Y_i^{(k)}$$

which gives the value of X_i.

3 A Parallel Iterative method

In this paragraph it will be shown the implementation on a MIMD computer of an iterative method, the Parallel Asynchronous Gauss-Seidel algorithm (PAGS) [3]. Before starting a brief description of the PAGS algorithm it is convenient to remind some definitions. An algorithm is synchronous if it is mathematically equivalent to one for which the time of computation, of message transmission and message reception are fixed and given a priori. An algorithm is asynchronous if the time of computation, message transmission and message reception can vary in two different executions of the program with an effect on the result of the computation. The introduction of some kind of asynchronism can increase the parallelism of the algorithm; this observation is at the basis of the development of the PAGS algorithm.

Let P be the number of available processors and N the system size, with N >> P (this is the most common real situation); suppose, for sake of simplicity, that

N is a multiple of P. We split the coefficient matrix in a row-wise fashion so that each processor has assigned N/P rows, and calculates N/P elements of the vector of unknowns. Of course, this means that a domain decomposition technique has been adopted. Each iteration is divided into C cycles of length (N/(C.P)), where C=1,2,...,(N/P); in each cycle, a processor updates only a subset of size (N/(C.P)) of its components. Moreover, inside each cycle, as soon as the new values of the components are calculated they will be used for the evaluation of other components. At the end of each cycle, all the processors release the new values of their vector of unknowns for the other processors. These values are used for the calculation in the remaining cycles.

Note that, inside each cycle, the components are updated in a Gauss-Siedel fashion while at the iteration level the algorithm realizes the Jacobi method (in particular, for C=N/P we have the Gauss-Siedel method).

For a fixed value of C, by increasing the number of processors, the PAGS algorithm generally needs of a higher number of iterations in order to reach the convergence. In fact, for the updating of a component, at least O=(N-N/P) old values are used: this number rises according to the number of processors used. Instead, for a fixed value of P, by increasing the number of cycles C the convergence rate generally speeds up; however, the number of interprocessor communication increases according to C, so that the choice of the value of C must represent a good compromise between the estimated gain of computing time and the additional communication time.

The algorithm has been implemented on Meiko CS 1, a distributed memory MIMD computer with 128 T800 Transputer. The interconnection network topology used was a ring: even if this topology needs (P-1)/2 communication for an all-to-all broadcasting, it allows very simple and low overhead routing algorithms. Moreover, this topology is easy scalable. The PGAS algorithm has been used within a code for the solution of the flow field around a nacelle with panel methods[4]. From a logical point of view, a panel method can be decomposed in four main phases:

a) the calculation of the panels characteristics;

b) the calculation of the influence matrices;

c) the solution of a linear equations system;

d) the calculation of the flow quantities.

Clearly, the PAGS algorithm was used inside the c phase. An additional problem for the parallel implementation was the data distribution because, in the sequential version of the code, the calculation of the influence matrices was organized in a column manner. Instead of performing a matrix transposition between the b and c phase, it was preferred to reorganize part of the code for matching the data allocation constrain of the PAGS method. The results obtained for the PAGS on the Meiko have shown good performance: for a source panel problem with 2048 panels, the speed-up ranges from 15.5 (using 16 processors) to 96.4 (using 128 processors).

4 A SIMD Monte Carlo Implementation

Monte Carlo methods are inherently parallel and, consequently, a lot of applications using these methods on parallel machines have been developed. It should be noted that MIMD machines represent the "natural" hardware platform for implementing Monte Carlo methods; however, it seems to be interesting to investigate on the feasibility of parallel implementation on a SIMD and this for two main reasons:

- Monte Carlo methods are often used inside or in conjunction with more complex and large existing codes (usually written in FORTRAN); so, the easiness in programming SIMD machines makes their use very interesting;

- most of the MIMD machine of the new generation provide processing nodes equipped with vector processors; so, the parallelization strategies used for the SIMD machine can be efficiently used also on MIMD machines.

The SIMD computer used for the test was a Connection Machine CM 200, with 4k single-bit processors and 128 64-bit Floating Point Units; the codes were written by using the Fortran 90 language and the "slicewice" model[5]. For a general description of the Monte Carlo algorithm used for the tests see[6][7][8]

In a SIMD machine, the parallelism is exploited by performing operations on array structures (DATA Parallelism). This implies that if the problem dimension is equal to m (in other words the number of trajectories we would like to calculate is m) we define vectors of length m. instead of to iterate m times the realization of the scalar quantities.

Supposing p processors are available, we are able to calculate m realizations for each vector in $[m/p]$ steps. In other words, if we start from a sequential implementation of the algorithm, the main work to perform is the transformation of iterative operations on scalars in vectorial operations. The following problem arises whenever conditional statements have to be evaluated: for all the trajectories for which the condition is not verified , new realizations have to be calculated. This means that, we need to perform different operations on different elements of the same array and this is in contrast with the SIMD paradigm.

To overcome this problem the following strategy has been adopted. First of all, a "mask" vector of dimension m has to be declared. Then all the operations which lead to the calculation of the next point of the trajectory are performed for all the trajectories. For all the trajectories which do not satisfy some conditions, the corresponding elements of the mask array are set to $FALSE$ and the computation continues for all the trajectories. At the end of the iteration, all the trajectories with the corresponding value of mask equal $TRUE$ are taken into account; for the others, the evaluation of new trajectories begins. Alternative strategies to minimize this workload overhead (e.g. the compacting of the original structure into a smallest one) are not convenient because of the introduced communication overhead. Clearly, by adopting the above mentioned strategy, a certain amount of unnecessary operations have to be performed; in order to quantify the wasted work a parameter AE (Algorithm Efficiency) is defined as[]

$$AE = (N_{it}N_t - N_l)/N_{it}N_t$$

where N_{it} is the number of iterations performed, N_t is the number of trajectories computed at each iteration, Nl is the total number of discarded trajectories.

It should be pointed out that this definition of AE represents the "worst case", that is it includes the number of unnecessary work due to the algorithm itself. More precisely, in the sequential version of the code whenever a trajectory does not satisfy an IF statement it is immediately discarded (whereas in the parallel version the computation goes on); however, the operations (and the time) spent to reach the IF statement are "lost", but it is intrinsic of the Monte Carlo algorithm. The numerical test have been performed considering the following problem in the unit square

$$Lu = \Delta u = 0$$

with the boundary conditions

$$u = x_1, \quad \text{for} \quad x_2 = 0$$

$$u = x_2, \quad \text{for} \quad x_1 = 0$$

$$u = 1 + x_1, \quad \text{for} \quad x_2 = 1$$

$$u = 1 + x_2, \quad \text{for} \quad x_1 = 1.$$

Fig 1 shows the time spent of the Connection Machine (CM-time), the CM-utilization, that is the percentage of the overall computing time (Front End + Connection Machine) spent on the Connection Machine) and the time spent on a SUN Sparkstation 2. The results show that Monte Carlo methods can be efficiently implemented also on SIMD machines; the performances increase when the size of the problem grows.

5 Conclusions

In this paper some techniques for implementing numerical methods have been illustrated. Moreover, two applications on different types of parallel architectures have been also presented. The applications show that, even if the strategy to follow for the solution on a parallel architecture of a given problem is clear, there are many real aspects and difficulties to be taken into account.

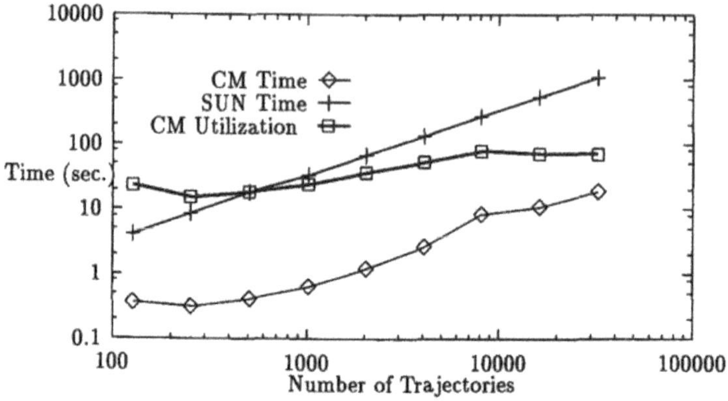

Figure 1: Monte Carlo Results

6 Acknowledgments

The author would like to thank miss M.G. Parrella for her helpful suggestions for the translation into English of this paper.

References

[1] J. J. Modi, *Parallel Algorithms and Matrix Computation*, Oxford University press, 1988

[2] D. I Moldovan, *Parallel Processing*, Morgan Kaufamann Eds, 1993

[3] C. de Nicola, G. De Pietro, G. Gallo, *Panel Methods in Transputer-Based Architecture*, Proc of PACTA' 92 Conference, IOS Press, Barcelona, 1992

[4] G. De Pietro, G. Gallo, R. Vaccaro, *A Parallel Asynchronous Gauss-Seidel Method for Solving Linear equation Systems*, EC Newsletter, Laterza Publishers, vol. 4 No. 1, 1992

[5] Thinking Machine Corporation, FORTRAN 90, Users Guide.

[6] I.Dimov, O.Tonev. Criteria Estimators for Speed-up and Efficiency of Some Parallel Monte Carlo Algorithms for Different Computer Architectures, Proc. WP & DP'91, Sofia, April 16-19, North-Holland Publ. Co., Amsterdam, 1992, 243-262.

[7] I.Dimov, O.Tonev. Random Walk on Distant Mesh Points Monte Carlo Method. Journal of Statistical Physics, vol 70,Nos 5/6, 1993, pp. 1333-1342.

[8] I. Dimov, O. Tonev. Random walk on distant mesh points Monte Carlo methods, *Journal of Statistical Physics*, vol. 70, Nos. 5/6, 1993, pp. 1333 –1342.

[9] G. M. Baudet *Asynchronous Iterative Methods for Multiprocessors*, Journal of the Association for Computing Machinery, Vol.25, No.2, April 1978, pp.226-244.

IRSIP
National Research Council of Italy
Via P. Castellino 111
Napoli
Italy

A Comparison of Three Parallel Matrix Product Algorithms

F. Desprez *† M. Pourzandi‡

Abstract

In this paper, we present results of the study of matrix multiplication algorithms on the Intel Paragon and the Archipel VolVox IS-860 respectively with 1-port wormhole and 4-port store & forward communication protocols. We compare and analyze three algorithms on two dimensional torus/grid for both machines.

1 Introduction

The matrix product is the routine GEMM (GEneral Matrix Matrix product) of the level 3 BLAS (Basic Linear Algebra Subroutines) [6]. The matrix product is a basic routine involved in a lot of linear algebra algorithms which can be rewritten as a sequence of calls to this routine. First designed to sequential and parallel shared memory machines, the BLAS have to be recoded for distributed memory machines [3, 4] in order to get parallel versions of basic libraries such as LAPACK (Linear Algebra PACKage) [1, 3].

In this paper, we compare three basic algorithms on a two dimensional torus. The first one, designed by Fox and al. [7] for the Caltech hypercube, is based on horizontal broadcast. The second one, based on Cannon's algorithm [2], requires only neighbor to neighbor communications, but preskewings of two of the three matrices before and after execution ($C = C+AB$). Snyder's algorithm requires a transposition of matrix B before and after execution, vertical rollings of this matrix and global summations over the rows at each step [9]. These algorithms have already been implemented and compared on the Intel Delta machine by Choi, Dongarra and Walker in [4].

*On leave from LIP, CNRS URA 1398, ENS Lyon, 46 allée d'Italie, 69364 Lyon Cedex 07, France.

†This work was supported in part by the National Science Foundation Science and Technology Center Cooperative Agreement No. CCR-8809615, CNRS-NSF grant 950.223/07, PRC C^3, Archipel SA and MRE under grant 974, and DRET.

‡This work was supported in part by CNRS-NSF grant 950.223/07, PRC C^3, Archipel SA and MRE under grant 974, and DRET.

We have implemented and compared them on the ARCHIPEL Volvox IS-860 and Intel Paragon XPS/5. These two machines have different communication protocols, respectively *Store & forward* 4-ports and *Wormhole* 1-port but have the same kind of computation processor, i.e. the Intel i860. On the Paragon machine, the i860 is the latest version of this processor (i860 XP) and on the Volvox machine, this is the previous version (i860 XR). This explains the computation performances differences between the two machines. However, they have the same kind of behavior. Our goal is to compare matrix product algorithms and to highlight the differences in the implementations and thus the performances due to the different models of communication. The resulting matrix product algorithms can then be used for the distributed version of level 3 BLAS library. Results of experiments are given for both machines.

For each algorithm, we assume that we start the execution with a natural full-block distribution of the matrices on the logical torus of processors. We assume that we have to compute the matrix product $C = C + AB$ where A, B and C are $N \times N$ matrices. If we store matrices A, B and C on a $P \times P$ two dimensional torus, then submatrices A_{ij}, B_{ij} and C_{ij} of size $N/P \times N/P$ are located on processor which coordinates are (i, j) in the torus. If the algorithm needs a redistribution of the matrices before execution starts, we take into account this communication overhead. And we assume that after computations, the submatrices must return to their original distribution on the processors for further computation.

2 Description of the algorithms

In this section, we discuss the necessary communications for each algorithm. We present the communications patterns of the algorithms in table 1.

Algorithm	Cannon	Fox	Snyder
Pre and post processing	skewing of A and B	None	Transposition of B
Computation of C	locally	locally	horizontal global sums
Movements of A	horizontal rollings	horizontal broadcast	none
Movements of B	vertical rollings	vertical rollings	vertical rollings

Table 1: Communication patterns of the three algorithms.

2.1 Fox's algorithm

The algorithm, also called broadcast-multiply-roll, described by Fox in [7, 8] has been first developed for the Caltech hypercube. It broadcasts horizontally a diagonal of the submatrices A_{ij} through the processors, at each step of the execution. The communication pattern of the submatrices B_{ij} is vertical rollings.

This algorithm requires a natural mapping of the submatrices, i.e. submatrices $[A, B, C]_{ij}$ are stored on the processors which coordinates in the grid are (i, j). The algorithm is given in Figure 1 and his two first steps are given in Figure 4.

```
/* No pre treatment */

/* Computation of the matrix product */
Broadcast of the first diagonal of A
for k = 1 to √P − 1 in // do
   GEMM(A, B, C)
   Vertical rollings of B
   Broadcast of the next diagonal of A
endfor
in // do
   GEMM(A, B, C)
   Vertical rollings of B
fin

/* No post treatment */
```

Figure 1: Broadcast-multiply-roll algorithm.

2.2 Cannon's algorithm

In order to have only neighbor-to-neighbor communications, submatrices A_{ij} and B_{ij} are distributed by diagonals through the rows of processors. The submatrices on the diagonal of A are distributed on the first column of the processors and the first diagonal submatrices of B are distributed on the first row of the processors (Figure 5). Because of this repartition, we have to redistribute submatrices A_{ij} and B_{ij} after the computation of the matrix product in order to put them back to their original location. Therefore, during the computation phase, we perform vertical and horizontal rollings according to the coordinates of the processor in the two dimensional torus. Figure 2 gives Cannon's algorithm.

2.3 Transpose algorithm

This algorithm has been presented by Snyder [9]. The submatrix A_{ij} is stored on processor (i, j) while the submatrix B_{ij} is stored on processor (j, i), i.e.

```
/* Pre treatments */
Rollings of A and B

/* Computation of the matrix product */
for k = 1 to √P in // do
   GEMM(A, B, C)
   Vertical rollings of B
   Horizontal rollings of A
endfor

/* Post treatment */
Rollings of A and B
```

Figure 2: Cannon's algorithm.

matrix B is transposed before the computation phase. The algorithm requires a horizontal global sum at each step on each row of processors. The communication pattern of the sub-matrices B_{ij}, as for the Fox algorithm, is vertical rollings. We also have to redistribute submatrices B_{ij} after the matrix computation by transposing matrix B backward. Figure 3 gives Snyder's algorithm for node (i, j). Figure 6 shows the two first steps of this algorihm. The submatrices C_{ij} in bold are the ones which are computed at this step.

3 Experimental results

In this section, we present the performances of the three algorithms on both machines. We show that the most efficient algorithm changes according to the parallel machine used. On both machines, we try to maximize the performances by using overlap of the rollings by the computations. This can be done easily on both machines using asynchronous communications. Broadcast and global sums are not overlap because they need many communication steps and thus their overlap will imply a reduction of the grain of computation and thus a reduction of the performances of the calls to the level 3 BLAS routine DGEMM. This has already been studied in [5]. The sequential DGEMM routine is assembly coded in order to obtain the best performances.

We first compare the three algorithms with pre and post treatments on a 6×6 torus on the Archipel Volvox machine (Figure 7). Despite of the pre and the post-processing stages, Cannon's algorithm remains the most efficient. The difference with the two other algorithms is more important if we do not redistribute matrices at the beginning and at the end of the computation (Figure 8). Notice that the costs of the broadcast and global sums are almost the same and the difference between Fox's and Snyder's algorithms comes from the pre and

```
/* Pre treatment */
transpose matrix B

/* Computation of the matrix product */
in // do
   GEMM(A, B, C)
   Vertical rollings of B
endfor
for k = 1 to √P − 1 in // do
   Global sums on the rows of processors for C_{i,(i+k−1)mod√P}
   GEMM(A, B, C)
   Vertical rollings of B
endfor
Global sums on the rows of processors for C_{i,(i+√P−1)mod√P}

/* Post treatment */
transpose matrix B
```

Figure 3: Snyder's algorithm for node (i, j).

post-transpositions.

On the Intel Paragon, communication startups are smaller and the broadcast algorithm is very efficient using wormhole routing (Figures 9 and 10). Because of the 1-port assumption, the Intel Paragon seems very sensitive to the algorithms with intensive communications even if they are grid-shaped. Fox's algorithms is thus a bit more performant on that machine. If we do not pre and post-redistribute matrices, Cannon and Fox are almost equivalent. Notice also that, due to an heavy operating system, Paragon curves are more "chaotic".

On both machines, Snyder's algorithm is the most expensive one. It requires a horizontal global sum at each step, with a pre- and post-transposition.

Finally, we have compared both machines using a 6×6 torus/grid (Figures 11 and 12). Due to a limitation in the memory space (the operating system is using a large amount of memory), we stopped our experimentations on the Paragon for 1800×1800 matrices. With its more performant processor and communication network, the Intel Paragon is more performant for his algorithms. However, we can see that the two machines have two different behaviors. Cannon's algorithm is always more performant on the Archipel machine and less performant on the Intel machine.

4 Conclusion

In this paper we show clearly that the cost of the communication algorithm is not a simple function of the amount of data to be communicated but also depends on the machine's features which fit the communication schema of the algorithm well or are able to overlap the communications by the computations.

It seems that the 4-ports parallel machines are particularly adapted to Cannon's algorithm. This algorithm, using the communications on the all 4 ports of the 2D-torus topology in parallel, squeezes the best performance out of the Archipel VolVox machine. Neither Fox's algorithm nor Snyder's one, achieve the performances as well as Cannon's algorithm on these machines. This is not the case for 1-port parallel machines. These ones do not benefit from the simultaneous communications.

Actually, a broadcast based algorithm with a distance free wormhole protocol without communications intensive pre and post-treatments, is more efficient for matrix product algorithms. According to our results, we can predict Cannon's algorithm to be the most performant on wormhole 4-port machines using the i860 XP.

References

[1] C.H. Bischof. Fundamental Linear Algebra Computations on High-Performances Computers. Technical report, Mathematics and Computer Science Division - Argonne National Laboratory, 1988.

[2] L.E. Cannon. *A Cellular Computer to Implement the Kalman Filter Algorithm.* PhD thesis, Montana State University, 1969.

[3] J. Choi, J.J. Dongarra, and D.W. Walker. The Design of Scalable Software Libraries for Distributed Memory Concurrent Computers. In J.J. Dongarra and B. Tourancheau, editors, *Environments and Tools For Parallel Scientific Computing.* Elsevier, 1992.

[4] J. Choi, J.J. Dongarra, and D.W. Walker. PUMMA: Parallel Universal Matrix Multiplication Algorithms on Distributed Memory Concurrent Computers. Technical Report ORNL/TM-12252, Oak Ridge National Laboratory, April 1993.

[5] F. Desprez. A Library for Coarse Grain Macro-Pipelining in Distributed Memory Architectures. In *IFIP 10.3 Conference on Programming Environments for Massively Parallel Distributed Systems.* Birkhaeuser Verlag AG, Basel, Switzerland, 1994.

[6] J.J. Dongarra, J. Du Croz, I. Duff, and S. Hammarling. A Set of Level 3 Basic Linear Algebra Subprograms. *ACM Transaction on Mathematical Software*, 16(1):1–17, 1990.

[7] G. Fox, M. Johnson, G. Lycenza, S. Otto, J. Salmon, and D. Walker. *Solving Problems on Concurrent Processors - General Techniques and Regular Problems*, volume 1. Prentice Hall, 1988.

[8] G.C. Fox and S.W. Otto. Matrix Algorithms on a Hypercube I: Matrix Multiplication. *Parallel Computing*, 4:17–31, 1987.

[9] C. Lin and L. Snyder. An Algorithm of Choice for Solving QR Factorization. In E.D. Brooks, B.J. Heston, K.H. Warren, and L.J. Woods, editors, *The 1992 MPCI Yearly Report: Harnessing the Killer Micros*, pages 276–289. Lawrence Livermore National Laboratory, 1992.

F.Desprez
Computer Science Department
The University of Tennessee
107 Ayres Hall
Knoxville, TN 37996-1301, USA

M. Pourzandi
CNRS - Laboratoire LIP, URA 1398
ENS Lyon
46, Allée d'Italie
69364 LYON Cedex 07, France

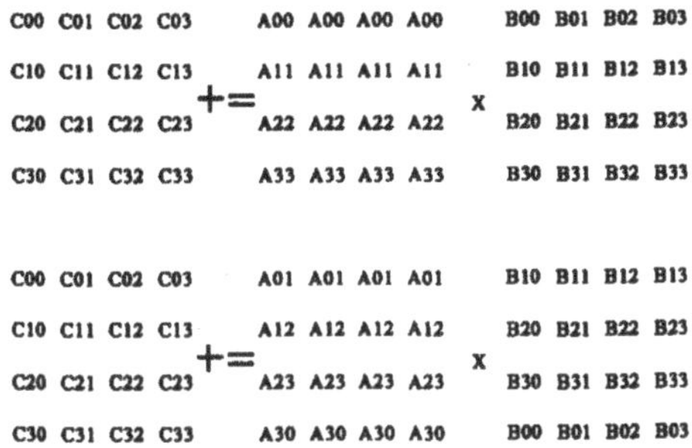

Figure 4: The two first steps of the broadcast-multiply-roll algorithm.

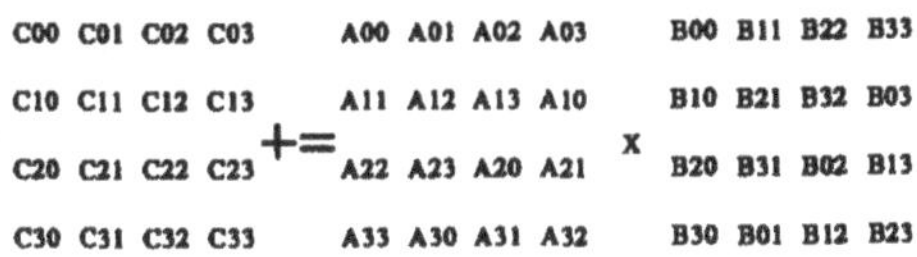

Figure 5: Data distribution for Cannon's algorithm.

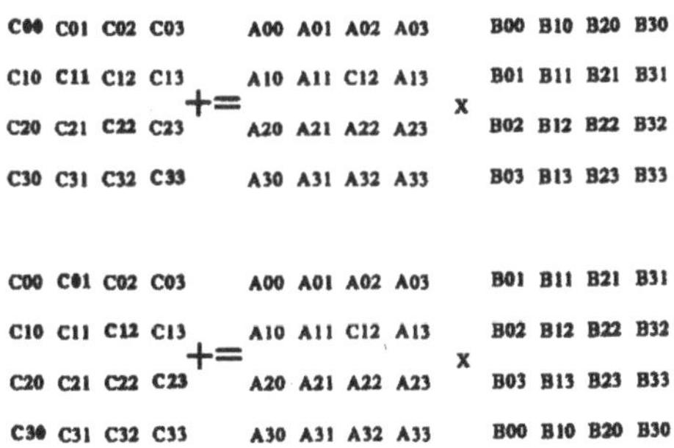

Figure 6: The two first steps of Snyder's algorithm.

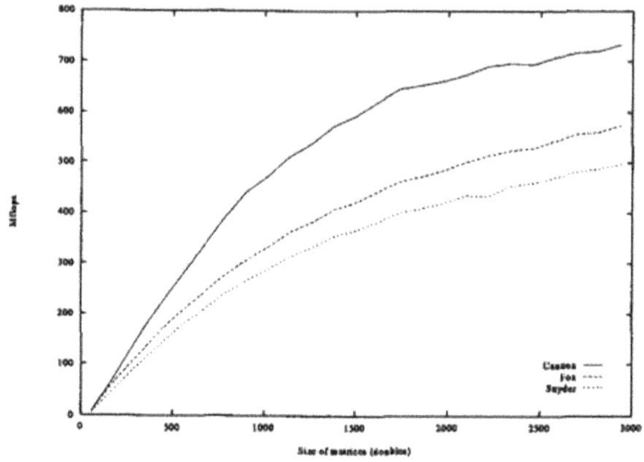

Figure 7: Performances of the matrix product with pre(post)treatments in Mflops according to the matrix size on VolVox IS-860 for a 6×6 torus

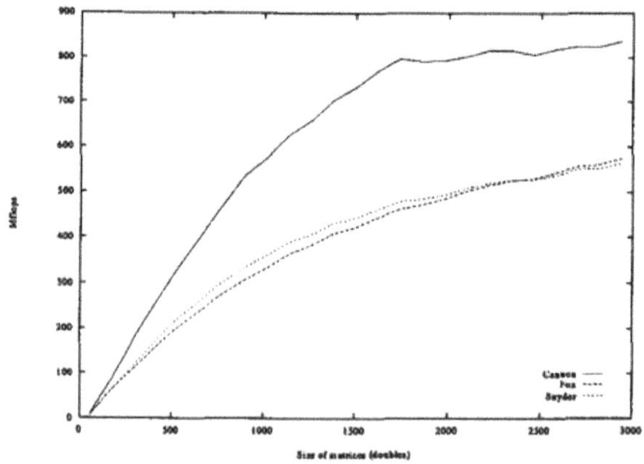

Figure 8: Performances of the matrix product without pre(post)treatments in Mflops according to the matrix size on VolVox IS-860 for a 6×6 torus

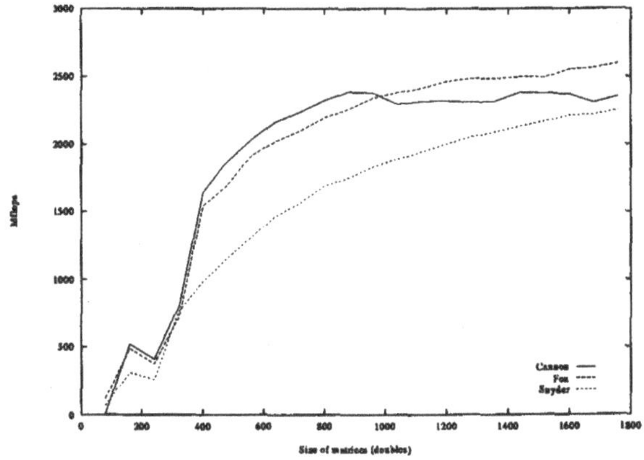

Figure 9: Performances of the matrix product with pre(post)treatments in Mflops according to the matrix size on Intel Paragon XPS/5 for a 8×8 grid

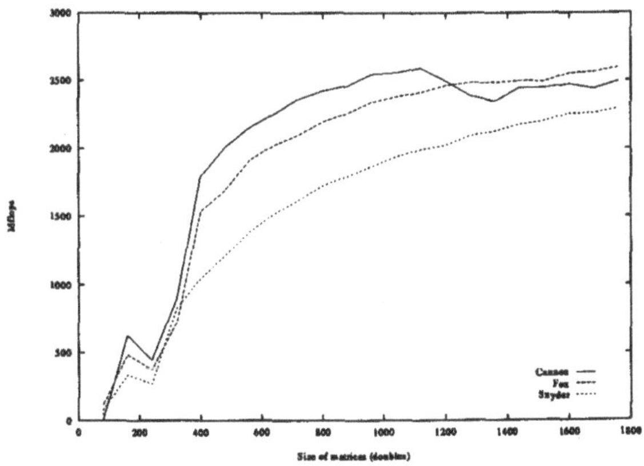

Figure 10: Performances of the matrix product without pre(post)treatments in Mflops according to the matrix size on Intel Paragon XPS/5 for a 8×8 grid

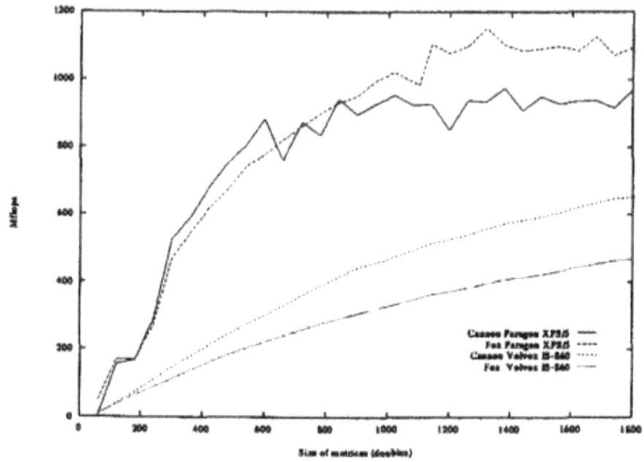

Figure 11: Performances of the matrix product with pre(post)treatments in Mflops according to the matrix size on VolVox IS-860 and Paragon XPS/5 for a 6×6 torus/grid

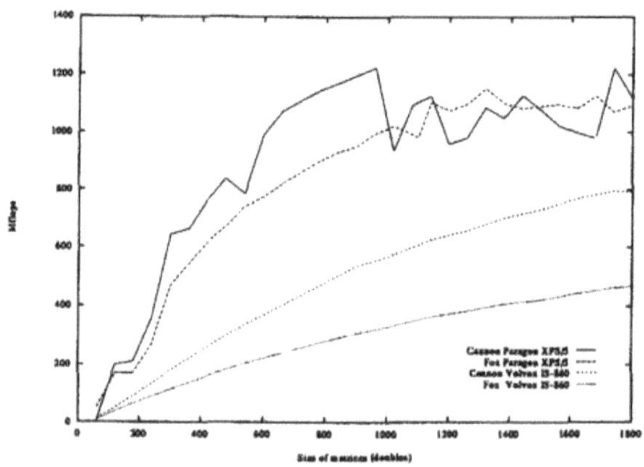

Figure 12: Performances of the matrix product without pre(post)treatments in Mflops according to the matrix size on VolVox IS-860 and Paragon XPS/5 for a 6×6 torus/grid

A Data Parallel Functional Language for Distributed Memory Machines

Herbert Kuchen and Holger Stoltze*

Abstract

Current Implementations of functional languages give insufficient means to control parallelism and communication. Our approach is to support data parallelism by providing a set of primitives on arrays which allow the user to control the parallelism and communication on a high level, disabling problems like deadlocks . Only one unique version of an array may be referenced at a time. This restriction allows arrays to be updated in place and enables the user to control the space requirements of the program. Experimental results demonstrate the efficiency of our data parallel functional language.

1 Introduction

MIMD machines with distributed memory offer high computation power at relatively low costs, but they are still hard to program. Typically, imperative languages, e.g. Occam 2 [7], are used which include features for controlling parallelism and communication at a low level. Non-trivial programs are likely to contain deadlocks, which are hard to detect, especially since errors can hardly be reproduced due to the non-determinism in the language.

Functional languages avoid this problem. Here, the implementation can decide to exploit implicit parallelism and to introduce the required communication automatically. The user specifies an application without caring about parallelism and communication at all. In a functional language, the value of an expression only depends on its denotation and not on the order of computations. This property is called *referential transparency.* It ensures that the subexpressions of an expression $f\ e_1 \dots e_n$ can be evaluated in any order, especially in parallel. Hence, the implementation can spawn processes for the evaluation of $e_1, \dots, e_n$, which inform the process for $f\ e_1 \dots e_n$ when they are ready.

Several parallel implementations of functional languages have been developed. While an efficient implementation on shared memory machines seems to be possible [1], implementations on distributed memory machines (see e.g. [11, 9]) could not

*This work was supported by Graduiertenkolleg Informatik und Technik, RWTH Aachen.

reach the desired speed. The reason is that too much is done automatically by some heuristics, which does not take advantage of knowledge about the application and the data structures used. Such heuristics are, for example, used for the choice of the granularity of processes, for load balancing, and for the data distribution.

It is possible to solve these problems by adding some annotations to the program which allow to control the behavior of these heuristics. This leads to some kind of para-functional programming [5]. Nevertheless, some problems remain. For example, the information is lost, which data are connected from the point of view of the application. Thus, for instance, the elements of an array have to be requested element by element and cannot be transmitted in one message. Some of the previous problems have been solved in the implementation of Concurrent Clean [13]. It allows to control the distribution of processes and data, and it can avoid request messages. However, Concurrent Clean allows a level of programming which very well enables the construction of deadlocks.

Moreover, an important drawback of functional languages remains, namely that the typical data structure used in these languages, the list, is not appropriate for parallel implementations since it forces sequential accesses to their elements.

A key point of our new approach is that we offer arrays and a couple of operations on them, which allow to handle their parts in parallel and move them between different processes (communication). Moreover, it is crucial for an efficient parallel implementation to ensure that at each time only one "version" of an array exists and is referred to. This allows to update them in place, which in turn enables their elements to be accessed in constant time. This "uniqueness" property will be checked by the compiler (see [10, 14] for details).

In contrast to parallel imperative languages, the presented functional language is deterministic. Parallel applications can be roughly divided into real-time applications and super computing. Real-time applications require a fast reaction to a number of inputs. Non-determinism is needed here for reacting to the first available input. However, our approach is aiming at super computing. Here, we have applications which can in principle be handled sequentially, and where the only reason for using a parallel machine is its higher computation power. Non-determinism is not needed for such applications. On the contrary, it would unnecessarily complicate the program development, especially the debugging. Since we are aiming at super computing, we support data parallelism, the typical kind of parallelism in such applications.

The rest of this paper is organized as follows. In Section 2, we discuss some problems for an efficient implementation of functional languages due to their applicative data structures. Section 3 shows a couple of operations which can be used to write parallel programs working on arrays at a high level of abstraction. Section 4 contains experimental results, and in Section 5 we conclude and point out future work.

2 Applicative Data Structures

The typical data structure in functional programs is the *list*. While lists are convenient to specify many applications and have the advantage of enabling easy pattern matching, they have several disadvantages for a parallel implementation. First of all, it is not possible to access the elements in constant time, in contrast to arrays in imperative languages. Hence, lists are slower than arrays w.r.t. random accesses to their elements. But such accesses are needed in a parallel implementation, since we would like several processes to work in parallel at different places of a data structure.

But even if random accesses to the elements can be avoided and the application admits to scan them from left to right, spawning a process at every element, this takes linear time (in the length of the list). This is especially severe, if the computation at each element is small compared to the length of the list, and this is often the case.

Moreover in a distributed environment, every access to an element of the list has to pass the first cons-cell first. The processor storing this cell has to answer all requests for this cons-cell and it is another source of sequentialization. If other data structures, e.g. arrays, are used, this alone does not solve the problem, since in a functional language updates cannot (in general) be done in place, but for every update a new version of the data structure has to be maintained. This has to be done, since the old version may still be needed elsewhere. [4, 15] propose a method to discover, where updates can be done in place. In fact, this is not enough, since we want to be sure, that every update can be done in place, and we want the user to change the program, if this is not the case. He knows best what can be done for some given application.

The usual way to handle different versions is to introduce so-called *modification cells* [10]. Such a cell represents an updated version of a data structure. It contains the value of some particular updated element of the data structure and a pointer to some older version, where the values of all other elements can be found.

The problem with modification cells is similar to that with lists. The elements of a data structure cannot be accessed in constant time but all modification cells may need to be passed before the desired element is found, and, on a distributed memory machine, all accesses to the data structure have to pass the first modification cell. Thus, the processor storing it has to answer all requests for this cell and sequentializes the accesses.

Alternatively to modification cells, the data structure (list, array, ...) can be stored as a tree [8, 11]. A new version of the data structure will be created by adding a new path from a new node to the updated leaf and the rest of the old tree can be shared with the new tree. But again, the elements cannot be accessed in constant time (but logarithmic time). Moreover the root of each tree now gets a bottleneck. All accesses have to pass it and are sequentialized by the processor storing it.

An alternative is to forbid individual updates and to force arrays to be updated as a whole (monolithically). This avoids modification cells and the bottlenecks caused by them. However, this style of programming is not adequate for all applications, e.g. for the histogram problem or quicksort on arrays it is hard to get around individual updates. Moreover maintaining several versions of the whole array leads to significant storage costs.

We can conclude that, if updates are not performed in place but require the management of different versions of a data structure, then either the accesses to the data structure are sequentialized or the style of programming is restricted and a huge amount of storage is required. Hence, the only way to enable efficient parallel algorithms is by guaranteeing that updates can safely be done in place. Our approach is the following. We force the user to program in such a way that only one version of an array is needed at a time. The imperative programmers will do this anyway.

One way to reach this aim is to use monads [16] or mutable abstract data types [6]. Unfortunately, they are too restrictive for parallel programming where we would like several processes (corresponding to different subexpressions) to access the same data structure at the same time as long as this is safe. In our case, the compiler will check the *uniqueness* of arrays. The main idea is that parallel "reading" accesses are allowed. If there is an "updating" access, parallel or subsequent accesses to the old data structure are forbidden (see [10] for details).

3 Operations on Arrays

Almost all interesting applications for large distributed memory machines work on arrays. Thus, we will offer them in our language with the restriction that each instance of such an array has to be unique, i.e. only one version of it can be referred to at a time.

We require all functions and constructors to be strict in arguments "containing" array operations. The rest of the language can be lazy. The reason for this restriction is that functions which are lazy in such arguments would not allow to control the order of array accesses as required in order to fulfill the uniqueness restriction. Dangling references to an already updated array might survive.

We assume the binary type constructor Array to be predefined. Array a b is the type of arrays with index type a and component type b. Moreover, we assume a predefined type

$$\text{Annotations} = \text{Divide [Int]} \mid \text{Allocate [Int]} \rightarrow \text{Int} \mid \text{Cyclic} \mid \ldots$$

When creating an array, a list of annotations may specify the data division and distribution. An empty list of annotations indicates that a default distribution shall be used. Possible annotations are e. g.

Divide [parts] specifies that parts partitions have to be produced.

Divide [rowparts,colparts] specifies that rowparts partitions in horizontal direction and colparts partitions in vertical direction shall be produced. This makes of course only sense for arrays which are at least two dimensional. All partitions have equal size in order to enable an easy exchange between processors.

Allocate f maps partition a_i of the new array a to processor $f\ i$. For many applications, such a general and runtime dependent allocation scheme is not needed. Thus, we also provide some simpler allocation schemes. Let us just show one of them:

Cyclic leads to a cyclic distribution of partitions to processors.

3.1 Basic Operations

Let us now describe the basic operations which we offer on arrays. Here, we will not give a complete list of array operations but rather concentrate on the most important ones. The reader can imagine that more specific or more general variants of some of these operations can also be defined.

3.1.1 Creation and Basic Accessing Operations

create:: (Ix a) ⇒ (a → b) → (a,a) → [Annotations] → Array a b

create $f\ (l,u)$ *atn* generates a new array a, where element a_i has the value $f\ i$ (for $l \leq i \leq u$). The list of annotations *atn* determines the division and distribution of data as explained above. Moreover, it determines the distribution of work. Basic operations working on a partition of an array will be executed where this partition is located (note that this works since all the basic operations only access one array). *create* is executed in parallel.

(!):: (Ix a) ⇒ Array a b → a → b

$a!i$ delivers element a_i of array a.

upd:: (Ix a) ⇒ Array a b → a → b → Array a b

upd $a\ i\ x$ updates array a in place by replacing the old value of a_i by x. *upd* can only be used in an expression which does not cause any concurrent or subsequent accesses to a. We only show *upd* for completeness and recommend to use wherever possible the operations shown below, which update a whole array at a time, since they provide a higher level of programming and work in parallel on the elements of an array (in contrast to a sequence of individual updates).

3.1.2 Monolithical Computation Operations

$$\text{map:: (Ix a)} \Rightarrow \text{(a} \to \text{b} \to \text{c)} \to \text{Array a b} \to \text{Array a c.}$$

map f a replaces in parallel (and in place) every element a_i of array a by $f\ i\ a_i$. Note that the new value of a_i not only depends on a_i but also on its position i, since this is required by many practical applications.

$$\text{fold:: (Ix a)} \Rightarrow \text{(b} \to \text{b} \to \text{b)} \to \text{Array a b} \to \text{b}$$

fold $(\oplus)$ a computes in parallel $\bigoplus_{i=l}^{u} a_i$ where $\oplus$ is an associative and commutative binary operation. The reason for defining *fold* in this unconventional way (i.e. without some initial value and without a direction of folding) is that we want to evaluate *fold* in parallel. The classical *foldl* and *foldr* force a sequential execution. A disadvantage is that we require the user to ensure that the first argument supplied to *fold* is always an associative and commutative function. Otherwise, *fold* will behave non-deterministically.

3.1.3 Communication Operations

The following operations will be used to exchange partitions of an array between processors. It is ensured that each processor receives exactly one new partition, i.e. each partition is replaced by another one. All movements of partitions are done in parallel and update the argument array in place. Thus, these operations provide a powerful means of *communication*. Note that a deadlock is not possible, since there are no individual messages but rather a coordinated group of communications.

$$\text{rotate\_rows: (Ix a)} \Rightarrow \text{(Int} \to \text{Int)} \to \text{Array (a,a) b} \to \text{Array (a,a) b}$$

rotate_rows f M rotates the i-th row of matrix M cyclically by $f\ i$ partitions to the right, i.e. partition $M_{i,j}$ (not to be confused with element $m_{i,j}$) is moved to the old location of partition $M_{i,(j+f\ i\ mod\ colparts)}$ $(0 \leq i \leq colparts,\ 0 \leq j \leq rowparts)$. Analogously, rotate_cols rotates the columns.

3.1.4 How to Fulfill the Uniqueness Property

If it is not possible to fulfill the uniqueness property in order to program some specific application, the user can either sequentialize parts of the program by using the letseq-construct (see below) which evaluates its arguments sequentially, or a physical copy of the array can be created. In an expression

$$\text{letseq } x_1 = e_1, \ldots, x_n = e_n \text{ in } e_{n+1}$$

e_i will be evaluated before e_{i+1} $(i = 1, \ldots, n)$. Moreover, e_{i+1} may refer to $x_1, \ldots, x_i$ in order to access the results of $e_1, \ldots, e_i$. The value of the whole letseq-expression is the value of e_{n+1}.

copy:: (Ix a) ⇒ Array a b → [Annotations] → Array a b.

copy a l creates a physical copy of *a*. Both, the original and the copy, can subsequently be used independently, including updates in place. When copying, the data distribution may differ from the original according to the list *l* of annotations given as the 2nd argument of *copy*. *copy* works in parallel. The access to its first argument *a* is counted as a reading one.

3.1.5 Referring to the Data Distribution

In addition to the previous operations, there is an operation, which only refers to the (permanent) data distribution of an array. When checking the uniqueness of an array this operation (and its arguments) can be ignored.

local_bounds:: (Ix a) ⇒ Array a b → a → (a,a)

local_bounds a i gives a pair with the bounds of the partition of array *a*, which contains the element with index *i*. Example: if a matrix *m* with indexes in $[0,\ldots,99] \times [0,\ldots,99]$ is divided into 4×4 partitions, *local_bounds m* (57, 35) is ((50,25),(74,49)), indicating that the desired element is contained in the submatrix with indexes in $[50,\ldots,74] \times [25,\ldots,49]$.

3.2 Example: Matrix Multiplication

Now, we have a basic set of operations on arrays which allow us to program a broad range of applications efficiently and elegantly. Let us consider an example and see, how matrix multiplication can be programmed using our operations (see Fig. 1). This program is based on the algorithm of Gentleman [3]. This program is almost as short as a typical automatically parallelized functional program, but it behaves like a parallel imperative program (see [10] for details). Note that there is one situation where we needed to sequentialize the computation using a letseq-expression in order to preserve the uniqueness property. We have forced the local multiplication to take place before the rotations. The corresponding non-sequentialized expression

```
repeat n (rotate_rows (const 1) a)(rotate_cols (const 1) b)(localmult a
b c)
```

is rejected by the compiler, since it contains reading accesses to `a` and `b` in `(localmult a b c)` and writing accesses to them in `(rotate_rows (const 1) a)` and `(rotate_rows (const 1) b)`, respectively.

4 Experimental Results

In order to compare the efficiency of the (implementations of the) different parallel languages, we have measured the runtimes of matrix multiplication on a $\sqrt{p} \times \sqrt{p}$

```
n = 100;  p = 4;  const x y = x
repeat 0 a b c = c
repeat (n+1) a b c = letseq c2 = localmult a b c
                     in repeat n (rotate_rows (const 1) a)
                                 (rotate_cols (const 1) b)
                                 c2
localmult a b c =
   map sprod c
   where  sprod (i,j) x =
             let sum = x
                ((fr,fc),(lr,lc)) = local_bounds a (i,j)
             in for  k <- [0..lr-fr]
                     next sum = sum + a!(i,fc+k) * b!(fr+k,j)
                in sum
matmult a b =
   repeat (n/p) (rotate_rows (negate) a)
                (rotate_cols (negate) b)
                (create (const 0) ((0,0),(n-1,n-1))
                                  [Divide [p,p], Cyclic])
```

Figure 1: Matrix multiplication in the data parallel functional language.

torus of T800 transputers (20 MHz). The programs in Occam, Helios-C [12], and our data parallel functional language (DPFL) are based on Gentleman's algorithm [3]. The implicitly parallel functional program (IPF) uses the usual three nested loops without referring to parallelization and communication explicitly. Occam, Helios-C, and DPFL provide a linear speedup, while the implicitly parallel program shows an impressive slowdown due to the communication overhead. Note moreover that due to its huge space requirements, it was not even possible to multiply 500×500 matrices (but only 30×30 matrices)! Taking into account that our prototype implementation was not carried out with the same level of sophistication than the commercial Occam and Helios-C compilers, it is not surprising that we cannot reach the efficiency of the code produced by them. However, we could approach Occam and Helios-C to a factor

	Occam	Helios-C	DPFL	IPF	D/O	D/C
$p \setminus n$	500	500	500	30	500	500
4	244.93	362.49	1253.36	38.05	5.11	3.46
25	38.54	57.96	214.25	38.30	5.56	3.70

Table 1: Runtimes for $n \times n$ matrix multiplication on p processors in seconds.

of about 5.5 and 3.5 respectively (see Table 1, columns D/O and D/C), which shows that, in spite of the much higher level of programming, our data parallel functional language can be efficiently implemented. Experiments with other applications, e.g. the numerical solution of partial differential equations, show similar results.

5 Conclusions and Future Work

Current parallel implementations of functional languages on distributed memory machines are too slow, since they handle too much automatically and since applicative data structures cause some implementation overhead. Our solution is to provide unique arrays together with a set of operations on them which allow to control the parallelism and the communication on a high level. The problems of low level communication with individual messages, i.e. mainly deadlocks, are avoided. The determinism of the functional language is preserved, which clearly facilitates debugging. Our operations on arrays can be understood as specific algorithmic skeletons [2], i.e. parallelization patterns. It remains to be investigated, how far it is possible to automatically translate more general skeletons to our operations.

References

[1] L. Augustsson, T. Johnsson: *Parallel Graph Reduction with the ν-G-machine*, FPCA'89, ACM, 1989, pp. 202-213.

[2] J. Darlington: *Parallel Programming Using Skeleton Functions*, Proceedings of PARLE '93, LNCS 694, 1993, pp. 146-160.

[3] W.M. Gentleman: *Some Complexity Results for Matrix Computations on Parallel Processors*, Journal of the ACM, Vol. 25, 1978, pp. 112-115.

[4] P. Hudak, A. Bloss: *The Aggregate Update Problem in Functional Programming Systems*, ACM Symp. on Principles of Programming Languages, 1985, pp. 300-314.

[5] P. Hudak: *Para-functional Programming in Haskell*, in B.K. Szymanski (ed.): *Parallel Functional Languages and Compilers*, Addison-Wesley, 1991.

[6] P. Hudak: *Mutable Abstract Datatypes or How to Have Your State and Munge It Too*, Yale Research Report YALEU/DCS/RR-914, 1993.

[7] G. Jones, M. Goldsmith: *Programming in Occam 2*, Prentice Hall, 1988.

[8] H. Kuchen, G. Geiler: *Distributed Applicative Arrays, Workshop Parallel Implementation of Functional Languages*, Proceedings appeared as Technical Report no. 90-16, University of Nijmegen, The Netherlands, 1990, pp. 223–235.

[9] H. Kingdon, D. Lester, G.L. Burn: *The HDG-machine: a highly distributed graph-reducer for a transputer network*, Computer Journal, Vol.34(4), 1991, pp. 290-301.

[10] H. Kuchen, R. Plasmeijer, H. Stoltze: *Efficient Distributed Memory Implementation of a Data Parallel Functional Language*, PARLE'94, LNCS, Springer, 1994.

[11] R. Loogen, H. Kuchen, K. Indermark, W. Damm: *Distributed Implementation of Programmed Graph Reduction*, PARLE '89, LNCS 365, 1989, pp. 136-157.

[12] Perihelion Software Ltd.: *The Helios Parallel Operating System*, Prentice Hall, 1991.

[13] R. Plasmeijer, M. van Eekelen: *Functional Programming and Parallel Graph Rewriting*, Addison-Wesley, 1993.

[14] J. Smetsers, E. Barendsen: *Conventional and uniqueness typing in graph rewrite systems*, 13th Conf. on Foundations of Software Technology and Theoretical Computer Science, Bombay, India, LNCS, Springer, 1993.

[15] A.V.S. Sastry, W. Clinger, Z. Ariola: *Order-of-evaluation Analysis for Destructive Updates in Strict Functional Languages with Flat Aggregates*, FPCA'93, ACM, 1993, pp. 266-275.

[16] P. Wadler: *The essence of functional programming*, 19th Symp. on Princ. of Programming Languages, ACM, 1992, pp. 1-14.

Address of Authors:
RWTH Aachen, Lehrstuhl für Informatik II
D-52056 Aachen
Germany
email: {herbert,holger}@zeus.informatik.rwth-aachen.de

A Fixed Sized Regular Array for Matrix Inversion by Monte Carlo Method

G.M. Megson, V.N. Aleksandrov, I.T.Dimov

Abstract

A fixed sized systolic array for inverting a matrix using a Monte Carlo method is presented. The array requires $m_2 + 2m_1 + T + N - 1$ time steps and $(m_2 + 2)TN$ cells for a single row of the inverse where $m_1 m_2 = n^2$, N is the m.e. of the number of chains required to reach given precision, T is the m.e. of chains length in the stochastic process and n is the matrix order. When $m_2 = m$ and $m_1 = n^2/m$ for $m > 0$ a constant the matrix can be restructured to produce a run-time of $n^2/m + 2m + T + N - 1$ and $(m+2)TN$ cells. A simple re-timing produces a multi-pass organisation requiring only $(m+2)T$ cells with the effective length of $m_2 = Nn^2/m$. To compute the complete inverse we can replicate the fixed sized architecture n times. Or re-use a single copy of the architecture n times. The latter has advantages where the matrices are toeplitz or only a few elements of the matrix are required.

1 Introduction

The problem of inverting a real $n \times n$ matrix is of unquestionable importance in the development of many scientific and engineering applications over a range of disciplines. For example real time speech coding, digital signal processing, communications, stochastic modeling as well as many physical problems involving partial differential equations. The direct methods of solution require $O(n^3)$ sequential steps when some elimination or annihilation scheme (e.g non-pivoting Gaussian Elimination, Gauss-Jordan methods) are employed [1]. As a result the computation time for very large problem size or real-time problems is prohibitive and prevents the use of many established algorithms and results. To over come these limitations the recent advances in VLSI manufacturing techniques and the development of so-called regular array synthesis techniques based on manipulation of data dependencies have offered to produce highly efficient, cost effective and massively parallel architectures [10, 11].

We concentrate on Monte Carlo method applied to matrix inverse (MI), since only $O(NT)$ steps are required to find an element of inverse matrix (N is the m.e. of the number of chains and T is a measure on the chains length in the stochastic process, which are independent of n) if pre-processing of matrices is not necessary, and because of the inherent parallel nature of the sampling process for stochastic methods. Thus for sufficiently large n the Monte Carlo approach is theoretically more efficient [3].

Regular arrays for matrix inverse that have been derived from affine and uniform recurrences using linear timing schedules have received considerable attention in the literature [6, 7, 9, 13]. But most of the methods are based on variations of the Gauss-Jordan algorithm or triangularization with the assumption that the matrix being inverted is diagonally dominant or positive definite so that no pivoting is required. If no data is duplicated in the array then approximately $5n$ steps and $O(n^2)$ processors are required for inversion. If data duplication is allowed then $4n$ steps and $O(n^2)$ processors is possible. In [12] the authors demonstrated the existence of an array which required $O(n^4)$ cells and a time bounded above by $4n$. In fact an array with $O(n^2NT)$ cells and time $3n/2 + N + T$ was constructed. In the current paper we extend these results to derive a fixed size array independent of n, N.

2 Stochastic methods and matrix inversion

First we require some definitions. Let the system of linear algebraic equations (SLAE) be presented in the form:

$$x = Ax + \varphi \tag{1}$$

where $x = (x_1, x_2, ..., x_n)^t$ is a $1 \times n$ solution vector and $\varphi = (\varphi_1, \varphi_2, ..., \varphi_n)^t$ is a given vector. Matrix A is a square $n \times n$ matrix and satisfies the condition $\max_{1 \le i \le n} \sum_{j=1}^{n} |a_{ij}| < 1$.
Further suppose that we have Markov chains with n - states. The random trajectory (chain) T_i of length i starting in state k_0 is defined as follows: $k_0 \to k_1 \to \cdots \to k_j \to \cdots \to k_i$, where k_j means the number of the state chosen, for $j = 1, 2, \cdots, i$. Assume that $P(k_0 = \alpha) = p_\alpha$, $P(k_j = \beta | k_{j-1} = \alpha) = p_{\alpha\beta}$ where p_α is the probability that the chain starts in state α and $p_{\alpha\beta}$ is the transition probability to state β after being in state α. Probabilities $p_{\alpha\beta}$ define a transition matrix P. We require that $\sum_{\alpha=1}^{n} p_\alpha = 1$, $\sum_{\beta=1}^{n} p_{\alpha\beta} = 1$, $\alpha = 1, 2, ..., n$ and that the distribution $(p_1, ..., p_n)^t$ is acceptable to vector g and similarly that the distribution $p_{\alpha\beta}$ is acceptable to A [15]. Now consider the problem of evaluating the inner product $(g, x) = \sum_{\alpha=1}^{n} g_\alpha x_\alpha$ of a given vector g with the vector solution of (1). It is known [15] that the mathematical expectation of r.v. $\Theta^*[g]$ is:

$$E\Theta^*[g] = (g, x) \text{ where } \Theta^*[g] = \frac{g_{k_0}}{p_{k_0}} \sum_{j=0}^{\infty} W_j \varphi_{k_j} \text{ and } W_0 = 1,\ W_j = W_{j-1} \frac{a_{k_{j-1} k_j}}{p_{k_{j-1} k_j}} \tag{2}$$

To find the inverse C of some matrix B we must first compute the elements of matrix $A = I - B$, where I is the identity matrix. Thus $C = \sum_{i=0}^{\infty} A^i$, which converges if $\|A\| < 1$. To find the element $c_{rr'}$ of the inverse matrix C we must set $g = e(r) = (0,...,0,1,0,...,0)$, where the one is in the r-th place, and $\varphi = e(r')$ [15]. Thus the Monte Carlo method becomes $c_{rr'} \approx \frac{1}{N}\sum_{s=1}^{N}[\sum_{(j|k_j=r')} W_j]_s$ where $W_j = W_{j-1}a_{k_{j-1}k_j}/p_{k_{j-1}k_j}$, $(j|k_j = r')$ means that only W_j for which $k_j = r'$ are included in the sum and N is the number of chains. Since W_j is included only into the corresponding sum for $r' = 1,2,...,n$ then the same set of N chains can be used to compute a single row of the inverse [15].

Define the statistical error of the method, assuming that we have the random variable (r.v.) θ with mathematical expectation $E\theta$ and average $\bar{\theta}$. Suppose that we have N independent realizations of θ. Then it is known [15] that $\bar{\theta}$ has a normal distribution with the same average $E\theta$. The so called **probable error** , which gives the order of error, is $r_N = 0.6745\sqrt{D\theta/N}$, where $P\{|\bar{\theta} - E(\theta)| < r_N\} \approx 1/2 \approx P\{|\bar{\theta} - E(\theta)| > r_N\}$.

Applying Monte Carlo methods with **optimal** or **almost optimal** frequency function [4, 12] we reduce the variance (and N respectively) for a given probable error. An example showing the algorithmic efficiency of the Monte Carlo Method with Almost Optimal frequency function (MAO) in comparison with the Usual (Monte Carlo) Method (UM) (with uniformly distributed frequency function) is presented in [12]. There is shown that N is about 10 times less for MAO in comparison with UM.

3 Uniform Recurrences

Following the work in [12] which derived a systolic array for matrix inversion we derive a fixed sized array (i.e an array independent of the matrix order n). Our starting point is a review of the uniformized algorithm in [12] together with a description of the computation as a convex polyhedral domain in 4-D. For convenience let us assume that $t = 1(1)T$ and $r = 1(1)N$ are indexes representing the current observation and the current trial of the process. Also define the $n \times n$ matrices $E = E_{pq} = a_{pq}/p_{pq}$ and $C = (c_{ij})$ as the input and inverse matrices respectively. By the sequence of program transformations in [12] we derived the uniformized for-loop program for computing c_{ij} the elements of the matrix inverse shown in Fig 1.

The loops involving p and q remove explicit use of run-time dependent variables [12]. The order of p and q is not significant. Before we can apply an array synthesis method the program has to be fully indexed and uniformized so that a regular array is guaranteed. Provided the full indexing is consistent [14] any choice can be used. We choose the ordering (p,q,t,r) to make projections of the architecture more intuitive. Thus with $k_{0,0,t,r} = k_{0,r}$, $w_{0,0,0,r} = w_{0,r}$, and $E_{p,q,0,0} = E_{p,q}$ which demonstrates that the Monte-Carlo algorithm can be represented geometrically as a composite convex polytope in four dimensions with the assignment statements mapped to integer co-ordinate positions. The variables $R_{p,q,0,r} = R_{p,q} = p$ and $C_{p,q,0,r} = C_{p,q} = q$ which represent matrices of row and column indices together with $\bar{w}$ to remove the non-uniformity between T and w are introduced. This last transformation produces an essentially uniform program except that it requires knowledge of j and $c_{i,j}$. The latter can be treated as a simple variable (rather than an indexed one) because i and j are constant and not part of the computation domain of our nested loop program. However for the sake of completeness we require variables which will chart the movement of the $c_{i,j}$ value through the space defined by (p,q,t,r). In addition we know (from the previous section) that any of our programs can be used for any $j = 1(1)n$. Consequently a whole row of the inverse matrix can be computed in parallel. To exploit this fact we introduce the variables v and J to pipelined the elements in row i of the inverse and the values of j for this row. Letting $E_{p,q,0,0} = E_{p,q}$, $R_{p,q,0,0} = R_{p,q}$, $C_{p,q,0,0} = C_{p,q}$ as before and $k_{0,0,0,0} = i$, $v_{m_2+1,q,0,0} = c_{i,j} = 0$ and $J_{m_2+1,q,0,0} = j$ for $q = m_1 + 1(1)2m_1$ we produce the program in (Fig. 1). The functions h_1, h_2, h_3 are introduced for convenience and compute the conditional statements of the algorithm.

Clearly the assignments can be allocated to unique integer co-ordinate positions in the 4-D hypercube D defined by the vertices

$$\begin{array}{llll}
(0,1,1,1) & (0,2m_1,1,1) & (m_2+1,1,1,1) & (m_2+1,2m_1,1,1) \\
(0,1,T,1) & (0,2m_1,T,1) & (m_2+1,1,T,1) & (m_2+1,2m_1,T,1) \\
(0,1,1,N) & (0,2m_1,1,N) & (m_2+1,1,1,N) & (m_2+1,2m_1,1,N) \\
(0,1,T,N) & (0,2m_1,T,N) & (m_2+1,1,T,N) & (m_2+1,2m_1,T,N)
\end{array}$$

where $m_1 = n$, $m_2 = n$, which itself is composed of small convex regions associated with the separate nested loops and assignments. Thus we have a system of recurrence equations defined over different domains inside D so that the data dependencies are piece-wise linear over the whole domain. Indeed by treating the subscripts as vectors (or co-ordinates) and subtracting variable indexes on the right of assignment from those on the left the dependencies for the whole computation can be summarized by the data dependency vectors

$$(0,0,0,1) \quad (0,1,0,0) \quad (1,0,0,0) \quad (0,0,1,0)$$

which are simply the direction vectors of the 4-D space. Thus the data dependency graph (DDG), formed by connecting co-ordinates in D according to vectors given above, must be acyclic.

Observe that once the point $z = (p,q,t,r)$ has been computed at time $time(z)$ we can immediately compute the functions at points $(p+1,q,t,r)$, $(p,q+1,t,r)$, $(p,q,t+1,r)$ and $(p,q,t,r+1)$ on time step $time(z)+1$. We have four non-collinear points which must lie on the same (timing) hyperplane thus the normal to the plane is $\lambda = (1,1,1,1)$. In addition we require that the vertex $(0,1,1,1)$ of the domain is computed on the first time step so that $\alpha = 2$. Consequently

$$time((p,q,t,r)) = \lambda^t z - \alpha = p + q + t + r - 2. \qquad (3)$$

and it is easy to see that the maximum degree of parallelism is obtained from this family of timing planes. It follows straightforwardly that any linear allocation of the DDG nodes in D onto a processor network which is conflict-free (i.e

```
for r = 1 to N do
  {
    k_{0,0,0,r} = k_{0,0,0,r-1}; w_{m_2+1,n,0,r} = 1;
    for q= 1 to m_1 do
      {
        for p = 1 to m_2 do
          {
            E_{p,q,0,r} = E_{p,q,0,r-1};
            R_{p,q,0,r} = R_{p,q,o,r-1}; C_{p,q,0,r} = C_{p,q,0,r-1} ;
          }
        v_{m_2+1,m_1+q,0,r} = v_{m_2+1,m_1+q,0,r-1}, J_{m_2+1,m_1+q,0,r} = J_{m_2+1,m_1+q,0,r-1};
      }
    for t = 1 to T do
      {
        k_{0,0,t,r} = MP(k_{0,0,t-1,r}); \bar{k}_{0,0,t,r} = k_{0,0,t,r-1};
        \bar{w}_{n+1,0,t,r} = 1;
        for q = 1 to m_1 do
          {
            k_{0,q,t,r} = k_{0,q-1,t,r}; \bar{k}_{0,q,t,r} = \bar{k}_{0,q-1,t,r};
            T_{0,q,t,r} = 0;
            for p = 1 to m_2 do
              {
                k_{p,q,t,r} = k_{p-1,q,t,r}; \bar{k}_{p,q,t,r} = \bar{k}_{p-1,q,t,r};
                E_{p,q,t,r} = E_{p,q,t-1,r};
                R_{p,q,t,r} = R_{p,q,t-1,r}; C_{p,q,t,r} = C_{p,q,t-1,r};
                T_{p,q,t,r} = h_1(T_{p-1,q,t,r}, E_{p,q,t,r}, R_{p,q,t,r}, C_{p,q,t,r}, k_{p,q,t,r}, \bar{k}_{p,q,t,r} );
              }
            \bar{w}_{m_2+1,q,t,r} = h_2(T_{m_2,q,t,r}, \bar{w}_{m_2+1,q-1,t,r});
          }
        w_{m_2+1,m_1,t,r} = w_{m_2+1,m_1,t-1,r}*\bar{w}_{m_2+1,m_1,t,r};
        for q = n+1 to 2n do
          {
            w_{m_2+1,q,t,r} = w_{m_2+1,q-1,t,r};
            k_{m_2+1,q,t,r} = k_{n,q-1,t,r};
            J_{m_2+1,q,t,r} = J_{m_2+1,q,t-1,r};
            v_{m_2+1,q,t,r} = h_3(v_{m_2+1,q,t-1,r}, w_{m_2+1,q,t,r}, k_{m_2,q,t,r}, J_{m_2+1,q,t,r});
          }
      }
    for q = m_1+1 to 2m_1 do
      v_{m_2+1,q,T,r} = v_{m_2+1,q,T,r-1} + v_{m_2+1,q,T,r};
  }
```

Figure 1: Uniform Algorithm for a single row of the inverse

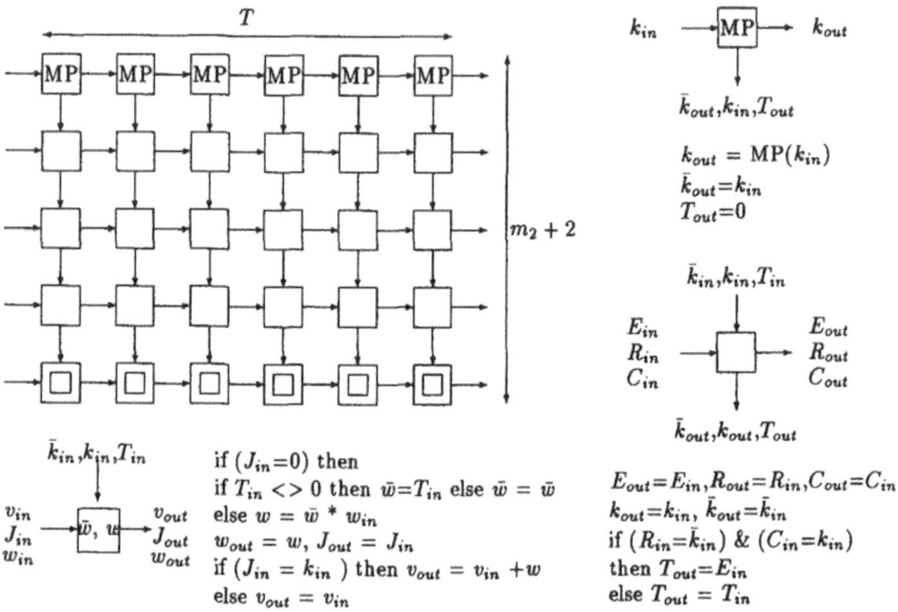

Figure 2: Systolic Array for Monte Carlo Inversion ($r = 1$)

no two computations with the same time step mapped to the same processor) has the total computation time

$$time((m_2+1, 2m_1, T, N)) = m_2+1+2m_1+T+N-2 = m_2+2m_1+T+N-1 \quad (4)$$

where a step is the cost of the most complex function at a DDG node. Notice that this bound on the time is independent of the allocation of computations to processors which has yet to be determined.

4 Fixed Sized Systolic Arrays

In [12] we considered a simple allocation defined by

$$s = (0, 1, 0, 0)^t \quad (5)$$

which projects D onto a 3-D processor topology by removing the q direction. Notice that the dimensions of this array are $m_2 + 2$, T and N giving a total of $(m_2 + 2)TN$ cells for finding a single row of the inverse. Indeed the choice of projection prevents the range $q = 1(1)2m_2$ from contributing to the cell count thus saving $2m_2$ cells.

The array structure for the case when $N = 1$ is given in Fig 2. The full array is produced by replicating this design in the r direction N times, observe that the recurrences in Fig 1 already deal with the necessary pipelining of the matrix input and accumulation of the inverse row values in each of the processor planes (see Fig 3).

Our task is to place bounds on the size of the array. Observe that the loops involving p ad q are introduced into the program artificially to index the elements of the input matrix. The order of indexing is completely arbitrary. In [12] we chose $m_1 = m_2 = n$ for an $n \times n$ matrix however it should be clear that any choice satisfying $m_1 m_2 = n^2$ is admissible. Since the array size depends on p and not on q it is sensible to trade-off the sizes of m_1 and m_2 to minimize hardware. Suppose we choose $m > 0$ and set $m_2 = m$ then $m_1 = n^2/m$ and the input matrices $E_{p,q}$, $C_{p,q}$, $R_{p,q}$ etc must be modified accordingly. For example, when $m = 1$, $m_2 = 1$ and $m_1 = n^2$ so that the input data is a 1-D set of elements with length n^2, where element (p,q) is found in the $(p-1)*n+q$ position of the input. For instance in the 3×3 case

$E_{p,q} \mapsto$	E_{11}	E_{12}	E_{13}	E_{21}	E_{22}	E_{23}	E_{31}	E_{32}	E_{33}
$R_{p,q} \mapsto$	1	1	1	2	2	2	3	3	3
$C_{p,q} \mapsto$	1	2	3	1	2	3	1	2	3

It follows that the array requires $3TN$ cells which is independent of n and n^2+N+T steps In general we have $(m+2)TN$ cells and $n^2/m+N+T+m-1$ steps where m is chosen arbitrarily.

Next consider N and T. The array in Fig 2 illustrates the case for $N = 1$. When $N > 1$ the design is three dimensional- Fig 3 with copies of Fig 2 in each plane in the r direction. Clearly the design can be reduced to 2-D by projecting out the r-axis. Unfortunately this projection creates data conflict and requires substantial re-timing of the data. In particular we pipeline N copies of the data through the 2-D array one after another this makes the effective length of $m_1 = Nn^2/m$ so that mT cells compute a single row of the inverse in $Nn^2/m + T + m$ steps.

5 Parameter Estimation

Now let us outline some estimates on N and T. Using an almost optimal frequency function and according to the principal of collinearity of norms [4] we choose $p_{\alpha\beta}$ proportional to the $|a_{\alpha\beta}|$. Depending on estimates of the convergence of the Neuman series one of the following stopping rules can be selected to drop Markov chains: when $|W_i| < \delta$ [15] or when a chain enters an absorbing state [2, 3, 8]. In case of **Monte Carlo method without absorbing states (MWA)** ($\varphi_{k_j} = \delta_{k_j\beta}$ in (2) if $\alpha\beta$-th entry of inverse matrix is computed) the bounds on T and $D\Theta^*$ are $T \leq \frac{|\log\delta|}{|\log\|A\||}$ and $D\Theta^* \leq \frac{\|\varphi\|^2}{(1-\|A\|)^2} \leq \frac{1}{(1-\|A\|)^2}$. Consider the **Monte Carlo methods with absorbing states (MA)** [2, 3, 8], where $\eta^*[g]$ means r.v. $\eta_T[g]$ (T is the length of the chain when absorption take place) taken over infinitely long Markov chain. The bounds on T and $D\eta^*[g]$ [2, 3] if the chain starts in state $r = \alpha$ and $p_{\alpha\beta} = |a_{\alpha\beta}|$, for $\alpha, \beta = 1, 2, ..., n$ are

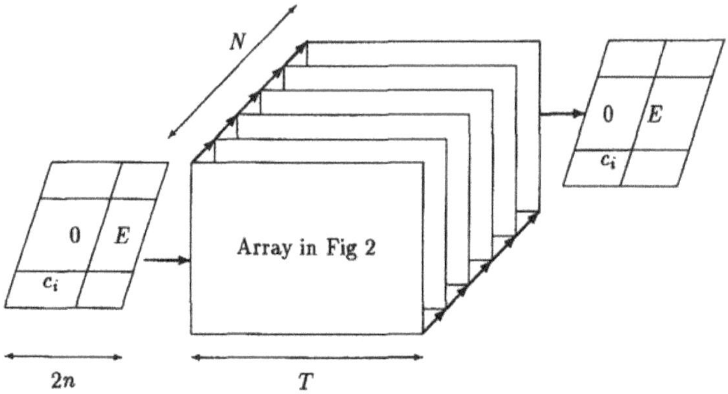

Figure 3: Full Array for $r = 1(1)N$

$E(T|r = \alpha) \leq \frac{1}{(1-\|A\|)}$, and $D\eta^*[g] \leq \frac{1}{(1-\|A\|)^2}$. According to the Central Limit Theorem $N \geq \frac{0.6745^2}{\epsilon^2} D\eta^*[g]$ for the given error ϵ. Thus $N \geq \frac{0.6745^2}{\epsilon^2} \frac{1}{(1-\|A\|)^2}$ is a lower bound on N. Accepting low precision solutions (e.g. $10^{-2} < \epsilon < 1$) it is clear that $n >> N$ as $n \mapsto \infty$. Consider N and T as functions of $\frac{1}{(1-\|A\|)}$. Thus in both methods T is bounded by $O(\sqrt{N})$, since in MWA $T < \sqrt{N}\frac{\epsilon|\log \delta|}{0.6745}$ and in MA $T \leq \sqrt{N}\frac{\epsilon}{0.6745}$. Results in [5] show the fact that for sufficiently large N, $T \approx \sqrt{N}$. Results of experiments for different matrix norms $\|A\|$ and precision ϵ are outlined in Table 1.

ϵ	δ	$\|A\|$	T_{min}	T_{max}	N	$\|A\|$	T_{min}	T_{max}	N
0.4	0.1	0.9	7	16	284 - 2500	0.5	0	4	11 - 100
0.3	0.1	0.9	6	17	505 - 4444	0.5	0	4	20 - 177
0.2	0.1	0.9	6	17	1137 - 10000	0.5	0	4	45 - 399
0.1	0.1	0.9	7	17	4549 - 40000	0.5	0	4	181 - 1599
0.05	0.05	0.9	10	21	18198 - 160000	0.5	0	5	727 - 6398

Table 1: Method without absorbing

Next denote by $\mathcal{N}$ the maximum of N and by $\mathcal{T}$ the maximum of T. Thus the time to evaluate a row of the inverse is bounded by $\mathcal{N}\ n^2/m + \mathcal{T} + m$ steps on $m\mathcal{T}$ cells. Fortunately $\mathcal{T} = O(\sqrt{\mathcal{N}})$ and thus the time for a row of inverse is determined by: $\mathcal{N}\ n^2/m + \sqrt{\mathcal{N}} + m$. The run-time for full inverse is the same but $mn\mathcal{T} = O(mn\sqrt{\mathcal{N}})$ cells are required.

Remark 1: If in the Monte Carlo method without absorbing states the stopping rule is $|W_i| < \delta$, the small change in the program of the cells in last row on Fig. 2, i.e. only these W_i with $|W_i| \geq \delta$ are added to the sum for the computed element of inverse. Therefore taking $T = \mathcal{T}$ we ensure the worst case appearance.

Remark 2: In case when $\|A\| > 1$ new efficient approaches [16] based on the solutions of Boundary Value Problems can be applied to the MI.

6 Conclusion

We have derived a fixed sized systolic array for inverting a matrix using a Monte Carlo method with the following parameters. The array requires $m_2 + 2m_1 + T + N - 1$ time steps and $(m_2 + 2)TN$ cells for a single row of the inverse where $m_1 m_2 = n^2$, N is the m.e. of the number of chains, T is the m.e. of chains length and n is the matrix order. When $m_2 = m$ and $m_1 = n^2/m$ for $m > 0$ a constant the matrix can be restructured to produce a run-time of $n^2/m + 2m + T + N - 1$ and $(m + 2)TN$ cells. A simple re-timing produces a multi-pass organisation requiring only $(m + 2)T$ cells with the effective length of $m_2 = Nn^2/m$. Notice also that N and T depend only on the matrix norm and precision and do not depend on the matrix size.

Following the above discussion to compute the complete inverse we can replicate the fixed sized architecture n times in which case the time is bounded by $n^2/m + 2m + \mathcal{T} + \mathcal{N} - \infty$ on $(m+2)n\mathcal{T}\mathcal{N}$ cells. Or re-use a single copy of the architecture n times. The run-time then is bounded by $n^3/m + 2m + \mathcal{T} + \mathcal{N} - \infty$ on $(m + 2)\mathcal{T}\mathcal{N}$ cells.

In particular it should be noted that the Monte-Carlo method is well suited to large problems where other solution methods are impractical or impossible for computational reasons or when a few elements or one row of inverse matrix are required to be computed.

References

[1] Bertsekas D.P. and Tsitsiklis , *Parallel and Distributed Computation* , Prentice Hall, 1989

[2] Curtiss J.H., Monte Carlo methods for the iteration of linear operators. *J. Math Phys.*, vol 32, No 4 (1954), pp.209–232.

[3] Curtiss J.H. *A Theoretical Comparison of the Efficiencies of two classical methods and a Monte Carlo method for Computing one component of the solution of a set of Linear Algebraic Equations.* ,Proc. Symposium on Monte Carlo Methods , John Wiley and Sons, 1956, pp.191-233.

[4] Dimov I. *Minimization of the Probable Error for Some Monte Carlo methods.* Proc. Int. Conf. on Mathematical Modeling and Scientific Computation, Varna, 1991

[5] Dimov I. and O.Tonev, Random Walk on Distant Mesh Points Monte Carlo Method. *Journal of Statistical Physics,* Vol. 70, Nos.5/6, pp. 1333–1342,1993.

[6] Delosme J.M., *A parallel algorithm for the algebraic path problem* Parallel and Distributed Algorithms, eds. Cosnard et all, Bonas, France, 1988, pp67-78.

[7] El-Amawy A., A Systolic Architecture for Fast Dense Matrix Inversion, *IEEE Transactions on Computers,* Vol.C-38, No3, pp.449-455, 1989.

[8] Forsythe G.E. and Leibler R.A. *Matrix Inversion by a Monte Carlo Method,* MTAC Vol.4 , 1950 , pp. 127-129.

[9] Kung S.Y., S.C.Lo and P.S.Lewis, Optimal Systolic Design for the Transitive Closure and shortest path problems, *IEEE Transactions on Computers* , Vol.C-36, N05, 1987, pp.603-614.

[10] Kung S.Y., *On Supercomputing with systolic wavefront array processors* , IEEE Proc. Vol.72, pp.867-884, 1984.

[11] Kung H.T. and C.E.Leiserson, *Systolic Arrays for VLSI* , in sparse matrix proceeding 1978, SIAM, 1979, pp256-282.

[12] G.M. Megson, V.N. Aleksandrov, I.T. Dimov, *Systolic Matrix Inversion Using a Monte Carlo Method*, To appear in Journal of Parallel Algorithms and Applications, Vol 4, No 1.

[13] Megson G.M., A Fast Faddeev Array, *IEEE Transactions on Computers*, Vol.41, N012, December 1992, pp.1594-1600.

[14] Rapanotti L. , G.M.Megson, *Pre-processing in SADE*, TR-431, Dept. Computing Sci., Univ. of Newcastle upon Tyne.
Algebraic

[15] Sobol' I.M. *Monte Carlo numerical methods.* Moscow, Nauka, 1973 (Russian)(English version Univ. of Chicago Press 1984).

[16] Sabelfeld K.K., *Monte Carlo methods for Boundary Value Problems* , Nauka , Novosibirsk, 1989.

Address of the Authors:
Department of Computing Science,
University of Newcastle
Claremont Road Newcastle-Upon-Tyne NE1 7RU UK
Bulgarian Academy of Sciences
Acad.G.Bonchev str. Bl.25-A, 1113 Sofia
Bulgaria

Automatic parallelization of numerical computations in fields and lattices *

Nathalie Revol Jean-Louis Roch
Institut IMAG - Laboratoire LMC
46, avenue F. Viallet
38031 Grenoble cedex, France
Nathalie.Revol@imag.fr, Jean-Louis.Roch@imag.fr

Abstract

We present a new approach of automatic parallelization, using the semantic of operations. The basic tool is an algorithm for the parallel evaluation of arithmetic circuits with lattice computations. It can also be used for semi-ring, ring and field computations ([2, 5, 9, 10, 8]). After having detailed this algorithm, we present its usage in automatic parallelization. Finally we present experiments giving tight bounds in the case of infinite precision arithmetic as well as some linear algebra problems, and trivial bounds on a $\mathcal{P}$-complete problem.

1 Introduction

Our goal is the measurement of the fine grain parallelism of a sequential algorithm; we determine if there is a parallel program corresponding to this algorithm, working in polylogarithmic time, with a polynomial number of processors. This study is based on the algebraic properties of the operations involved in the program, like fields, in the realm of numerical computations. In section 2, we define the class of programs called arithmetic circuits and the fundamental evaluation operations. A general class of programs is straight-line programs, which corresponds to the arithmetic circuit model proposed by von zur Gathen [4]. We could transform every circuit into arithmetic expressions, but this would lead to an exponential number of operations; we thus work directly on the DAG (Direct Acyclic Graph) representing the

*This work is supported by the French Council of Scientific Research CNRS and the Ministery of Research and Education MRES

arithmetic circuit, instead of the equivalent arithmetic forest.
In section 3, we detail a new algorithm for lattice computations, based on the Eval, Shunt and Group procedures defined before, and we sketch the proof of its parallel complexity, which is of time $\mathcal{O}(\log(n)\log(nd))$ if n is the number of nodes of the circuit and d is its algebraic degree, with a polynomial number of processors.
In section 4 we sketch how this algorithm can be used in order to automatically parallelize a sequential program. Finally, in the last section, we present some predictions for the parallel complexity of some numerical problems, as well as our experimental results.

Why computing in a lattice? A lattice is an algebraic structure occurring in some problems such as graph theory (computing a shortest path involves $(\mathbb{R}^+, max, +, 0, 0)$ lattice), as well as production modelling or performance evaluation (timed Petri nets).
A lattice of particular interest is the Boolean algebra, since in practice programs are based on this algebra. We expect to extract more parallelism by working at this low level: indeed Boolean operations can be better combined than integer or floating-point operations. From a theoretical point of view, an open question is whether $\mathcal{NC} \subsetneq \mathcal{P}$ or $\mathcal{NC} = \mathcal{P}$ ([11]). $\mathcal{P}$ is the set of problems solvable with time and space polynomial in the entry size. $\mathcal{NC}$ is the set of log-space uniform parallel programs that compute within a polylogarithmic time with a polynomial number of processors.
A reference $\mathcal{P}$-complete problem, the MVCP (monotone circuit value problem, cf [6]), is a Boolean problem, and studying its parallelization leads to general results for $\mathcal{P}$ and $\mathcal{NC}$.
In what follows, we shall consider semi-rings, rings, fields and distributive lattices, with a greatest and a smallest elements (it is always possible to add dummy elements in order to meet this condition).

2 Arithmetic circuits

2.1 Arithmetic circuits

In the following, arithmetic circuits are represented by DAGs defined as follows (cf [4]): given $(S, +, \times, 0, 1)$ a commutative algebraic structure (semi-

ring, ring, field or lattice), an arithmetic circuit is a DAG satisfying [8]:
1. edges are weighted by functions over S.
2. each node is labelled as one of the three types: a leaf (associated with a value $\in S$), a $+$ node, a $\times$ node.
3. the in-degree of a leaf is 0, the in-degree of $+$ and $\times$ nodes is ≥ 1.
4. every edge is directed bottom-up (leaving the leaves).
The number of nodes is denoted by n. An algorithm is said to have parallel complexity equal to $CREW(f(n), g(n))$ if its computational time on a CREW (Concurrent Read Exclusive Write) PRAM is $t(n) = \mathcal{O}(g(n))$, using $\mathcal{O}(f(n))$ processors (cf [7]).

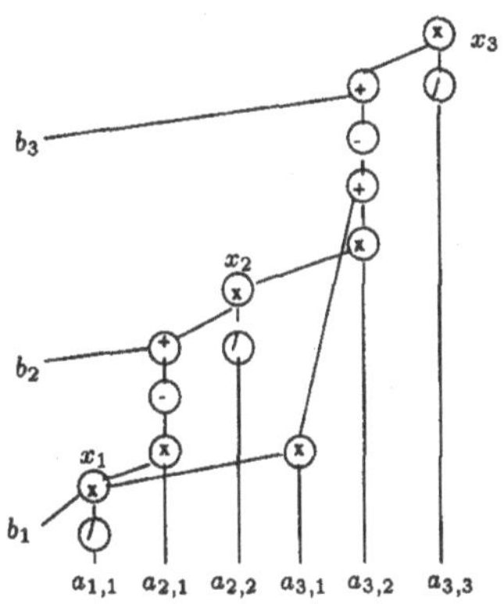

Figure 1: *Solving of a linear lower triangular* 3×3 *system.*

For the solving of a linear triangular 3×3 system, classical parallelization would lead to a parallel execution time of $\mathcal{O}(n)$, which is the length of the critical path, whereas we predict a $CREW(n^{2.37}, (\log^2(n))$ complexity.

2.2 Fundamental procedures

The important properties we use are the associativity and the distributivity of $+$ over $\times$ in finite sets, and also commutativity in non finite sets (otherwise tables can be used, instead of functions).

Eval An obvious procedure is the **Eval** one: an operation is performed if every operand is evaluated. Since any internal node has its arity bounded by n, it has at the most n children, and computing the result of an n-ary operation is equivalent to an iterated product; performing n iterated products has parallel complexity $CREW(n^2, \log n)$, if we assume that any binary $+$, $\times$ operation can be performed in unit time.

Shunt The **Shunt** (cf [2, 5, 9, 8]) exploits the algebraic properties: the edges are weighted with simple functions (affine functions for computations on semi-rings or rings, $\frac{ax+b}{cx+d}$-like functions for computations on fields). When a grandfather operation has one grandson which is a leaf, it shunts his son (fig. 2).

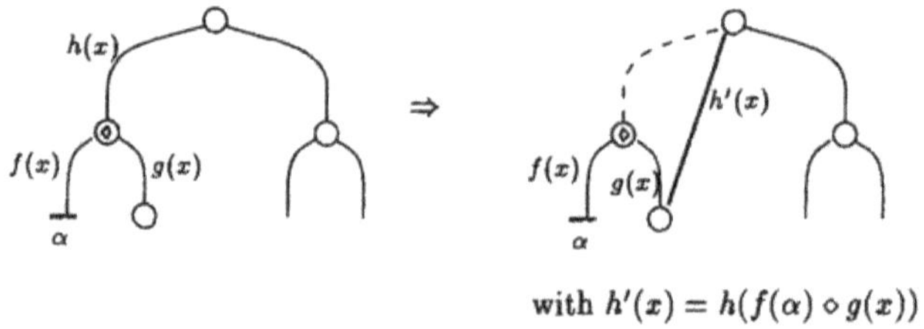

Figure 2: **Shunt** *operation.*

The functions weighting the edges are stable under such transformations, i.e. h' still has a simple form.
Because of the variable arity, only an internal node having one single non leaf child can be shunted, and to prepare itself it computes its partial value (i.e. evaluate the sum or the product of its leaf children) and put it on the edge to one of its leaf children, so as to have only one leaf child. During the **Shunt** it computes the number of non leaf children. Each computation can be considered as n simultaneous iterated products. Then every internal node looks for "shuntable" children and shunts them; this updating step requires only one processor per child. If several functions occur for the same child, another iterated product is required. Hence, the overall complexity of **Shunt** is $CREW(n^2, \log(n))$.

Group We have extended the usual definition of arithmetic DAGs, internal nodes can have in-degree ≥ 1, i.e. they can have one, two or more operands.

We then group + and × nodes in order to fully exploit their variable arity (cf [8]). For this purpose, we define four auxiliary matrices from the adjacency matrix U, the coefficients of which represent the weighting functions:
- U^{++} (resp. $U^{\times\times}$) is the matrix of edges between two + (resp. ×) nodes
- $U^{+\cdot}$ (resp. $U^{\times\cdot}$) is the matrix of edges from a + (resp. ×) node to a × (resp. +) node or to a leaf
The grouping operation can be written as matrices operations:

$$\begin{cases} U^{++} & \longleftarrow & U^{++}.U^{++} \\ U^{+\cdot} & \longleftarrow & U^{++}.U^{+\cdot} + U^{+\cdot} \end{cases} \text{and} \begin{cases} U^{\times\times} & \longleftarrow & U^{\times\times}.U^{\times\times} \\ U^{\times\cdot} & \longleftarrow & U^{\times\times}.U^{\times\cdot} + U^{\times\cdot} \end{cases}$$

The **Group** procedure has the same complexity as a matrix product: $CREW(M(n), \log(n))$, where $M(n)$ is the minimal number of processors required to multiply two $n \times n$ matrices in time $\mathcal{O}(\log(n))$ ([1]).

Remark

It is possible to treat programs with few tests: the weighting functions then become piecewise affine functions, and if the number of tests is small and bounded it is possible to compose these functions in constant time.

3 A general algorithm for lattice computations

Our algorithm consists in applying successively the three procedures **Eval**, **Shunt** and **Group**.

Algorithm Lattice Straight-Line Program
/* Input: the DAG representing the arithmetic circuit */
/* Output: a set of non connected nodes containing their value */
/* The functions created by the Shunt mechanism are stored on the edges */
while (there is a non evaluated node) **loop**
Phase
Grouping: matrix multiplication to group + and × nodes
Eval: every + or × node having all its children being leaves computes its value
and disconnects itself from its children
every internal node computes its partial value

[1] this number is roughly bounded between $\mathcal{O}(n^2)$ and $\mathcal{O}(n^3)$, the best known bound is $\mathcal{O}(n^{2.37})$ [3]

Shunt: every node with one operation child having only one non leaf child shunts this child

end_loop

Complexity. If we use a kind of height, looking like Miller, Ramachandran and Kaltofen's one, defined by $h(v) = min(h_{+,\times}, h_{\times,+})\times$ the number of nodes in U_v the sub-circuit computing the node v, we can prove by induction on the size of U_v that this height is divided by a constant factor at each step. Hence the number of steps is $\mathcal{O}(\log(h))$, and [8] have shown that $h \leq 1/2n^2d + n^2$, where $d = min(d_{+,\times}, d_{\times,+})$. Each step having a $CREW(M(n), \log(n))$ complexity, we obtain a global $CREW(M(n), \log(n)\log(nd))$ complexity.

Special cases When the algebraic underlying structure is a semi-ring, a ring or a field, we can simplify the algorithm: if the $\times$ nodes are binary nodes, and if they are not grouped, we obtain Miller, Ramachandran and Kaltofen's algorithm ([8]).

If we have an arithmetic expression instead of a general circuit, we gain a $\log(n)$ factor by avoiding any grouping. Avoiding grouping leads to good results on some particular arithmetic circuits (where a $+$ node is often followed by a $\times$ node, and thus the group is quite useless). For trees, this algorithm has a $CREW(n, \log n)$ complexity. We do not know its complexity when used on DAGs, but we obtain for instance a $CREW(n, \log n)$ complexity with the parallel prefix problem and the addition or the multiplication of two n-bits integers.

4 Using these algorithms in a parallelizing compiler

We assume we work at the lowest granularity, to work on arithmetic circuits (i.e. we do not treat neither conditional jumps nor loops of unknown length). To parallelize a program, we firstly build its complete execution graph, then we apply symbolically the previous algorithm of parallel evaluation to this DAG to build the execution graph of the parallel evaluation. This graph corresponds to the same computations, but it is shallower, i.e. it contains more potential parallelism.

This graph is quite particular: it is both UET (Unit Execution Time) and UCT (Unit Communication Time), and therefore it is well suited to be well scheduled. An other possibility is firstly to cluster it, since its tasks have a very fine grain, and then to schedule it.
This is very different from the classical techniques of automatic parallelization, that work on the dependence graph (scheduling or nested-loops parallelization) without changing the critical path length, and of the techniques used to design new efficient parallel algorithms, such as a "divide-and-conquer" approach.

5 Results

Infinite precision arithmetic This algorithm has been applied to the infinite precision addition circuit. Since there is no trick using the actual values of the inputs, only one test with any inputs is sufficient. The algebraic degree of this graph is linear in the number of nodes, and thus the predicted complexity is $CREW(M(n), \log^2(n))$. We can see on figure 3 that this complexity is achieved, and the multiplicative constants are small. The same results occur for the infinite precision multiplication.

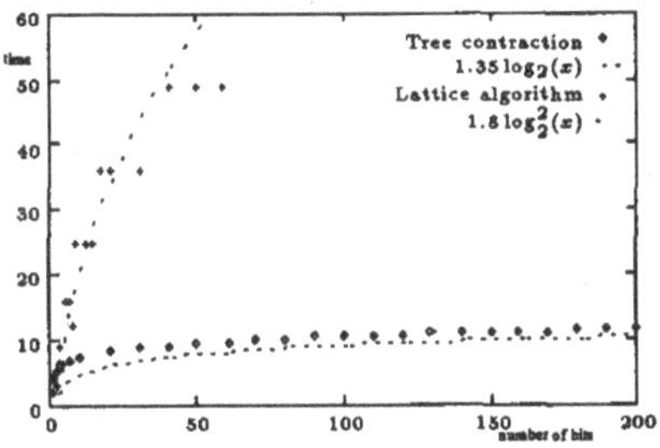

Figure 3: *Parallel time for the addition of two n-bits numbers*

Linear algebra When we solve a linear triangular system, (a frequent operation in numerical computations!), we perform operations in a field. The algebraic degree is in this case linear in the number of nodes, and thus we obtain a $CREW(M(n), \log^2(n))$ complexity, which is the best effective complexity, even if the lower bound established for this problem is $\mathcal{O}(\log(n))$.

Lexicographic maximal independent set This problem is $\mathcal{P}$-complete, which is easier to study than the general MCVP[2]. The theoretical complexity for this problem $CREW(M(n^2), n * \log(n))$ (the corresponding DAG has n^2 nodes). The experiments confirm this result (we simulate a CRCW-PRAM, thus we gain a $\log(n)$ factor).

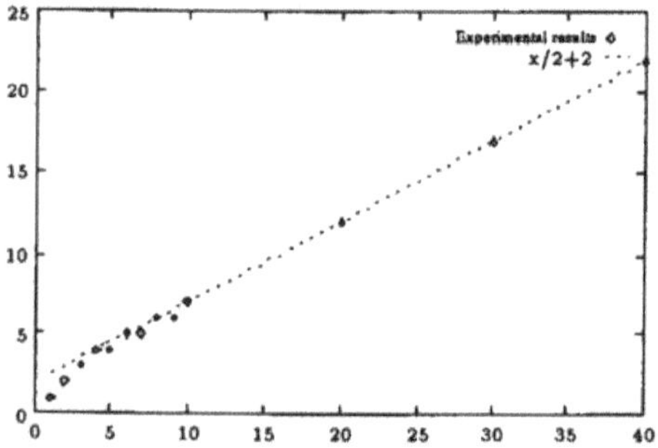

Figure 4: *Parallel time for the lexicographic maximal independent set.*

We expect to improve this bound by the use of additional lattice's properties, such as the absorption law.

6 Conclusion

In this paper we have presented an algorithm for the parallel evaluation of arithmetic circuits over a lattice. It generalizes the algorithms of tree contraction and of semi-ring DAG contraction. We have sketched how it can be used as a parallelizing tool, and we have presented experiments giving good complexity estimation. We have also presented our results on a problem which is difficult to parallelize.
Our research perspectives are on the one hand the test and practical experiments of this tool as a compiler, and on the other hand a further work on the Boolean lattice, for the study of complexity theory and of circuit design.

[2]Without the lexicographic order constraint, this problem can be solved efficiently in $\mathcal{O}(\log^2(n))$ parallel time.

References

[1] E.G. Coffman and P.J. Denning. *Operating systems theory.* Series in Automatic Computation. Prentice-Hall, NJ, 1973.

[2] R. Cole and U. Vishkin. Optimal parallel algorithms for expression tree evaluation and list ranking. *LNCS*, 319:91–110, 1988.

[3] D. Coppersmith and S. Winograd. On the asymptotic complexity of matrix multiplication. *SIAM J. of Computing*, 11(3):472–492, August 1982.

[4] J. von zur Gathen. Parallel arithmetic computations : a survey. In *Proc.* 12th *Int. Symp. Math. FOCS, Bratislava*, LNCS, 233, 1986.

[5] A. Gibbons and W. Rytter. Optimal parallel algorithms for dynamic expression evaluation and context-free recognition. *LNCS*, 319:32–45, 1988.

[6] L. M. Goldschlager. The monotone and planar circuit value problems are log space complete for P. *SIGACT News*, 9(2):25–29, 1977.

[7] R. M. Karp and V. Ramachandran. *Handbook of Theoretical Computer Science, J. van Leeuwen, editor*, chapter Parallel Algorithms for Shared-Memory Machines, pages 869–941. Elsevier Science Publishers, B.V., 1990.

[8] G. L. Miller, V. Ramachandran, and E. Kaltofen. Efficient parallel evaluation of straight-line code and arithmetic circuits. *LNCS*, 227, July 1986.

[9] G. L. Miller and J. H. Reif. Parallel tree contraction and its application. 26th IEEE Symp. on FOCS, pages 478–489, 1985.

[10] G. L. Miller and S. H. Teng. Dynamic parallel complexity of computational circuits. *J.ACM*, pages 254–263, 1987.

[11] David S. Johnson. *Handbook of Theoretical Computer Science, J. van Leeuwen, editor*, chapter A Catalog of Complexity Classes, pages 67–161. Elsevier Science Publishers, B.V., 1990.

A parallelized placement algorithm for printed circuit board design[1]

W. Schade*, V. Jossifov**, I. Koepp**
*GFaI Berlin, **IIEF Berlin
Rudower Chaussee 5, Geb. 13.7; D-12484 Berlin
Tel.: 030-63921603 Fax: 030-63921602
e_mail: schade@gfai.fta-berlin.de

Abstract

Higher integration of printed circuit boards as well as the consideration of physical disturbance effects during the PCB design lead to higher development times. This paper presents a new placement algorithm for PCB design. We achieve an acceleration of the placement procedure by parallelizing parts of the algorithm.

1 Introduction

The automatical design of printed circuit boards (PCB) is based on very time consuming algorithms. As the PCB design is an interactive process between designer and design tool there is a demand for the speed up of the used algorithms.
For this reason we parallelism several steps of the placement procedure. We implement the parallelized algorithms described in this paper on a transputer system Parsytec MC3 DE containing eight processors T805.

1.1 Floorplaning

The automatical placement of the modules of a PCB is divided into three steps: global placement, slicing and final placement. The global placement compute an optimal arrangement of the modules with the criterion of shortest net lengths. In this step a module is considered as a point representing the centre of the enclosing rectangle. So the computed placement is not free of overlappings. The slicing procedure builds sets of modules which can be handled by final placement algorithms. The final placement has to eliminate all overlappings.

The coded application of the floorplaning–procedure can be formalized as follows:

actual_tree_node = PCB_area
while **exist area_to_be_cut**
   global-placement procedure
   while **same_tree_level**

[1]This research is supported by grant Nr. 444/93 of the Arbeitsgemeinschaft industrieller Forschungsvereinigungen (AiF)

slicing procedure **:parallel section (TREE)**
actual_tree_node = *next-node-of-same-tree-level function*
endwhile
actual_tree_node = next_tree_level_node
endwhile
final-placement procedure **:parallel section (HYPERCUBE)**

2 Slicing

The slicing algorithm is based on the divide–and–conquer paradigm. During each slicing step sections of the area are divided into two subsections. The modules of a sliced section are assigned to the son sections corresponding to their positions computed during the global placement procedure [SHJK_94].
The elements in each son segment are positioned by a global placement step applied to these areas [SJ_94].

2.1 Parallelization of the slicing algorithm

As the hierachical order of the sliced sections corresponds to a binary tree, the nodes of the transputer cluster are mapped to such a topology.
In each node two new subsections are created. The data describing a subsection are transmitted to the corresponding son node where the subsection is cut. This process is broken of if a subsection does not need to be sliced.

The circuit board used for the processing time measurements described below contains 181 components (fig.8).
The diagram (fig.1) shows a run time comparison of the sequential slicing program with the parallel program. The sequential version runs on a single node the parallel version on seven nodes of the transputer cluster. After the sections belonging to the second tree level are sliced the process is stopped. For the parallel program the portion of the run time used up only for node-to-node communication is marked.

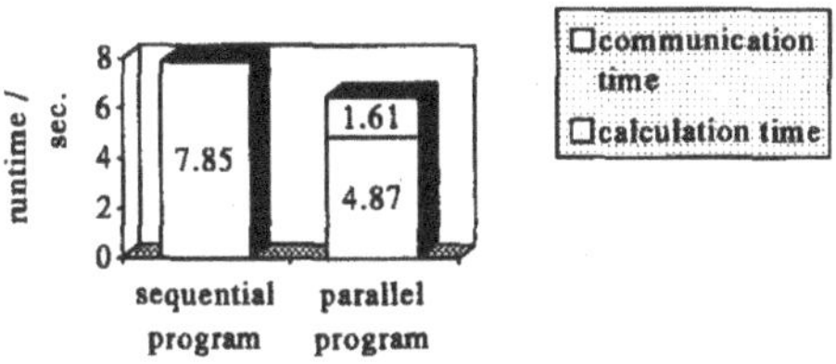

fig.1 Processing times of the sequential and the parallel slicing program.

The calculation time of the parallel program is about 30 percent higher than theoretically expected. This delay is caused by synchronization processes of communicating nodes.

The diagram in fig.2 shows which components the communication time consists of. The seek time is used up by searching for data to be sent. The packing up of several data into buffers (during the pack time) leads to a decrease of the communication time by the factor 2.5. During the transmition time the data from the buffers are sent.

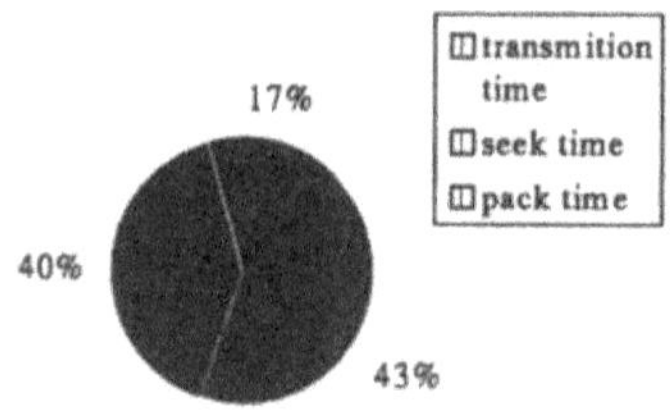

fig.2 Components of the communication time.

3 Final placement

A geometrical algorithm called 'wheels' (fig.3) is implemented and improved [WW_90].

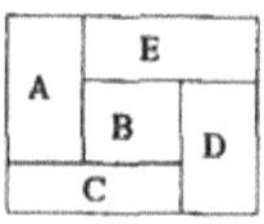

fig.3 Five components are arranged in shape of a wheel.

A *final placement* for m modules consists of an enveloping rectangle subdivided into m non-overlapping rectangles, which have to be large enough to accommodate the module assigned to it.
A *block* is a connected set of at least one rectangle. There are *rectangular* and *L-shaped* blocks (R- and L-blocks). Each block has a finite number of possible implementations. An implementation of an L-block is given by a 4-tupel of its widths and heights (w_1, w_2, h_1, h_2). An R-block can be considered as a special type of an L-block with

$w = w_1 = w_2$ and $h = h_1 = h_2$ (fig.4).

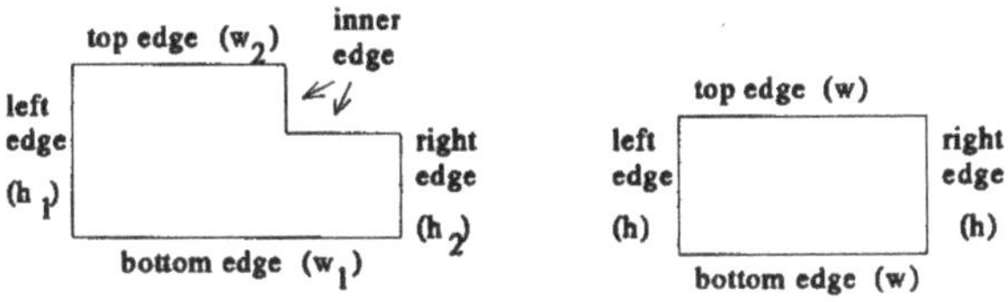

fig.4 Terms to denote the various sides of rectangular and L-shaped blocks.

Implementations of one block are joined to at least one list (*R-list*, *L-list*) which elements are sorted by decreasing widths w_1 as well as increasing heights h_1 and h_2.

The positions of the blocks A and B (left and right $\alpha-$rectangle) are calculated by the $\alpha-$procedure. The blocks C, D and E ($\beta-, \gamma-, \delta-$rectangle) are placed by the $\beta-, \gamma-$ and $\delta-$procedure, respectively (fig.3).
The time complexity and the total number of all stored implementations of the $\alpha-$procedure is $O_\alpha(|R_A||R_B|)$. The $\beta-$ and $\delta-$procedure have complexity of $O_{\beta/\delta}((\sum_{i=1}^{p} |L_{\alpha/\gamma i}|) + p|R_{C/E}|)$. $\gamma-$procedure has complexity of $O_\gamma((\sum_{i=1}^{p} |L_{\beta i}|)|R_D|)$. L_x is the L-list constructed by procedure x, R_X is the R-list corresponding to the block X, p is the number of L-lists to be processed [WW_90].

3.1 Improvement of the wheel-algorithm

The algorithms 'Module shifting' and 'Placement on unused areas' proposes a better utilization of unused areas and are improvements of the 'wheel' algorithm.

Module shifting
Module shifting allows the placement of components in the unused area computed by the $\beta-$ or $\gamma-$procedure (fig.5). If $h_\beta \geq h_\gamma$ the $\gamma-$rectangle is left shifted. The $\delta-$rectangle is shifted downwards if $w_\gamma \geq w_\delta$.

fig.5 Shifting of components to unused areas.

The pseudocode for the moduleshifting procedure is shown in fig.6.

$\gamma-$procedure:

if $h_\beta \geq h_\gamma$
  $w_{1\gamma} = maximum(w_\beta + w_\gamma, w_{1\beta})$
endif
else
  $w_{1\gamma} = w_{1\beta} + w_\gamma$
endelse

$\delta-$procedure:

if $(w_{1\gamma} - w_{1\alpha m}) \geq w_\delta$
  $h_{1\delta} = h_{2\delta} = maximum(h_{1\gamma}, h_\gamma + h_\delta)$
endif
else
  $h_{1\delta} = h_{2\delta} = maximum(h_{1\gamma}, h_{2\gamma} + h_\delta)$
endelse

fig.6 Pseudocode of the moduleshifting procedure

Module placement on unused areas

Each rectangle is given an identification number (ID) and each L-implementation contains a list (ID-list) in which the IDs of all rectangles the L-implementation consists of are stored.

After the $\beta-, \gamma-$ and $\delta-$procedure each R-lists containing the rectangles to be placed is scanned for R-implementations which are small enough to be accommodated by the unused area computed by the $\beta-, \gamma-$ or $\delta-$procedure respectively. If a suitable rectangle is found the corresponding ID-list is checked for the rectangle's ID. The rectangle is placed in the unused area if its ID was not found in the ID-list. In this case the remaining unused area is calculated anew before the scanning procedure of the R-list for further suitable rectangles is continued (fig.7).

Obviously, this feature leads to an increase in time complexity of $O'_\beta : |R_C| - 3 - k_\beta$, $O'_\gamma : |R_D| - O'_\beta - 1 - k_\gamma$ and $O'_\delta : |R_E| - O'_\gamma - 1 - k_\delta$, k_x is the number of utilized elements in the unused area of procedure x.

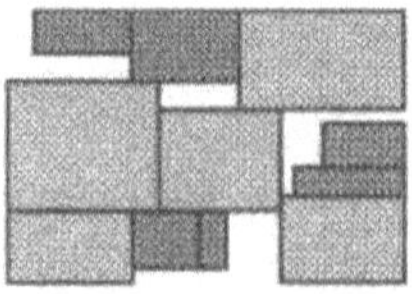

fig.7 The dark modules are placed in unused areas.

3.2 Parallelization of the wheel-algorithm

As a result of the slicing procedure the circuit board is divided into several regions. So the rectangles belonging to the same region are assembled to the same wheel. Consequently a rectangle's position on the PCB results from its position within the region as well as the region's position on the circuit board. Hence the placement inside one region has no influence on the placement result of other regions. So it is possible to apply a coarse grained parallelism with a very small demand for communication.

For test purposes with virtual data sets the floorplaning for a circuit board containing 900 modules and sliced into 168 regions was implemented. The 168 regions are mapped as virtual processors in a hypercube- or ring- topology, respectively.
The diagram (fig.9) shows the run times for various numbers of transputers in comparison with the sequential run on a SUN workstation. and global placement.

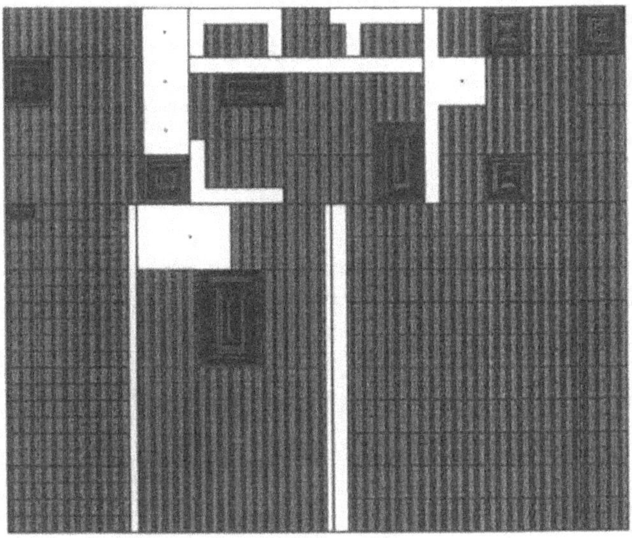

SUN workstation and hyper cube the exact runtimes are signed at the bars top.
fig.8 Final placement after slicing and global placement.

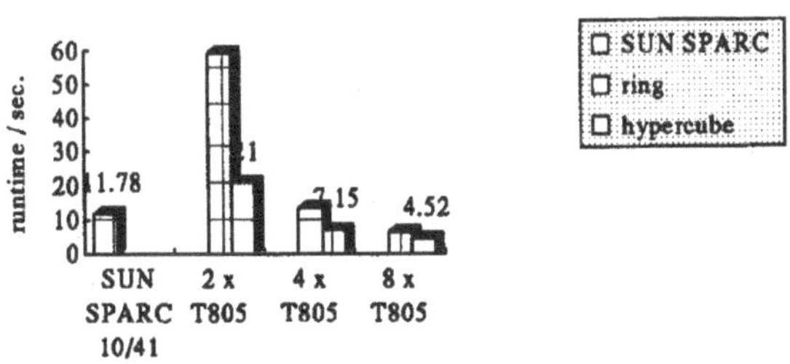

fig.9 Runtime comparison on various topologies and processor quantities - for SUN workstation and hyper cube the exact runtimes are signed at the bars top.

4 Bibliography

[Sigl_92] Sigl, Georg, *Plazierung der Zellen bei der Layoutsynthese mittels Partitionierung und quadratischer Optimierung.* Dissertation, Lehrstuhl für Rechnergestütztes Entwerfen, Technische Universität München (1992).

[WW_90] Wang, Ting-Chi, Wong, D.F., *An Optimal Algorithm for Floorplan Area Optimization.* ACM/IEEE Proceedings 27th Design Automation Conference (1990), pp.180-186.

[SHJK_94] Schade, W., Herrmann, H., Jossifov, V., Koepp, I., *Parallelized placement tools for printed circuit board design.* 6th International Workshop on Parallel Processing by Cellular Automata and Arrays PARCELLA (1994), submitted paper.

[SJ_94] Schade, W., Jossifov, V., *Parallelisierung von Verfahren zur Plazierung von MOdulen auf Leiterplatten unter Berücksichtigung von quasi on-line bereitgestellten Simulationsergebnissen.* GFaI - internal report (1994).

Parallel Processing of Communication Protocols

Otto Spaniol, Christian Engel

Lehrstuhl für Informatik IV, RWTH Aachen, Germany

Abstract

Current and future communication applications like multimedia, distant conferencing, and open distributed conferencing require data transfer rates of 100 MBit/s and more. New physical and media access layers like ATM, FDDI2, DQDB are about to support the demanded performance. Now the higher protocol levels have to offer appropriate efficiency to bring this performance to the user. Traditional protocols, their architecture, the implementation are overcome. Parallel protocols and efficient concurrent implementation is required. To transfer concurrence from specification to implementation a proper and expressive formal system is required. The need for an environment taking all stages of protocol development into account is obvious. This paper will present a strategy based on higher Petri nets. An XTP implementation by Petri nets and some performance evaluations are presented. An outline of the performance evaluation is given.

1 Introduction

In response to increasing requirements to performance and functionalities of communication protocols efficient high speed implementations have to be gained. Several different approaches have been contributed. Some suggest new protocol designs supporting optimized and enhanced facilities. Other efforts deal with hardware for the implementation and its programming. Common to both is the evidence of parallelism. Concurrent execution of a protocol with high inherent parallelism promises high performance. But this is not sufficient. The platform and the programming language used by the implementor have to fit the form of parallelism supported by the protocol specification as well. Moreover, the environment used for the protocol design has to support a method to transfer the parallelism from specification to the implementation level. Thereby the implementation is enabled to exploit concurrence thoroughly. Drawbacks put down to formal system transfer have to be avoided.

A powerful protocol design and implementation method based on Petri nets has been developed and researched in the PICOM project[1] at our department. Among others the eXpress Transfer Protocol (XTP) has been redesigned and implemented by it.

[1] PICOM (**Parallelism in Communication** Protocols) is a project sponsored by the DFG (Deutsche Forschungsgemeinschaft = German Foundation for Research)

The specification and the implementation has been evaluated and compared to other concurrent protocol implementations.

This paper will deal with parallelism in protocol design and implementation. The next section presents approaches to design a protocol and enchant parallelism by means of Petri nets. The environment for protocol design and implementation by Petri nets is outlined in section 3. Section 4 presents a programming environment and a multiprocessor platform for concurrent protocol execution. The evaluation of the obtained XTP protocol implementation is outlined in section 5. The paper ends with some conclusions.

2 Parallelism and Protocol Design

Different formal systems have been suggested for protocol specification. Their suitability for the individual forms of parallelism is different. Some of these formal systems have been applied for formal specification languages and implementation methods and are under wide use. The most famous examples are finite state machines (EFSM) and process algebras.

Extended finite state machines:

EFSMs are the basic formal system of popular specification languages like SDL and Estelle. But the suitability of EFSMs to parallelism is limited. Every EFSM for itself is strictly sequential. Concurrence only occurs by processing independent instances of EFSMs. This restricts parallelism to grain granularity. But the power of EFSMs is reduced even more by the need of at least partial independence of the individual tasks. Two instances of EFSMs may share a common global variable. In this case they can't be processed concurrently. Since dependencies like this are hidden by the formal system the protocol developer has to uncover them.

Process algebras:

The formal specification language LOTOS is suitable for concurrence. Parallel composition is supported and a list of the synchronization points called gates must be given explicitly every time. Therefore parallelism and dependencies are obvious. Descending even to fine granularity is possible. There are several implementation methods and tools based on algebraic process algebras.

This paper won't be another paper about these commonly used methods. It will present a different approach based on Petri nets.

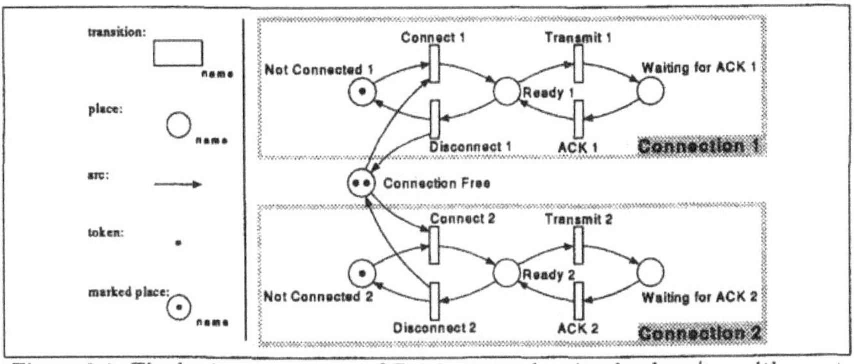

Figure 2.1: The basic components of Petri nets and a simple place/transition net for the control of two concurrent transfer connections

Petri nets:

Petri nets consist of *places*, *transitions*, and *arcs* (see figure 2.1). Places may be considered as data locations and transitions as actors or functions. Between a transition and a place an arc exists iff the transition depends on the contents of a place or modifies it. The execution of a transition (called firing) is atomic. Concurrent firing of a subset of two or more transitions is possible if the firing of one transition of this set does not affect the firing of another transition from this set. In figure 2.1 this might arise for any pair of transitions one from Connection1 and one from Connection2. This example shows that Petri nets comprise parallelism in a very natural way. The rule explained above yields concurrency in a fashionable and implicit manner.

Petri nets have proven themselves as ideal for analysis, verification, simulation, modeling, and performance evaluation. In the next section a versatile protocol development environment is presented. It makes use of these methods and the properties of Petri nets.

In comparison to EFSMs Petri nets are superior in the support of parallelism. Compared to process algebras the capability of both formal systems is similar. There are only minor advantages and disadvantages. When the decisions for formal specification systems were made parallelism was only a minor aspect. This may be the reason for the low popularity of Petri nets. The theory of Petri nets is well researched. The mathematical model is very powerful.

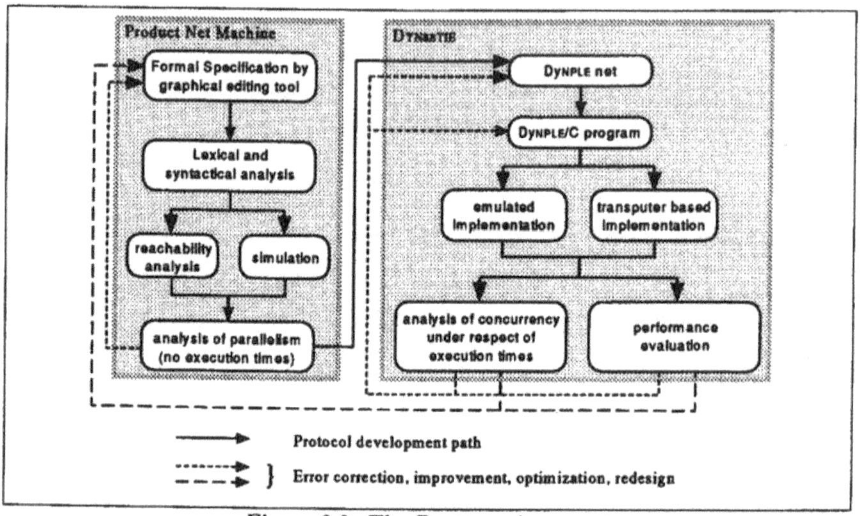

Figure 3.2: The DyNE environment

3 Protocol Design comprising Concurrency

In [7] a computer aided environment for modular protocol design is introduced. It is called DyNE[2]. Its construction has been driven by four main goals:

- to use one common formal system on all stages of protocol development from specification down to implementation;
- to support and exploit all forms of parallelism including fine granularity for both protocol specification and implementation;
- to analyze, validate, verify, test, and evaluate protocols;
- to achieve a high performance protocol implementation on a multiprocessor platform.

See figure 3.2 for the structure of the environment and the path of protocol development. DyNE makes use of product nets, a kind of higher algebraic Petri nets with individual tokens [2]. They have been designed for protocol specification, analysis, and verification [11, 9]. A suitable protocol design tool, the product net machine, is available [10]. It is integrated in the specification phase of DyNE. Several protocols including XTP have been specified, analyzed, and validated with this tool.

[2]DyNE= Dynamic Nets Environment

From a product net specifying a protocol a DyNPLE net is derived. It is a Petri net with programming language characteristics. Because product nets and DyNPLE nets are related this step can focus on implementation decisions and efficiency. Since detailed implementation informations are not due to specification this step can be supported only partially by automation. For the treatment of DyNPLE nets a graphical tool is available.

The remaining steps towards implementation are very simple and mechanical. They can be automated. DyNPLE/C[3] is a high level parallel programming language [5] comprising Petri nets. The DyNPLE net is translated to a program in this language. Then a DyNPLE/C compiler generates the machine code program. It is downloaded to the target machine which conforms an abstract multiprocessor controller architecture called DyNARC[4] [6].

4 Multiprocess Protocol Implementation

The reflections in the previous section lead to the approach to use Petri nets as basic formalism for specification and implementation. Moreover, a controller which executes a Petri net at machine level can be constructed. Transitions will be executed concurrently by several processors. The process deciding the firing of transitions can be isolated and distributed to some other processors.

Consider the structure of the abstract communication controller architecture DyNARC depicted in figure 4.3. DyNARC is a master-slave architecture. The processors of DyNARC are grouped in four component types: firing machines, evaluators, activators, and deactivators. The firing machines are the slaves. They execute the actions taking place when a transition fires. They are controlled by the masters in control unit, consisting of evaluators, activators, and deactivators. The master components

- evaluate the conditions under which transitions have to be fired
- synchronize the firing of transitions and ensure memory consistency.

The data in the places of the Petri net, called marking, are held in the token memory. Global memory, interfaces and special components like DMA and timer complete the construction.

Essentially the code of the program is distributed to firing machines and evaluators. All component types have to execute some overhead like distribution, synchronization, and communication between the components.

[3]DyNPLE/C = Dynamic Nets

[4]DyNARC = Dynamic Nets Architecture for Communication Controllers

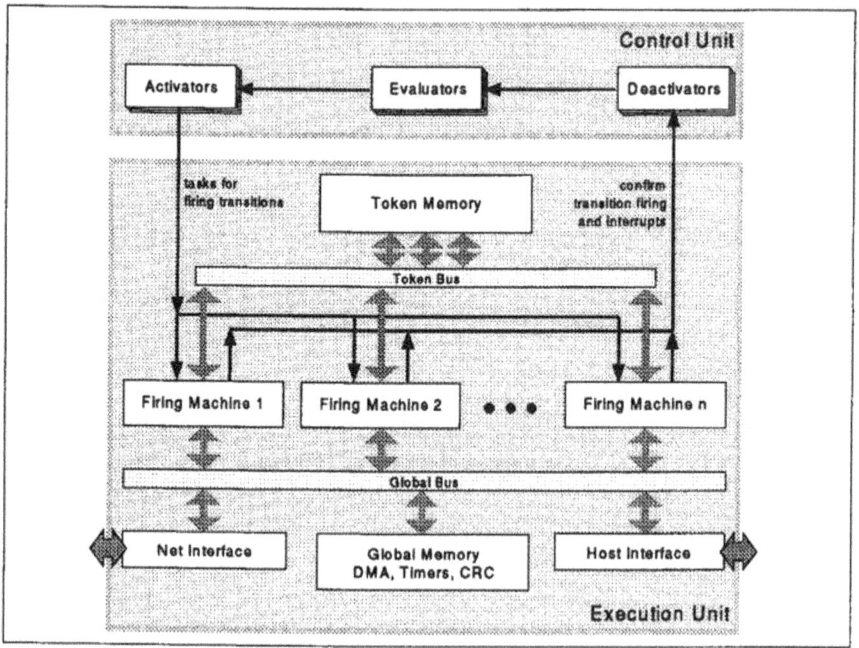

Figure 4.3: Structure of the DyNARC communication controller

It is a remarkable property of DyNARC that every component type can be configured by an arbitrary multitude. There is no unique processor with a special and central task. A processor like this would tend to be a bottleneck.

DyNARC has been implemented twice.

- An emulator on Unix systems is in use for some time and well approved. It allows emulation of a large scope of scenarios and configurations.
- A new implementation on Parsytec transputer clusters (inmos transputer T800) with an arbitrary number of up to 256 processors is in alpha test phase.

The final goal is to construct a single board high speed communication controller. Universal processors and chips could be used, but special VLSI, GALs, and rather primitive special processors should optimize the performance.

For the programming of the DyNARC controller a special high level programming language called DyNPLE/C[5] has been designed. It is an extension of the programming

[5]DyNPLE/C = **Dynamic Nets**

language ANSI C. Essentially the syntax for transitions and places has been added. The semantics are dictated by the underlying Petri nets. The net incorporated by a Program is dynamic. A subnet of the Petri net is considered as a class which may be constructed and destructed at runtime. Thereby several instances of the same transition may occur and execute concurrently. Transitions may consist of a single instruction or complex subprograms.

For more details on the language DyNPLE/C and the controller architecture DyNARC see [5, 6, 7].

5 Evaluation of an XTP Implementation

XTP (eXpress Transfer Protocol) [12] has been designed for data transfer rates of 100 MBit/s to 1GBit/s. It is a transfer system, i.e. according to the ISO/OSI reference model it is assigned to the network and transport layers. Primarily XTP offers connection oriented service but connectionless data transfer is supported as well. XTP comprises addressing modes of OSI, Internet, and other standards. The error control is suited to modern physical media with low error probabilities. It supports *Go Back N* and *Selective Resent* error control. The overhead for error, rate and flow control is decreased evidently compared to traditional protocols like TCP/IP. Other aspects of optimization for new physical and media access layers influenced the protocol design. For instance, in most protocol implementations the receiver has become apparent as a bottleneck. Consequently in XTP some parts have been moved from the receiver to the sender, as far as possible and useful.

Originally XTP was designed for VLSI implementation. It was intended for a chipset with individual electronic circuits working concurrently. Apart from this different approaches for efficient implementation have been made, mostly founded on the construction and programming of a multiprocessor architecture.

In the PICOM project the protocol has been redesigned by Petri nets and implemented on the DyNARC controller. The transitions and evaluation of net conditions are distributed to processors. Not only different parts of the protocol like sender and receiver are able to run concurrently. Very often multiple instances of parts of the protocol are replicated to work independently and in parallel. For instance, different connections at the same time make use of this performance improvement.

As sketched in section 3 the specification is transferred to a DyNPLE net. In this step the parallelism has been paid increased attention. Places commonly used by several transitions are of overriding importance. Places that are incident to a majority of transitions with high probability of firing have been avoided. Transitions with a common incident place usually take effect on each other. Therefore, they cannot be

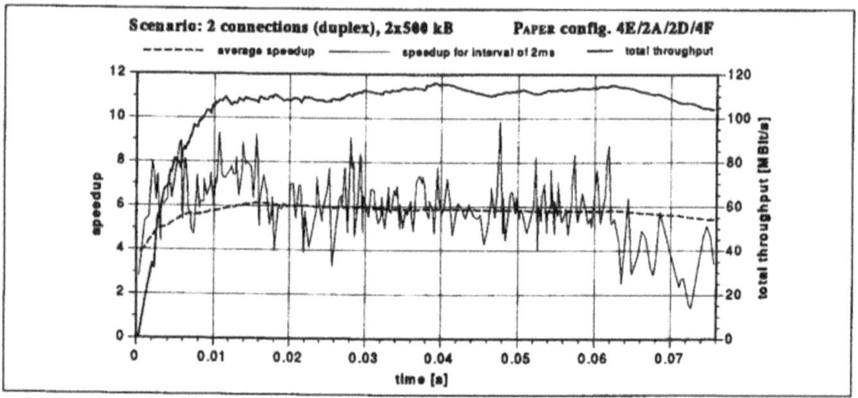

Figure 5.4: Performance in duplex data transfer

fired concurrently and lead to loss of parallelism. This drawback can be decreased by rare use of variables of overall importance. Splitting or distributing such a variable to several places avoids the problem. Petri net theory provides such techniques. Unlike some other formal systems like EFSMs the Petri nets uncover the described bottlenecks.

The resulting XTP implementation has been tested on both completed platforms. Several configurations and scenarios have been evaluated. The performance values given in this section have been measured with the emulator on an intel processor 80486 with 50 MHz clockrate and memory with 70 ns access. We got comparable values on a simple Sun Sparcstation 10 and a HP Apollo 700. The values of the throughput will improve if hardware with better performance, special VLSI circuits, GAL chips, or special processors are used. We expect a performance improvement by factor 4.

In figure 5.4 the performance curves for a duplex (bidirectional) connection oriented data transmission of 500 kBytes in each direction are shown. The configuration of the DyNARC emulator was 4 evaluators, 2 activators, 2 deactivators, and 4 firing machines. The speedup is given for one station, the throughput is the sum of both connections.

The speedup is evaluated by computing the fraction of the time it takes a single processor controller and the multiprocessor controller to execute the program. The average speedup in this scenario is more than 5.5, the peaks top at 10. This is a satisfying result in comparison to the forecast given in [3]. In the context with more connections, other XTP functionality, and other protocols in the protocol stack a much higher speedup will result.

The final throughput for one connection in the duplex scenario is 53 MBit/s. For

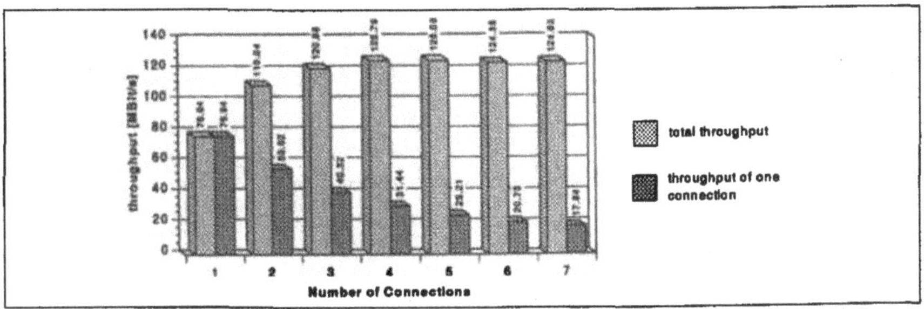

Figure 5.5: Number of connections affecting the throughput of one connection

a single single simplex connection it is almost 77 MBit/s. This is the end-to-end performance of the XTP implementation by Petri nets. Remind that no other protocol layers are attached. These value will only be achieved if the lower layers support a performance of at least 100 MBit/s.

In comparison to a simplex transmission the speedup for duplex connection is nearly twice the value, but the throughput is not. Certain places in one controller must be shared by all connections. Although the two connections can almost be executed in parallel and the speedup increases, the throughput of one connection is affected and decreased by another connection. This effect cannot be avoided and occurs in all implementations. There are always certain resources which have to be shared, e.g. interfaces, synchronization data, schedulers, service access points.

In comparison to other XTP implementations [4, 1] the throughputs above show the presented implementation to be similar or superior. This result holds for other scenarios not presented in this paper.

Consider figure 5.5 to see how the number of connections affects the throughput. While the total performance increases due to the available concurrency the per-connection throughput decreases. In the 4/2/2/4 configuration we reach saturation for 5 and more connections. The total throughput doesn't increase anymore. There are two possible reasons for this effect:

- All processors or at least one type of components has gained full load. The addition of some more processors of this type will solve the problem and increase the performance.
- The concurrency of the implementation is completely exploited. Addition of more processors won't have any effect. Improvement of performance is only possible if more parallelism can be uncovered and exploited in the implementation. In this case redesign and optimization of the protocol and its implementation will improve the situation. Otherwise, the implementation is considered to be optimal and no more performance improvement can be achieved.

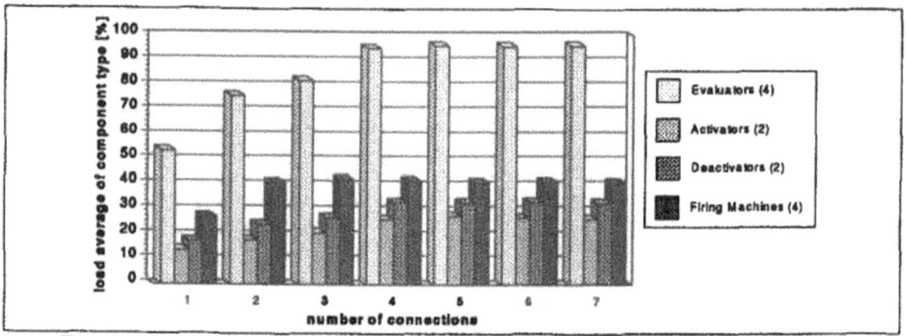

Figure 5.6: load of the components by increasing number of connections

Figure 5.6 confirms the second explanation. For 4 and more connections the 4 evaluators are almost at full load. An increase of the number of evaluators would help, but not for all situations. There is a point where the first reason for saturation arises. It has been proved by other measurements that this point is reached for 7 connections and a total throughput of more than 200 MBit/s. Below this point the expected maximal number of connections with full load at the same time dictates the number of processors required for maximal throughput.

It has to be remarked that some parts of the DyNPLE/C compiler still have to be optimized. Currently the run time system for the evaluation scheduler is rather primitive. Many transitions are re-evaluated although there is no need for it. It has been shown that a simple data flow analysis at compile time will decrease the evaluation rate essentially. In the near future the DyNPLE/C compiler will be extended by a suitable optimizer. Furthermore, a special architecture instead of a universal processor will improve the performance of the evaluator. In our scenario the threshold for the need of more than four evaluators was five connections. After improving the compiler and the evaluator as suggested the threshold will increase.

Other evaluations have been made on the presented XTP implementation proving its high performance. It is beyond the scope of this paper to present them all. They will be presented in [7].

6 Conclusions

This paper gave an outline of three basic formalisms applicable to protocol design. They were compared concerning their suitability for protocol design and concurrent protocol processing. The Petri net based method has been presented in more detail. Petri nets can be applied for distributed processing and a suitable multiprocessor hardware. Parallelism can be exploited substantially. If the protocol has been designed appropriately very high performance is achievable.

The resulting XTP implementation by Petri nets has been evaluated. Its efficiency has been proved. The proposed DyNE environment is a substantial approach to a protocol design and implementation strategy for high speed concurrent protocols.

References

[1] T. Braun, M. Zitterbart, *A parallel architecture for transport systems and gateways*, In [8], pages 321–329

[2] H.J. Burkhardt, P. Ochsenschläger, Product nets: A formal description technique for cooperating systems, GMD-Studien 165, Gesellschaft für Mathematik und Datenverarbeitung, Bonn, Germany, December 1989

[3] F. Conrads, *Formal Specification and Analysis of XTP with Product Nets* (in German), Master's thesis, Aachen University of Technology, Aachen, Germany, 1992

[4] C. Diot, V. Roca, *XTP/KRM implementation on a transputer network*, In [8], pages 310–320

[5] C. Engel, *DyNPLE/C - A language for concurrent programming of high speed communication protocols on a Petri net based multiprocessor controller architecture*, Aachener Informatik-Berichte, Aachen University of Technology, Aachen, Germany, 1994, (to be released) [anonymous ftp: ftp.informatik.rwth-aachen.de, /pub/reports/1994]

[6] C. Engel, *DyNARC: A multiprocessor high speed communication controller based on dynamic Petri nets*, Aachener Informatik-Berichte, Aachen University of Technology, Aachen, Germany, 1994, (to be released) [anonymous ftp: ftp.informatik.rwth-aachen.de, /pub/reports/1994]

[7] C. Engel, *Implementation of Communication Systems by Dynamic Petri nets* (in German), PhD thesis, Aachen University of Technology, Aachen, Germany, 1994

[8] *16th Annual IEEE Conference On Local Computer Networks*, Minneapolis, Mn, USA, October 1991

[9] U. Nitsche, *Reachability Analysis in Product Nets* (in German), Technical Report GMD-Studien No 521, Gesellschaft für Mathematik und Datenverarbeitung, Bonn, Germany, April 1991

[10] P. Ochsenschläger, *The Product Net Machine* (in German), Technical Report GMD-Studien No 505, Gesellschaft für Mathematik und Datenverarbeitung, Bonn, Germany, January 1991

[11] P. Ochsenschläger, *Module Homomorphisms II* (in German), Technical Report GMD-Studien No 597, Gesellschaft für Mathematik und Datenverarbeitung, Bonn, Germany, November 1991

[12] Protocol Engine Incorporated, *XTP Protocol Definition Revision 3.6, PEI 92-10.*, October 1992

Parallel domain decomposition algorithms for time-dependent problems of mathematical physics

P.N. Vabishchevich

1 Introduction

Domain decomposition methods are considered at present time as the most perspective ones for parallel computer systems. Nowadays the main success is obtained in approximate solving classic boundary value problems for elliptic equations of second order [1–3]. Two main methods are used for these equations – with and without overlapping.

As for time-dependent problems of mathematical physics, there are in common use approaches based on ordinary implicit schemes and implemented via iterative methods of the domain decomposition. The alternative technique is considered in the present work and includes non-iterative schemes (region-additive schemes) [4–9]. On the basis of the general theory of additive schemes a wide class of difference schemes [10,11] (alternative directions, locally one- dimensional, factorized schemes, summarized approximation schemes, etc.) is investigated in application to the domain decomposition technique.

Three various types of decomposition operators,corresponding to the Dirichlet and Neumann conditions on the subdomain boundaries, are considered in thiswork. General stability conditions are obtained for the region-additive schemes. The emphasis is on studying the accuracy of domain decomposition schemes. In particular, the dependence of the convergence rate on the width of subregion overlapping is investigated as the primary property. New classes of vector additive domain decomposition schemes are introduced which are the most suitable for the parallel computer architecture and demonstrate a very high accuracy.

2 Model problem

A two-dimensional rectangular domain Ω with parallel to the coordinate axises sides is considered. The solution of the parabolic equations is sought in the domain Ω:

$$\frac{\partial u}{\partial t} - \sum_{\alpha=1}^{2} \frac{\partial}{\partial x_\alpha}(k_\alpha(x) \frac{\partial u}{\partial x_\alpha}) = f(x), \quad x = (x_1, x_2) \in \Omega, \quad t > 0. \tag{1}$$

Equation (1) is supplemented by the homogeneous boundary condition (the Dirichlet problem)

$$u(x,t) = 0, \quad x \in \partial\Omega, \quad t > 0 \tag{2}$$

and the initial condition

$$u(x,0) = u_0(x), \quad x \in \Omega. \tag{3}$$

Let us introduce in the domain Ω the uniform grid x_α with the fixed spacing $h_\alpha, \alpha = 1,2$. Let ϖ be the set of the internal nodes. Passage to the more common problems in context of this paper is carried the editing character. Let us define a difference operator A on the set of functions $y \in H$ such as $y(x) \equiv 0, x \notin \varpi$,

$$Ay = -\sum_{\alpha=1}^{2}(a_\alpha(x) y_{\bar{x}})_x, \quad x \in \varpi. \tag{4}$$

Here the standard index-free notation of the difference schemes theory [10] is used, for example:

$$a_1(x) = 0,5(k_1(x) + k_1(x_1 - h_1, x_2)), \quad a_2(x) = 0,5(k_2(x) + k_2(x_1, x_2 - h_2)).$$

Let us introduce a scalar product and a norm in the Hilbert space H as follows

$$(y,v) = \sum_{x \in \varpi} y v h_1 h_2, \quad \|y\| = \sqrt{(y,y)}.$$

Note, that in Hoperator A is selfadjoint and positive definite [10], i.e. $A = A^* > 0$.

From equation (1)-(3) for the given $y(x,0), x \in \varpi$ we pass the following equation

$$\frac{dy}{dt} + Ay = 0, \quad x \in \varpi. \tag{5}$$

For the last equation difference schemes of domain decomposition type are constructed. Numerical implementation of these schemes is based on the solution of problems in separate subdomains of the calculation domain Ω at every time-level.

3 Spatial decomposition

Let the domain Ω consists from p separate subdomains:

$$\Omega = \Omega_1 \cup \Omega_2 \cup \ldots \cup \Omega_p.$$

The separate subdomains may have overlapping. The work of any of the elementary processors is directed to solving the problem in an individual subdomain or on a set of the isolated subdomains. We assume that the namely splitting of the whole domain onto separate subdomains could lead to the effective usage of a parallel computer. The domain decomposition methods make possible to adapt the numerical algorithms of the approximate solving nonstationary problems on the arbitrary parallel computer architecture and also for the arbitrary memory configurations and commutation of the different computer units.

Let ϖ_α be the subsets of nodes ϖ, lying in subdomains $\Omega_\alpha, \alpha = 1, 2, ..., p$. Let us construct the decomposition difference schemes similar to presented in [7] on the basis of the unit splitting for domain Ω. Let us define the following functions

$$\chi_\alpha(x) = \begin{cases} > 0, & x \in \Omega_\alpha, \\ 0, & x \notin \Omega_\alpha, \end{cases} \qquad \alpha = 1, 2, \ldots, p \tag{6}$$

where

$$\sum_{\alpha=1}^{p} \chi_\alpha(x) = 1, \quad x \in \Omega. \tag{7}$$

We shall consider the class of decomposition schemes, where for the operator A the following additive presentation takes place:

$$A = \sum_{\alpha=1}^{p} A_\alpha, \tag{8}$$

here operators $A_\alpha, \alpha = 1, 2, ..., p$ are associated with isolated subdomains and also with splitting (6),(7) and with the solution of the individual subproblems in subdomains $\Omega_\alpha, \alpha = 1, 2, ..., p$.

Simplest difference decomposition scheme is defined via the definition of operators $A_\alpha, \alpha = 1, 2, ..., p$ in the following way [5],[6]:

$$A_\alpha = \chi_\alpha A, \quad \alpha = 1, 2, ..., p. \tag{9}$$

The following presentation for the decomposition operators can be used:

$$A_\alpha = A\chi_\alpha, \quad \alpha = 1, 2, ..., p. \tag{10}$$

Clearly, that for the such a splitting operator A is not selfadjoint, i.e.$A_\alpha \neq A_\alpha^*, \alpha = 1, 2, ..., p$.

It is naturally to use in this case the symmetric splitting (7),(8) (see, for example, [4],[8]), where

$$A_\alpha y = -\sum_{\beta=1}^{2} (a_\beta^\alpha(x) y_{\bar{x}_\beta})_{x_\beta}, \quad x \in \varpi, \quad \alpha = 1,2,...,p. \tag{11}$$

Grid operators $A_\alpha, \alpha = 1,2,...,p.$ are approximated via difference degenerating elliptic operators

$$-\sum_{\beta=1}^{2} \frac{\partial}{\partial x_\beta}(k_\beta \chi_\alpha(x) \frac{\partial u}{\partial x_\beta}), \quad \alpha = 1,2,...,p,$$

where coefficients a_β^α are defined like a_β . With expressions (7),(8),(11) we obtain $A_\alpha = A_\alpha^* \geq 0, \quad \alpha = 1,2,...,p.$

4 Region-additive schemes

First, simplest case of decomposition of domain Ω by means of the straight lines $x_1 =$ const will be demonstrated. In this case the constructing of the difference schemes can be oriented on the usage of the difference splitting schemes with $p = 2$, where Ω_1 and Ω_2 are defined as a combination of corresponding subdomains. While splitting on two operators classical alternating direction schemes would be appropriate for transition from previous time level to the next one.

Unconditionally stable difference schemes for the solving equation (6) with the corresponding initial condition are easily constructed via splitting (7),(8) with self-adjoint and positive definite operators $A_\alpha, \alpha = 1,2$ (with decomposition operator (11)). Of special interest is the accuracy of the approximate solution, its dependence on the width of subdomains overlapping and also from functions $\chi_\alpha(x), \alpha = 1,2$. Usage of the schemes with the asymmetric decomposition operators requires individual investigation.

Let us select among of unconditionally steady factorized schemes the scheme of the stabilizing correction similar to the classical Douglas-Rechford scheme. Let y^n be the difference solution at the time moment $t^n = n\tau$, where $\tau > 0$ is the time-step. The transition from the previous time-level to the next one is performed in accordance with the following expressions

$$\frac{y^{n+1/2} - y^n}{\tau} + A_1 y^{n+1/2} + A_2 y^n = 0,$$

$$\frac{y^{n+1} - y^{n+1/2}}{\tau} + A_2(y^{n+1} - y^n) = 0. \tag{12}$$

It is well known from [10] that the scheme (12) is unconditionally stable, if $A_\alpha = A_\alpha^* \geq 0, \quad \alpha = 1,2.$

Difference scheme (12) with decomposition operators (9),(10) belongs to the symmetrized difference schemes class. Similar schemes can be obtained via using of usual difference schemes for the parabolic equation with varying weights [12].

First,consider difference scheme (12) with splitting (8),(9), i.e.

$$\frac{y^{n+1/2}-y^n}{\tau}+\chi_1(x)Ay^{n+1/2}+\chi_2(x)Ay^n=0,$$

$$\frac{y^{n+1}-y^{n+1/2}}{\tau}+\chi_2(x)A(y^{n+1}-y^n)=0. \tag{13}$$

Let $A^{1/2}y^{n+\alpha/2}=v^{n+\alpha/2}$ for $\alpha=1,2$. In this case multiplying expression (13) on $A^{1/2}$, we have

$$\frac{v^{n+1/2}-v^n}{\tau}+\tilde{A}_1v^{n+1/2}+\tilde{A}_2v^n=0,$$

$$\frac{v^{n+1}-v^{n+1/2}}{\tau}+\tilde{A}_2(v^{n+1}-v^n)=0,$$

where

$$\tilde{A}=A^{1/2}\chi_\alpha(x)A^{1/2},\quad \widetilde{A_\alpha}=\widetilde{A_\alpha^*}\geq 0,\quad \alpha=1,2.$$

Scheme (8),(10),(12) can be symmetrized in a similar way.

For the difference decomposition scheme (8),(12) with selection of operators $A_\alpha,\alpha=1,2$ according to (9),(10) and (11) the following estimate for stability in respect to the initial data is valid:

$$\left\|(E+\tau A_2)y^{n+1}\right\|_D\leq\left\|(E+\tau A_2)y^0\right\|_D,$$

where $D=A,A^{-1}$ and E respectively.

Investigation of decomposition method convergence leads to the following estimates

$$\left\|(E+\tau A_2)z^{n+1}\right\|_D\leq M((1+\|\chi_2\|_A)\tau+|h|^2). \tag{14}$$

Moreover, accuracy of decomposition schemes depends on the width of subdomains overlapping Ω_1 and Ω_2 (see term at $\|\chi_2\|_A$ in estimate (14)).

While constructing of difference schemes for the parallel computer it should be oriented on the decomposition with minimum overlapping of domains, i.e. on the minimization of exchange between individual processors. At minimal overlapping of subdomains (width of subdomains overlapping equal $O(|h|)$) we have from estimate (14) that the convergence rate is $O(|h|^{-1/2}\tau+|h|^2)$.

Under the more general domain decomposition (in this case the question is set of straight lines $x_i=\text{const},\ i=1,2$) grid operators A in equation (5) have the form (7) with number of operators $p>2$. For such problems difference schemes of summarized approximation [10] have some advantages. Investigation of difference schemes of summarized approximation shows that these schemes have low accuracy at spatial domain approximation. For designing parallel numerical methods, it should be more suitable to use vector additive difference schemes with full approximation, besides these numerical schemes are unconditionally stable at any p. Similar difference schemes were proposed in [13] and considered from general positions of theory of difference schemes stability.

Additive difference schemes for equations (5),(7) are constructed on the basis of splitting the main problem on the set of p subproblems, each of them is connected with separate grid operator $A_\alpha, \alpha = 1, 2, ..., p$. Let's define vector $Y = y_1, y_2, ..., y_p$. The following set of equations should be solved for calculation of each component of this vector:

$$\frac{dy_\alpha}{dt} + \sum_{\beta=1}^{p} A_\beta y_\beta = 0, \tag{15}$$

$$y_\alpha(0) = y_0, \quad \alpha = 1, 2, ..., p. \tag{16}$$

As long as from equations (15),(16) follows that $y_\alpha(t) = y(t)$, then arbitrary component of vector $Y(t)$ may be chosen like a solution of the main problem for equation (5).

The following scheme is the example of unconditionally stable schemes for set of equations for (15),(16)

$$(E + \sigma\tau A_\alpha)\frac{y_\alpha^{n+1} - y_\alpha^n}{\tau} + \sum_{\beta=1}^{p} A_\beta y_\beta^n = 0, \quad \alpha = 1, 2, ..., p$$

for case $\sigma \geq p/2$. Realization of this scheme is connected with the inversion of operators $E + \sigma\tau A_\alpha, \quad \alpha = 1, 2, ..., p$ at every time-level in just the same way as in case with general (scalar) difference additive schemes. These schemes may be considered like difference schemes with weights where the weight σ is larger then unit ($\sigma > 1$). Investigation of accuracy of vector difference schemes leads to estimates similar to estimates (14). Three-level vector difference schemes can be used In order to construct parallel numerical algorithms with the same advantages.

References

1. G.I. Marchuk. *Methods of computational mathematics.* Nauka, Moscow, 1989 (in Russian).

2. V.I. Agoshkov. Methods of domain decomposition for mathematical physics problems. In: *Computational processes and systems.* Nauka, Moscow, 1991, pp.3-51 (in Russian).

3. A. Quarteroni. Domain decomposition and parallel processing for the numerical solution of partial differential equations.*Surveys on Mathematics for Industry.* 1991. v.1, pp.75-118.

4. Yu.M. Laevskii. Methods of domain decomposition for the numerical solution of two-dimensional parabolic equations. In: *Variational difference methods for problems of numerical analysis.* Novosibirsk, 1987, pp.112-128 (in Russian).

5. P.N. Vabishchevich. Region-additive difference schemes for unsteady problems of mathematical physics. In: *Computational mathematics and cybernetics.* Moscow University, 1989, v.3, pp.61-63 (in Russian).

6. P.N. Vabishchevich. Difference schemes of domain decomposition for the numerical solution of nonstationary problems. *J. of Computational Mathematics and Mathematical Physics*, 1989, v.29, 12, pp.1822-1829 (in Russian).

7. Yu.M. Laevsky. Quadratic elements in splitting methods. *Sov. J. of Numer. Anal. and Math. Model.*, 1990. v.5, pp.244-249.

8. M. Dryja. Substructuring methods for parabolic problems. *Proc. IV Intern. Simp. on Domain Decomposition Methods for Partial Differential Equations.* Philadelphia, SIAM, 1990, pp.264-271.

9. Yu.M. Laevsky. About one algorithm of domain decomposition without domain overlapping for the numerical solution of parabolic equations. *J. of Computational Mathematics and Mathematical Physics.* , 1992, v.32, 11, pp.1744-1755 (in Russian).

10. A.A. Samarckii. *Theory of difference schemes.* Nauka, Moscow, 1983 (in Russian).

11. A.A. Samarckii, A.B. Gulin. *Stability of difference schemes.* Nauka, Moscow, 1973 (in Russian).

12. P.N. Vabishchevich, P.P. Matus. Two-level difference schemes with varying weights. *Reports of Belorussian Academy of Sciences*, 1993, v.37, 6.

13. V.N. Abrashin. On a variant of alternating direction method for the numerical solution of multi-dimensional problems of mathematical physics. 1. *Differential equations*, 1990, v.26, 2, pp.314-323.

Institute for Mathematical Modeling RAS,
Moscow, Russia

Two classes of efficient hypercube orderings for SVD

Marián Vajteršic

Abstract

The paper presents formulations of parallel ordering schemes for computing the singular values of a real rectangular matrix by the block variant of the two-sided Jacobi method. The schemes are developed for multiprocessor systems with the hypercube architecture. The schemes represent two classes of orderings the cardinality numbers of which equal to the number of processors in the hypercube system. Each of these orderings can be considered as an optimal one because of its minimal interprocessor communication requirements.

Key words: parallel linear algebra, block Jacobi methods, hypercube multiprocessors

1 Introduction

The singular value decomposition (SVD) of a real matrix M of the size $m \times n (m \geq n)$ is given by

$$M = USV^T$$

where U and V respectively are orthogonal matrices of order m and n and S is an $m \times n$ diagonal matrix. The diagonal elements of S represent the singular values of M and the columns of U and V are the left and right singular vectors, respectively.

Due to the importance of this procedure today no one doubts about its crucial role in scientific and engineering computing. Effective methods have been developed for sequential SVD computation (e.g. [3,4]). Among procedures amenable to parallel computing, valuable candidates are the one-sided Jacobi approach [5] and the two-sided method [6]. For multiprocessors with coarse-grained parallelism a promising attempt represents the block variant of the two sided Jacobi approach [2]. There exist also orderings for implementing the permutations required by Jacobi-type SVD methods which support systolic of these methods on massively parallel arrays [7,8].

In this paper, we will formulate new parallel orderings for an efficient version solution of the SVD problem by the block two sided Jacobi method on hypercube multiprocessor systems. We omit a mathematical description of the method which can be found e.g. in [1,2]. We recall here only those basic features of the method which are needed to understand the ordering formulations.

Let us consider an $l \times l$ block-structure of the matrix M. The blocks $M_{i,j}(i,j = 1,2,...,l)$ of M are of the size $(m/l) \times (n/l)$. The $m \times (n/l)$ sized column blocks of M are denoted by $M_j(j = 1,2,...,l)$. The parameter l is chosen according to the number p of processors in the hypercube system under consideration. If a hypercube of a dimension k, $k > 1$ is considered then $p = 2^k$ and $l = 2 \times p = 2^{k+1}$. Hence, the matrices occurring in the two-sided Jacobi approach are divided into $2p$ block columns. Most operations of the method can be computed on the block columns parallelly without any transportation steps needed. Only the operation of updating the matrix A by the left singular vectors arising from the solution of subproblems has to be performed on block rows. This computational operation requires across processor communication [1]. The computational process proceeds in sweeps whereby in each sweep $(l(l-1)/2)$ subproblems of the size $(2m/l) \times (2n/l)$ have to be solved with corresponding block-row and block-column updates.

The parallel solution requires to find such an order of subproblem solving which divides the computation of one sweep into $l-1$ stages whereby in each of them $l/2$ subproblems are solved on p processors concurrently. The aim is to formulate optimal orderings, i.e. when in each stage distinct pairs of block rows and columns are treated by the processors without doubling or omitting some of the subproblem solutions. For hypercube multiprocessors there is also a requirement to organize the data transportations among directly connected processors only.

In the subsequent sections of the paper some optimal parallel orderings schemes will be presented. A class CO of orderings with a regular exchange scheme is described in section 2. Section 3 brings a formulation of the second ordering class DO. Each of these classes represent p different orderings, the patterns of which depend on the processor numbers they are related to. All these orderings are optimal for a hypercube system with processors numbered by Gray code.

2 The ordering scheme CO

The ordering scheme CO will define a class of orderings $CO(1), CO(2), ..., CO(p)$ the number of which corresponds to the number of processors. At the beginning of the SVD algorithm for each of these orderings each processor $P(j)\, j = 1,2,...,p$ holds in memory locations $A(j)$ and $B(j)$ respectively the blocks M_{2j-1} and M_{2j} of the original matrix M. In the formulation of the orderings, a transfer of a data block located on the processor $P(r)$ at the memory place $L1(r)$ to a memory place $L2(t)$ of a processor $P(t)$ will be denoted as $L2(t) \leftarrow L1(r)$ where $L1, L2 \in \{A, B\}$. Then for a given $i \in 1,2,...,p$ the ordering $CO(i)$ can be defined in the following manner:

ORDERING $CO(i), i = 1,2,\ldots,p$.

```
for j = 1, 2, ..., p
location of M_{2j-1} and M_{2j} in respectively positions A(j) and B(j) on a processor P(j)
for j = 1, 2, ..., p − 1
        if i = 1 then
                A(p) ← B(1) else                          (1)
                A(i − 1) ← B(i)
        if i = p then
                B(p) ← A(1) else                          (2)
                B(i) ← A(i + 1)
        if i ≠ 1 ∧ i ≠ p then                             (3)
                A(p) ← A(1)
        for s = 2, 3, ..., p
                if s ≠ i ∧ s ≠ i + 1 then                 (4)
                   A(s − 1) ← A(s)
        for s = 1, 2, ..., p − 1
                B(s + 1) ← B(s)                           (5)
        B(1) ← B(p)
```

For each $j = 1, 2, \ldots, p-1$ there are defined two stages of the ordering (steps 1 - 4 and step 5 respectively). The algorithm will be illustrated for an ordering applied to a 16×16 block matrix M. The number of processors is 8 (i.e. a cube of the size 3 is considered). The ordering $CO(5)$ for $i = 5$ is obtained as given in Table 1 (the block columns of the matrix M which meet at memory locations $A(i), B(i)$ on processors $P(i), i = 1, 2, \ldots, 8$, in order to solve the corresponding subproblems, are specified by their indices given in parenthesis):

P(1) (A(1),B(1))	P(2) (A(2),B(2))	P(3) (A(3),B(3))	P(4) (A(4),B(4))	P(5) (A(5),B(5))	P(6) (A(6),B(6))	P(7) (A(7),B(7))	P(8) (A(8),B(8))
(1,2)	(3,4)	(5,6)	(7,8)	(9,10)	(11,12)	(13,14)	(15,16)
(3,2)	(5,4)	(7,6)	(10,8)	(9,11)	(13,12)	(15,14)	(1,16)
(3,16)	(5,2)	(7,4)	(10,6)	(9,8)	(13,11)	(15,12)	(1,14)
(5,16)	(7,2)	(10,14)	(8,6)	(9,13)	(15,11)	(1,12)	(3,14)
(5,14)	(7,16)	(10,2)	(8,4)	(9,6)	(15,13)	(1,11)	(3,12)
(7,14)	(10,16)	(8,2)	(6,4)	(9,15)	(1,13)	(3,11)	(5,12)
(7,12)	(10,14)	(8,16)	(6,2)	(9,4)	(1,15)	(3,13)	(5,11)
(10,12)	(8,14)	(6,16)	(4,2)	(9,1)	(3,15)	(5,13)	(7,11)
(10,11)	(8,12)	(6,14)	(4,16)	(9,2)	(3,1)	(5,15)	(7,13)
(8,11)	(6,12)	(4,14)	(2,16)	(9,3)	(5,1)	(7,15)	(10,13)
(8,13)	(6,11)	(4,12)	(2,14)	(9,16)	(5,3)	(7,1)	(10,15)
(6,13)	(4,11)	(2,12)	(16,14)	(9,5)	(7,3)	(10,1)	(8,15)
(6,15)	(4,13)	(2,11)	(16,12)	(9,14)	(7,5)	(10,3)	(8,1)
(4,15)	(2,13)	(16,11)	(14,12)	(9,7)	(10,5)	(8,3)	(6,1)
(4,1)	(2,15)	(16,13)	(14,11)	(9,12)	(10,7)	(8,5)	(6,3)
(2,1)	(16,15)	(14,13)	(12,11)	(9,10)	(8,7)	(6,5)	(4,3)

Table 1

The initial column location is given by the first ordering row. The last row (below the line) is the starting position for the next sweep. (It results from the above algorithm if we would run the steps 1 - 4 for j = p.)

Although the neighbouring matrix blocks with indices $(2j-1, 2j)$ are not located on a processor $P(j), j = 1, 2, \ldots, p$, this initial distribution allows to perform the next sweep of the block SVD method by applying the above ordering algorithm without any complication.

The ordering scheme CO is more transparent from respectively Figure 1 and Figure 2 where the exchange pattern with a specification of individual steps 1 - 5 is illustrated for $CO(5)$ and $CO(8)$.

As seen for each ordering $CO(i)$ the matrix block initially located at the position $A(i)$ on given processor $P(i), i = 1, 2, \ldots, p$ is kept fixed during the exchange process. Since the exchange steps are performed chronologically from 1 to 5 , two entering arrows (for 2 and 5) at the position $B(i)$ indicate that the transfer 5 follows on a new matrix block which resulted from the transfer 2.

From the analysis of the orderings $CO(i), i = 1, 2, \ldots, p$ it follows, that the exchanges of matrix blocks are executed between processor pairs $P(j), P(j+1), j = 1, 2, \ldots, p-1$ and $P(1), P(p)$. If the processors are numbered by the Gray code (see Figs.1,2) then these transfers are performed between processors which are directly connected in the hypercube. Since only one matrix block is transferred between processors, we may conclude that CO represents a class of p optimal orderings for a hypercube.

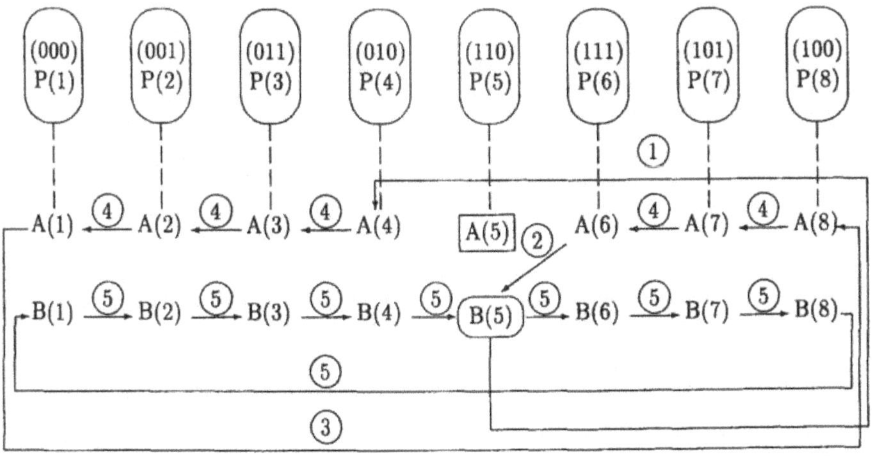

Figure 1

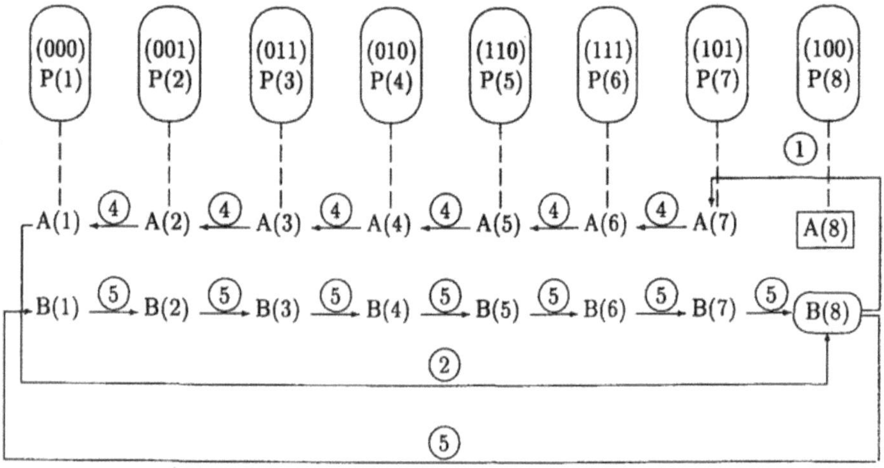

Figure 2

3 The orderings scheme DO

The ordering scheme DO defines a class of p different ordering schemes $DO(1), DO(2), \ldots, DO(p)$ the number of which coincides with the number of processors in the hypercube. For the formulation of these schemes analogous tools are used for the formulation of the CO schemes.

An ordering $DO(i)$ is defined for a given $i = 1, 2, \ldots, p$ as follows:

for $j = 1, 2, \ldots, 2p - 1$
if $i = p$ then
  $A(1) \leftarrow B(p)$ else
  $A(i+1) \leftarrow B(i)$ ①
if $i = 1$ then
  $B(p) \leftarrow A(p)$ else
  $B(i-1) \leftarrow A(i-1)$ ②
if $i \neq 1$ then
  $B(p) \leftarrow B(1)$ ③
if $i \neq 1 \wedge i \neq p$
  $A(1) \leftarrow A(p)$ ④
 for $s = 2, 3, \ldots, p$
  if $s \neq i \wedge s \neq 1$ then
   $A(s) \leftarrow A(s-1)$ ⑤
 for $s = 1, 2, \ldots, p - 1$
  if $s \neq i - 1$ then
   $B(s) \leftarrow B(s+1)$ ⑥

As an illustration, the scheme $DO(4)$ is shown by Table 2 for a 16×16 blocking of the input matrix.

P(1) (A(1),B(1))	P(2) (A(2),B(2))	P(3) (A(3),B(3))	P(4) (A(4),B(4))	P(5) (A(5),B(5))	P(6) (A(6),B(6))	P(7) (A(7),B(7))	P(8) (A(8),B(8))
(1,2)	(3,4)	(5,6)	(7,8)	(9,10)	(11,12)	(13,14)	(15,16)
(15,4)	(1,6)	(3,8)	(5,1)	(7,9)	(11,14)	(12,16)	(13,2)
(13,6)	(15,8)	(1,10)	(3,9)	(5,7)	(11,16)	(14,2)	(12,4)
(12,8)	(13,10)	(15,9)	(1,7)	(3,5)	(11,2)	(16,4)	(14,6)
(14,10)	(12,9)	(13,7)	(15,5)	(1,3)	(11,4)	(2,6)	(16,8)
(16,9)	(14,7)	(12,5)	(13,3)	(15,1)	(11,6)	(4,8)	(2,10)
(2,7)	(16,5)	(14,3)	(12,1)	(13,15)	(11,8)	(6,10)	(4,9)
(4,5)	(2,3)	(16,1)	(14,15)	(12,13)	(11,10)	(8,9)	(6,7)
(6,3)	(4,1)	(2,15)	(16,13)	(14,12)	(11,9)	(10,7)	(8,5)
(8,1)	(6,15)	(4,13)	(2,12)	(16,14)	(11,7)	(9,5)	(10,3)
(10,15)	(8,13)	(6,12)	(4,14)	(2,16)	(11,5)	(7,3)	(9,1)
(9,13)	(10,12)	(8,14)	(6,16)	(4,2)	(11,3)	(5,1)	(7,15)
(7,12)	(9,14)	(10,16)	(8,2)	(6,4)	(11,1)	(3,15)	(5,13)
(5,14)	(7,16)	(9,2)	(10,4)	(8,6)	(11,5)	(1,13)	(3,12)
(8,16)	(5,2)	(7,4)	(9,6)	(10,8)	(11,13)	(15,12)	(1,14)
(1,2)	(3,4)	(5,6)	(7,8)	(9,10)	(11,12)	(13,14)	(15,16)

Table 2

As seen from the row below the line, the nice property of DO schemes is the original location of matrix blocks on the processors at the beginning of each sweep.

The diagrams on Figure 3 and Figure 4 for respectively i = 8 and i = 4 make the above given algorithmic formulation more transparent.

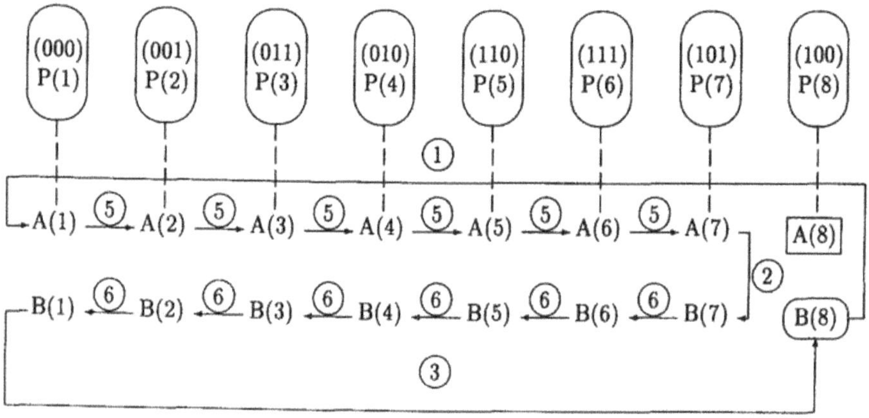

Figure 3

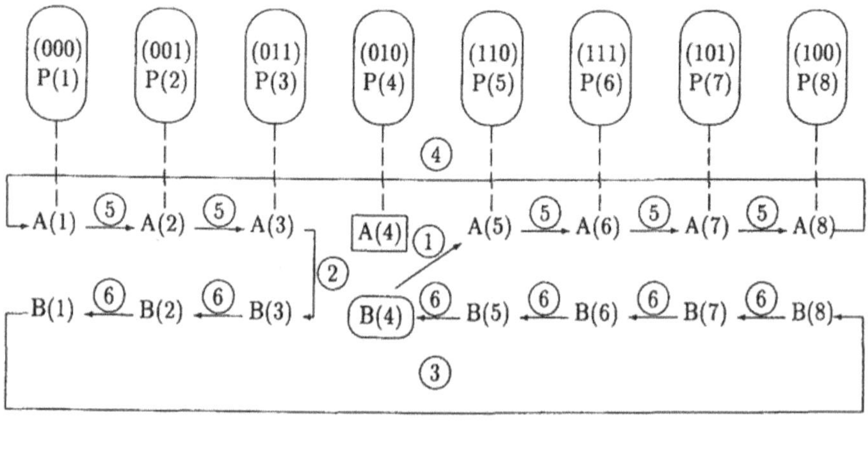

Figure 4

What concerns the communication pattern of the DO schemes, an analogous statement as for the CO schemes holds. Namely, the matrix-block exchanges according to the individual steps 1 - 6 follow between neighbouring processors in the hypercube. (Also in this case we assume the enumeration of processors by the Gray code.) Thus each of the scheme from the DO class satisfies the conditions put on optimal parallel ordering schemes.

4 Conclusions

Parallel ordering schemes for the block Jacobi method have been developed. The schemes allow evaluation of the singular values in a fully parallel manner. The schemes are applicable to those parallel architectures which can be embedded into the hypercube topology. From the point of view of communication and memory requirements, the algorithms based on these schemes are optimal when considering the MIMD distributed memory computational regime.

It is important to note that another actual problem with Jacobi methods is convergence. As shown by experience, a relatively large amount of sweeps are needed to achieve quadratic convergence [9]. A matter of further research will be to classify the new schemes from the above mentioned respect.

References

[1] C.H.Bischof: Computing the singular value decomposition on a distributed system of vector processors. Techn. Rep. 87-869, Dept. of Comp. Science, Cornell University,1987

[2] C.H.Bischof,C.F.Van Loan: Computing the singular value decomposition on a ring of array processors. In:J.Culumn and R.A.Willoughby(eds.):Large Scale Eigenvalue Problems. Amsterdam: North-Holland 1986, pp.51-66

[3] T.Chan: An improved algorithm for computing the singular value decomposition. ACM Trans. Math. Software 8, 72-83 (1982)

[4] G.H.Golub, C.Reinsch: Singular value decomposition and least squares solutions. Numerische Mathematik 14, 403-420 (1970)

[5] .R.Hestenes: Inversion of matrices by biorthogonalization and related results. J. Soc. Indust. Appl. Math. 6, 51-90 (1958)

[6] E.Kogbetliantz: Solution of linear equations by diagonalization of coefficient matrices. Quart. Appl. Math. 13, 123-132 (1955)

[7] F.T.Luk: A triangular processor array for computing the SVD. Linear Algebra Applic. 77, 259-274 (1986)

[8] F.T.Luk, H.Park: On parallel Jacobi orderings. SIAM J. Sci. Stat. Comp. 10 ,18-26 (1989)

[9] G.Shroff, R.Schreiber: On the convergence of the cyclic Jacobi method for parallel block orderings. SIAM J. Sci. Stat. Comp. 10, 326-346 (1989)

Institute of Control Theory and Robotics,
Slovak Academy of Sciences
Dúbravská 9, 842 37 Bratislava,
Slovak Republic

Optimizing large air pollution models on high speed computers

Zahari Zlatev, Ivan Dimov, Krassimir Georgiev *

Abstract

Long range transport of air pollution is an important environmental problem. Therefore the interest in this phenomenon has been increasing during the last decade. On the other hand, it is rather difficult to treat numerically the mathematical models describing long range transport of air pollutants because these models lead, after some kind of discretization, to very huge systems of ordinary differential equations (up to order of 10^6). The use of modern parallel and/or vector machines is a necessary condition for a successful handling of big air pollution models. However, this is often far from sufficient. One should also optimize the code in order to be able to exploit fully the power of the modern computers. The code is optimized for several different types of high speed computers (vector machines, parallel computers with shared memory and parallel computers with distributed memory; including in the last group the massively parallel computers). Results obtained by using representative computers from each of these three groups will be presented.

1 Mathematical description of large air pollution models

The long range transport of air pollutants over Europe is studied by a system of partial differential equations **PDE's**:

$$\begin{aligned}\frac{\partial c_s}{\partial t} &= -\frac{\partial(uc_s)}{\partial x} - \frac{\partial(vc_s)}{\partial y} + \frac{\partial}{\partial x}\left(K_x\frac{\partial c_s}{\partial x}\right) + \frac{\partial}{\partial y}\left(K_y\frac{\partial c_s}{\partial y}\right) \\ &\quad + E_s + Q_s(c_1, c_2, \ldots, c_q) - (\kappa_{1s} + \kappa_{2s})c_s, \qquad s = 1, 2, \ldots, q\end{aligned} \tag{1}$$

*This project was supported in part by NATO-ENVIR.CRG930449, the Danish Natural Sciences Research Council and the National Environmental Research Institute (Denmark)

The number of equations is equal to the number of species that are involved in the model and varies in different studies. The largest number of equations used until now was 35 (but the physicists in the **Danish Environmental Research Institute** would like to increase this number if the model so obtained could be handled numerically). The different quantities that are involved in the mathematical model have the following meaning: (**i**) the concentrations are denoted by c_s; (**ii**) u and v are wind velocities; (**iii**) K_x and K_y are diffusion coefficients; (**iv**) the sources are described by the function E_s; (**v**) κ_{1s} and κ_{2s} are deposition coefficients; (**vi**) the chemical reactions are described by $Q_s(c_1, c_2, \ldots, c_q)$.

The chemical reactions are of the form:

$$Q_s(c_1, c_2, \ldots, c_q) = -\sum_{i=1}^{q} \alpha_i c_i + \sum_{i=1}^{q}\sum_{j=1}^{q} \beta_{ij} c_i c_j \tag{2}$$

This is a special kind of non-linearity, but it is not clear how to exploit this fact during the numerical treatment of the model.

It is clear from the above description of the quantities involved in the mathematical model that all five physical processes (advection, diffusion, emission, deposition and chemical reactions) can be studied by using the above system of **PDE's** . The most important processes are the advection (the transport) and the chemical reactions. Kernels for these two parts of the model will sometimes be discussed in more details.

2 Numerical treatment of the model

It is difficult to treat the system of **PDE's** (1) directly. This is the reason for using different kinds of splitting. A simple splitting procedure, based on ideas proposed in [5] and [6] can be defined, for $s = 1, 2, \ldots, q$, by the following sub-models:

$$\frac{\partial c_s^{(1)}}{\partial t} = -\frac{\partial (u c_s^{(1)})}{\partial x} - \frac{\partial (v c_s^{(1)})}{\partial y} \tag{3}$$

$$\frac{\partial c_s^{(2)}}{\partial t} = \frac{\partial}{\partial x}\left(K_x \frac{\partial c_s^{(2)}}{\partial x}\right) + \frac{\partial}{\partial y}\left(K_y \frac{\partial c_s^{(2)}}{\partial y}\right) \tag{4}$$

$$\frac{dc_s^{(3)}}{dt} = E_s + Q_s(c_1^{(3)}, c_2^{(3)}, \ldots, c_q^{(3)}) \tag{5}$$

$$\frac{dc_s^{(4)}}{dt} = -(\kappa_{1s} + \kappa_{2s}) c_s^{(4)} \tag{6}$$

The discretization of the spatial derivatives in the first system, (3), leads to q independent systems of ordinary differential equations (ODE's). The discretization can be carried out both by using pseudospectral expansions and by applying finite elements; [8], [9].

The discretization of the second-order spatial derivatives in the second system, (4), is quite similar. Both pseudospectral expansions and finite elements can be used. The number of independent systems of ODE's is again q; the same as in the previous case, the important fact here is that different numerical algorithm may be used in the advection and diffusion parts; [8], [9].

The third and the fourth systems do not contain partial derivatives. These systems are directly expressed by ordinary differential equations. The third system contains N small systems of ODE's (N being the number of grid-points, while each system contains q ODE's). The fourth system contains Nq independent ODE's.

The four systems (3) -(6) are solved successively at each time-step. The initial conditions are used to couple the four systems. Consider a given time-step n ($n = 1, 2, ...$). The initial condition used in the solution of the first system is the vector containing the concentrations at the grid-points which were obtained during the previous time step. The approximation obtained after the solution of each of the first three systems is used as an initial condition in the solution of the second system. The approximation obtained after the solution of the second system is used as an initial condition in the solution of the third system. Finally, the approximation obtained after the solution of the third system is used as an initial condition in the solution of the fourth system. The approximation obtained after the solution of the fourth system is considered as an acceptable approximation of the concentrations at step n. It is clear that everything is ready to start the computations at a new step, time-step $n + 1$, after the solution of the fourth system.

Predictor-corrector methods with several different correctors are used in the solution of the first and the second systems. The correctors are carefully chosen so that the stability of the method is enhanced; [7].

Several different methods have been tried in the solution of the third system; [3]. The QSSA (quasi-steady-state algorithm) is simple and relatively stable but not very accurate (therefore it has to be run with a small time-stepsize). The classical numerical methods (Backward Euler Method, the Trapezoidal Rule and a second-order Runge-Kutta algorithm) lead to the solution of non-linear systems of algebraic equations and, therefore, they are more expensive. On the other hand, these methods can be incorporated with an error control and perhaps with larger time-steps. The extrapolation methods are also promising. It is easy to achieve an error estimation and to carry out the integration with large time-steps when these algorithms are used. However, it is difficult to implement such methods in an efficient way. The experiments with different integration methods for the chemical sub-model (5) are continuing. The QSSA will be used in the experiments described here.

The fourth system, (6), contains Nq independent ODE's. Moreover, all these ODE's are linear. Therefore they are solved exactly during the numerical treatment of the model.

3 Need for high speed computers

The size of the systems that arise after the space discretization and the splitting procedures used to treat numerically (1) is enormous. Let us assume that the model is discretized on a (96x96) grid (such a grid is used in the Danish Eulerian Model after 1993). Then the number of equations in each of the four systems of ODE's (3)-(6) is 322560. The time-stepsize used in the advection step is 15 min. The chemical sub-model (5) cannot be treated with such a large time-stepsize (because it is very stiff; especially when photochemical reactions are involved). Therefore six small time-steps are carried out for each advection time-step (this means that the chemical time-stepsize is 2.5 min.). From this description it is clear that in fact nine systems of ODE's (each of them containing 322560 equations) are to be treated per advection step. Assume that a one-month run is to be carried with the model. This will result in 3456 advection time-steps (taking here into account that it is necessary to use five extra days in order to start up the model).

It is clear that such a large problem can be solved only if a new modern high speed computers are used. Moreover, it is necessary to select the right numerical algorithms (which are most suitable for the high speed computers available) and to perform the programming work very carefully in order to exploit fully the great potential power of the vector and/or parallel computers.

4 Vector processors

Vector processors (and especially the latest CRAY products) are still among the most power computers available on the market. The runs with the Danish Eulerian Model on different vector processors started in 1980, and continuing. Most of the numerical methods were adapted for runs on such computers.

During the advection part one must be careful when the pseudospectral method is vectorized. The use of multiple FFT's (fast Fourier transforms) is very efficient. One should also take care when finite elements are used. In the latter case, banded matrices appear in the computations and, thus, efficient algorithms for computations with such matrices are necessary.

The chemical part is also critical (the treatment of the chemical reactions will require more than 95 % of the computing time when this part is not vectorized).

The other two parts are not very critical. The diffusion part, (4), can be treated in the same way as the advection part. The deposition part, (6), can be vectorized in a very easy way.

It is commonly accepted to measure the speed of computations achieved on a vector processor in MFLOPS (millions of floating point operations, additions and multiplications, per second). The speeds of computations (achieved in the different parts of the Danish Eulerian Model when it is run on a CRAY Y-MP computer) are given in **Table 1**. More results can be found in [2], [3], [9], [10], [11].

Part of the code	MFLOPS	Time	Percent
Advection	331	448	16
Diffusion	337	285	10
Chemistry	405	1500	54
Deposition	321	483	17

Table 1. Results obtained in run of the model on CRAY Y-MP. The times are given in seconds. Percent refer to the percentages of the different parts of the model with regard to the total time for the run (the time for the input and output operations, approximately 3%, are not taken into account in this table).

It is seen that all parts are optimized rather well for vector computations. The speed of computations of the chemical part is very high. However, this part is nevertheless the most time-consuming part of the model, and improvements are needed here.

5 Parallel computers with shared memory

As a rule, parallel computers with shared memory have only a few but tightly connected processors. Therefore it is natural to parallelize the advection and the diffusion parts according to the pollutants (the number of independent ODE systems is equal to the number of pollutants for these two parts).

To parallelize the deposition part is trivial. All Nq ODE's in this part are independent. This means that there are Nq independent tasks which must be distributed among the processors. If there are p processors, then every processor has to perform approximately Nq/p simple tasks.

The parallelization of the chemical part is also easy. The number of independent ODE systems is N (each of them containing q equations). Thus, if there are p processors, then every processor will receive approximately N/p tasks.

If the parallel processor available has a cache memory, then the requirement for "data locality" must be satisfied, i.e. the computer must work as long as possible with the same data and the amount of these data should not be too large (so that it can be kept in cache).

If the parallel computer with shared memory has processors with vector capabilities, then one can optimize the code on two levels: (i) each processor must work as long as possible in a vector mode and (ii) all processors must work as long as possible concurrently.

When splitting procedures are used, it is relatively easy (in principle, at least) to optimize the large air pollution models for runs on parallel machines with shared memory. However, there are a lot of technical difficulties, which are caused, among other things, by the fact that the codes are very long.

Many experiments were carried out on several parallel machines with shared memory using modules of the models containing the most difficult parts; the advection part and the chemical part. Some results obtained on two ALLIANT computers are shown in **Table 2**. The efficiency is also given in **Table 2**. Let t_1 be the computing time achieved on one processor. Let t_p be the computing time achieved when p processors are used. Then the efficiency is defined by $100t_1(pt_p)$.

Number of processors	Alliant computer	
	FX/80	FX/2800
1	309	219
2	173 (89%)	120 (91%)
4	104 (74%)	69 (79%)
8	71 (54%)	45 (61%)
10		33 (66%)
14		32 (49%)

Table 2. Computing times (in seconds) on Alliant FX/80 and FX/2800 model 200PMs for advection + chemistry. The best compiler options were used.
The efficiency achieved when more than one processor is in use is given in brackets.

6 Massively parallel computers

All experiments described in this section were performed on a **CM-200** computer. **CM-200** is a massively parallel computer that contains up to **65536** bit serial processors that work asynchronously. The particular computer used in the experiments contains **8196** bit serial processors. Each processor has up to 1 Mbit of local memory and is able to access other processors through a general purpose high-speed communication router, which utilizes a hypercube communication network of an appropriate dimension. Physically, **16** bit serial processors reside on a single chip. Each pair of chips (or, in other words, each group of **32** processors) shares either a **32** or **64** bit floating point unit. Therefore one can consider a **CM-200** with **8196** bit serial processors as a machine configured by **8196/32=256** computational units that are able to work concurrently and to communicate during the computational process.

The theoretical peak performance of the **CM-200** is rather high. In the case when it is configured with **8196** bit serial processors and, when double precision is used, the peak performance is about **2.56 GFLOPS** (billion floating point operations, additions or multiplications, per second) see, for example, [1]. However, the speed of computations that is achieved in real application packages is normally much lower. The speed is degraded ([4]): **(i)** by communication costs (when information must be sent to other processors), **(ii)** by reduction costs (when, for example, norms of vectors and/or matrices are to be calculated), **(iii)** by the fact that different processors have to carry out different calculations (when maskings or the "where" statements are used), **(iv)** by some other reasons.

Let us start the discussion with **the first part** of the computational work. Communications are needed in this part. The simple chapeau finite elements were chosen in order to reduce the number of communications. When this choice is made, each processor is communicating only with two neighbours. Thus, the communication cost is nearly minimal. Nevertheless, the speed of computations during this part is considerably lower than that for the second part (see below) when a 32x32 grid is in use. Refining the grid results in improving the performance; see **Table 3**. Reductions are also needed during the first part; because the systems of linear algebraic equations (that arise there) are solved by iterative methods and vector norms are needed in the stopping criteria. The main effects of the reduction are similar to those caused by the communications: (i) the reductions degrade the speed of computations and (ii) the relative cost of the reductions is reduced when the grid is refined (which leads to an increase of the speed of computations); this is illustrated in **Table 3**.

Consider now the second part of the computational process sketched above. No communications and no reductions are needed because simple explicit methods for solving **ODE's** are used in this part. Moreover, the loops are very simple; the same kind of arithmetic operations are used for all components of the vectors that are involved in this part (i.e. no masking is needed and no constructions with **where** statements are used). The performance of **CM-200** is very high when the second part is executed; it is seen in **Table 3**.

The third part of the computational work is not very expensive (in comparison with the other parts). It is included here only because it is crucial for the numerical treatment of the chemical sub-model. The concentrations that are to be used as input to the chemical model must be non-negative. However, the advection sub-model produces negative concentrations in sub-domains where the exact concentrations are zeros (or very small numbers). Therefore the concentrations in such sub-domains (if such sub-domain exist) must be smoothed. It is clear from this explanation that smoothing is to be performed only in some parts of the space domain (and these parts must be determined). The subroutine that performs smoothing contains all kinds of calculations that delay the speed of computation: communications, reductions, masking and constructions of loops with the **where** statement. This means that one should expect the speed of computations to be very low when this subroutine is executed on **CM-200**. The results in **Table 3** illustrate this effect. It is seen also that the speed of computations becomes greater when the grid is refined. It must be mentioned here that this routines does not perform very well on other high-speed computers either.

The fourth part is easy for **CM-200**. There are no communications and no reductions in this part. However, the code contains some loops that are constructed by using **where** statements. This delays the computations and the speed of computations in this part is not as high as the speed of computations in the second part; see

Table 3.

Numerical results concerning the speed of computations that could be achieved in the different parts of the third case (advection + chemistry) of the class of test problems introduced in Section 4 are shown in **Table 3**. Five different grids, (32x32), (64x64), (128x128), (256x256) and (512x512), were used in the experiments. It is seen from **Table 3** that the performance of the machine is nearly always improved when the grid is refined. The case where the finest grid, (512x512), is used is sometimes an exception. The reason for such a phenomenon is probably the fact that there is not sufficient memory to store all data that will be needed by a processor in its local memory.

Part of work	GRID - SIZE				
	(32x32)	(64x64)	(128x128)	(256x256)	(512x512)
Part 1	64	169	320	480	580
Part 2	332	685	910	961	850
Part 3	13	47	136	248	312
Part 4	352	510	560	577	579

Table 3. Speed of computations measured in MFLOPS.
Part 1 - space discretization of the advection terms.
Part 2 - time integration of the advection sub-model.
Part 3 - smoothing to avoid negative concentrations.
Part 4 - treatment of the chemical sub-model.
For all grids the number of time-steps is 3456.

One can consider the total time spent on **CM-200** (it will be called the CM total time) as a sum of three major components: **(i)** the CM CPU time, **(ii)** the CM communication time, **(iii)** the CM reduction time. It is interesting to investigate the relations between these times. An experiment has been performed to get information about the behaviour of the major components of the computing time when the grid is refined. In this experiment the above problem has been run on five grids: (32x32), (64x64), (96x96), (128x128) and (160x160). The computing times (the total time, the CPU time, the communication time and the reduction time) obtained in this experiment are given in **Fig. 1**.

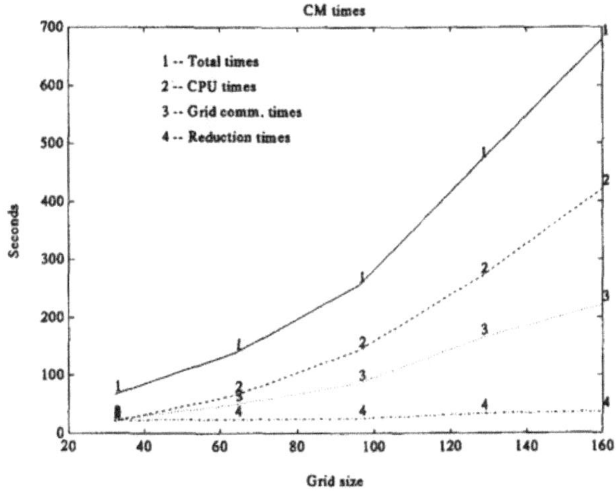

Figure 1. Increasing the CM times with refining the grids.

Figure 2. Percentages of the most important CM times (as parts of the total CM time) when the grids are refined.

It is seen that the first three times, drawn on Fig.1, are steadily increasing when the grids are refined, while the fourth time is nearly constant.

It is interesting to investigate the relations between these times. An experiment has been performed to get information about the behaviour of the major components of the computing time when the grid is refined. In this experiment the above problem has been run on five grids: (32x32), (64x64), (96x96), (128x128) and (160x160). The computing times (the total time, the CPU time, the communication time and the reduction time) obtained in this experiment are given in **Fig. 1**. It is seen that the first three times are steadily increasing when the grids are refined, while the fourth time is nearly constant.

The relative parts of the three major components of the total time are shown, as percentages of the total time, in **Fig. 2**. The results that are presented in **Fig. 2** are very illustrative. It is clearly seen that the following four conclusions can be drawn by using these results: **(i)** The relative part of the CM CPU time grows when the number of grid-points is increased. **(ii)** The relative part of the CM communication time stays nearly constant when the number of grid-points is increased. **(iii)** The relative part of the CM reduction time decreases when the number of grid-points is increased.

The main conclusion is that if one intends to run the model on a very fine grid (because of the accuracy requirements), then one should not bother too much about

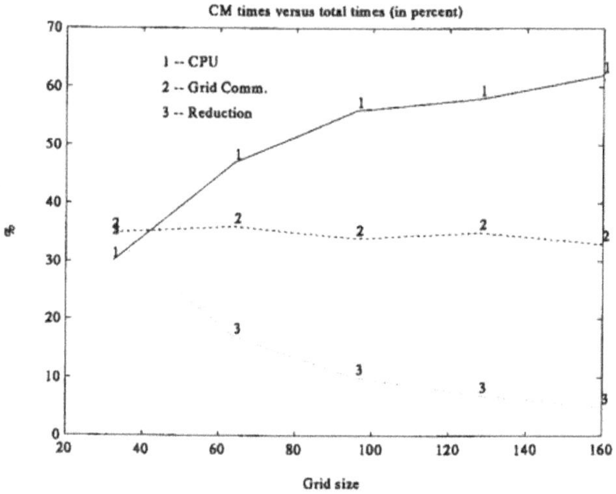

the reduction time. The communication time is the more important factor in this situation. However, the most important factor in the case where a very fine grid is to be used is the CPU time, and the major efforts in the attempts to optimize the code must be made to reduce the CPU time. If the grid is not very fine, then one must attempt to reduce both the CPU time and the communication time. Finally, if the grid is rather crude, say (32x32), then one must attempt to reduce all three major components of the computing time; i.e. both the CPU time, the communication time and the reduction time.

7 Concluding remarks

There are still many tasks which are to be solved in the attempts to improve the performance of the Danish Eulerian Model. Two major types of improvements should be emphasized: (i) improvements of the physical procedures in the model and (ii) the optimization of the model to the new more complicated modern computers (heterogeneous computers, PVM, etc.)

Improvement of the physical procedures. The development of a three dimensional model is one of the most important tasks among the tasks in this group. A three dimensional model describes much better the reality than a two dimensional model. A three dimensional model is in process of development now at the National Environmental Research Institute in collaboration with scientists from the Bulgarian Academy of Sciences.

Some other tasks (as, for example, improvements of the chemical modules, of the deposition modules, etc.) must also be accomplished in the near future.

Use of more complicated architecture. More complicated architectures, in which different principles are implemented, are now coming on the market. This process will probably continue in the future. Therefore it is necessary to adapt continuously the model to the last achievements in the supercomputer technology. The use of heterogeneous computers is attractive.

PVM (Parallel Virtual Machine) has been rather popular in the last two-three years. Some modules of the model have been tested on PVM. This work continues.

References

[1] **J. J. Dongarra:** *"Performance of various computers using standard linear equations software"*. Report No. CS-89-85. Computer Science Department, University of Tennessee, Knoxville, TN 37996-1301, USA, **1991**.

[2] **R. Harrison, Z. Zlatev and C. J. Outley:** *"A comparison of the predictions of an Eulerian atmospheric transport-chemistry model with experimental measurements over the North Sea"*. Atmos. Envinron., **29(1994)**, 497–516.

[3] **Ø. Hov, Z. Zlatev, R. Berkowicz, A. Elliassen and L. P. Prahm:** *"Comparison of numerical techniques for use in air pollution models with non-linear chemical reactions"*. Atmos. Environ., **23(1989)**, 967-983.

[4] **C. Levit:** *"Grid communication on the Connection Machine: Analysis, performance and improvements"*. In: "SCIENTIFIC APPLICATIONS OF THE CONNECTION MACHINE" (H. D. Simon, ed.), pp. 316-322. World Scientific Publishing, Singapore, **1989**.

[5] **G. I. Marchuk:** *"Mathematical modeling for the problem of the environment"*. Studies in Mathematics and Applications, No. 16, North-Holland, Amsterdam, **1985**.

[6] **G. J. McRae, W. R. Goodin and J. H. Seinfeld:** *"Numerical solution of the atmospheric diffusion equations for chemically reacting flows"*. J. Comp. Phys., **45(1984)**, 1-42.

[7] **Z. Zlatev:** *"Application of predictor-corrector schemes with several correctors in solving air pollution problems"*. BIT, **24(1984)**, 700-715.

[8] **Z. Zlatev:** *"Mathematical model for studying the sulphur pollution over Europe"*. J. Comput. Appl. Math., **12(1985)**, 651-666.

[9] **Z. Zlatev:** *"Treatment of some mathematical models describing air pollution problems"*. Parallel Computing, **6(1988)**, 87-99.

[10] **Z. Zlatev, J. Christensen and A. Eliassen:** *"Studying high ozone concentrations by using the Danish Eulerian Model"*. Atmos. Environ., **27A(1993)**, 845–865.

[11] **Z. Zlatev, J. Christensen and Ø. Hov:** *"An Eulerian air pollution model for Europe with nonlinear chemistry"*. J. Atmos. Chem., **15(1992)**, 1-37.

Addresses of the Authors:

Z. Zlatev
National Envinronmental Research Institute
Frederiksborgvej 399, P.O. Box 358,
DK–4000 Roskilde, DENMARK

I. Dimov, K. Georgiev
Bulgarian Academy of Sciences
Acad.G.Bonchev str. Bl.25-A,
1113 Sofia, BULGARIA

Part III

Applications of Numerical Methods

Metal Forming Processes Simulated by the Finite Element Method

N. Bontcheva, R. Iankov, M. Datcheva *

1 Introduction

Numerical simulation of metal forming processes gives the possibility to predict the exact form of the end product of free forging processes, depending on the technological parameters, as well as to determine the conditions under which cracks may appear inside the deformed body. In the case of compaction and forming processes of powder based materials the developement of density distribution during the technological process is obtained.

The application of the finite element method to simulating of forming and compaction processes of powder based materials will be considered. As a particular case metal forming of dense metals will be obtained.

2 Mechanical model

Metal forming processes are characterized with large plastic strains. Elastic strains are negligible. That is why the latter will be omitted and the rigid–plastic model will be applied. We assume that a yield condition of the form ([1]):

$$F \equiv AII_s^2 + BI_1^2 + CII_s + DI_1 - \sigma_p^{*2} = 0 \,. \tag{2.1}$$

exists, where

$$I_1 = \sigma_{ij}\delta_{ij} = \sigma_{kk} \,, \quad II_s^2 = \frac{1}{2}s_{ij}s_{ij} \,, \quad s_{ij} = \sigma_{ij} - \frac{1}{3}I_1\delta_{ij} \tag{2.2}$$

and σ_{ij} is the Cauchy stress tensor.

The existence of the first stress invariant I_1 in (2.1) gives the possibility to take into account volume changes during the deformation process.

*This work was supported in part by the National Foundation "Scientific Investigations" under grant MM–41

In (2.1) A, B, C, D and σ_p^* are material functions. depending on the relative density of the material

$$\rho = \frac{V_b}{V}, \tag{2.3}$$

where V is an arbitrary small volume of the porous material, consisting of the volume V_b of the base material and the volume V_{por} of the pores:

$$V = V_b + V_{por}. \tag{2.4}$$

In the case of fully dense material the material functions B, C and D vanish and the well known Mises yield condition, valid for metals is obtained.

The density change is connected with the volume change $\dot{\varepsilon}_v$ that takes place during the deformation process [2, 3, 4, 5]:

$$\frac{\dot{\rho}}{\rho} = -\dot{\varepsilon}_v, \quad \dot{\varepsilon}_v = \dot{\varepsilon}_{kk}. \tag{2.5}$$

In (2.5) $\dot{\varepsilon}_{ij}$ is the strain rate tensor of the porous material:

$$\dot{\varepsilon}_{ij} = \frac{1}{2}(v_{i,j} + v_{j,i}) \tag{2.6}$$

and v_i is the velocity vector.

The material function σ_p^* depends on the yield limit of the base (dense) material σ_p^b:

$$\sigma_p^* = \beta\sigma_p^b, \tag{2.7}$$

where $\beta = \beta(\rho)$ is a material function. The yield limit of the base material depends on the equivalent strain rate of the base material

$$\dot{\bar{\varepsilon}}_b = \int_0^t \dot{\varepsilon}_b dt, \tag{2.8}$$

e.g. hardening effects and strain-rate sensitivity may be described.

We assume the equivalent strain rate energy principle

$$\dot{W}_p = \sigma_{ij}\dot{\varepsilon}_{ij} = \bar{\sigma}\dot{\bar{\varepsilon}}, \tag{2.9}$$

where $\bar{\sigma}$ is the equivalent stress and $\dot{\bar{\varepsilon}}$ is the equivalent strain rate. We assume that the strain rate energy of the porous material is connected with the strain rate energy of the base material by means of the following expression:

$$\dot{W}_p = \bar{\sigma}\dot{\bar{\varepsilon}} = \rho\bar{\sigma}\dot{\bar{\varepsilon}}_b, \tag{2.10}$$

which means that the deformation of the base material is responsible for the plastic work.

Experiments show ([7]) that the surface roughness changes during the deformation process, de[ending on the density developement. According to [7] we assume the following kinetic equation for the surface roughness parameter evolution:

$$\frac{dR_max}{d\bar{\varepsilon}} = C_R\rho^\gamma\,, \tag{2.11}$$

where $C_R = \kappa G$, G is the mean crystal grain diameter and κ and γ are material constants.

The normality rule has the form:

$$\begin{aligned} \dot{\varepsilon}_{ij} &= \dot{\lambda}\left(\tfrac{\partial F}{\partial I_1}\delta_{ij} + \tfrac{\partial F}{\partial II_s^2}s_{ij}\right)\,, \\ \dot{\varepsilon}_v &= \dot{\varepsilon}_{kk} = 3\dot{\lambda}\tfrac{\partial F}{\partial I_1}\,, \end{aligned} \tag{2.12}$$

and taking into account (2.1) we obtain:

$$\begin{aligned} \dot{\varepsilon}_{ij} &= \dot{\lambda}\left[(2BI_1 + D)\,\delta_{ij} + \left(A + \tfrac{1}{2II_s}C\right)s_{ij}\right]\,, \\ \dot{\varepsilon}_v &= \dot{\varepsilon}_{kk} = 3\dot{\lambda}\,(2BI_1 + D)\;. \end{aligned} \tag{2.13}$$

The mean stresses in the uniaxial stress state are:

$$\begin{aligned} \sigma_I &= \bar{\sigma}\,, \quad \sigma_{II} = \sigma_{III} = 0\,, \\ s_I &= \tfrac{2}{3}\bar{\sigma}\,, \qquad s_{II} = s_{III} = -\tfrac{1}{3}\bar{\sigma} \end{aligned} \tag{2.14}$$

and according to (2.12) the effective strain rate is:

$$\dot{\bar{\varepsilon}} = \dot{\varepsilon}_I = \dot{\lambda}\left[\left(\frac{\partial F}{\partial I_1}\right)_{equiv} + \frac{2}{3}\left(\frac{\partial F}{\partial II_s^2}\right)_{equiv}\bar{\sigma}\right]\,, \tag{2.15}$$

where $()_{equiv}$ denotes the values, corresponding to the equivalent uniaxial state. Taking into account (2.12) and (2.14) we obtain the following expression for $\dot{\lambda}$:

$$\dot{\lambda} = \frac{\dot{\bar{\varepsilon}}}{2\left(B + \frac{1}{3}A\right)\bar{\sigma} + \left(D + \frac{1}{\sqrt{3}}C\right)}\,, \tag{2.16}$$

Introducing (2.16) and (2.13) into (2.9) we obtain:

$$\dot{W}_p = \frac{\dot{\bar{\varepsilon}}\left(AII_s{}^2 + BI_1^2 + \frac{1}{2}CII_s + \frac{1}{2}DI_1\right)}{\left(B + \frac{1}{3}A\right)\bar{\sigma} + \left(D + \frac{1}{\sqrt{3}}C\right)} = \bar{\sigma}\dot{\bar{\varepsilon}} \tag{2.17}$$

which leads to the expression for the equivalent stress:

$$\bar{\sigma} = \frac{-\frac{1}{2}\left(D + \frac{1}{\sqrt{3}}C\right) \pm \sqrt{\frac{1}{4}\left(D + \frac{1}{\sqrt{3}}C\right)^2 + 4\left(B + \frac{1}{3}A\right)R}}{2\left(B + \frac{1}{3}A\right)}\,, \tag{2.18}$$

where due to (2.1)

$$R = \sigma_p^{*2} - \frac{1}{2}(CII_s + DI_1) \ , \tag{2.19}$$

or

$$R = \frac{1}{2}\left(\sigma_p^{*2} + AII_s^2 + BI_1^2\right) . \tag{2.20}$$

The flow rule (2.13) leads to the expression for the stress tensor:

$$\begin{aligned} \sigma_{ij} &= \frac{1}{\dot\lambda\left(A + \frac{1}{II_s}C\right)}\left(\dot\varepsilon_{ij} - \frac{1}{3}\dot\varepsilon_v\delta_{ij}\right) + \frac{1}{18\dot\lambda B}\left(\dot\varepsilon_v - 3\dot\lambda D\right)\delta_{ij} \\ &= \frac{1}{\dot\lambda}\left[\frac{1}{A + \frac{1}{2II_s}C}\dot\varepsilon_{ij} + \frac{1}{3}\left(\frac{1}{6B} - \frac{1}{A + \frac{1}{2II_s}C}\right)\dot\varepsilon_v\delta_{ij}\right] - \frac{1}{6}\frac{D}{B}\delta_{ij} . \end{aligned} \tag{2.21}$$

Introducing (2.21) into (2.9) we obtain the expression for the effective strain rate:

$$\dot{\bar\varepsilon} = -\frac{1}{12}\frac{D}{B}\frac{\dot\varepsilon_v}{\bar\sigma} + \sqrt{Q} \, , \tag{2.22}$$

where

$$Q = \frac{1}{\bar\sigma}\left\{\frac{S_3}{S_2}\dot\varepsilon_{ij}\dot\varepsilon_{ij} + \frac{1}{3}\left[\frac{1}{48}\left(\frac{D}{B}\right)^2\frac{1}{\bar\sigma} + S_3\left(\frac{1}{6B} - \frac{1}{S_2}\right)\right]\dot\varepsilon_v^2\right\} , \tag{2.23}$$

The following notations are used in (2.23):

$$\begin{aligned} S_2 &= A + \frac{1}{2II_s}C \ , \\ S_3 &= 2\left(B + \frac{1}{3}A\right)\bar\sigma + \left(D + \frac{1}{\sqrt{3}}C\right) . \end{aligned} \tag{2.24}$$

3 Application of the finite element method

The application of the finite element method is based on the functional ([5]):

$$\pi = \int_V \bar\sigma\dot{\bar\varepsilon}dV + \int_{S_f} \tau_t v_i dS - \int_{S_p} p_i v_i dS \ , \tag{3.1}$$

where V is the volume of the deformed body, S_p is the part of its surface S on which tractions are prescribed and S_f is the part of S, where friction takes place. τ_t is the friction stress, directed opposite to the relative tangential velocity at points on S_f. We assume that the friction stress τ_t is obtained according to the Siebel law:

$$\tau_t = \frac{1}{\sqrt{3}}m\sigma_p^b \qquad x_i \in S_f \ , \tag{3.2}$$

where σ_p^b is the yield limit of the basic material and m is the friction factor.

The true values of the velocity field bring stationarity of the functional (3.1):

$$\delta\pi = \int_V \bar\sigma\delta\dot{\bar\varepsilon}dV + \int_{S_f} \tau_t\delta v_i dS - \int_{S_p} p_i\delta v_i dS = 0 \, . \tag{3.3}$$

The velocity vector v_i is:

$$\{v\} = [N]^T \{q^e\} \,, \tag{3.4}$$

where q^e is the nodal velocities vector of a finite element and $[N]^T$ is the matrix of the shape functions. The strain rate vector is:

$$\{\dot{\varepsilon}\} = [B]\{q^e\} \,, \tag{3.5}$$

where $[B]$ is the strain rate matrix. The volumetric strain rate is:

$$\dot{\varepsilon}_v = \{\dot{\varepsilon}\}^T \{e\} = \{q^e\}^T [B]^T \{e\} = \{e\}^T [B] \{q^e\} = \{C\}^T \{q^e\} \,, \tag{3.6}$$

where

$$\{e\}^T = \{1,1,0,1\} \,, \quad \{C\}^T = \{e\}^T [B] \,. \tag{3.7}$$

The effective strain rate according to (2.23) and (3.5) is:

$$\dot{\bar{\varepsilon}} = -\frac{1}{12}\frac{1}{\bar{\sigma}}\frac{D}{B}\{C\}^T \{q^e\} + \sqrt{\{q^e\}^T [P] \{q^e\}} \,, \tag{3.8}$$

where

$$[P] = [B]^T [D^*] [B] \,, \qquad [D^*] = [D_1] + [D_2] + [D_3] \,, \tag{3.9}$$

$$\begin{aligned} [D_1] &= \tfrac{S_3}{S_2\bar{\sigma}}\left([I_1] - \tfrac{1}{3}[I_2]\right) = \left(\tfrac{1}{3}A + B + \tfrac{D}{2\bar{\sigma}} + \tfrac{C}{2\sqrt{3}\bar{\sigma}}\right)\left[\tfrac{2}{S_2}\left([I_1] - \tfrac{1}{3}[I_2]\right)\right] \,, \\ [D_2] &= \tfrac{S_3}{18B\bar{\sigma}}[I_2] = \left(\tfrac{1}{3}A + B + \tfrac{D}{2\bar{\sigma}} + \tfrac{C}{2\sqrt{3}\bar{\sigma}}\right)\tfrac{1}{9B}[I_2] \,, \\ [D_3] &= \tfrac{1}{144\bar{\sigma}^2}\left(\tfrac{D}{B}\right)^2 [I_2] \,, \end{aligned} \tag{3.10}$$

$$[I_1] = \begin{bmatrix} 1 & 0 & 0 & 0 \\ 0 & 1 & 0 & 0 \\ 0 & 0 & \frac{1}{2} & 0 \\ 0 & 0 & 0 & 1 \end{bmatrix} , \qquad [I_2] = \begin{bmatrix} 1 & 1 & 0 & 1 \\ 1 & 1 & 0 & 1 \\ 0 & 0 & 0 & 0 \\ 1 & 1 & 0 & 1 \end{bmatrix} . \tag{3.11}$$

According to (2.22) the stress vector is:

$$\{\sigma\} = \frac{\bar{\sigma}}{\dot{\bar{\varepsilon}}}[D]^T \{\dot{\varepsilon}\} - \frac{1}{6}\frac{D}{B}\{e\} \,, \tag{3.12}$$

where

$$[D] = [D_1] + [D_2] \,. \tag{3.13}$$

Taking into account (3.4) and (3.8), the functional (3.1) takes the form:

$$\begin{aligned} \pi &= \int_V \bar{\sigma}\left(-\frac{1}{12}\frac{D}{B}\frac{1}{\bar{\sigma}}\{C\}^T \{q^e\} + \sqrt{\{q^e\}^T [P] \{q^e\}}\right) dV \\ &+ \int_{S_f} \{\tau\}^T [N] \{q^e\} \, dS - \int_{S_p} \{p\}^T [N] \{q^e\} \, dS \,. \end{aligned} \tag{3.14}$$

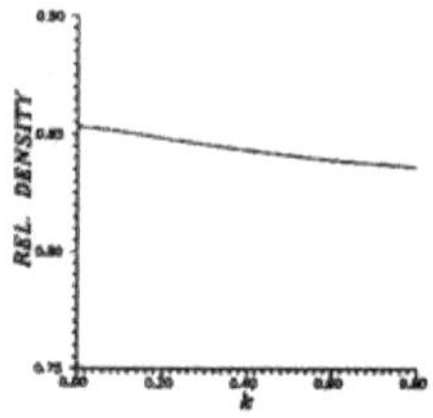

Figure 4.1:

The variation of (3.14) according to the velocity field leads to:

$$\int_V \frac{\bar{\sigma}}{\dot{\bar{\varepsilon}} + \frac{D}{12B}\dot{\varepsilon}_v} \{q^e\}^T [P]\, dV - \int_V \frac{D}{12B} \{C\}^T dV$$
$$+ \int_{S_f} \{\tau\}^T [N]\, dS - \int_{S_p} \{p\}^T [N]\, dS = 0\,. \tag{3.15}$$

We denote

$$[K] = \int_V \frac{\bar{\sigma}}{\dot{\bar{\varepsilon}} + \frac{D}{12B}\dot{\varepsilon}_v} [P]\, dV\,. \tag{3.16}$$

Equation (3.15) takes then the form:

$$[K]\{q^e\} = \{F\} - \{F_\tau\} - \{F_V\}\,, \tag{3.17}$$

where $[K]$ is the element stiffness matrix and the nodal forces are:

$$\begin{aligned} \{F\} &= \textstyle\int_{S_p} [N]^T \{p\}\, dS\,, \\ \{F_\tau\} &= \textstyle\int_{S_f} [N]^T \{\tau\}\, dS\,, \\ \{F_V\} &= \textstyle\int_V \frac{D}{12B} \{C\}\, dV\,. \end{aligned} \tag{3.18}$$

It is seen that the existence of the term DI_1 in the yield condition (2.1) influences the element stiffness matrix (3.16) and leads to an additional nodal force $\{F_V\}$ in the basic equation of the finite element method (3.17). This term is dependent on the relative density ρ and is determined during the iteration procedure ([5]), like the friction nodal force $\{F_\tau\}$.

4 Numerical results

On the basis of the relations obtained in part 3 the program system PORO was created. Four–node isoparametric finite elements were used.

By means of the program PORO the process of free forging was simulated. Porous material with the following material characteristics was considered ([2]):

$$\begin{aligned} A &= 2+\rho^2\,, & B &= 1-\tfrac{1}{3}A\,, \\ C &= 0\,, & D &= k\sqrt{\tfrac{1}{3}A+B\sigma_p^*}\,, \\ \sigma_p^* &= \sqrt{2\rho^2-1}\sigma_p^b\,, & \sigma_p^b &= \sigma_{p0}^b\,(1+0.42857\bar{\varepsilon})\,. \end{aligned} \tag{4.1}$$

Numerical examples for different values of k were calculated. Cylinders with diameter $2d = 80mm$ and hight $2h = 100mm$ were considered. Due to the axial symmetry, the problem is two–dimensional and a forth of the axial cross section was considered. The cylinders undergo a reduction of 30% in hight. The initial relative density of the material is $\rho_0 = 0.75$. Friction between die and workpiece was considered. The friction factor was taken $m = 0.6$. The die velocity v_d is prescribed. A finite element mesh consisting of 100 elements was applied.

Fig.4.1 shows how the relative density ρ in the centre of the workpiece changes with the variation of k in the interval $0 \le k \le 0.8$. As a second example, compaction of powder–metal cylindrical preforms, $15mm$ in height and $20mm$ in diameter was considered. The following material functions were used ([6]):

$$\begin{aligned} A &= 3\rho^{-2N}\,, & B &= \tfrac{\rho^{2N}}{9\eta}\,, \\ \eta &= \tfrac{1}{c(1-\rho)^d}\,, & C &= 0\,, \\ D &= 0\,, & \sigma_p^* &= \sigma_p^b\,, \\ \sigma_p^b &= \sigma_{p0}^b\,(1+0.42857\bar{\varepsilon})\,. \end{aligned} \tag{4.2}$$

The material constants, appearing in (4.2) are: $c = 2.8, d = 0.58, N = 2.7$. The material constants, taking part in the roughness evolution equation (2.11) are: $\kappa = 1.2, \gamma = -3.4$. The mean particle size is taken $G = 98\mu m$.

Fig.4.2 shows the distribution in hight of R_{max}, the equivalent strain $\bar{\varepsilon}$ and the relative density ρ at 21% reduction.

References

[1] Betten J., H.-G. Frosch, and M. Borrmann. Pressure-dependent yield behaviour of metals and polymers. *Materials Science and Engineering*, 56:233–246, 1982.

[2] Doraivelu S. M., H.L. Gegel, J.S. Gunasekera, J.C. Malas, J.T. Morgan, and J.F. Thomas. A new yield function for compressible p/m materials. *Int. J. Mech. Sci.*, 26(9/10):527–535, 1984.

[3] Duszczyk J. New yield function for forming processes of porous materials. *Adv. Technology of Plasticity*, II:969–977, 1987.

[4] Kim K. T. and J. Suh. Elastic-plastic strain hardening response of porous metals. *Int. J. Engng. Sci.*, 27(7):767–778, 1989.

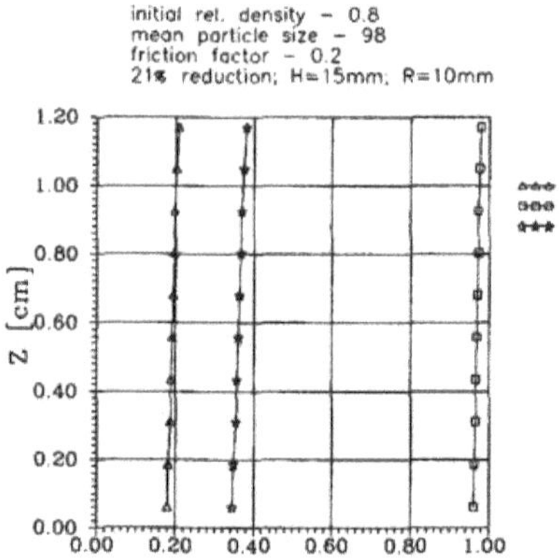

Figure 4.2:

[5] Kobayashi S., S.-I. Oh, and T. Altan. *Metal Forming and the Finite Element Method.* Oxford University Press, New York Oxford, 1989.

[6] Shima S. and M. Oyane. Plasticity theory for porous metals. *Int. J. Mech. Sci.*, 18:285–291, 1976.

[7] Takakura N., Yamaguchi K., and FukudaM. Surface roughness development in upsetting of sintered powder metals. *Advanced Technolojy of Plasticity*, 2:903–908, 1990.

Address of Authors:
Institute of Mechanics, Bulg. Academy of Sci.
Sofia
Bulgaria

LIQUID JET DISINTEGRATION ANALYSIS : DIRECT NUMERICAL SIMULATION

M. Chacha, R. Occelli, L. Tadrist
I.U.S.T.I. - U.A 1168, U.P - Centre de Saint Jérôme, case 162
Av. Escadrille Normandie Niemen
13397 Marseille Cedex 20 FRANCE

and S. Radev
I.M.B.M. Bulgarian Academy of Sciences
ul. Georgi Bonchev, bl. 4, 1113 Sofia BULGARIA

Abstract

The capillary instability of an infinite axisymmetric viscous liquid column in a gaseous or viscous liquid medium is investigated. The process of disintegration is simulated numerically using a (second order) finite difference method in conjunction with a streamfunction-vorticity formulation of the Navier-Stockes equations. All nonlinear and viscous terms are included. Nonlinear terms are found to be responsible of the formation of satellite drops. The basic characteristics of liquid jet disintegration such as drop sizes and breakup times are in good agreement with those obtained by previous investigators. **Key words:** Nonlinear capillary instability, Liquid-fluid systems, Streamfunction-vorticity formulation.

1 INTRODUCTION

Problems in which the solution of a (partial) differential equation has to satisfy certain conditions on the boundary of a specified domain are referred to as boundary-value problems. In many cases however, the boundary of the domain is not known in advance but, has to be determined as a part of the solution : Jet flows constitute good examples of such problems. Because of the presence of a mobile interface boundary (liquid-liquid or liquid-gas) one encounters here the so-called capillary phenomena due to the existence of a surface tension.

In recent years the capillary jet has been used in many practical applications such as printing, particle sorting, dispersing liquids, fibers spinning, etc...

Rayleigh [1] was the first to make a theoretical study of the inviscid liquid column in vacuum and to establish the possibility of droplet formation. He showed that the capillary jet is stable for all purely non-axisymmetric disturbances and, the jet is stable or unstable to axisymmetric disturbances, depending on whether the wave length is less or greater than the circumference of the undisturbed cylinder.

Other authors, among them Weber [2], studied the condition of axi-symmetrical breakup of a viscous liquid jet.

In addition to the viscosity, Tomotika [3] took into account the density and the viscosity of the medium surrounding the jet.

Neither of the aforementioned theories predicted the formation of droplets of different diameters after breakup of a liquid jet. It was assumed as a starting point of their analysis that a spectrum of infinitesimal disturbances of the form of eq. (1) is initiated on the surface of stationary liquid column as a result of pressure fluctuations :

$$\delta = \delta_0 \exp(\mathrm{Q}t + i\mathrm{K}z) \tag{1}$$

Nonlinear theories were conducted first by Yuen [4]. He developed a solution for an inviscid jet. He carried out the solution up to 3rd order for a monochromatic initial displacement of the free surface. In general nonlinear theories lead to good qualitative results.

One of the goals of the present work is to predict numerically, the experimentally observed satellite drops as reported for example by Rutland and Jameson [5].

A complete (numerical) solution of the full Navier-Stockes equations should lead to both qualitative and quantitative results. It is not easy to carry out numerical computation of the problem due to the free surface. In the field of numerical simulation of liquid jet disintegration, the work of Shokoohi [6] appears to be of particular importance.

This paper presents a direct simulation of the process of disintegration of a viscous liquid jet in a liquid or gaseous medium. The applicability of the numerical procedure is verified in comparison with the results of Shokoohi [6] and temporal linear stability analysis, in the case of a jet in vacuum. Drop sizes are also compared to those obtained experimentally by Rutland and Jameson [5] and Lafrance [7]. For a jet in viscous liquid phase, the applicability is verified according to the linear theory analysis of Tomotika [3].

2 PHYSICAL MODEL AND EQUATIONS

The geometry considered in the present work implies what has been called "infinite jet". That is, the jet has no beginning nor end. We consider a system of two incompressible Newtonian fluids initially at rest. Gravity effects are neglected.

Figure 1 shows a periodic section of the axisymmetric jet with a single wavelength disturbance (fundamental disturbance). The continuous phase is limited at R_{max} with a rigid cylinder. This ensures a physical external boundary to the computational domain. This domain is bounded elsewhere by the axis of symmetry Oz and two jet cross-sections at a distance one wavelength of the prescribed disturbance.

Owing to the symmetry of the problem the hydrodynamic equations are written in a cylindrical coordinate system. The undisturbed jet radius R_N and the capillary time $t_0 = \sqrt{\rho_2 R_N^3/\sigma}$ are chosen to be the characteristic length and characteristic time respectively. Hereafter two new dependent variables are introduced to use the so-called streamfunction-vorticity formulation : ψ (the stream function) and ω (the vorticity). These variables are defined in both the jet and continuous phases as : $V = \frac{1}{r}\frac{\partial\psi}{\partial z}$, $U = -\frac{1}{r}\frac{\partial\psi}{\partial r}$ (2) and, $\omega = \frac{\partial V}{\partial z} - \frac{\partial U}{\partial r}$ (3) where U and V are the axial and radial components of the velocity respectively.

To overcome the difficulty related to the interface location, a coordinate transformation is then used as follows :

$$\eta_j = \frac{(j-1)^* R_{max} - r}{(j-1)^* R_{max} - R_s} \tag{4}$$

j = 1 (respectively j = 2) refers to the dispersed phase (respectively the continuous phase).

The resulting forms of the governing equations and boundary conditions are :

2.1 Equations

$$\frac{\partial\omega}{\partial\tau} = a_1\frac{\partial^2\omega}{\partial\eta^2} + a_2\frac{\partial\omega}{\partial\eta} + a_3\frac{\partial^2\omega}{\partial\eta\partial\xi} + a_4\frac{\partial\omega}{\partial\xi} + a_5\frac{\partial^2\omega}{\partial\xi^2} + a_6\omega \tag{5}$$

$$b_1\frac{\partial^2\psi}{\partial\eta^2} + b_2\frac{\partial\psi}{\partial\eta} + \frac{\partial^2\psi}{\partial\xi^2} + b_3\frac{\partial^2\psi}{\partial\eta\partial\xi} - r\omega = 0 \tag{6}$$

$$V = g_1\frac{\partial\psi}{\partial\xi} + g_2\frac{\partial\psi}{\partial\eta} \qquad U = g_3\frac{\partial\psi}{\partial\eta} \tag{7}$$

The above-mentioned equations are to be written for the jet and the continuous phase.

2.2 Boundary conditions

At the centerline : $\omega_1 = 0$ (8) $\psi_1 = 0$ (9) At the interface :

a) The continuity of the tangential stress gives :

$$\left[c_1\frac{\partial^2\psi}{\partial\eta^2} + c_2\frac{\partial\psi}{\partial\eta} + c_3\frac{\partial^2\psi}{\partial\eta\partial\xi} + c_4\frac{\partial^2\psi}{\partial\xi^2}\right] = 0 \tag{10}$$

b) The continuity of the normal stress leads after some transformation to :

$$\left[\sum_{k=1}^{k=9} d_k D_k^{\eta,\xi}\psi\right] = s \qquad \left(or \left[\left(\frac{\partial P}{\partial\xi}\right)\right]_{\eta=1}^{CN} - \left[\left(\frac{\partial P}{\partial z} + \tan\alpha\frac{\partial P}{\partial r}\right)\right]_{\eta=1}^{Mvt} = 0\right) \tag{11}$$

CN means that $\left[\frac{\partial P}{\partial\xi}\right]_{\eta=1}$ is expressed using the normal stress condition. Mvt means that the same partial derivative is expressed using the momentum equations; i.e : $\left[\left(\frac{\partial P}{\partial\xi}\right)\right]_{\eta=1}^{Mvt} = \left[\left(\frac{\partial P}{\partial z} + \tan\alpha\frac{\partial P}{\partial r}\right)\right]_{\eta=1}^{Mvt}$ which is exactly $\left[\left(\frac{\partial P}{\partial\xi}\right)\right]_{\eta=1}^{CN}$ This leads to the elimination of the pressure term in the surface tension boundary condition. c) The kinematic condition becomes (for both phases): $\frac{\partial R_s^2}{\partial \tau} = 2\frac{\partial \psi}{\partial \xi}$ (12) ***Notation :*** $[G] = G_1 - G_2$ At $r = R_{max}$:

$\omega_2 = bw\frac{\partial^2 \psi_2}{\partial {\eta_2}^2}$ (13) and $\psi_2 = Cste\ (= 0)$ (14) At both ends of one-wavelength jet

segment : $F(0, \eta_j, \tau) = F(\lambda, \eta_j, \tau)$ (15)

where F is any flow variable or its derivatives and λ the wavelength of the fundamental disturbance.

3 NUMERICAL PROCEDURE

An exact solution to the resulting nonlinear equations is not known. Of course this set of equations belongs to the class of (partial) differential equations for which numerical methods have been extensively studied. But the presence of an unknown interface with no condition on the interface vorticity reveals the difficulty encountered when solving this type of problems; especially using streamfunction-vorticity formulation.

At zero time we impose upon the surface of the cylinder, an axisymmetric cosinusoidal disturbance of dimensionless wavenumber K and amplitude δ_0. The initial solution is derived from the linear temporal analysis. When there is a viscous continuous phase, R_{max} is chosen in such a way to ensure a dimensionless growth rate Q practically equal to that obtained from Tomotika [3] dispersion relation. -Tomotika [3] worked with an unbounded continuous phase-. Here follows the (iterative) numerical procedure :

step 1 : Increase the time.

step 2 : Compute a new surface profile Rs using eq. (12).

step 3 : Differentiate Rs (3rd order splines are used).

step 4 : Solve eq. (5) for the vorticity in the jet (ADI method is used).

step 5 : Solve eq. (5) for the continuous phase (ADI method is used).

step 6 : Approximate the vorticity on the rigid wall using eq. (13)

step 7 : Solve eq. (6) for the stream function in the jet (ADI method is used after introducing a fictitious time derivative of ψ).

step 8 : Solve eq. (6) for the continuous phase (ADI method is used after introducing a fictitious time derivative of ψ).

step 9 : Compute the velocity components values everywhere using eq. (7).

step 10 : Determine the jet interface vorticity using eq. (10).

step 11 : Approximate eq. (6) on the interphase boundary for the continuous phase interface vorticity.

step 12 : Determine the interface stream function value using eq. (11).

step 13 : Go to step 2 if the convergence criterion is not satisfied (Reduce the time step if necessary).

Go to step 1 if not too small time step.

stop

4 RESULTS

Our criterion for the jet breakup is satisfied when the time step becomes very small ($\Delta t \approx 10^{-5}$). At this stage in general, the radius of the jet at breakup points is less than 0.05.

Numerical solutions have been conducted for a water jet in vacuum and a water jet in dodecan surrounding medium. To validate the present numerical approach, numerical results are compared with the linear theory solutions of Weber; and with Shokoohi [6] results for the first system. They are also compared in term of drop sizes with the results of Rutland and Jameson [5] and to those of Lafrance [7]. For the liquid-liquid system, Tomotika linear solutions are used for comparison. Entering parameters for a column of water in vacuum are :

$\rho_1 = 1\text{g/cm}^3$ $\mu_1 = 0.01\text{po}$ $\sigma = 72.5\text{dynes/cm}$ $\delta_0 = 0.02$

Those for a column of water in dodecan continuous phase are :

$\rho_1 = 1\text{g/cm}^3$ $\mu_1 = 0.01\text{po}$ $\mu_2 = 1.35\ 10^{-2}\text{po}$ $\rho_2 = 0.748\text{g/cm}^3$ $\sigma = 47\text{dynes/cm}$ $\delta_0 = 0.02$

In the case of the liquid-gas system, the growth rates of one half of the difference between the swell and the neck are compared against the growth rates obtained from linear analysis (Fig. 2). The close agreement obtained in the linear range manifests the correctness of the computation. The jet contour keeps a sinusoidal form during the early stages of the destabilisation (fig. 2, 3). In the later stages, because the slope of the contour becomes large, the linear analysis is no longer valid (hence more realistic flow dynamics can only be studied through the nonlinear equations). Figure 3 shows the amplification of the perturbation corresponding to the neck and the swell, for three different nondimensional wavenumbers. For a short wavelength, the neck contracts faster than the swell grows. For a long wavelength, the contrary becomes true.

The basic characteristics of jet break-up are the duration of the interval it remains unbroken and the size of the drops that are formed : After the two breakup points have been identified, the volume of liquid contained in each drop is calculated. The drop volumes are converted to the drop radii assuming that the resulting drops will be spherical. These calculations have been performed for different unstable dimensionless wavenumbers. Predicted and measured drop sizes are represented in fig. 4. The accordance is very satisfying. The breakup times are in good agreement with those obtained by Shokoohi [6] (Fig. 5). The fastest growth rate is found for $K \approx 0.7$ which is in agreement with the linearized theory. At moderate wavelengths ($0.3 < K < 0.8$), linear theory predicts relatively good breakup time. Nonlinear breakup times are shorter than those of the linear theory.

Figure 6 shows the interface profiles before breakup. The satellite drop size increases with the wavelength. At short wavelengths, the crest of the satellite and the crest of the main drop are spaced one-half wavelength apart. This indicates that the second harmonic component is mainly responsible for the formation of the satellite in the range of moderate wavelengths. With long wavelengths, the possibility of the formation of two (or more) satellites exists : Here, the satellite drop is a long ligament with a "neck". ***Concerning the liquid-liquid system,*** conclusions similar to those in the previous part can be drawn. For example the fastest growth rate has been found near $K = 0.7$ (not shown). In fig. 7 the growth rates of one half the difference between the swell and the neck are shown together with those obtained from linear analysis of Tomotika. Breakup times are longer and

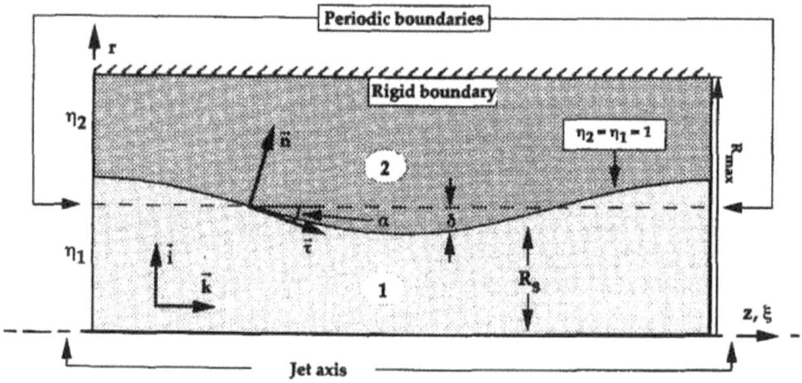

Fig. 1 Geometry of the flow under consideration

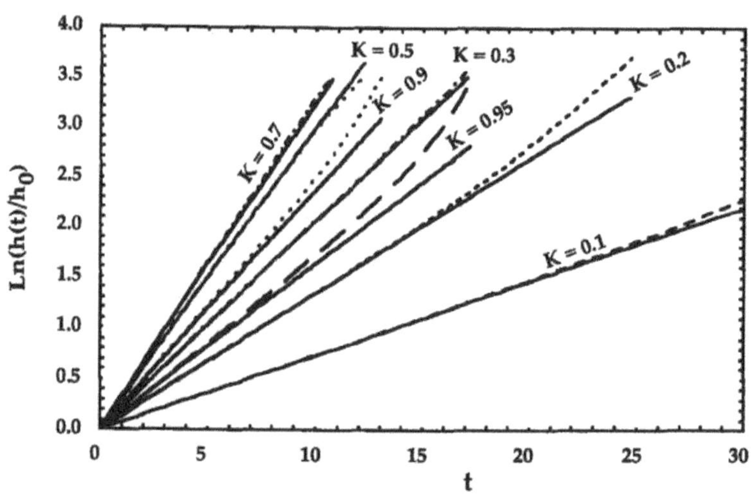

Fig. 2 Growth of disturbance vs. time for water-vacuum system ($h(t) = R_s(0, t) - R_s(\lambda/2, t)$; $h_0 = 2^* \delta_0$)

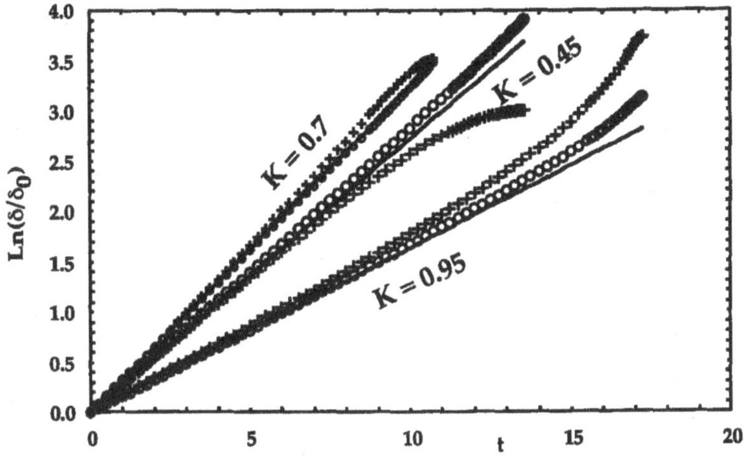

Fig. 3 Growth of the disturbance in time. —— Weber's solution
Numerical solution : + neck; o swell.

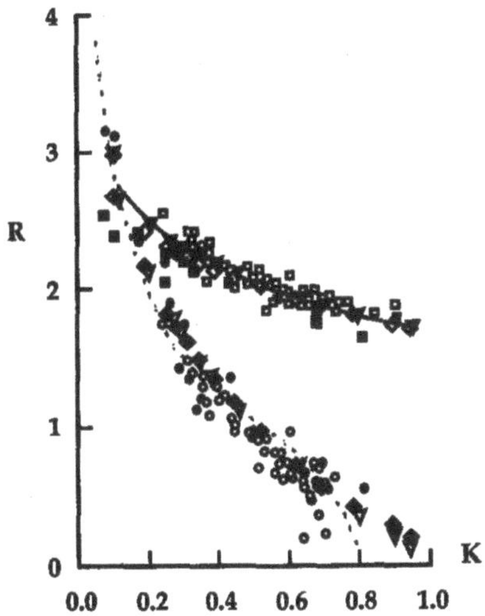

Fig. 4 Drop radii vs. wavenumber for a jet of water:
Comparison between the present and the previous works.

Theory :	Lafrance,	— main drop;	- - satellite drop.
Numerical :	Shokoohi,	▼ main drop;	▽ satellite drop;
	Present work,	◇ main drop;	◆ satellite drop.
Experiment :	Lafrance,	▫ main drop;	o satellite drop;
	Rutland and Jameson,	▪ main drop,	• satellite drop.

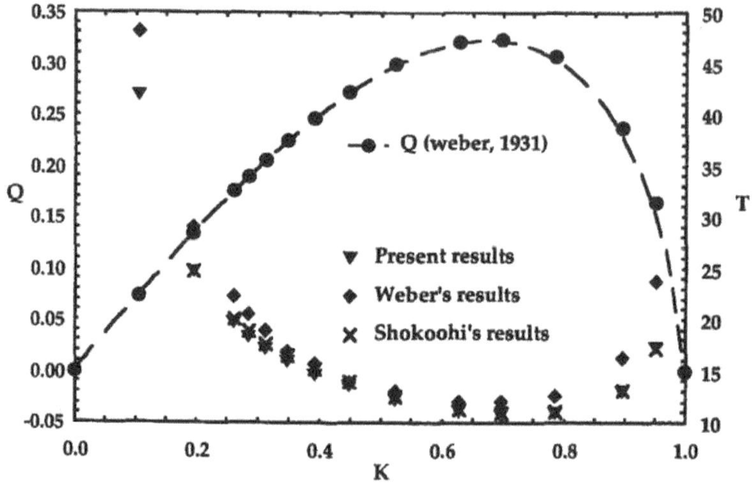

Fig. 5 Nondimensional growth rate Q and breakup time T vs. nondimensional wavenumber K

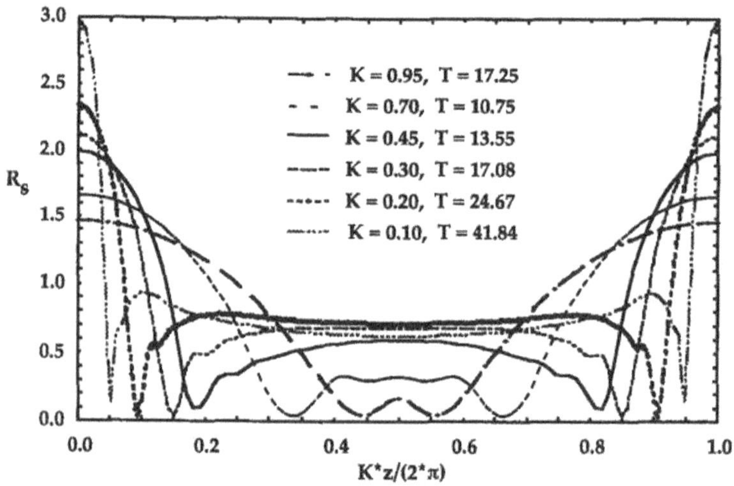

Fig. 6 Interface profiles before breakup ($\delta_0 = 0.02$)

nonlinear effects occur much earlier in this system than in the first one. And the differences between linear and nonlinear breakup times are much larger. Hydrodynamic effects of the continuous phase increase the speed of nonlinear effects. Satellites are obtained for all unstable wavelengths. However they appear smaller than in the former system (Fig. 8). Figure 8 compares the drops radii obtained in water-dodecan system and water-vacuum system.

5 CONCLUSION

Nonlinear effects appear to be important for they are the cause of the small satellite drops that develop between the large drops. Nonlinear breakup times are found to be shorter than linear ones assuming the same initial amplitude of perturbation. Satellite drops exist for all unstable wave numbers in the two flow systems that we studied. In the liquid-liquid system, the breakup time is larger than that obtained for a jet disintegration in vacuum assuming the same initial condition (K and δ_0). Hydrodynamic effects of the continuous phase reduce the size of the satellite drops and accelerate the amplification of the perturbations. The calculation code is still subjected to further validations.

References

[1] L. Rayleigh, On the instability of jets, *Proc. Lond. Math. Soc.*, **10**; 4, 1878.

[2] C. Weber, Zum Zerfall eines Flüssigkeitsstrahles, *Z. Angew, Math. Meh.*, **11**, 1931, pp. 136–154.

[3] S. Tomotika, On the instability of a cylindrical thread of a viscous liquid surrounded by another viscous fluid, *Proc. R. Soc. Lond.* **Cl A**, 1935, pp. 322–337.

[4] M. C. Yuen, Nonlinear capillary instability of a liquid jet, *J. Fluid Mech.* **33**, 1, 1968, pp. 151–163.

[5] D. F. Rutland, G. J. Jameson, Theoretical prediction of the size of drops formed in the breakup of capillary jets, *Chem. Eng. Sci.* **25**, 1970, pp. 1689–1698.

[6] F. Shokoohi, Numerical investigation of the desintegration of liquid jets, Ph.D. Thesis, 1976, Colombia University, New York.

[7] P. Lafrance, Nonlinear breakup of a laminar liquid jet, *The Phys. of Fluids*, **18**, 4, 1975, pp. 428–432.

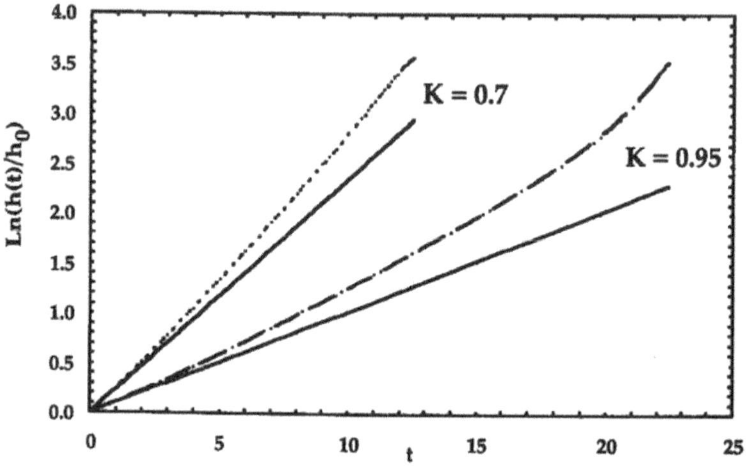

Fig. 7 Growth of disturbance vs. time for water-dodecan system ($h(t) = R_g(0, t) - R_g(\lambda/2, t)$; $h_0 = 2^*\delta_0$)

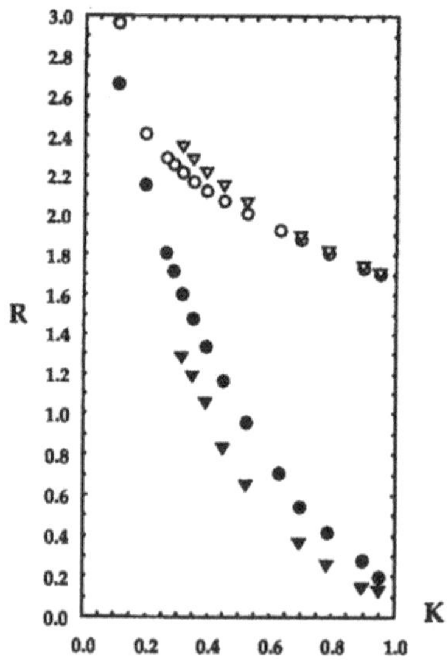

Fig. 8 Drop radii vs. wavenumber (Present work) :
Effects of the continuous phase on drop size.
Water-vacuum system : ○ main drop; ● satellite drop.
Water-dodecan system: ▽ main drop; ▼ satellite drop.

Splitting Scheme for Iterative Solution of Bi-Harmonic Equation. Application to 2D Navier–Stokes Problems

C. I. Christov* A. Ridha

Abstract

A ficititious time is introduced into the bi–harmonic equation rendering it to a higher-order (generalized) parabolic equation. The method of splitting of operator is used and shown that the splitting scheme approximates the generalized parabolic equation. In the half-time steps the splitting scheme reduces to 1D problem with five-diagonal band matrix. The test problem of steady viscous flow in lid-driven cavity is considered and results are obtained for the range $0 < Re < 40000$. The characteristics of the flow obtained here are in good agreement with the results of other authors.

The bi-harmonic equation appears in different fields of mechanics of continua, e.g., elasticity and viscous flows and developing efficient techniques for its numerical solution are of primary practical importance. It goes beyond the scope of the present paper to give extensive review of the literature on numerical approaches to Navier–Stokes equations. We refer the reader to the comprehensive review [3]. Here we concern ourselves with developing a fully implicit cost-effective coordinate-splitting algorithm directly for the fourth–order equation for stream function. Thus the splitting method is applied as iterative procedure for solving the bi-harmonic boundary-value problems.

1 Posing the Problem

The 2D Navier–Stokes equations in terms of stream function reduce for steady flows to the nonlinear bi–harmonic problem

$$Re\left(\frac{\partial\psi}{\partial y}\frac{\partial\Delta\psi}{\partial x}-\frac{\partial\psi}{\partial x}\frac{\partial\Delta\psi}{\partial y}\right)=\Delta\Delta\psi \tag{1}$$

*On leave from the National Institute of Meteorology and Hydrology, Bulgarian Academy of Sciences, Sofia, BULGARIA

where $Re = UL/\nu$ is the Reynolds number based on the characteristic length scale L and velocity U. The dimensionless form of (1) is obtained after scaling the independent spatial variables by L, the stream function - by UL. The b.c. for an internal flow have the form

$$\psi = G(x,y) \, , \quad \frac{\partial \psi}{\partial n} = F(x,y) \, , \qquad \text{at} \qquad (x,y) \in \partial D \tag{2}$$

Here n stands for the outward normal to the region D and G, F are given functions, specified by the velocity components at the boundary. For stationary problems the boundary is a stream line and then without loosing the generality one may set $G \equiv 0$.

In the literature, the stream-function formulation was treated numerically by introducing the vorticity function $\omega = -\Delta\psi$ and rendering the problem to a coupled system of two second-order equations for ψ, ω. Although apparently more tractable, the $\psi - \omega$ formulation is inherently explicit, because of the lack of boundary condition for the vorticity function ω. Here we do not use the vorticity-stream function formulation, but rather solve directly the bi-harmonic equation for ψ. Thus we not only tackle the genaral problems of numerical solution of bi-harmonic equations, but also provide a method to overcome the inherent difficulties of the $\psi - \omega$ formulation.

2 Case Study: The Lid-driven Cavity Flow

As a rule, the new schemes and algorithms for solving Navier-Stokes equations are tested on certain test cases, most frequently on the case study of flow in a rectangular cavity driven by the steady horizontal motion of the lid. The advantage of this test case is that its geometry is the simplest possible. The disadvantage is that there are singularities at the points where the lid touches the vertical walls, but as it turns out the discontinuous boundary condition does not poses much difficulties to the algorithms.

The Chorin-Harlow method (MIC or fractional-step method) was applied to the lid-driven cavity flow in [6]. The latest progress along these lines can be found in [8] where the 3D problem of lid-driven cavity is treated. This kind of methods are by definition semi-explicit. The finite-element method was applied to the lid-driven cavity flow in [4] and results were obtained up to $Re = 5000$. In [11] the finite-element method was applied on adaptive grids and systematic results were obtained for $Re \leq 15000$.

The higher-accuracy methods in terms of vorticity-stream function were tested on the lid-driven cavity flow in [5] and [7], but due to the intrinsic instability of the said methods, reliable results were obtained in [5] only for $Re < 1000$ although the $Re = 2000$ is also mentioned, while in [7] the problems started showing up as early as for $Re = 300$.

The most-closely related to ours is the work of Schreiber & Keller [9] where the non-linear bi-harmonic equation (1) was solved. They did not use neither the staggered mesh nor the coordinate splitting. They reported results for Re as high as 10000 but the higher Reynolds number were far from the values which are being now accepted as the solution to the problem. Their results compared reasonably with the other works only after Richardson extrapolation was applied. Yet, we believe, the work [9] is very important because it shows that the the bi-harmonic problem is a good basis for numerical treatment of Navier–Stokes problems.

As a benchmark numerical paper for the flow under consideration is considered [1] where the $\psi - \omega$ formulation was used with explicit implementation of the condition on vorticity and upwind differences for the convective terms. Reliable results were obtained for $Re \leq 50000$ while the $Re = 10000$ was reached only after a smoothing operator was applied.

The square cavity occupies the situated in the domain $0 \leq x \leq 1, 0 \leq y \leq 1$ with the lid coinciding with the plane $y = 1$ and moving to the right in the said plane. Then the boundary conditions adopt the form

$$\psi = \frac{\partial \psi}{\partial x} = 0 , \quad \text{for } x = 0,1 ; \qquad \psi = \frac{\partial \psi}{\partial y} = 0, \text{ for } y = 0 , \quad \psi = 0, \ \frac{\partial \psi}{\partial y} = 1, \text{ for } y = 1 .$$

3 Difference Scheme

We add artificial time (say, s) in the equation for the stream function (1) rendering it into a generalized higher-order diffusion equation, namely

$$\frac{\partial \psi}{\partial s} = Re \left(\frac{\partial \psi}{\partial y} \frac{\partial \Delta \psi}{\partial x} - \frac{\partial \psi}{\partial x} \frac{\partial \Delta \psi}{\partial y} \right) - \Delta\Delta\psi \tag{3}$$

Since in the case under consideration the time is artificial, one need not worry about a second-order approximation with respect to it. The sole concern remains the stability of the scheme. In this instance we generalize the so-called "second Douglass scheme" or the scheme of "stabilizing correction" (see, e.g., [13] for terminology) to fourth-order operators as follows

$$\frac{\tilde{\psi}_{ij} - \psi_{ij}}{\sigma} = -\Lambda_{11}\tilde{\psi}_{ij} - \Lambda_{22}\psi^n_{ij} - \Lambda_{12}\psi^n_{ij} + F^n_{ij} \qquad \frac{\psi^{n+1}_{ij} - \tilde{\psi}_{ij}}{\sigma} = -\Lambda_{22}\left(\psi^{n+1}_{ij} - \psi^n_{ij}\right) \tag{4}$$

where σ is the increment with respect to the fictitious time s. Here $\Lambda_{11}, \Lambda_{22}$ and Λ_{12} are respectively the difference approximations of the fourth order derivatives with respect to x, y and the mixed fourth order derivative (see (9)), while F^n is a difference approximation of the nonlinear term (see (9)). Here the superscript n refers to the time stage: n is the current stage (the "old" one) and $n+1$ is the "new" one.

Denoting by E the unitary operator we recast the scheme as follows

$$(E + \sigma\Lambda_{11})\tilde{\psi} = (E - \sigma\Lambda_{22})\psi^n + (-\Lambda_{12}\psi^n + F^n)\sigma , \qquad (E + \sigma\Lambda_{22})\psi^{n+1} = \tilde{\psi} + \sigma\Lambda_{22}\psi^n . \tag{5}$$

After the fractional-step quantity $\tilde{\psi}$ is excluded one gets

$$(E + \sigma\Lambda_{11})(E + \sigma\Lambda_{22})\psi^{n+1} = (E + \sigma^2\Lambda_{11}\Lambda_{22})\psi^n + (-\Lambda_{12}\psi^n + F^n)\sigma \tag{6}$$

or which is the same

$$(E + \sigma^2\Lambda_{11}\Lambda_{22})\frac{\psi^{n+1} - \psi^n}{\sigma} = -(\Lambda_{11} + \Lambda_{22})\psi^{n+1} - \Lambda_{12}\psi^n + F^n \tag{7}$$

The last equality shows that the scheme in half-time steps (4) is equivalent to an implicit scheme in full-time steps. It is worth mentioning that the splitting scheme appears to be even more stable than an implicit scheme in full steps because of the presence of operator $\Lambda_{11}\Lambda_{22}$ which is positive definite (for commuting Λ_{11} and Λ_{22}) and hence the norm of the operator $(E+\sigma^2\Lambda_{11}\Lambda_{22})^{-1}$ is always lesser than unity.

An important feature of (7) is the explicit implementation of the nonlinear convective term and the mixed derivative term. As already mentioned, our main purpose is to propose an effective approximation of the viscous terms (the bi-harmonic operator). In such a case the most important is to have the diffusion terms represented implicitly, rather than the convective ones and this notion predestines our choice. In addition, in using the explicit scheme for convective terms we are spared the difficulties of their approximation on the "new" time stage notorious for generating artificial "scheme" viscosity, etc. In a sense, our treatment of convective term is a "purest" one that does not additionally pollute the overall $O(h_x^2+h_y^2)$ approximation of the scheme

In order not to obscure the main ideas we do not use here non-uniform meshes or other sophistications usually needed to improve the accuracy in the technological applications. The grid in both x- and y-directions is taken uniform, namely

$$x_i = -\frac{1}{2}h_x + (i-1)h_x\,, \quad i=1,\ldots,M+1 \qquad y_j = -\frac{1}{2}h_j + (j-1)h_x\,, \quad j=1,\ldots,N+1$$

where h_x and h_y are the spacings in x- and y-directions, respectively. On the uniform grid we take the standard central-difference approximations:

$$\begin{aligned}
\Lambda_{11}\psi &\equiv \frac{1}{h_x^4}\left(\psi_{i+2,j}-4\psi_{i+1,j}+6\psi_{ij}-4\psi_{i-1,j}+\psi_{i-2,j}\right)\\
\Lambda_{22}\psi &\equiv \frac{1}{h_y^4}\left(\psi_{i,j+2}-4\psi_{i,j+1}+6\psi_{ij}-4\psi_{i,j-1}+\psi_{i,j-2}\right)\\
\Lambda_{12}\psi &\equiv \frac{2}{h_x^2h_y^2}\left[\psi_{i+1,j+1}-2\psi_{i,j+1}+\psi_{i-1,j+1}-2\left(\psi_{i+1,j}-2\psi_{i,j}+\psi_{i-1,j}\right)\right.\\
&\qquad\left.+\psi_{i+1,j-1}-2\psi_{i,j-1}+\psi_{i-1,j-1}\right]
\end{aligned} \tag{8}$$

$$F_{ij} = \frac{Re}{4h_xh_y}\left[\left(\psi_{i,j+1}-\psi_{i,j-1}\right)\left(\omega_{i+1,j}-\omega_{i-1,j}\right)\left(\psi_{i+1,j}-\psi_{i-1,j}\right)\left(\omega_{i,j+1}-\omega_{i,j-1}\right)\right] \tag{9}$$

where

$$\omega_{ij} \equiv \frac{1}{h_x^2}\left(\psi_{i+1,j}+\psi_{i-1,j}-2\psi_{ij}\right)+\frac{1}{h_y^2}\left(\psi_{i,j+1}+\psi_{i,j-1}-2\psi_{ij}\right)\,, \tag{10}$$

is introduced only for the sake of convenience of notations.

At each half-time step the scheme reduces to a linear algebraic system with five-diagonal matrix. For instance, for the line $y=y_j$ one has to solve the following system for the vector

$\psi_{1j}, \ldots, \psi_{Mj}$:

$$\begin{pmatrix} 1 & 0 & 0 & & & & & & \\ 0 & 1 & 0 & 0 & & & & & \\ 1 & -4 & 6+\frac{h_x^4}{\sigma} & -4 & 1 & & & & \\ \cdots & \cdots & \cdots & & & \cdots & \cdots & \cdots & \\ & & 1 & -4 & 6+\frac{h_x^4}{\sigma} & -4 & 1 & & \\ \cdots & \cdots & \cdots & & & \cdots & \cdots & \cdots & \\ & & & & 1 & -4 & 6+\frac{h_x^4}{\sigma} & -4 & 1 \\ & & & & & 0 & 0 & 1 & 0 \\ & & & & & & 0 & 0 & 1 \end{pmatrix} \begin{pmatrix} \psi_{1j} \\ \psi_{2j} \\ \psi_{3j} \\ \cdots \\ \psi_{ij} \\ \cdots \\ \psi_{M-1,j} \\ \psi_{Mj} \\ \psi_{M+1,j} \end{pmatrix} = \begin{pmatrix} 0 \\ 0 \\ b_{3j} \\ \cdots \\ b_{ij} \\ \cdots \\ b_{M-1,j} \\ 0 \\ 0 \end{pmatrix} \qquad (11)$$

The iterations (time steps in fictitious time) are conducted until the satisfaction of the uniform norm criterion: $\max|\psi^{n+1} - \psi^n| < \epsilon$. After the calculations are completed the solution is "returned" to the non-staggered mesh according to the formula $\Psi_{ij} = \frac{1}{4}(\psi_{ij} + \psi_{i+1,j} + \psi_{i,j+1} + \psi_{i+1,j+1})$ and it is Ψ which is presented graphically

4 Results and Discussion

As has already been above mentioned, the main objective of the present work is to provide a fully implicit treatment of the bi-harmonic problem (the viscous terms) then the relevant case are the small and moderate Reynolds numbers.

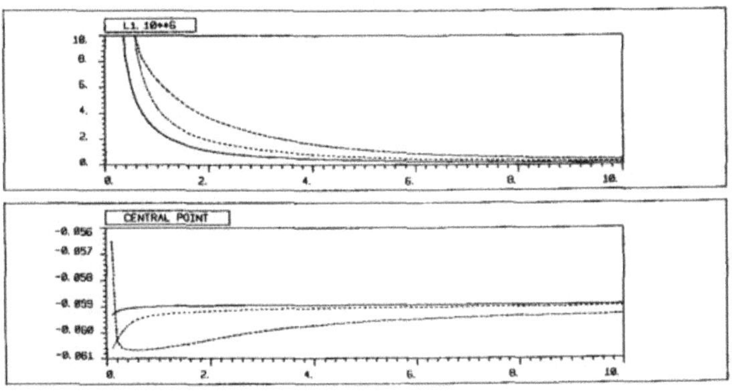

Figure 1: The evolution of solution and L_1-norm with the iterations for Re=1 for three different time increments: 1. $\sigma = 0.002$ (solid line); 2. $\sigma = 0.02$ (short dashes; 3. $\sigma = 0.2$ (long dashes)

First of all, we show that the approximation of the steady solution does not depend of the fictitious-time increment of the splitting scheme. Theoretically it follows from eq. (7)

and provides a good check for the correctness of the algorithm if it is respected in practice. We calculated the flow with $Re = 1$ with three different fictitious-time increments: $\sigma = 0.002, 0.02, 0.2$. The evolution of the solution in the centre of the cavity (point $x = 0.5, y = 0.5$) is presented in Fig. 1 for a spatial resolution of 82×82 points and juxtaposed it to the evolution of the L_1 norm. We deliberately selected the case $\sigma = 0.2$, because in the virtue of $O(\sigma)$ approximation of the scheme one expects very rough approximation for large σ. Indeed, it is the case for the transient, but not for the steady state solution. In fact the cases $\sigma = 0.002$ and $\sigma = 0.2$ differ by less than 0.1% for the same number of iterations. It is clear then that the overall approximation of the steady problem does not depend on the time increment. The general feature here is the initial "overshooting" of the maximal value and consecutive monotonous convergence "from above".

The second important verification is the spatial approximation of the scheme. The case of $Re = 1$ is a trivial in this sense, because there is almost no influence from the convective term. We did obtain also a solution with 162×162 points for the same time increment $\sigma = 0.02$ and the difference was indeed of order of $h^2 \approx 4.10^{-5}$. Much more interesting is the case of higher Reynolds numbers, because we claim that our scheme has no artificial viscosity due to the explicit central-difference representation of the convective term. For that reason we treated the case $Re = 2000$ with $\sigma = 0.2$ with the two different spatial meshes and the difference between the two solutions was strictly of the second order.

These tests suffice to claim the good convergence of the scheme and its adequate spatial approximation. A general feature that distinguish it from the convergence methods for second–order diffusion equations is the somewhat "nonlinear" convergence in the sense that on the initial stages the algorithm descends rapidly to the sought solution, but when it gets in a closer vicinity of the latter the rate of convergence decreases and the final refinement takes considerable number of iterations. Obviously, the fourth-order diffusion is faster than the second-order one when it comes to dissipating the initial condition (when the gradients of the residual are large), but slower when the gradients are small (when the algorithm is near the sought solution). As a result the algorithm is slower in arriving to the final, steady–state stage. It is clear that this is the price that must be paid for the other advantages gained by the present approach to the problem, e.g., full implicitness and strong stability features.

There is consensus now (see, e.g., [11]) that the first sights of the third secondary vortex in the left-upper corner of the cavity appear for $Re = 2000$. That is another reason for solving that case with two different spatial meshes (resolutions). In Fig. 2 are shown the flow patterns for $Re = 1, 1000, 2000, 5000$ as calculated with mesh size 82×82. It is seen that the embryo of the third vortex does indeed appear for $Re = 2000$ and it is present there for the high–resolution calculations too.

The highest Re for which we were able to reach a truly steady solution was $Re = 5000$. In Fig. 3 we present the evolution of the solution in the centre of the cavity and the behaviour of the norm with the number of iteration for $Re = 5000$. After an initially violent evolution the norm gradually subsides to the order of 10^{-9} and the amplitudes of pulsations of solution decrease while their period increases. This means that the solution is approaching to the steady state (the iterations converge). Then the pattern of the flow presented in the last of

Figs. 2 pertains indeed to a steady solution. In Table 1 we present the comparison of our results (denoted by $C-R$ to the literature results. We select the most robust and in a sense, most representative characteristics – the value of streamfunction in the center of the primary vortex and its coordinates. Table 1 gives the comparison. It is interesting to note the general tendency of our results being off the closest reference results by 0.4%. This can be attributed first to the absence of of an artificial viscosity in our scheme and second – to the fact that we solve directly a problem with fourth derivatives that is much more flexible than a problem with second derivatives. In fact, here the lowest-order artificial terms comming from the difference approximation are the sixth–order derivatives, while in $\psi - \omega$ formulations, the lowest–order are the fourth derivatives, save for the second derivatives which arise from the convective terms when approximated with upwind differences. It is natural that a fourth derivative is "stiffer" than a sixth one, i.e., the profile of stream function ψ is more flexible in our case and its maximum/minimum is expected to be higher/lower than in the case of vorticity–stream function formulation.

The high-Reynolds-number flows could not be the primary concern for a scheme aimed to provide correct treatment of the diffusion terms. It is also clear than one can never have a scheme performing equally well for the whole interval of Re starting from zero and going up to the realm of turbulent régimes. Yet, because of its strong implicitness, the algorithm developed here turns out very efficient also for high Reynolds numbers. Surprisingly enough, for large Re the stability is achieved with very large increments of the fictitious time. In other words, in order to tackle the high-Re cases one need to jump during a given time step as farther ahead in time as possible. This way, the destabilizing effect of the convective term is diminished. It would have been a sheer paradox, if the progress made by the scheme during one time step were proportional to the size of the time increment. However, it is not. Even for time steps of order of hundreds, the relative progress of the solution toward the steady state is very slow and in some cases the convergence requires millions of time steps. The general scenario for very large Re is as follows. The rough solution to the problem is obtained within several hundreds of iterations (fictitious-time steps) and then the algorithm oscillates around the sought solution. The differences then are so slight that are hardly discernible. We were able to discover this kind of behaviour only after we went for L_1-norms of order of 10^{-7}. It is quite deceiving to stop the iterations when the norm reaches, say 10^{-6}, although it might look perfectly natural to do that. This sophistication is needed because of the nature of the investigated flow which exhibits a prolonged (in Re) laminar region and delayed transition to turbulence. In fact, the "weakly chaotic" solutions we encounter for $10000 < Re < 40000$ are very far from what is nowadays known as fully developed turbulence.

5 Conclusions

In the present paper we have studied numerically the steady 2D Navier–Stokes equations in terms of stream function. A fictitious time is added in the fourth–order equation for the stream function and coordinate splitting is applied directly to the bi–harmonic diffusion operator. It is shown that the splitting scheme does approximate an implicit scheme in full–

Table 1: The maximal value of $-\psi(x,y)$ and the positions x_{max} and y_{max} where it occurs

Re	1	100	400	1000	2000	3200	5000
[1]		0.10342 0.6172 0.7344	0.11391 0.5547 0.6055	0.11793 0.5513 0.5625		0.12038 0.5165 0.5469	0.11897 0.5117 0.5352
[9]	0.10006 0.50000 0.76667	0.10330 0.61667 0.74167	0.11297 0.55714 0.60714	0.11603 0.52857 0.56419			
[6][1]	0.099	0.103	0.112	0.116[2]		0.115	0.112
[12][3]		0.1034 0.6188 0.7375	0.1136 0.5563 0.6000	0.1173 0.5438 0.5625	0.1116 0.5250 0.5500		0.0920 0.5125 [4] 0.5313
[5][5]	0.100027 0.5 0.775	0.1035 0.625 0.75	0.1128 0.55 0.60	0.1115 0.525 0.575			
[10][6]		0.1034		0.1176[7]		0.1065	0.1181
[2]			0.1198	0.1136			
[11]					0.123 0.531 0.553		
[7]	0.0991 0.5 0.775	0.1054 0.633 0.748	0.1098 0.556 0.633	0.1092 0.556 0.595	0.1068 0.5 0.556		
C–R	0.100181 0.5017 0.7650	0.10397 0.6198 0.7369	0.11432 0.5533 0.6021	0.11871 0.5346 0.5645	0.11223 0.5228 0.5507		0.12661 0.5062 0.5232

[1] 65×65, [2] 97×97, [3] 321×321, [4] 161×161, [5] fourth order, [6] 128×128, [7] 256×265.

time steps. Thus no vorticity function is needed and the problem of boundary conditions is solved directly in an implicit manner at each half-time step imposing both the boundary conditions for the stream function.

The test problem of lid-driven cavity flow is treated by the new scheme as a featuring example. Results are obtained for Reynolds numbers in the interval $0 \leq Re \leq 40000$. For the higher Reynolds numbers $10000 < Re < 40000$ a kind of chaotic bifurcation is observed and the solution becomes pulsatile, exhibiting small but not decaying aperiodic oscillations around the mean flow.

Acknowledgment A Sabbatical Fellowship form the Spanish Ministry of Science and Education is gratefully acknowledged by CIC. CIC also acknowledges the hospitality of Laboratoire de Mécanique, Université de Caen.

References

[1] U. Ghia, K. N. Ghia, and C. T. Shin. High-*Re* solutions for incompressible flow using the Navier–Stokes equation and a multigrid method. *J. Comp. Phys.*, 48:387–411, 1982.

[2] J. W. Goodrich and W. Y. Soh. Time–dependent viscous incompressible Navier–Stokes equations: The finite-difference galerkin formulation and streamfunction algorithms. *J. Comp. Phys.*, 84:207–241, 1989.

[3] Ph. M. Gresho. Incompressible fluid dynamics: Some fundamental formulation issues. *Ann. Rev. Fluid Mech.*, 23:413–453, 1991.

[4] Ph. M. Gresho, S. T. Chan, C. Upson, and R. Lee. A modified finite element method for solving the time–dependent, incompressible Navier–Stokes equations. part 1: Theory. part 2: Applications. *Int. J. Numer. Methods Fluids*, 4:557–598, 619, 1984.

[5] M. M. Gupta. High accuracy solutions of incompressible Navier–Stokes equations. *J. Comp. Phys.*, 93:343–359, 1991.

[6] J. Kim and P. Moin. Application of fractional–step method to incompressible Navier–Stokes equations. *J. Comp. Phys.*, 59:308–323, 1985.

[7] S. J. Liao. Higher-order streamfunction–vorticity formulation of 2d steady–state Navier–Stokes equations. *Int. J. Numer. Methods Fluids*, 15:595–612, 1992.

[8] M. Rosenfeld, D. Kwak, and M. Vinokur. A fractional step solution method for the unsteady incompressible Navier–Stokes equations in generalized coordinate system. *J. Comp. Phys.*, 94:102–137, 1991.

[9] R. Schreiber and H. B. Keller. Driven cavity flows by efficient numerical techniques. *J. Comp. Phys.*, 49:310–333, 1983.

[10] M. C. Thompson and J. H. Ferziger. An adaptive multigrid tecvhnique for the incompressible Navier–Stokes equations. *J. Comp. Phys.*, 82:94–121, 1989.

[11] S. Turek. Tools for simulating non-stationary incompressible flow via discretely divergence–free finite element models. *Int. J. Numer. Methods Fluids*, 18:71–105, 1994.

[12] S. P. Vanka. Block–implicit multigrid solution of Navier–Stokes equations in primitive variables. *J. Comp. Phys.*, 65:138–158, 1986.

[13] N. N. Yanenko. *Method of Fractional Steps.* Gordon and Breach, 1971.

C. I. Christov
Instituto Pluridisciplinar, Universidad Complutense,
Paseo Juan XXIII, No 1,
28040, Madrid, SPAIN

A. Ridha
Dept. of Mathematique et Mécanique,
Université de Caen, Caen 14200
FRANCE
Laboratoire de Modélisation en Méchanique, CNRS-URA229,
Université Pierre et Marie Curie,
Tour 66, 4 Place Jussieu, 75252 Paris, FRANCE

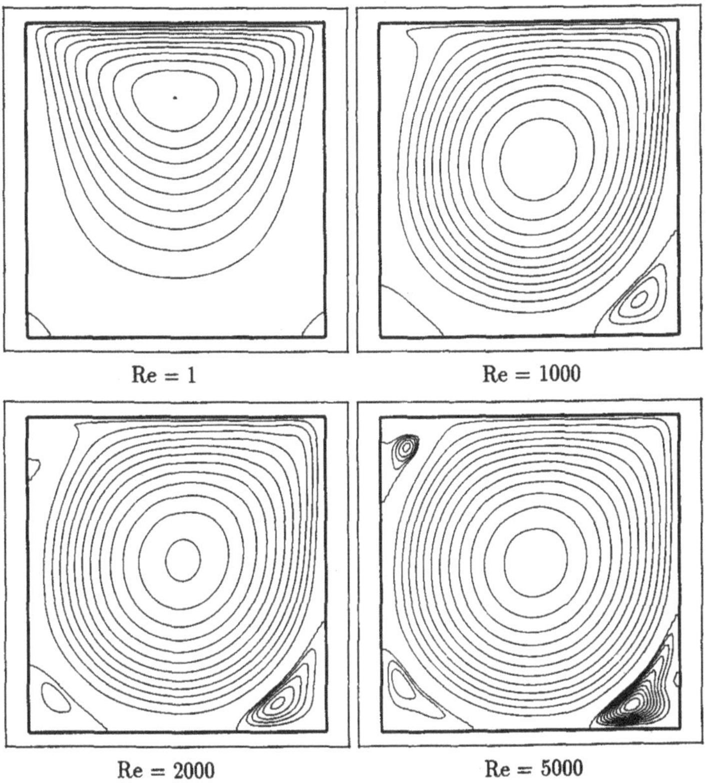

Figure 2: The stream lines for: a) $Re = 1$; b) $Re = 1000$; c) $Re = 2000$; d) $Re = 5000$. The negative isolines (the main vortex) of ψ are evenly spaced by -0.01, and the positive isolines (secondary vortices) - by $+0.0005$

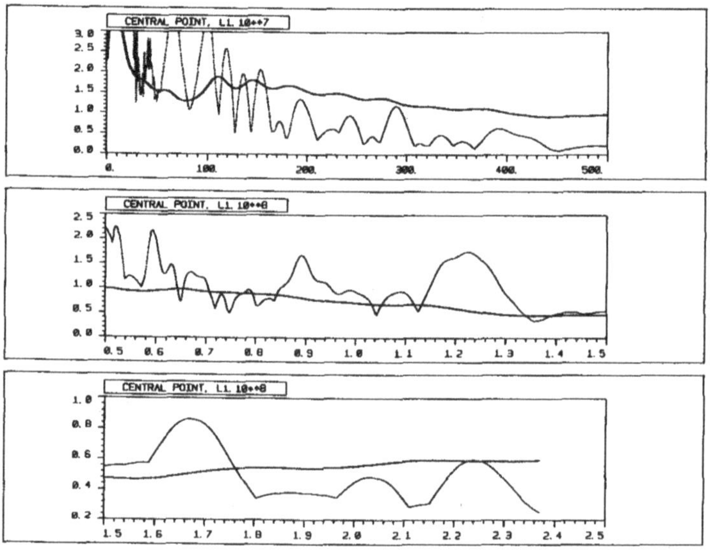

Figure 3: The evolution of solution and L_1-norm with the iterations for Re=5000. Solid line represents a scaled value of the stream function in the geometric center. The dashed line - scaled value of the L1-norm. In the upper figure the abscissa is measured in thousands of iterations; in lower two figures - in milions of iterations.

Reliability Estimation by optimal Monte Carlo Method*

E. Dimova, S. Trifonova

Abstract

The paper deals with the reliability of the information systems. The method of minimum path/cut sets is applied for evaluation of system reliability. The reliability function is presented as a solution of a linear system of equations. The problem is solved by Monte Carlo method with optimal transition density function.

1 Problem formulation

The problem of the reliability of the information systems is considered. The purpose is to outline a key importance of the reliability analysis in the design of systems with defined effectiveness.

The function of the operational effectiveness is [1]:

$$Y(R) = W(R) - C(R) , \tag{1}$$

where R is function of system reliability; $W(R)$ is function of system efficiency at the reliability level R; $C(R)$ is function of cost at the reliability level R.

Consider a system $S = (V, E)$, where $V = \{v_1, v_2, \ldots, v_n\}$ is a set of vertices and $E = \{e_1, e_2, \ldots, e_m\}$ is a set of edges. Vertices does not fail and have fixed positions. Edges can fail and have reliability parameters. System reliability is defined as a probability of paths existing among polar subset vertices. The system structure and minimum paths set $A = \{A_1, A_2, \ldots, A_p\}$ for evaluation of complex system reliability is used [6].

The system reliability function is

$$R = P(\bigcup_{i=1}^{p} A_i) = \sum_{i=1}^{p} P(A_i) - \sum_{i<j} P(A_i, A_j) + \ldots, (-1)^{p-1} P(A_1, A_2, \ldots, A_p) \tag{2}$$

*This work is partialy supported under Grant N:NIMU22IN/93.

In general the system reliability model for different polar subsets of vertices depends on dinamic system structure A and is of kind

$$R = AR + \varphi, \tag{3}$$

where φ is related to structure transformations.

The system reliability R is a random function $f(x_1, x_2, \ldots, x_m)$ of m independent random variables $x_1, x_2, \ldots, x_m$.

It is known nominal values of random variables x_1^{nom}, $i = 1, 2, \ldots, m$ and maximal deviation $\pm\delta x_1$, $i = 1, 2, \ldots, m$.

The function $R = f(x_1, x_2, \ldots, x_m)$ can be considered as a linear function

$$f(x_1, x_2, \ldots, x_m) \approx f(x_1^{nom}, x_2^{nom}, \ldots x_m^{nom}) + \sum_{i=1}^{m} \frac{\partial f(x_i^{nom})}{\partial x_i} \delta x_i, \tag{4}$$

where

$$\delta x_i = x_i - x_i^{nom}, \tag{5}$$

The value of standard deviation σ_R can be estimated from the relation:

$$\sigma_R^2 = \sum_{i=1}^{m} \left[\frac{\partial f}{\partial x_i} \sigma x_i \right]_l^2 + \sum_{i=1}^{m} \left[\frac{\partial f}{\partial x_i} \sigma x_i \right]_g^2 \tag{6}$$

where: σx_i is standard deviation of x_i random variable.

Subscript "l" and "g" shows that these deviations must be considered as local and global.

In the most case the distribution of R is sufficiently near normal for $m \geq 6$.

It is assumed that

$$K x_i = \frac{\partial f}{\partial x_i} \frac{\partial x_i}{\partial x} = \frac{\partial f}{\partial x_i} \sigma x_i \tag{7}$$

where $K x_i$ is sensitivity coefficient.

Equation for σ_R, can now be rewritten as:

$$\sigma_R^2 = \sum_{i=1}^{m} (K x_i)_l^2 + \sum_{i=1}^{m} (K x_i)_g^2. \tag{8}$$

It is convenient to calculate average partial derivatives by the finite differences in $x_i - \delta x_i \leq x_i \leq x_i + \delta x_i$:

$$\frac{\partial R}{\partial x_i} = \frac{R_i^{nom} - R_i}{\delta x_i} \tag{9}$$

2 The Monte Carlo method

Usually, Monte Carlo method is used for estimation inner products

$$J(f) = (g, f) = \int g(x) f(x)\, dx\,. \tag{10}$$

From now on the integration will mean the integration in the domain G. The function $g(x)$ belongs to the space $L_2(G)$ and $x \equiv (x^{(1)}, x^{(2)}, \ldots, x^{(d)}) \in G \subset R^d$ is a point in the Euclidean space R^d.

The function $f(x)$ is:

(i) "i"-order iterative function with a kernel $k(x, x_1)$:

$$f(x) = K^i\varphi(x) = \int \ldots \int k(x, x_1), k(x_1, x_2), \ldots k(x_{i-1}, x_i)\varphi(x_1) dx_1 \ldots dx_i; \tag{11}$$

(ii) a solution of the second-order integral equation:

$$f(x) = \int k(x, x_1) f(x_1) dx_1 + \varphi(x) \tag{12}$$

$$(f = Kf + \varphi)\,.$$

It is assumed that the Neuman series

$$f(x) = \sum_{i=0}^{\infty} K^i \varphi(x) \tag{13}$$

converges.

It is known [2], that the mathematical expectation of the random variable (r.v.) $\Theta_{i[g]}$ is equal to (g, f):

$$M\Theta_{i[g]} = (g, f) \tag{14}$$

where $\Theta_{i[g]} = \frac{g(\xi_0)}{p(\xi_0)} W_i \varphi(\xi_i)$, $W_0 = 1$; $W_j = W_{j-1} \frac{k(\xi_{j-1}, \xi_j)}{p(\xi_{j-1}, \xi_j)}$ for $j = 1, \ldots, i$.

The points ξ_j $(j = 0, 1, \ldots, i)$ are random points in the domain G with a density function $p(x_0)$ (if $j = 0$) and with a transition density function $p(x_{j-1}, x_j)$ (if $j = 1, \ldots, i$).

The random trajectory $T_i = (\xi_0 \to \xi_1 \to \ldots \to \xi_i)$ may be interpreted as a random point in the space $G \times \ldots \times G = G^{i+1}$ with a density

$$p_i(t^i) = p_i(x_0, x_1, \ldots, x_i) = p(x_0) \prod_{j=1}^{i} p(x_{j-1}, x_j) \tag{15}$$

Then, the series $\frac{1}{N}(\sum_{s=1}^{N} \Theta_{i[g]}) \approx (g, f)$ is an estimate of the functional (10); $(\Theta_{i[g]})_s$ is the value of $\Theta_{i[g]}$ for the "s" random trajectory; N is the number of the random trajectory.

In this case the probable error is [3]

$$r_N = c\sigma(\Theta_{i[g]})N^{-1/2}, \tag{16}$$

where $c \approx 0.6745$ and $\sigma^2(\Theta_{i[g]})$ is the standard deviation. The problem of optimization of the Monte Carlo algorithms consists in the minimization of $\sigma^2(\Theta_{i[g]})$.

We shall use the method for minimization of the dispersion $\sigma^2(\Theta_{i[g]})$ proposed by Iv. Dimov [4] by means of a suitably chosen of the density function $p_i(t^{(i)})$.

The next functions are considered in [3]:

$$\psi(x_j) = \left(\int \frac{k^2(x_j, x_{j+1})}{p(x_j, x_{j+1})} \psi^2(x_{j+1}) dx_{j+1}\right)^{1/2}, \quad j = 0, 1, \ldots, i-1 \tag{17}$$

$$\hat{\psi}(x_j) = \int \mid k(x_j, x_{j+1})\psi(x_{j+1}) \mid dx_{j+1}, \; j = 0, 1, \ldots, i-1 \tag{18}$$

where $\psi(x_i) = \hat{\psi}(x_i) = \varphi(x_i)$ and $p(x_j, x_{j+1})$ is a tolerant density function to the kernel $k(x, x_1)$.

We will use the following statement, proved in [4]:

Statement 1. The density function

$$\hat{p}_i(x_0, \ldots, x_i) = \hat{p}(x_0) \prod_{j=1}^{i} \hat{p}(x_{j-1}, x_j) = c_0 \mid g(x_0) \mid \prod_{j=1}^{i} \mid k(x_{j-1}, x_j) \mid \varphi(x_i)$$

where $c_0 = \left(\int \mid g(x_0)\hat{\psi}(x_0) \mid dx_0\right)^{-1}$ minimizes the second moment of the r.v. $\Theta_{i[g]}$ – $M\Theta^2_{i[g]}$.

Because

$$\sigma^2(\Theta_{i[g]}) = M\Theta^2_{i[g]} - (M\Theta_{i[g]})^2 \tag{19}$$

it is easy to show, that the density function $\hat{p}(x_0, \ldots . x_1)$ minimizes the standard deviation $\sigma(\Theta_{i[g]})$. It is so, because $M\Theta_{i[g]} = const$ for any tolerant density function.

It is shown that a density function

$$\bar{p}_i(x_0, \ldots, x_i) = c_0 \mid g(x_0) \mid \prod_{j=1}^{i} \mid k(x_{j-1}, x_j) \mid \tag{20}$$

is almost optimal [4].

Consider a minimization of the probable error for calculation of inner product of given function with solution of the second-order integral equation.

In this case, the r.v. $\Theta^*_{[g]}$ for which $M\Theta^*_{[g]} = (g, f)$ is

$$\Theta^*_{[g]} = \frac{g(\xi_0)}{p(\xi_0)} \sum_{i=0}^{\infty} W_i \varphi(\xi_i)$$

The next statement is valid [4]:

Statement 2. If $\varphi(x_i)K^{k-i}\varphi(x_i) \geq 0$ for any $x_i \in G$ and $k > i$, then the density function

$$p_k(x_0, \ldots, x_i) = c_0|g(x_0)| \prod_{j=1}^{i} |k(x_{j-1}, x_j)||[\varphi(x_i)K^{k-i}\varphi(x_i]^{1/2} \tag{21}$$

minimizes $M(\Theta_{i[g]}\Theta_{k[g]})$.

Now we will show how these results can be used for solving systems of linear algebraic equations.

Consider a method for estimation the inner product of a given vector and a vector-solution f of a linear algebraic system:

$$(g, f) = \sum_{\alpha} g_\alpha f_\alpha, \tag{22}$$

where g and f are vectors with components g_α and f_α respectively.

Consider $G \equiv [0, m) \subset R^1$ and separate G into subintervals G_α so that $x \in G_\alpha \Leftrightarrow \alpha - 1 \leq x < \alpha$.

It is obvious, that if

$$k(x, x_1) = k_{\alpha\beta}, \quad x \in G_\alpha, \quad x_1 \in G_\beta, \tag{23}$$

$$\varphi(x) = \varphi_\alpha, \quad x \in G_\alpha, \tag{24}$$

the integral equation (12), described for $x \in G_\alpha$ (where $\alpha = 1, 2, \ldots, m$) leads to a linear algebraic system:

$$f_\alpha = \sum_{\beta} k_{\alpha\beta} f_\beta + \varphi_\alpha, \quad \alpha = 1, \ldots, m, \tag{25}$$

where $f_\alpha = f(x)$, for $x \in G_\alpha$, $\alpha = 1, \ldots, m$. (We may note, that if (22) and (23) occur and $x \in G_\alpha$, then $f(x)$ does not depend on x and $f(x)$ is constant, i.e. $f(x) = f_\alpha$, because for any $x \in G_\alpha$ the next equality is held: $f(x) = \sum_\beta k_{\alpha\beta} \int_{G_\beta} f(x)dx + \varphi)$ [5].

The system (24) can be written in the next form:

$$f = Kf + \varphi, \tag{26}$$

where $f = (f_1, \ldots, f_n)^T$, $\varphi = (\varphi_1, \ldots, \varphi_n)^T$;

$$K = \begin{bmatrix} k_{11} & k_{12} & \ldots & k_{1n} \\ k_{21} & k_{22} & \ldots & k_{2n} \\ \vdots & & & \\ k_{n1} & k_{n2} & \ldots & k_{nn} \end{bmatrix}$$

This observation permit us to use all results obtained before for minimization of the probable error for calculation of the inner products (21).

Let us consider the case, when

$$f_\alpha = \sum_\beta k_{\alpha\beta}\varphi_\beta.$$

This case corresponds to the case (i) for the function $f(x)$. The r.v. $\eta_1[g]$ for which $M\eta_1[g] = (g, K\varphi)$ is

$$\eta_1[g] = \frac{g_{k_0}}{p_{k_0}}\frac{k_{k_0k_1}}{p_{k_0k_1}}\varphi_{k_1}, \tag{27}$$

where $p_{k_0} = P(\alpha = k_0)$, $p_{k_0k_1} = P\{\beta = k_1|\alpha = k_0\}$; $\sum_\alpha p_\alpha = 1$; $\sum_\beta p_{\alpha\beta} = 1$ for $\forall\alpha = 1,\ldots,m$.

The random trajectory $T_i = (\xi_0 \to \ldots \to \xi_i)$ is degenerated to a Marcov chain with two states $(k_0 \to k_1)$.

The vectors ψ and $\hat{\psi}$ with components ψ_α and $\hat{\psi}_\alpha$ are presented analogically to the functions $\psi(x_0)$ and $\hat{\psi}(x_0)$

$$\psi_\alpha = \left(\sum_\beta \frac{k^2_{\alpha\beta}\varphi^2_\beta}{p_{\alpha\beta}}\right)^{1/2}$$

$$\hat{\psi}_\alpha = \sum_\beta |k_{\alpha\beta}\varphi_\beta| = \|k_\alpha\varphi\|.$$

Consider the case of calculation the inner product (21) when f is a vector-solution of the system (25). In this case we must consider an infinitely long Marcov chain

$$T_\infty = (k_0 \to k_1 \to \ldots \to k_i \to \ldots)$$

A corresponding r.v. is

$$\eta^*[g] = \frac{g_{k_0}}{p_{k_0}}\sum_{i=0}^{\infty}\frac{k_{k_0k_1}\ldots k_{k_{i-1}k_i}}{p_{k_0k_1}\ldots p_{k_{i-1}k_i}}\varphi_{k_i}\,. \tag{28}$$

Then

$$M\eta^{*2}[g] = \sum_{i=0}^{\infty}\sum_\alpha\cdots\sum_{\alpha_i}\frac{g^2_{\alpha_1}}{p_{\alpha_0}}p_iW_i^2\varphi^2_{k_i} + 2\sum_{i=0}^{\infty}\sum_{k=i+1}^{\infty}M\left[\frac{g^2_{k_0}}{p^2_{k_0}}W_i\varphi_{k_i}W_k\varphi_{k_k}\right], \tag{29}$$

where $p_i = p_{\alpha_0}p_{\alpha_0\alpha_1}p_{\alpha_1\alpha_2}\cdots p_{\alpha_{i-1}\alpha_i}$.

It is possible to show that the minimization density function for $M(\eta_i[g]\eta_k[g])$ is

$$\hat{p}_{k,i} = c_0|g_{\alpha_0}|\prod_{j=1}^{i}|k_{\alpha_{j-1}\alpha_j}|[\varphi_{\alpha_i}K^{(k-i)}\varphi_{\alpha_i}]^{1/2} \tag{30}$$

for any i and k and $\varphi_{\alpha_i} K^{(k-i)} \varphi_{k_i} \geq 0$.

Consider Monte Carlo algorithms with null probable error.

These algorithms have unique features. They can be very important for solving some problems, which need a large-scale computations.

Let us write the expression for $\sigma^2(\Theta_{1[g]})$, where $f(x)$ is integral operator for the frequency function $\hat{p}_1(x_0, x_1)$:

$$\sigma^2(\Theta_{1[g]}) = \left(\int |g(x_0)\hat{\psi}(x_0)|dx_0\right)^2 - \left(\int g(x_0)f(x_0)dx_0\right)^2, \tag{31}$$

where $\hat{\psi}(x_0) = \int |k(x_0, x_1)\varphi(x_1)|dx_1 = \|k(x_0)\varphi\|$;
$f(x_0) = \int k(x_0, x_1)\varphi(x_1)dx_1$.

If for any $x_0 \in G$ the function $k(x_0, x_1)\varphi(x_1)$ does not change his sign, then $\hat{\psi}(x_0)$ and $f(x_0)$ either coincide for any $x_0 \in G$ or they are reflection in a plane x_0. If in this case the product of the functions $g(x_0)f(x_0)$ is of constant sign, then the variance is zero.

The necessary and sufficient conditions for the zero probable error for the optimal frequency function are [4]:

1. For any $x_0 \in G$ the function $k(x_0, x_1)\varphi(x_1)$ does not change his sign;
2. The product $g(x_0)f(x_0)$ is of constant sign.

Let us formulate the conditions for obtaining the algorithms with null probable error for the case when $f(x)$ is "i"-order iterative function:

1. For any $x_j \in G(j = 0, \ldots, i)$ the functions $k(x_j, x_{j+1})K\varphi(x_j)$ (where $K\varphi(x_j) = \int k(x_j, x_{j+1})dx_{j+1}$) are of constant sign;
2. The product $g(x_0)f(x_1)$ is of constant sign.

As the above conditions must be fulfilled in the case when $f(x)$ is a solution of the second-order integral equation (12) for obtaining of null probable error algorithms. Monte Carlo method described above is applied for investigation of the system reliability.

To calculate

$$Y_i = \left(e_{(i)}, Y\right) = \sum_{\alpha=1}^{m} e_\alpha^{(i)} Y_\alpha \tag{32}$$

it is necessary to use algorithm (26) of Monte Carlo method for the case when vector g is the unit vector $e_{(i)} = (0, \ldots, 0, 1, 0, \ldots, 0)$, in which only "i" component is equal to 1.

The problem is that the system (9) is very large and often it is necessary to know only one or a small number of solutions components.

For estimation of the derivatives (8) for every parameter "i" it is necessary to solve the following systems of the type (9)

$$Y^{(i)+} = K^{(i)+}Y^{(i)+} + \varphi$$

$$Y^{(i)-} = K^{(i)-}Y^{(i)-} + \varphi,$$

where the matrix $K^{(i)+}$ corresponds to the maximum value of the parameter $x_i = x_i + \delta x_i$ and $K^{(i)-}$ corresponds to the minimum value of the parameter $x_i = x_i - \delta x_i$. In this case an almost optimal Monte Carlo algorithm consists of random walk on elements of matrix K by transition probability (29) in the Marcov chain.

The method is used for calculation the system reliability of the control system.

Acknowledgment

We wish to thank Ivan Dimov for the useful discussions and consultations, which improved this work significantly.

References

[1] Reliability and Engineering Efficiency. Handbook, vol.5, Mashinostroyeniye, Moscow, 1989.

[2] Sobol I.M., Monte Carlo numerical methods. **Moscow, Nauka**, 1973 (in Russian).

[3] Ermakov S.M., Mikhailov G.A. Statistical simulation. **Moscow, Nauka**, 1982 (in Russian).

[4] Iv. Dimov. Minimization of the probable error for Monte Carlo method. Appl. of Math. in Technology. Differential Equations, Varna 1986, 1987, pp. 161–164.

[5] Curtis J.H. Monte Carlo methods for the iteration of linear operators. **J. Math. Phys., vol 32, No 4**, (1984), pp. 209–232.

[6] Agrawal A., R. Barlow A survey of network reliability and domination theory. **Operational Research, vol. 32, No 3**, (1984), pp.478–492.

Address of the Authors:
Central Laboratory of Control Systems
Bulgarian Academy of Sciences
Acad.G.Bontchev St., Bl.107, PBox 79
1113 Sofia, BULGARIA Email: sltriff@bgcict.bitnet

Seismic Waves in Layered Soil : Boundary Integral Equation Method

Petia Dineva *

Abstract

The one and two- dimensional problems of SH, P, SV- seismic wave propagation in an multilayered half-space is studied by the DBIEM (Direct Boundary Integral Equation Method) for steady state and transient cases. A hybrid formulation is used on the base of DBIEM and finite difference procedure applied to the time variable in the transient problem solution. The present hybrid formulation employs the fundamental solution depending neither on the frequency nor on the time variable. Time records of surface responses of the half-space are computed and compared with numerical evaluation of exact analytical solutions to this problem.

1 Introduction

Elastodynamics was the oldest subject of elasticity. It has originated with Navier's theory of motion of elastic bodies and later it has been adopted by Cauchy and Poisson, Lamme, Reileigh and Love. The publication of Kolsky [1], has marked the revival of research on elastic waves. By the 1970s, the monographs by Nowacki [2], Achenbach [3], Graff [4], Eringen and Suchubi [5] has appeared. The research of elastic waves is active in the field of geophysics as well.

The mathematical theory of elasticity is formulated in terms of body force f, surface force (traction) T, stress tensor σ_{ij}, strain tensor ε_{ij}, and displacement vector u_i of an elastic body. Isotropic materials are characterized by material constants, such as Young's modulus E, shear modulus G, Poisson's ratio ν and Lamme's constants λ and μ. The theory of elastodynamics especially considers in the case when all fields are time-dependent.The mass density ρ becomes an important parameter in measuring the inertia. The linear theory of elastodynamics is embodied in the following set of equations for a body of volume V enclosed by a surface $S = S_1 + S_2$.

EQUATIONS OF MOTION

*The author express his gratitude to the National Science Foundation, grant N TH-364/93 for financial support.

$$\sigma_{ij,j} + f_i = \ddot{u}_i \quad x \in V \tag{1.1}$$

PHYSICAL EQUATIONS

$$\sigma_{ij} = 2\mu(\nu/(1-2\nu))\delta_{ij}\varepsilon_{kk} + 2\mu\varepsilon_{ij} \quad x \in V \tag{1.2}$$

EQUATIONS OF GEOMETRY

$$\varepsilon_{ij} = 0,5\left(\frac{\partial u_i}{\partial x_j} + \frac{\partial u_j}{\partial x_i}\right) \tag{1.3}$$

BOUNDARY CONDITIONS

$$u(x,t) = \overline{u} \quad x \in S_1; \quad \sigma_{ij}n_j(x,t) = \overline{T} \quad x \in S_2 \tag{1.4}$$

INITIAL CONDITIONS

$$u(x,t) = u_0 \quad \dot{u}(x,0) = v_0 \quad x \in V \tag{1.5}$$

Eqs. 1.1, 1.2 and 1.3 can be combined into one vector equation for $u(x,t)$: Navier- Cauchy equation of motion. The main aim of this paper is to show and numerically illustrate the solution of the problem:SH, P, SV- steady state and transient waves in multilayered media on the base of BIEM.

2 SH- seismic waves propagation (anti-plane problem) in layered soil

"Anti-plane problem "occurs when the displacement vector is perpendicular to the direction of propagation of the displacement waves. For this reason this motion is also called shear antiplane or horizontally polarized shear motion.

2.1. STEADY - STATE PROBLEM is considered when the time dependence is harmonic, i.e. $u(x,t) = u_0(x)exp(-iwt)$, and the media is free of sources if f in V and $\overline{T}$ on S vanish. The initial conditions 1.5 is omitted. Eq. 2.1 asquires the form:

$$\frac{\partial^2 u}{\partial x^2} + \frac{\partial^2 u}{\partial y^2} + k^2 u(x,y) = 0 \tag{2.1}$$

The geometry of a multilayered medium with nonparallel boundaries between layers is shown in Fig.1. The incident wave is:

$$u_z^i = exp[ik(y\cos\theta - x\sin\theta)]exp(-i\omega t) \tag{2.2}$$

here : θ is the incident angle, $k = \frac{w}{V}$; w is frequency,the V_{SH} — the wave velocity. Thus, we have the "anti-plane" problem in plane (x,y) (see Fig.1):

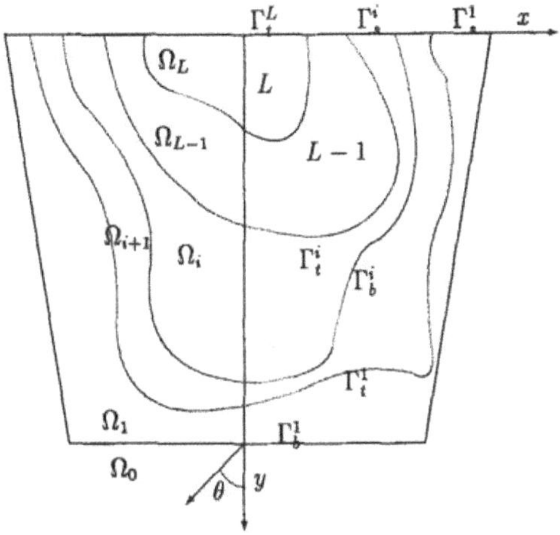

Figure 1: Fig.1

$$\begin{aligned} \sigma_{xz} &= \mu\frac{\partial u_z}{\partial x} \quad \sigma_{yz} = \mu\frac{\partial u_z}{\partial y} \quad T_z = \mu\frac{\partial u_z}{\partial n} \\ u_x &= u_y = 0 \quad u_z = u \end{aligned} \tag{2.3}$$

where: n is the unit normal vector.
The boundary conditions are:

$$\begin{aligned} &\tfrac{\partial u}{\partial n} = 0 \text{ for } (x,y) \in \Gamma_s, \Gamma_t \\ &u^i(x,y) = u^{i+1}(x,y) \quad \sigma^i_{kl}n^i_l = \sigma^{i+1}_{kl}n^{i+1}_l \\ &\text{for}(x,y) \in \Gamma^i_{t,b} \\ &u^0_b = u_0 + u^0 = u^L_b;\ \sigma^0_{ij}n^0_j = \sigma^L_{ij}n^L_j \\ &\text{for}(x,y) \in \Gamma^L_b \end{aligned} \tag{2.4}$$

where:u^0 -free field motion, i.e. the displacement obtained as a result of the interference of the incident and the reflected by the free surface waves, not accounting for the multilayered region. u_0 and u^0_b represent the scattered field in the half-space Ω_0 and Ω_L resp.; Ω_0 is the infinite region under Ω_L (Fig.1). In addition u^0_b has to satisfy Sommerfeld radiation condition.

2.1.1. Using BIEM for the solution of boundary-value problem: eqs.2.1, 2.4

Applying BIEM for the above described problem, the boundary integral equations is obtained for the i-th layer with its upper Γ_t^i, lower Γ_b^i, and lateral Γ_s^i bounds:

$$
\begin{aligned}
u^i(x,y,\omega)c_{zz} = & \int_{\Gamma_t^i}\left[u^i(x_0,y_0,\omega)\frac{\partial U^*(x,y,x_0,y_0,\omega)}{\partial n} - U^*(x,y,x_0,y_0,\omega)\frac{\partial u^i(x_0,y_0,\omega)}{\partial n}\right]d\Gamma_t + \\
& \int_{\Gamma_b^i}\left[u^i(x_0,y_0,\omega)\frac{\partial U^*(x,y,x_0,y_0,\omega)}{\partial n} - U^*(x,y,x_0,y_0,\omega)\frac{\partial u^i(x_0,y_0,\omega)}{\partial n_0}\right]d\Gamma_b + \\
& \int_{\Gamma_s^i}\left[u^i(x_0,y_0,\omega)\frac{\partial U^*(x,y,x_0,y_0,\omega)}{\partial n} - U^*(x,y,x_0,y_0,\omega)\frac{\partial u^i(x_0,y_0,\omega)}{\partial n_0}\right]d\Gamma_s
\end{aligned}
\tag{2.5}
$$

where: $U^*(x,y,x_0,y_0,\omega) = (i/4)[H_0^{(2)}(kr)+H_0^{(2)}(k\bar{r})]$ is Green function of the set problem, $H_0^{(2)}$ is the Hankel function of a second type, zero order; $r = [(x-x_0)^2 + (y-y_0)^2]^{1/2}$; $\bar{r} = [(x-x_0)+(y+y_0)]^{1/2}$; r, $\bar{r}$ are the position field vector and the source point vector.The integral singularities at $r \to 0$ are calculated by Banaugh [6]. Eq.2.5 is solved by BEM discretization using constant boundary elements. Numerical results for a concrete geological column with complex geometry, nonparallel boundaries between layers solving the steady state and transient SH-wave propagation problem are given in [10].

3 Two-dimensional transient seismic wave propagation problem (in-plane problem) in soil

Since the problem is two-dimensional and of plane-strain, the displacement vector in the coordinate system (x,y) in Fig.1 is $u = [u_x, u_y]$. The governing equation is:

$$(c_p^2 - c_s^2)u_{p,pi} + c_s^2 u_{i,pp} - \ddot{u}_i = -\rho^{-1}f_i(t) \tag{3.6}$$

where $c_p^2 = [(\lambda+2\mu)/\rho]^{1/2}$ and $c_s = (\mu/\rho)^{1/2}$
The application of the Wilson-Θ method [7] to $\dot{u}_i$ and $\ddot{u}_i$ yelds to equation 3.7.

$$
\begin{aligned}
(c_p^2 - c_s^2)u_{p,pi}(t+\Theta\Delta t) + c_s^2 u_{i,pp}(t+\Theta\Delta t) - k^2 u_i(t+\Theta\Delta t) = -X_i \\
k^2 = (6/\Theta^2\Delta t^2);\ X_i = \rho^{-1}f_i(t+\Theta\Delta t) + k^2u_i(t) + (6/\Theta\Delta t)\dot{u}_i(t) + 2\ddot{u}_i(t)
\end{aligned}
\tag{3.7}
$$

Equations 3.7 are elliptic linear partial differential equations at time $\Theta\Delta t + t$. Further the formulation of DBIEM is applied to equation 3.7. The discretization by constant BE is done. As a result, the following system of discretized BIE is obtained:

$$c_{ij}u_j(x^p,y^p) = A_i(x^p,y^p) - B_i(X^p,y^p) + \rho f(t)C_i(x^p,y^p) + D_i(x^p,y^p) \tag{3.8}$$

Here: $f(t)$ denotes the time signature of the wave source, $c_{ij} = \delta_{ij}0.5$

$$
\begin{aligned}
A_i(x^p, y^p) &= \sum_{k=1}^{N} P_j(x^k, y^k) \int_{\Gamma_k} U^*_{ij}(x^p - x^k, y^p - y^k) d\Gamma_k \\
B_i(x^p, y^p) &= \sum_{k=1}^{N} u_j(x^k, y^k) \int_{\Gamma_k} P^*_{ij}(x^p - x^k, y^p - y^k) d\Gamma_k \\
C_i(x^p, y^p) &= -U^*_{ij,j}(x^p - x_I, y^p - y_I) \\
D_i(x^p, y^p) &= -\sum_{k=1}^{M} \alpha^k_j \int_{\Omega_k} U^*_{ij}(x^p - x^k, y^p - y^k) d\Omega_k; \quad i = x, y
\end{aligned}
\tag{3.9}
$$

where: $\alpha^k_i = k u_i(x^k, y^k, t) + (6/\Theta\Delta t)\dot{u}_i(x^k, y^k, t) + 2\ddot{u}_i(x^k, y^k, t)$

Ω_k is the k - th two-dimensional BE. N, M - the number of BE on Γ_k, Ω_k

Here x_I and y_I are the coordinates of the wave source, u_x, u_y, P_x, P_y - displacements and tractions (unknown), U^*_{ij} and P^*_{ij} denote the fundamental solutions of displacement and traction.In order to put equation 3.8 into the form convenient for the numerical computations, first the variables are rearranged and a part of integrals is analytically calculated, the remaining integrals are computed numerically using Gaussian quadrature. The integral singularities are weak:$U^*_{ij} \to \ln(1/r)$, $P_{ij} \to 1/r$ at $r \to 0$.As a result a linear complex algebraic system is solved at every time moment $t = \Theta\Delta t$, $t = \Delta t + \Theta\Delta t$, etc., where $\Delta t > 0$, $\Theta > 1$. The numerical procedure is:

(1) Solve Eq.3.7 to obtain u_i, P_i for $t = \Theta\Delta t$ at all boundary points (note that, for $t = 0$, $\alpha_i = 0$);

(2) For $t = \Theta\Delta t$ evaluate u_i, P_i, α_i at all interior points;

(3) By means of the Wilson- Θ method for $t = \Delta t$ calculate u_i, $\dot{u}_i$, $\ddot{u}_i$;

(4); Solve Eq.3.7 to obtain u_i, P_i for $t = \Delta t + \Theta\Delta t$ at all boundary points;

(5) For $t = \Delta t + \Theta\Delta t$ evaluate u_i, P_i, α_i at all interior points;

(6) By means of Wilson-Θ method, calculate u_i, $\dot{u}_i$, $\ddot{u}_i$ for $t = 2\Delta t$;

(7) Solve Eq.3.7 to obtain u_i, P_i for $t = 2\Delta t + \Theta\Delta t$ at all boundary points;

(8) For $t = 2\Delta t + \Theta\Delta t$ evaluate u_i, P_i, α at all interior points.

Based on this scheme, the evaluation of responses for $t = 3\Delta t + \Theta\Delta t$ and subsequent observation times is straightforward.

Figures 2 and 3 present the theoretical seismograms in time domain for u_x and u_y at $x = 4y_I$ and $x = 10y_I$ on free surface of a half space perturbed by a buried explosive seismic source at depth $y_I = 0,5$m. The time function of $f(t)$ is assumed to be the parabolic ramp function [9] which half rise time Δ is $0,1$s. The time step $\Delta t = 0,0001$s, $\Theta = 1,4$, $\tau_0 = 2y_I/c_S$, A is dimensionless coefficient [8]. Two space discretizations Δl_1 and Δl_2 are chosen as $\Delta l_1 = 0,007$m and $\Delta l_2 = 0,0069$m. Note an excellent agreement with the exact values provided by the method of generalized ray [9].

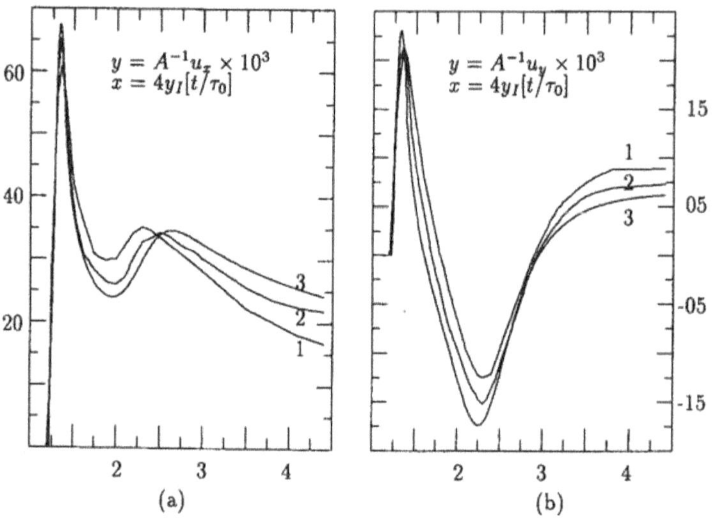

Figure 2: Theoretical seismograms on half-space perturbed by the buried line source ($A = 1/4\pi C_p^2$)

4 Conclusion

A two-dimensional formulation of the DBIEM has been proposed. First, by means of the Wilson-Θ method, the equation of motion are transformed into a set of equations of elliptic type. Then the BIEM is applied to the set of equations, employing the fundamental solution which depends neither on the frequency nor on the time variable. The mechano-mathematical models and numerical results for SH-,P-,SV- seismic wave propagation in soil is presented.

References

[1] Kolsky, H. Stress Waves in Solids. Clarendon Press, Oxford, 1953

[2] Nowacki, W. Dynamics of Elastic Systems Wiley, New York, 1963

[3] Achenbach, J., D. Wave Propagation in Elastic Solids North Holland, Amsterdam, 1973

[4] Graff, K.F. Wave Motion in Elastic Solids Ohio State University, Press, Columbus, 1975

[5] Eringen, A.C., Suhubi, E.S. Elastodynamics, vol.2 Academic Press, New York, 1975

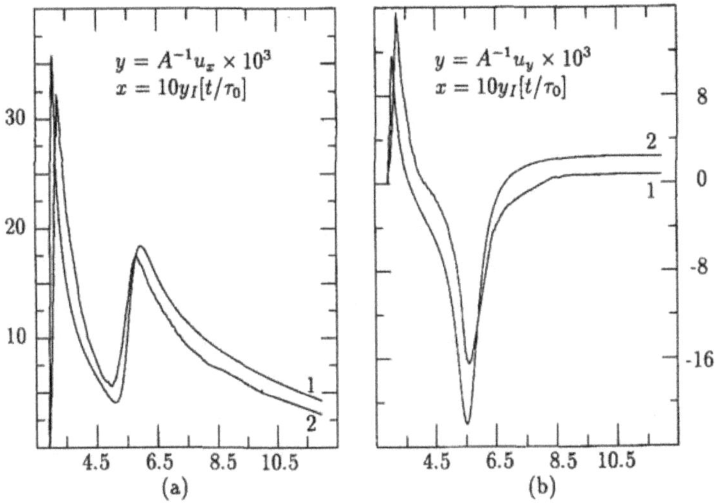

Figure 3: curve - by [9]; curve 2 - by OBIEM for $\Delta l_2 = 0.0069m$ curve 3 - by OBIEM for $\Delta l_1 = 0.017m$

[6] Banaugh, R.P. Diffraction of Steady Acoustic Waves by Surfaces of Arbitrary Shape The Journal of the Acoustic Soc. of Amer.,1963, vol.35, N10,pp.1590-1601

[7] Clough, R.W., Penzien, J. Dynamics of Structures, McGraw-Hall,Singapore, 1982

[8] Borejko, P., Ziegler, F. Transient elastic and viscoelastic responses of a half-space Proc. IUTAM Symp. on Elastic Wave Propagation and Ultrasonic Nondestructive Evaluation, Ed.S.K.Datta, Amsterdam,1990.pp. 401-402

[9] Borejko,P. Application of the Generalized Rays to Transient Waves in an Elastic Half-space due to aburied line source, Acta Mechanica 87, 1987,pp.79-90

[10] Dineva,P., L. Hadjikov, Ts, Rangelov On the solution of the problem of seismic wavespropagation in a multilayered nonelastic region by the BEM Int.J.of Soil Dynamics and Earthquake Engineering, 1991, vol.10,N6, pp.323-329.

Address of Author:
Institute of Mechanics, Bulg. Academy of Sci.
Sofia
Bulgaria

Monte Carlo modelling of quantum electron transport in semiconductor structures

Carlo Jacoboni
Dipartimento di Fisica, Universita di Modena ,Italy

Mihail Nedjalkov
CICT, Sofia, Bulgaria

1 Introduction

The main objective of device modelling is to provide theoretical and numerical descriptions of the physics of charge transport in microelectronic structures. In particular, the Monte Carlo (MC) method for charge-transport simulation is considered as the most reliable one, to which other techniques have to be compared in order to estimate their validity[1]. The MC method has been considered for more then 25 years as a direct simulation of the semiclassical transport, obtained by using the known probabilities for the elementary events taking part in the physical process. Its methods can also be used in situations unattainable by experiments, or even for the investigation of nonexisting materials and structures, thus having a scope somewhat similar to that of experimental techniques.

The recent device minituarization enhances the importance of rare physical events. Furthermore, so small space and time scales can be involved, that the applicability of semiclassical transport fail and a quantum description is required. How the MC algorithms may be evolved to meet this new challenge is the subject of this work.

2 Electron Transport in Semiconductors

The quantum theory of solids describes electrons by means of wavefunctions, with wave vector $\mathbf{k}$, travelling through the system under the action of the periodic crystal potential, the applied field $\mathbf{E}$ and scattering sources. In the semiclassical approximation, electrons are assumed to be described by wave-packets with well defined position $\mathbf{r}$ and momentum $\hbar\mathbf{k}$, changing with time according to Newton's law:

$$\hbar\frac{d\mathbf{k}}{dt} = e\mathbf{E}; \quad \frac{d\mathbf{r}}{dt} = \mathbf{v}; \quad \mathbf{v} = \frac{1}{\hbar}\nabla_k\epsilon(\mathbf{k}) \tag{1}$$

Here $\mathbf{v}$ is the electron velocity and $\epsilon(\mathbf{k})$ is the considered energy band that accounts for the perfect crystal potential. Scattering is due to periodicity imperfections such as impurities and lattice vibrations, whose quantized modes are called phonons. The interaction with the phonons consists of processes in which a phonon in mode $\mathbf{q}$ with momentum $\hbar\mathbf{q}$ and energy $\hbar\omega_q$ is absorbed or emitted so that the electron momentum and energy are increased or decreased by the amount $\hbar\mathbf{q}$ and $\hbar\omega_q$ respectively. The probabilities for phonon scatterings are obtained from quantum mechanical calculations.

The fundamental quantity in the semiclassical transport theory is the electron distribution function $f(\mathbf{k},\mathbf{r},t)$ defined as the electron density in the six-dimensional phase space $(\mathbf{k},\mathbf{r})$. The average electron quantities are then functionals of f. In absence of gradients and fields, f is the equilibrium distribution, commonly given by the Maxwell function. The electron mean velocity is zero in such a state and the mean energy ϵ_o appears as a definite function of the crystal temperature.

In presence of an electric field, each electron is accelerated according (1) until the time t_s of the scattering. At this time it changes instantaneously the value of $\mathbf{k}$ to some new value which, together with $\mathbf{r}(t_s)$ are the initial point for a new trajectory. Since $\epsilon(\mathbf{k})$ is an even function of $\mathbf{k}$, if the field is small, the average velocity is linear with respect to $\mathbf{E}$, while the electron energy maintains its equilibrium value (Ohm's law). In this case the transport properties of the system can be expressed by means of a few physical parameters, which are provided by theory or experiments.

If, on the contrary, the field is sufficiently high, the mean carrier energy increases together with the dissipation efficiency of the scattering mechanisms. We enter the nonlinear transport region, where "hot electron" effects have to be accounted for. The only adequate description of the electron system is then provided by the Boltzmann Equation:

$$\frac{\partial f}{\partial t} + \mathbf{v}.\nabla_r f + \frac{d\mathbf{k}}{dt}.\nabla_k f = \int d\mathbf{k}' f(\mathbf{k}',\mathbf{r},t)P(\mathbf{k}',\mathbf{k}) - f(\mathbf{k},\mathbf{r},t)\lambda(\mathbf{k}) \qquad (2)$$

with $\lambda(\mathbf{k}) = \int d\mathbf{k}' P(\mathbf{k},\mathbf{k}')$ and $P(\mathbf{k}',\mathbf{k})$ is the probability per unit time that the carrier is scattered from $\mathbf{k}'$ to $\mathbf{k}$. This equation can be derived by simply evaluating the change in the electron number in a small volume around $(\mathbf{k},\mathbf{r})$ due to the field, velocity and scattering contributions.

Present-day semiconductor devices can introduce so small time and space scales, that the semiclassical transport fail owing to the uncertainty principle. In fact the energy of an electron between two successive collisions is defined within an uncertainty $\Delta\epsilon \sim \hbar/\tau$, where τ is the time between the collisions. If we are willing to accept $\Delta\epsilon \leq 5meV$, then it must be $\tau \gg 10^{-14}s$. For fields as high as $10^6 V/cm$ as they may be present in very small devices, the time between collision may become smaller. At the same time the wavelength of an electron is of the order of several nm for typical electron energies and the space scale of these devices is already comparable to this figure, so that size quantization becomes important. As a consequence it is seen that a quantum approach to the problem is required.

3 MC simulation for Boltzmann equation

The MC method for transport in semiconductors, since its first application in 1966 [2], was based, for more then 20 years, on the imitation of the events taking place during the real electron motion. The natural probabilities for scattering processes are used in the simulation thus performing the particle number balance in any phase space point, as described by the BE. In that way a large variety of physical systems and phenomena have been investigated, accounting for full band structure, all types of phonons, impurity and carrier-carrier interaction[3]-[4]. The MC has also been extended to analyse space and time dependent problems and to include the effect of the exclusion principle[5]. Many transport quantities have been obtained such as mean velocity and energy, diffusivity, autocorrelation functions [6].

In devices with small dimensions, the local field can be very high. Some electrons can gain very high energy and participate in very rare processes, but having important effects (typical is the injection into the oxide in the MOS transistor; the fraction of electrons that participate in this process may be of the order of 10^{-6}), which cannot be provided from the traditional MC in a reasonable simulation time. To solve this problem, in the past years an alternative version of the MC has been proposed [7, 8, 9, 10]. It is based on the application of the numerical MC for solving integral equations.

Equation (2) can be transformed into an integral form, by expressing the solution in terms of its Green function [11].

$$f(\mathbf{k},\mathbf{r},t) = f_o(\mathbf{k}(0),\mathbf{r}(0))e^{-\int_0^t \lambda(\mathbf{k}(t'))dt'} + \int_0^t dt' \int d\mathbf{k}' e^{-\int_{t'}^t \lambda(\mathbf{k}(\tau))d\tau} P(\mathbf{k}',\mathbf{k}(t'))f(\mathbf{k}',\mathbf{r}(t'),t') \tag{3}$$

where f_o is the initial condition and $(\mathbf{k}(t'),\mathbf{r}(t'))$ are the values of $\mathbf{k},\mathbf{r}$ at time t', which will be in $(\mathbf{k}(t),\mathbf{r}(t))$ at time t. This equation has a straightforward physical interpretation: the distribution function in $Q = (\mathbf{k},\mathbf{r},t)$ has a contribution from the particles which at the initial time 0 are in the right trajectory and are not scattered away from it before time t, and a contribution from the particles from $Q' = (\mathbf{k}',\mathbf{r}',t')$, put on the right trajectory at time t' and not scattered away until time t. The kernel of (3) $\mathcal{K}(Q',Q)$ can be interpreted as a probability of reaching point Q if starting from Q'

The MC procedure for the evaluation of the ith term of the Neumann series of (3) can be chosen in an infinite number of ways. A point of the integrand, that is a possible electron trajectory, has to be randomly generated. The estimator is then given by the integrand function in the generated point divided by the probability density of generating that given point. Two possible estimators that correspond to generate the electron trajectory forward and backward in time respectively, are

$$\nu^f = \frac{f_o(Q_0)}{P(Q_0)}\frac{\mathcal{K}(Q_1,Q_0)}{PP(Q_0,Q_1)}\cdots\frac{\mathcal{K}(Q_i,Q_{i-1})}{PP(Q_{i-1},Q_i)}\delta(Q-Q_i)$$

$$\nu^b = \delta(Q - Q_0)\frac{\mathcal{K}(Q_0, Q_1)}{PP(Q_0, Q_1)} \cdots \frac{\mathcal{K}(Q_{i-1}, Q_i)}{PP(Q_{i-1}, Q_i)} f_o(Q_i)$$

Here P and PP are two arbitrary probability densities. P generates the initial point and $PP(Q', Q)$ generates the transition from Q' to Q. The mean value of ν^f or ν^b calculated for each trajectory gives the contribution of the ith term to the f value in Q. If the forward estimator is used with probability equal to the "natural" electron probabilities, the traditional ensemble MC is recovered. The weighted MC technique [7] is based on the forward estimator where the probabilities are used in such a way as to produce rare events more frequently than in nature and reduce the variance on the result of interest[12]

4 Quantum MC for the density matrix

Let us now consider a quantum formulation of an electron transport problem. The physical system under consideration is formed by a single electron and the phonon field with an applied, constant and uniform, electric field $\mathbf{E}$. The total hamiltonian is given by

$$H = H_e + H_p + H_E + H_i; \quad H_e = \frac{p^2}{2m} \quad H_p = \sum_q \hbar\omega_q a_q^\dagger a_q \quad H_E = e\mathbf{E}\cdot\mathbf{r} \tag{4}$$

where H_e is the free electron hamiltonian, H_p is the free phonon hamiltonian ,H_E is the contribution of the applied field and the electron-phonon interaction in terms of phonon creation, and annihilation operators $a_q^\dagger$, a_q is given by

$$H_i = \sum_q i\hbar F(q)\left\{a_q e^{i\mathbf{q}\cdot\mathbf{r}} - a_q^\dagger e^{-i\mathbf{q}\cdot\mathbf{r}}\right\} = H_a + H_e$$

where a refers to absorption and e to emission process.

We are interested in separating an "unperturbed" hamiltonian H_o: $H_o = H_e + H_p + H_E$, with a corresponding set of basis vectors: $|x,t\rangle = |\mathbf{k}_o, t\rangle|\{n_q\}, t\rangle$ where $\langle \mathbf{r}|\mathbf{k}_o, t\rangle = \frac{1}{\sqrt{V}} e^{i\mathbf{k}(t)\cdot\mathbf{r}} e^{-i\int_0^t \omega(t')dt'}$ $\mathbf{k}(t) = \mathbf{k}_o - \frac{e\mathbf{E}}{\hbar}t$, $\omega(t) = \hbar(\mathbf{k}(t))^2/2m$, and $|\{n_q\}, t\rangle$ is the time dependent H_p basis in the phonon number representation. We define the density matrix as $\rho(x, x', t) = \langle x, t|\rho|x, t\rangle$. Here ρ is the density operator whose time evolution is provided by the operator Liouville-von Neumann equation: $i\hbar\frac{\partial}{\partial t}\rho = H\rho - \rho H$.

In the chosen basis, the r.h.s. of the equation contains matrix elements only from $H' = H_i$: $\mathcal{H}'(x, x', t) = \frac{1}{i\hbar}\langle x, t|H'|x', t\rangle = -\mathcal{H}'^*(x', x, t)$ By integration with respect to time we obtain the integral equation for the density matrix:

$$\rho(x, x', t) = \rho(x, x', 0) + \int_0^t dt' \sum_{x''}\left\{\mathcal{H}'(x, x'', t')\rho(x'', x', t') - \rho(x, x'', t')\mathcal{H}'(x'', x', t')\right\} \tag{5}$$

The explicit forms of the absorption and emission components of $\mathcal{H}'(x, x', t)$ are given by:

$$\int d\mathbf{q}F(q)\delta(\mathbf{k} - \mathbf{k}' - \mathbf{q})e^{i\alpha(\mathbf{k}_o,\mathbf{k}'_o,t)}e^{-i\omega_q t}\sqrt{n'_q}\delta(n_q, n'_q - 1)\delta_{\{n_s\},\{n'_s\}}; \quad s \neq q \tag{6a}$$

$$\int d\mathbf{q}F(q)\delta(\mathbf{k} - \mathbf{k}' + \mathbf{q})e^{i\alpha(\mathbf{k}_o,\mathbf{k}'_o,t)}e^{i\omega_q t}\sqrt{n'_q + 1}\delta(n_q, n'_q + 1)\delta_{\{n_s\},\{n'_s\}}; \quad s \neq q \tag{6b}$$

$\alpha(\mathbf{k}_o, \mathbf{k}'_o, t) = (\omega_o - \omega'_o)t + \frac{e}{2m}(\mathbf{k}'_o - \mathbf{k}_o) \cdot \mathbf{E}t^2$. This is the integral equation to be solved with the Neumann expansion and a MC technique.

The whole information for the electronic subsystem is provided by the trace over the phonon coordinates: $\rho^e = \Sigma_{\{n_q\}}\rho(., \{n_q\}, \{n_q\}, .)$ Furthermore we are interested in diagonal elements only for the electron coordinates, since ρ^e is used for the evaluation of electron mean quantities. We assume that the density matrix at the initial time 0 is factorized into an electron part and a phonon part, and that they correspond to diagonal equilibrium distributions for the corresponding free Hamiltonians: $\rho_o(x, x') = Const.e\{-\frac{\epsilon(k_o)}{KT}\} \times P_{eq}(\{n_q\})\delta_{\mathbf{k}_o,\mathbf{k}'_o}\delta_{\{n_q\},\{n'_q\}}$. Since we start from diagonal initial ρ and each application of $\mathcal{H}'$ changes the phonon state by one unit we can conclude that:

(i) only even order terms of the Neumann expansion give contribution

(ii) a mode q which has been absorbed (or emited) by one argument must be absorbed (or emitted) by the other argument, or reemmited (or reabsorbed) by the same argument. Each couple of paired matrix elements (vertices) will be called a process and corresponds to a scattering event.

The numerical procedure starts with random selections with suitable arbitrary probabilities, of the order of the perturbative correction, the sequence of emission and absorbtion processes, the times of the processes in the sequence and the modes q of the involved phonons. Then, starting from the value $\mathbf{k}(t)$ at which ρ will be evaluated, both arguments of the density matrix are translated backwards in time until the time t_1 of the latest vertex. At this point the contribution of the corresponding matrix element $\mathcal{H}'$ is evaluated and the current value of $\mathbf{k}$ is changed accordingly. This step is repeated until all involved processes are considered and the initial values of the arguments of ρ are reached. Finally, according to the general theory, the estimator $\frac{\mathcal{H}'(t_1)...\mathcal{H}'(t_n)\rho(\mathbf{k}_{in},\{n_q\}_{in},0)}{\mathcal{P}}$ is then evaluated, where $\mathcal{P}$ is the product of the single probabilities used above. This estimator will contain a factor n_q or $n_q + 1$ respectively. We also assume that each mode q appears only once in a given perturbation term (no hot phonons), so that the corrections in ρ are linear with respect to n_q and the trace over the phonons results in replacing n_q with its equilibrium value $(exp(\hbar\omega_q/kT - 1))^{-1}$.

In this way we obtain the density matrix trough a random generation of quantum processes in the same way as the tradition ensemble MC obtains the distribution function through a random generation of electron trajectories.

5 Quantum MC for the Wigner function

Another way for quantum description of the hot electron transport is provided by the Wigner function. It is introduced as an extension of the concept of classical distribution function for the quantum case so that some of its properties have a direct probabilistic interpretation. That is why it is convenient to work in terms of Wigner picture when space boundaries are imposed in an inhomogeneous system.

If in the physical system considered above we replace the potential of the uniform electric field with the potential of a biased double barrier structure V_d, thus introducing space inhomogenyty, the total hamiltonian will be: $H = H_p + (H_e + V_d + H_i) = H_p + H'$ We choose as set of basis vectors $|\mathbf{k}\rangle|\{n_q, t\}\rangle$; where now $|\mathbf{k}\rangle$ does not include time dependence. The Wigner function is defined as the Weil-Wigner transform of the density matrix:

$$f_w(\mathbf{k}, \mathbf{r}, \{n_q\}, \{n'_q\}, t) = \int d\mathbf{k}' e^{i\mathbf{k}'.r} \rho(\mathbf{k} + \frac{\mathbf{k}'}{2}, \{n_q\}, \mathbf{k} - \frac{\mathbf{k}'}{2}, \{n'_q\}, t) \qquad (7)$$

By applying the Weil-Wigner transform to the corresponding Liouville-von Neumann equation we obtain the following integro-differential form of the Wigner equation:

$$\left\{\frac{\partial}{\partial t} + \frac{\hbar\mathbf{k}}{m}.\nabla_{\mathbf{r}}\right\} f_w(\mathbf{k}, \mathbf{r}, \{n_q\}, \{n'_q\}, t) =$$

$$\int d\mathbf{g} V_w(\mathbf{g}, \mathbf{r}) f_w(\mathbf{k} + \mathbf{g}, \mathbf{r}\{n_q\}, \{n'_q\}, t)\delta_{\{n_q\},\{n'_q\}} +$$

$$\Sigma_{\{n''_q\}} \int d\mathbf{g} e^{i\mathbf{g}\mathbf{r}} H_i(\mathbf{g}, \{n_q\}, \{n''_q\}, t) f_w(\mathbf{k} - \frac{\mathbf{g}}{2}, \mathbf{r}, \{n''_q\}, \{n'_q\}, t) -$$

$$\Sigma_{\{n''_q\}} \int d\mathbf{g} e^{i\mathbf{g}\mathbf{r}} f_w(\mathbf{k} + \frac{\mathbf{g}}{2}, \mathbf{r}, \{n_q\}, \{n''_q\}, t) H_i(\mathbf{g}, \{n''_q\}, \{n'_q\}, t) \qquad (8)$$

here the second term in the l.h.s. is obtained from the contribution of the free electron hamiltonian H_e, $V_w(\mathbf{g}, \mathbf{r}) = \frac{1}{i\hbar}\int d\mathbf{r}' e^{i\mathbf{g}.\mathbf{r}'} \left(V_d(\mathbf{r} + \frac{\mathbf{r}'}{2}) - (V_d(\mathbf{r} - \frac{\mathbf{r}'}{2})\right)$ gives the contribution of the device potential plus the applied field and $H_i(\mathbf{k} - \mathbf{k}', \{n_q\}, \{n'_q\}, t) = \mathcal{H}_i(\mathbf{k}, \{n_q\}, \mathbf{k}', \{n'_q\}, t)$ are obtained from (6) without the $e^{i\alpha t}$ term because of the different basis set.

The l.h.s. of the equation has the same form of the BE with a force term 0. Hence in a similar way it can be rewritten in an integral form by using the Newton phase trajectories. We suppose that the biased device potential and the interaction are switched on in the initial time t_o.

To the the iteration series of the resulting integral equation a numerical MC algorithm as described in the previous section can be applied. The major difference is that now $\mathbf{r}$ is moved during the "free flights" instead of $\mathbf{q}$. The Neumann expansion can be separated into 3 subseries:

(i) a d-series, due to H_d which gives the electron coherence interaction with the biased device potential . It yields the major quantum effects to transport properties. The series convergency together with the Monte Carlo algorithm for its simulation are discussed in [13].

(ii) i-series, due to H_i, accounts for the electron-phonon scattering events. All assumptions and conclusions done in previous section are valid also here.

(iii) a mixed series, which contains the ICFE and the interference effects of the barrier potential and phonon scattering.

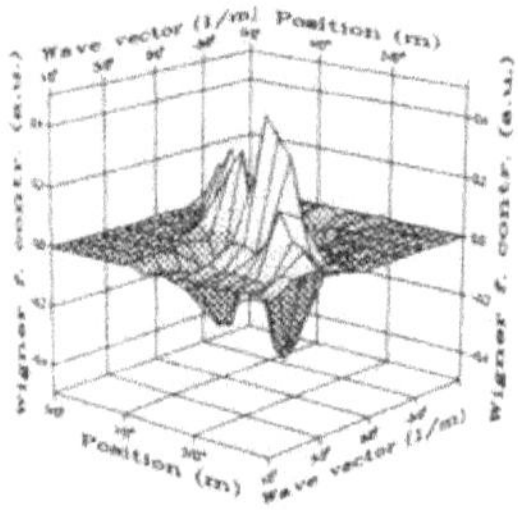

Fig. 1 : Structure, first order

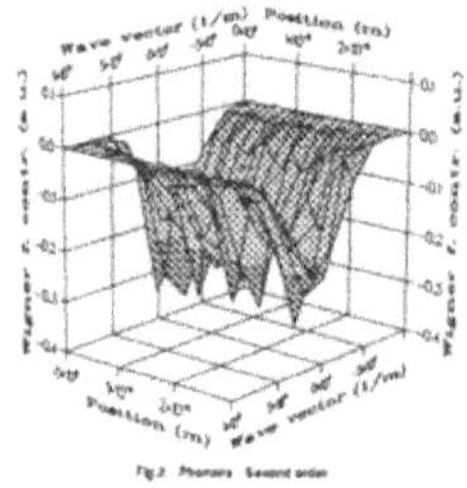

Fig.2 Phonons Second order

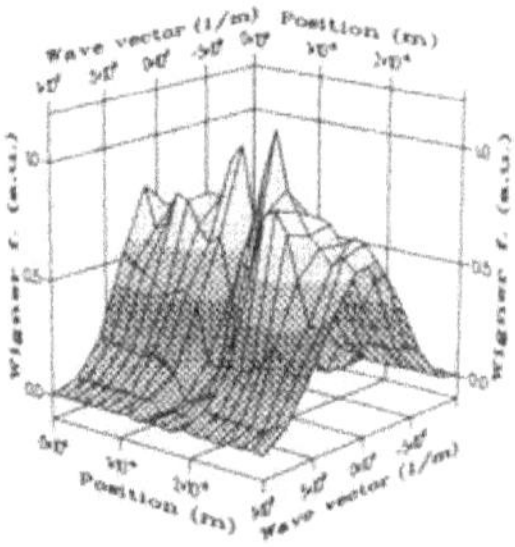

Fig.3 - Wigner f.

As an application we have studied a system with length $3 \times 10^{-8} m$ The potential profile is given by a symmetric double barrier. The distance between the barriers is $2 \times 10^{-9} m$ while the barrier size and height are 1×10^{-9} and $50 meV$ respectively. The applied potential is $10 meV$, and the lattice temperature is $300K$. A simplified semiconductor model characterized by effective mass $0.295 m_0$, density $2.33 \times 10^3 m^{-1}$ and nonpolar-optical phonons with temperature $450K$ and coupling constant $25 \times 10^{10} eVm^{-1}$ is considered. A spatial damping $\alpha = 4 \times 10^{-8}$ is used to improve the convergency [13]. Figs. 1 to 3 show some results obtained with the algorithm described above. Fig.1 shows the first order contribution from the device-structure field; Fig.2 shows the second order contribution due to phonons and Fig.3 the Wigner function as obtained by summing the results of the first two figures to the zero order (equilibrium) Wigner function.

6 Conclusions

The traditional MC technique used for electron transport simulation in semiconductors can be reformulated in terms of a more formal theory for the numerical MC solution of integral equation. In this form the method is also suitable for the solution of the quantum transport problems in mesoscopic structures, that appear to be the basis for future nanoelectronics. The computational effort is very large, and new variance reducing techniques should be devised in order to simulate electron "paths" with more than one or two scattering processes. Since, however, in mesoscopic system electron transport is typically quasi-ballistic, the method can be considered already suitable for practical application.

The authors wish to thank Dr. F. Rossi and Prof. I. Dimov for many useful discussions.

References

[1] C.Jacoboni and L.Reggiani *Rev. Mod. Phys. 55*, 1983, pp 645

[2] T.Kurosawa *J. Phys. Soc. Japan, Suppl. 24*, 1966, pp 424

[3] C.Jacoboni and P.Lugli *The Monte Carlo method for semiconductor device simulation, Springer-Verlag*, 1989

[4] R.Brunetti,C.Jacoboni,A.Matulionis and V.Dienys *Physica 134B*, 1985, pp 369 pp 424

[5] S.Bosi and C.Jacoboni *J. Phys. C*, 1976, pp 315 pp 424

[6] R.Brunetti and C.Jacoboni *in Semiconductors probed by ultrafast laser spectroscopy ed by R.Alfano Academic Press*, 1984

[7] C. Jacoboni, P. Poli and L. Rota, *Solid. St. Electr. 31*, 1988 pp 523

[8] P. Poli, L. Rota and C.Jacoboni *Appl. Phys. Lett. 55*, 1989 pp 1026

[9] L.Rota, C. Jacoboni and P. Poli, *Solid. St. Electr. 32*, 1989 pp 1417

[10] M.Nedjalkov and P.Vitanov *Solid. St. Electr. 32*, 1989 pp 893

[11] P. Vitanov, M. Nedjalkov, C. Jacoboni, F. Rossi and A. Abramo,*Advances in Parallel Algorithms, IOS Press, Amsterdam*, 1994 pp 117

[12] F.Rossi, P.Poli and C.Jacoboni *Semicond. Sci. Technol. 7* 1992, pp 1017

[13] M.Nedjalkov, I.Dimov F.Rossi and C.Jacoboni *submitted for publication*

NUMERICAL MODELING OF STATIONARY TWO-DIMENSIONAL MAGNETOSPHERE

M.D.Kartalev* M.S.Kaschiev** D.K.Koitchev***

*)Institute of Mechanics, BAS, Sofia, Bulgaria
**)Institute of Mathematics, BAS, Sofia, Bulgaria
***)Sofia University "St. Kl.Ohridsky", Sofia, Bulgaria

Abstract

A finite element numerical procedure is developed for two-dimensional modeling of the stationary magnetosphere. The whole magnetic field is supposed to be a sum of given dipole internal field and searched divergent-free and curl-free field of the magnetopause shielding current system. Some numerical implementations not only demonstrate the method abilities but also simulate some typical features of the magnetosphere magnetic field topology.

1 INTRODUCTION

It is well known that the solution and even the formulation of the complete self-consistence theoretical problem of finding the magnetospheric field and plasma topological features are not solved by now. That is why the efforts of studies are concentrated on the development of partially limited approaches, describing well enough different essential features of the problem. The selection of the model in each particular case depends on the pursued goals, existing information, estimation of the needed efforts.

The existing in the literature models of this kind are distinctive by the approaches, accuracy and physical conformity of the boundary conditions, as well as by the choice of the mathematical approaches used in the solution. A common aspect in the numerous mathematical formalisms used in the considerations of this problem in the literature is that the analytical solution is looked for mainly in the form of expansions of special functions: spherical harmonics, Fourier-Bessel integrals, then expanded over spherical harmonics with jointed Legendre polynomials [1], Bessel functions [2], parabolic harmonics [3], and so on.

The goal of this paper is a development of a numerical mathematical procedure for solving the problem for the magnetic field potential in a stationary magnetosphere.

The proposed here procedure is supposed to be more flexible and accurate in the following aspects:

- To avoid restrictions for the given shape of the magnetosphere boundary;
- To minimize restrictions for the given boundary conditions;
- To be universal enough as to permit a description of different astrophysical objects: different kind of magnetospheres as well as cavities of non-magnetic planetary bodies.

The two-dimensional consideration could lead to some quite essential physical conclusions. These could be some global features of the system magnetosphere-solar wind, important details of the magnetic field topology in the midday-midnight magnetosphere cross-section and so on.

2 FORMULATION OF THE PROBLEM

As a simplified geometry of the model magnetosphere, a two-dimensional fixed domain Ω with boundary $\partial\Omega = \gamma \cup \sigma$ (figure 1) is considered. Here Ω is bounded domain in $\mathbb{R}^2$; $\gamma = CALBD$ corresponds to the magnetopause - the border, separating magnetosphere from solar wind flow region (magnetosheath) and $\sigma = CD$ is a section, limiting the domain in which the solution is defined. The numerical approach will not impose any principal restrictions on the domain geometry. Following [2], a geometry, shown on figure 1 is chosen for the numerical implementations in this paper.

An assumption is made that the magnetic field inside Ω is stationary and divergent-free. The whole magnetic field $\mathbf{B}$ is a sum of different constituents:

$$\mathbf{B} = \mathbf{B}_s + \tilde{\mathbf{B}}_d \tag{1}$$

where $\tilde{\mathbf{B}}_d$ is given the Earth-dipole field. The sought part of the field $\mathbf{B}_s$ is "fitted" to this given field by conforming the boundary conditions in a way, which could be specific for each given field. $\mathbf{B}_s$ is supposed to be divergent-free and curl-free and therefore a corresponding harmonic scalar function, field potential U exists:

$$grad\ U = \mathbf{B}_s \tag{2}$$

$$\Delta U = 0 \tag{3}$$

The boundary condition for U on γ is based on the general supposition, that the distribution of the normal component of the whole magnetic field on γ could be given as function $\tilde{\psi}$, defined from additional physical reasons, which are not a subject of discussions in this paper. Referred to the searched solution $\mathbf{B}_s$, the condition turns to:

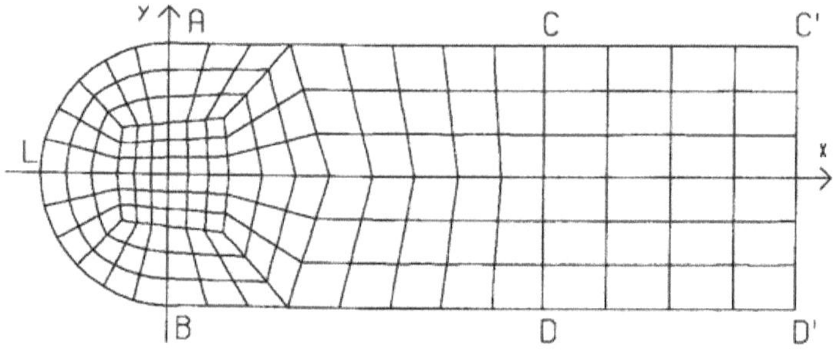

Figure 1: Accepted geometry of the basic computational domain $\Omega = CALBD$. The extended domain $\Omega^* = C'ALBD'$ is used in the realized computational procedures. The (x, y) coordinates of the typical points: $A(0,1)$, $B(0,-1)$, $L(-1,0)$, $C(3,1)$, $D(3,1)$, $C'(4.5,1)$, $D'(4.5,-1)$. Here ALB is semicircle with center $O(0,0)$, BD and AC are sections of a straight lines, CD is perpendicular to them straight section.

$$(\mathbf{B}_s, \mathbf{n}) = \frac{\partial U(x,y)}{\partial n} = \tilde{\psi}(x,y) - \tilde{B}_{d_n}(x,y), \quad (x,y) \in \gamma, \tag{4}$$

where $\mathbf{n}$ is the outward normal unit-vector of γ.

The boundary conditions on the boundary segment σ needs additional consideration and admits different variants. The numerical implementations in this paper use the Dirichlet boundary condition:

$$U(x,y) = \varphi(x,y); \quad (x,y) \in \sigma, \tag{5}$$

which is convenient mathematically and appears to be reasonable physically.

In this way the formulation of the problem for $\mathbf{B}_s$ (or for its potential U) is given by equations (3),(4),(5) and the whole field distribution is given by (1), (2).

3 FINITE ELEMENT DISCRETIZATION

First consider the weak formulation of the problem (3),(4),(5) (Galerkin scheme). Denote the set $H_E^1(\Omega) = \{w \in W_2^1(\Omega) : w|_\sigma = \varphi\}$ and the space $H_{0E}^1(\Omega) = \{w \in W_2^1(\Omega) : w|_\sigma = 0\}$. The weak formulation consists in seeking function

$$U \in H_E^1\ (\Omega), \quad a(v,U) = f(v) \ \text{ for each } v \in H_{0E}^1(\Omega), \tag{6}$$

where:

$$a(w_1, w_2) = \int_\Omega \nabla w_1 \nabla w_2 \, d\Omega, \text{ and } f(w) = \int_\gamma w\psi d\gamma$$

Represent Ω as $\Omega = \cup_{e \in \omega} e$, where e are curvilinear quadrangles, satisfying the conditions for consistency [4]. Here we designate as ω the totality of these quadrangles, which are the accepted here eight-node isoparametric cirendine finite elements [5]. A typical finite element grid is drawn on figure 1.

Denote $S^h(\Omega)$ and $S_0^h(\Omega)$ the finite element projections of $H_E^1(\Omega)$ and $H_{0E}^1(\Omega)$ respectively.

The problem (6) is considered in $S^h(\Omega)$, i.e. we are looking for

$$U^h \in S^h(\Omega), \quad a(v^h, U^h) = f(v^h) \text{ for each } v^h \in S_0^h(\Omega). \tag{7}$$

From this form we obtain a system of linear algebraic equations with semi definite matrix. Gauss three-points quadratures are used for the computation of the integrals over each finite element. The Cholezki expansion algorithm [5] is used in the numerical solving of the linear system, owing to the fact that its matrix is a sparse matrix.

The following error estimates are valid [4]:

$$\|U - U^h\|_{1,\Omega} \leq O(h^2)$$

where U is the solution of (6), U^h is the solution of (7); $h > 0$ is a majorant of the linear scale of the elements from ω.

4 DRAWING FIELD LINES OF THE VECTOR FIELD GIVEN IN FINITE ELEMENT APPROXIMATION

The vector lines of the whole magnetic field **B** (1) are given in parametric form: $x = x(t), y = y(t)$ by the system:

$$\frac{dx}{dt} = \frac{\partial U}{\partial x} + \tilde{B}_x, \quad \frac{dy}{dt} = \frac{\partial U}{\partial y} + \tilde{B}_y \tag{8}$$

where $\tilde{\mathbf{B}}_d^T = (\tilde{B}_x \ \tilde{B}_y)$ is the entire given part of the field.

It is important to note that **B** has singularities in Ω, which is the reason for considering in its subdomain $\Omega_1 \subset \Omega$, such that the field $\tilde{\mathbf{B}}_d(x, y)$ is limited in Ω_1. This leads to a corresponding partition of the Ω_1 to a set ω_1 of finite elements, generally not congruent to ω. A specific task in the considered here problem is the possible necessity of replacement of the finite element coverage ω by another one of the same kind ω_1. In terms of the next section for example, Ω_1 is the domain with circular cut around the singularity of the domain field. Practically the problem of transition into another grid set results in determination of the geometric locus of the given node (x_*, y_*) of ω_1 in ω, i.e. finding the element $e \in \omega$ containing this node, then resolving in (x_*, y_*) the local map on $e \in \omega$ $x = x(\xi, \eta), y = y(\xi, \eta)$ with respect to (ξ, η) by use of the Newton method and then determination from $v^h(x, y) \mid_e = v^h(x(\xi, \eta), y(\xi, \eta))$ the value of v^h in this node of ω_1.

Consider (8) locally over each element $e \in \omega_1$ as soon as the sought function $U^h \in S^h(\Omega)$ is not smooth on the boundary of the elements. Then over e, in variables (ξ, η), the system takes a form:

$$\frac{d\xi}{dt} = F_1^e(\xi, \eta), \quad \frac{d\eta}{dt} = F_2^e(\xi, \eta) \tag{9}$$

This system of differential equations is solved using Runge-Kutta method with error $0(|\tau|^4)$, where τ is a step to t. F_1^e and F_2^e are not defined analytically and their values in (ξ, η) have to be searched on each step from transformation of (8) in (9).

5 NUMERICAL RESULTS

The results of some numerical experiments expounded here, despite of the insufficiency of the two-dimensional approach, demonstrate nevertheless the perspectives of the developed scheme. The possibility is shown for quantitative investigation of some effects of the magnetospheres topology, obtained by qualitative reasons or from other model studings.

Under the presumption that the magnetic field vector tends asymptotically to be parallel to Ox downtail, the following scheme for settling the conditions on CD is adapted here:

The domain $CALBD = \Omega$ is extended to $C'ALBD' = \Omega^*$, obtained by lengthening BD and AC to BD' and AC'; $(D'C' \parallel DC)$ (figure 1). The domain Ω^* is divided into finite elements, maintaining the partition ω of the common domain Ω The finite elements of the $\Omega^* \setminus \Omega$ are constructed to be relatively larger. Than the considered mixed boundary value problem is solved numerically, concretizing the Dirichlet condition on $C'D'$ as $U = 0$ and putting the Neumann conditions for the segments CC' and DD' to be the same as determined in the points C and D respectively.

Having the latter solution, the corresponding distribution $\mathbf{B}_s$ on CD is obtained to be analyzed. Repeating the same procedure, taking larger extensions $CC' = DD'$, an asymptotic behavior of this distribution is found by means of appropriate norm estimation. In all of the considered here examples $CC' = AC$.

The first constituent of the whole field is two-dimensional dipole field $\tilde{\mathbf{B}}_d$ with potential $\tilde{U}_d = -(\mathbf{m}, \mathbf{r})/(2\pi\mathbf{r}^2)$ where $\mathbf{m}$ is the dipole moment and $\mathbf{r}$ is the radius-vector of (x, y).

The numerical examples, shown on figures 2, 3 correspond to the closed magnetosphere model ($\tilde{\psi} = 0$ over γ). The influence of the displacement O_eL of the dipole,

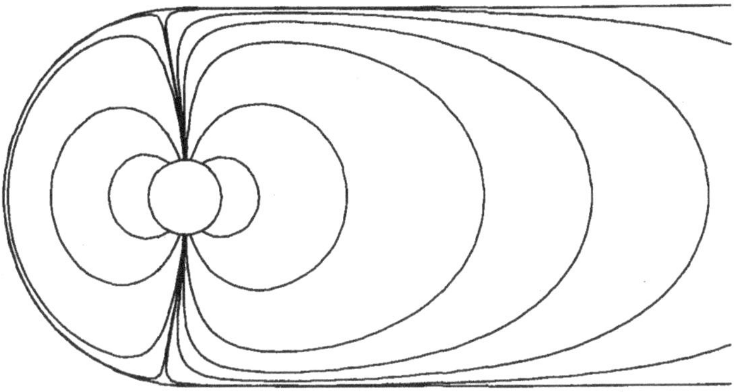

Figure 2: Closed magnetosphere simulation. The dipole center coordinates are $(0, 0)$. The obrained cusp inclination is 82.7^o.

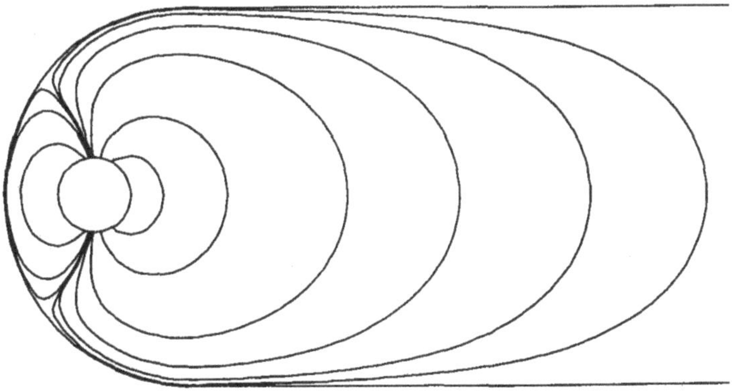

Figure 3: Closed magnetosphere with dipol, centered at the point $(-0.5, 0)$. The cusp inclination is 63.7^o.

centered at point O_e, from the magnetopause subsolar point L on the cusps inclination is demonstrated. In the accepted here geometry this is equivalent to the influence of the choice of the dayside magnetopause geometry. The distance O_eL in the case of figure 2 O_eL is equal to 1, (or $1/2AB$). The obtained cusps inclination is 82.7°. In the case of figure 3, $O_eL = 0.5$. The corresponding cusp inclination is 63.7°.

The next numerical example, reflected on figure 4 is an investigation of the expected influence of the interplanetary magnetic field (IMF), parallel to the solar wind velocity. An open magnetosphere is considered. Omitting discussions, it is quite natural to accept some nonzero normal magnetic field component on the dayside magnetopause only in this case. The following parameters are chosen: the dipole is centered in the point $(-0.2, 0)$, its moment is parallel to Oy and $\tilde{\psi} = 0.3(\mathbf{x}, \mathbf{n})\mathcal{B}_{L_T}$ over γ, where $\mathbf{n}$ is the outward normal unit-vector of γ, $\mathbf{x}$ is the unit vector of Ox and $\mathcal{B}_{L_T}$ is equal to the whole field at the subsolar point L (figure 1) found out in the approach of a vacuum closed magnetosphere. One can see on the figure 4 a strong influence of the x-IMF component on the tail region topology. Remind that this is only an idealized vacuum model.

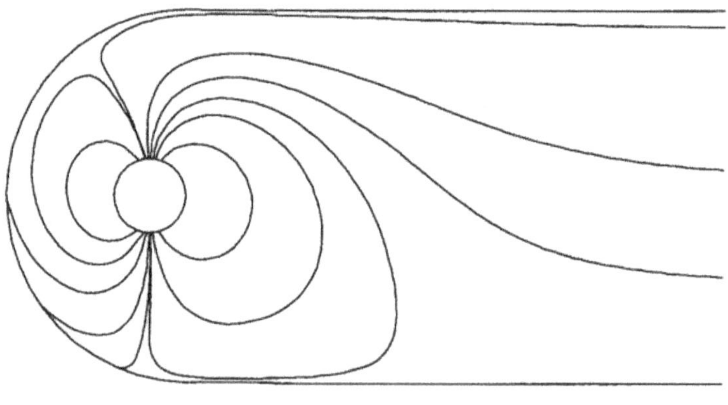

Figure 4: Simulation of the influence of the x-IMF on the vacuum magnetosprere topology.

6 SUMMARY

The simplified physical approach for magnetosphere magnetic field modelling, used in this paper, is well known and widely used in the theoretical magnetosphere studies. It seems, that no direct numerical approaches were used among different mathematical formalisms, solving this problem in the literature. An appropriate finite-elements numerical procedure is developed in this paper for solving the such obtained boundary value problem for the Laplace equation in two-dimensional approach. A procedure for finding the field lines is proposed.

Several numerical implementation in the paper are chosen rather to demonstrate some capabilities of the numerical procedure, than to obtain physical conclusions. Nevertheless, it becomes obvious, that even in this simplified two-dimensional approach, many typical features of the field topology are illuminated. An influence of the magnetopause geometry on the cusps inclination is demonstrated by replacing the dipole position, maintaining the remaining problem geometry. The cusps inclination increases with moving the dipole center toward the subsolar point L. A north-south magnetosphere tail asymmetry, caused by the interplanetary magnetic field X-component is shown in vacuum approach.

A development of the described study is planned in several directions. There are no any principal problems in including additional internal fields like cross-tail currents field for example. The two-dimensional approach here could be considered as a first step in the problem investigation before going to the three-dimensional problem.

Acknowledgements – This research was suppurted by the Bulgarian National Foundation "Scientific Research" under contracts: 428, F-55 and MM 207/92.

References

[1] I.I. Alekseev, and V.P. Shabansky. Planet. Space Sci., **20**, 117(1972).

[2] G.H. Voit. Planet. Space Sci. **29** 1 (1981).

[3] D.P. Stern. J.Geoph.Res. **90** 10851 (1985).

[4] G.Strang, G.J.Fix. An Analysis of the Finite Element Method, Prentice-Hall, Englewood Cliffs. NJ, 1973.

[5] K.J. Bathe, E.L. Wilson. Numerical Methods in Finite Element Analysis, Prentice-Hall, Englewood Cliffs. NJ, 1977.

Blow-up or Equilibria for Nonlinear Reaction Diffusion Processes in Plasmas

Marie-Noelle LE ROUX

Abstract

The purpose of this paper is to study the dynamic behaviour of the temperature profile of a fusion reactor plasma. This temperature is solution of a nonlinear heat equation with a source and sink terms. The solution may blow up in a finite time. We propose a scheme for discretization in time of that problem which has the same property. The convergence of the method is proved. Numerical results are also obtained.

1 Introduction

The purpose of this paper is to study the dynamic behavior of the temperature profile of a fusion reactor plasma. This temperature is solution of a nonlinear heat equation with source terms like alpha particle or ohmic heating and bremsstrahlung losses; the behavior of the solution is depending on the initial distribution and on the mutual weights and powers of the nonlinear reaction and diffusion processes. In particular the solution may blow-up in a finite time. So it is necessary to construct a scheme which has the same property. We shall study in detail the case with one source term, then we shall present the numerical results obtained with one or two source terms and one loss term, in particular, the number of possible equilibria of the solution.

Let Ω be a bounded domain in $\mathbf{R}^d (d \leq 3)$. The temperature u in the plasma is solution of the initial boundary value problem:

$$\begin{cases} u_t - \Delta u^{1+\delta} &= \alpha u^p, \ (x \in \Omega,\ t > 0) \\ u &= 0, \ (x \in \partial\Omega,\ t > 0) \\ u(x,0) &= u_0(x) > 0, \ (x \in \Omega) \end{cases} \tag{1}$$

with $\delta > 0, \alpha \geq 0, p \geq 1 + \delta$.

The problem (1) has the following properties: (Sacks [8]):

In the case $p = 1 + \delta$, if we denote by λ_1 the first eigenvalue of the Dirichlet problem : $-\Delta\rho = \lambda\rho$, $\rho = 0$ on $\partial\Omega$, we have:

ι) If $\alpha < \lambda_1$, the problem (1) has a global solution for $u_0 \in L^q(\Omega)$ $(q > 1)$ and this solution tends to zero as $t \to +\infty$.

ii) If $\alpha = \lambda_1$, the problem has again a global solution and as $t \to +\infty$, the solution of (1) tends to $\theta\rho_1^{1/(1+\delta)}$ for some constant θ depending on u_0 (ρ_1 is the eigenfunction corresponding to λ_1, satisfying $\rho_1 > 0$ and $\| \rho_1 \|_{L^1(\Omega)} = 1$).

iii) If $\alpha > \lambda_1$, there are no nonnegative solutions of (1) which exist for all time except $u = 0$. There exists $T > 0$ such that the problem (1) has a unique solution on $[0, T]$ and $\lim_{t \to T_-} \| u(.,t) \|_{L^\infty(\Omega)} = +\infty$.

In the case $p > 1 + \delta$, the theoretical results are far less important than for $p = 1 + \delta$. A sufficient blow-up condition has been obtained ([8]):

If the following inequality is satisfied:

$$\frac{1}{2}\int_\Omega | \nabla u_0^{1+\delta} |^2 \, dx - \frac{\alpha(1+\delta)}{1+\delta+p}\int_\Omega u_0^{1+\delta+p} dx < 0,$$

there exists $T > 0$ such that the problem (1) has a unique solution on $[0, T]$ and $\lim_{t \to T_-} \| u(.,t) \|_{L^\infty(\Omega)} = +\infty$.

For the sequel, it is more convenient to work with a transformed equation. Let $v = u^{1+\delta}$, $m = \frac{1}{1+\delta}$, $r = pm$, ($m \in]0,1[$, $r \in [1, +\infty]$); then v satisfies

$$\begin{cases} mv^{m-1}v_t + Av = \alpha v^r \ (x \in \Omega, t > 0) \\ v(x,0) = v_0(x) > 0, \ x \in \Omega \end{cases} \tag{2}$$

where A is the operator $-\Delta$, of domain $D(A) = H_0^1(\Omega) \cap H^2(\Omega)$.

In order to obtain a good approximation of the temperature, we have to construct a scheme which has the same property of blowup as the solution of the theoretical problem:

If v_n is the solution at the time level $t_n = n\Delta t$, the solution at the time level t_{n+1} is solution of the problem:

$$\frac{m}{1-m}(v_{n+1}v_n^{m-1} - v_{n+1}^m) + \Delta t A v_{n+1} = \alpha \Delta t v_{n+1} v_n^{r-1} \tag{3}$$

We prove the existence and the uniqueness of the solution of this scheme and study the properties of this solution. Then we prove the stability and we obtain the convergence for a class of initial conditions. Numerical results have been obtained in LeRoux-Wilhelmsson [7] for this scheme and we observe numerically that for any positive initial condition, the solution of the scheme converges to the exact solution.

2 Existence of the numerical solution

In this part, we prove that (3) has a unique positive solution if v_n is positive in Ω, continuous in $\bar{\Omega}$ and that the numerical solution exists during a positive time $T_1 = T_1(v_0)$.

We need several lemmas which have been proved in [5].

Lemma 2.1 *If the function v_n is positive in Ω, continuous in $\bar{\Omega}$ and satisfies the condition:*

$$\| v_n \|_{L^\infty(\Omega)} < \left(\frac{m}{(1-m)\alpha\Delta t} \right)^{1/(r-m)} \tag{4}$$

(3) has a maximal nonnegative solution $\hat{v}$ and if v is a solution , $v \in C^2(\bar{\Omega})$ and it satisfies:

$$0 \leq v \leq \hat{v}.$$

Lemma 2.2 *If (4) is satisfied, (3) has at most one positive solution $v \in C(\bar{\Omega}) \cap H_0^1(\Omega)$.*

It remains to prove that (3) admits one positive solution. This result is proved in the next lemma by using a minimization problem:

Lemma 2.3 *If (4) is satisfied, (3) has a positive solution v in $H_0^1(\Omega) \cap C^2(\bar{\Omega})$.*

The next theorem establish that (4) is satisfied during a positive number of time steps:

Theorem 2.1 *(3) has a unique positive solution v_n for n such that $t_n = n\Delta t < T_1$ where*

$$T_1 = \frac{m}{\alpha(r-m)} \| v_0 \|_{L^\infty(\Omega)}^{m-r} \tag{5}$$

Computation of the positive solution of problem (3). (3) may be written: $Av_{n+1} = f(v_{n+1})$ if f is the function defined by:

$$f(v) = (\alpha v_n^{r-1} - \frac{m}{(1-m)\Delta t} v_n^{m-1})v + \frac{m}{(1-m)\Delta t} v^m \tag{6}$$

If we define the sequence $(v_{n+1,j})$ by:

$$Av_{n+1,j} + \left(\frac{m}{(1-m)\Delta t} - \alpha v_n^{r-m} \right) v_n^{m-1} v_{n+1,j+1} =$$

$$= f(v_{n+1,j}) + \left(\frac{m}{(1-m)\Delta t} - \alpha v_n^{r-m} \right) = \frac{m}{(1-m)\Delta t} v_{n+1,j}^m,$$

we obtain a monotone sequence if the first iterate is a subsolution or a supersolution; it is the choice of $v_{n+1,0}$ which determines wether the sequence is decreasing or increasing . This sequence converges to the solution ([4]).

In order to compute v_1, we start with:

$$v_{1,0} = \frac{\| v_0 \|_{L^\infty(\Omega)}}{\left(1 - \alpha\Delta t \frac{1-m}{m} \| v_0 \|_{L^\infty(\Omega)}^{r-m} \right)^{1/m}}$$

since this constant is a supersolution.

If $n > 1$, a constant supersolution may be very large when t_n is close to T_1, so it is better to use for the initial iterate a subsolution $v_{n+1,0}$ of (3); it is the object of the next lemma to obtain such a subsolution:

Lemma 2.4 *For $n \geq 0$, we have the inequality:*

$$t_{n+1} v_{n+1}^{1-m} \geq t_n \frac{v_n^{1-m}}{1 + \Delta t K_n} \tag{7}$$

and $K_n = \alpha^{\frac{r-1}{m}} \max_{0 \leq j \leq n-1} \| v_j \|_{L^\infty_{(\Omega)}}^{r-m}$, $n \geq 2$; $K_1 = K_0 = 0$.

So it is possible to choose for the first iterate $v_{n+1,0}$ the function:

$$v_{n+1,0} = \left(\frac{t_n}{t_{n+1}} \frac{1}{1+\Delta t K_n} \right)^{1/(1-m)} v_n$$

and we obtain a monotone increasing sequence converging to $v_{n+1,j}$.

Numerically , we observe that if we choose $v_{n+1,0} = v_n$, the sequence $(v_{n+1,j})$ is not monotone, but it converges much more rapidly to the solution.

3 Properties of the numerical solution

Lemma 3.1 *For $n \geq 0$, we have the estimates:*

$$\begin{aligned} \frac{1-m}{m} \Delta t \frac{\int_\Omega (|\nabla v_{n+1}|^2 - \alpha v_{n+1}^2 v_n^{r-1}) dx}{\|v_{n+1}\|^2_{L^{m+1}(\Omega)}} & \leq \| v_{n+1} \|^{m-1}_{L^{m+1}(\Omega)} - \| v_n \|^{m-1}_{L^{m+1}(\Omega)} \\ & \leq \frac{1-m}{m} \Delta t \frac{\int_\Omega (|\nabla v_n|^2 - \alpha v_n^{r+1}) dx}{\|v_n\|^2_{L^{m+1}(\Omega)}} \end{aligned} \tag{8}$$

Lemma 3.2 *If $p = \delta + 1$, the sequence $(F(v_n))$ with*

$$F(v_n) = \frac{\int_\Omega (| \nabla v_n |^2 - \alpha v_n^2) dx}{\| v_n \|^2_{L^{m+1}(\Omega)}} \tag{9}$$

is decreasing.

From these two lemmas, we may deduce the following theorems on the behavior of the numerical solution.

Theorem 3.1 *In the case $p = \delta + 1$ and $\alpha \leq \lambda_1$, there exists a positive constant Δt_0 depending on v_0, p, α, Ω such for $\Delta t < \Delta t_0$, the numerical solution v_n exists for any $n \geq 0$. If $\alpha < \lambda_1$, this solution tends to zero when $n \to \infty$ and if $\alpha = \lambda_1$ this solution tends to $\theta \rho_1$ where θ is depending on the initial condition.*

Theorem 3.2 *In the case $p = \delta + 1$ and $\alpha > \lambda_1$, there exists T^* depending on Δt and on the initial condition v_0 such that the numerical solution v_n exists for $t_n < T^*$ and becomes infinite at T^*.*

Besides, we have:

$$\| v_n \|_{L^{m+1}(\Omega)} \leq \left(\frac{T^*}{T^* - t_n} \right)^{1/(1-m)} \| v_0 \|_{L^{m+1}(\Omega)} \tag{10}$$

If $F(v_0) < 0$, we have the estimate:

$$T^* < \frac{m}{1-m} \frac{\| v_0 \|^{m-1}_{L^{m+1}(\Omega)}}{-F(v_0)} \tag{11}$$

If $p > \delta + 1$, we obtain a sufficient condition for the blow-up of the solution. Let us denote by J^* the functional defined on $H^1_0(\Omega) \cap L^{r+1}(\Omega)$ by:

$$J^*(v) = \frac{1}{2} \int_\Omega | \nabla v |^2 \, dx - \frac{\alpha}{r+1} \int_\Omega v^{r+1} dx. \tag{12}$$

Theorem 3.3 *If $J^*(v_0) < 0$, there exists T^* such that v_n exists for $n\Delta t < T^*$ and becomes infinite at T^*.*

These estimates are the same as those obtained for the exact solution.

4 Convergence of the scheme

In this part we obtain estimates on the numerical solution; so we can extract by compacity a subsequence which converges to a function v. In order to prove that v is a weak solution, we need an hypothesis on the initial condition (this is due to the fact that we have a negative power in the scheme and we may not use a Holder inequality). If the initial condition does not satisfy the hypothesis, we observe numerically that the scheme again coverges and in fact this hypothesis is satisfied after a few number of time steps. Let us define a piecewise linear approximation $u_{\Delta t}$ of u by:

$$u_{\Delta t} = v_n^m + \frac{t - t_n}{\Delta t}(v_{n+1}^m - v_n^m), \; t_n \le t \le t_{n+1} \text{ and we denote } v_{\Delta t} = u_{\Delta t}^{1/m}.$$

We denote by $T_1^* = \inf_{0<\Delta t<\Delta t_0} T^*(\Delta t)$ if T^* is the existence time of the numerical solution and let $T \in [0, T_1^*[; (T_1^* \ge T_1)$.

Theorem 4.1 *The sequence $(v_{\Delta t})$ is uniformly bounded in $C(0,T; H^1_0(\Omega))$ and its derivative in time is uniformly bounded in $L^2(0,T; L^2(\Omega))$.*

Theorem 4.2 *Assume that the initial condition satisfies the hypotesis: there exists a positive constant C_0 such that:*

$$e - 4.1 A v_0 - \alpha v_0^r + C_0 v_0^m \ge 0. \tag{13}$$

There exists T_2 $(0 < T_2 < T_1^*)$ such that the sequence converges when $\Delta t \to 0$ to a function v, weak solution of (2) in $C(0,T; L^q(\Omega))$, ($q < \infty$ if $d = 2, q > 2d/(d-2)$ if $d > 2$) and in $C(0,T;\Omega)$ if $d = 1$).

5 Numerical results for temperature plasma

Generally, in physics plasmas, the temperature u is solution of a heat equation with one or two source terms and one loss term:

$$\begin{cases} u_t - \Delta u^{1+\delta} &= \alpha_1 u^{p_1} + \alpha_2 u^{p_2} - \gamma u^q, \ (x \in \Omega, \ t > 0) \\ u &= 0, \ (x \in \partial\Omega, \ t > 0) \\ u(x,0) &= u_0(x) > 0, \ (x \in \Omega) \end{cases}$$

with $\alpha_1, \alpha_2, \gamma \geq 0$.

To solve this problem, we have generalized the scheme used for (1). The behavior of the solution depends on the mutual weights and powers of the reactions terms and on the initial condition; this solution may blow-up in a finite time, tend to zero or to an equilibrium. Numerical results have been obtained on this equation, concerning in particular the number of equilibria in the case Ω is a ball of radius equal to one in $\mathbf{R}^2$, and u is a function of t and of the distance of x to the origin.

1– If $\alpha_2 = 0$, we have only one source term:

i) if $q < p_1 < 1+\delta$ or $1+\delta < p_1 < q$, there exists a critical value of α_1 (depending on the others parameters) such that if α_1 is larger than this value, there are two possible equilibria, the smallest is unstable and the larger is stable. If the initial condition is too small, the temperature tends to 0.

ii) if $q < 1+\delta \leq p_1$ or $1+\delta < q \leq p_1$ there exists one possible equilibrium which is unstable and the solution may blow-up in a finite time.

2– If $\alpha_2 > 0$ we obtain the following results:

i) if $q < p_2 < p_1 < 1+\delta$ or $1+\delta < p_2 < p_1 < q$, according the values of the coefficients, it is possible to have two equilibria , the first is unstable, the second is stable; the solution never blows up;

ii) if $p_2 < q < p_1 < 1+\delta$, or $p_2 < 1+\delta < p_1 < q$, it is possible to have three equilibria, the first and the third are stables and the second is unstable and there is no blow-up;

iii) if $p_2 < p_1 < q < 1+\delta$ or $p_2 < p_1 < 1+\delta < q$, there is one stable equilibrium;

iv) if $q < p_2 < 1+\delta < p_1$ or $1+\delta < p_2 < q < p_1$, it is possible to have three equilibria, the first and the third are unstable and the second is stable; the solution may blow up in a finite time;

v) if $p_2 < q < 1+\delta < p_1$ or $p_2 < 1+\delta < q < p_1$, there exists two possible equilibria, the first stable and the second unstable; the solution may blow up in a finite time;

vi) if $q < 1+\delta < p_2 < p_1$ or $1+\delta < q < p_2 < p_1$, there exists one possible unstable equilibrium and the solution may blow-up in a finite time.

References

[1] H.Amann *On the existence of positive solutions of nonlinear elliptic boundary value problems*, Indiana.Univ.Math.J. 21, No2(1971), pp.125-146.

[2] A.Friedman, B.Macleod, *Blowup of solutions of nonlinear degenerate parabolic equations*, Arch.Rational.Mech.Anal. 96,(1986),pp.55-80.

[3] V.A.Galaktionov, *A boundary value problem for the nonlinear parabolic equation* $u_t = \Delta u^{\sigma+1} + u^{\beta}$, Differential Equations, 17,(1981), pp.55-80.

[4] H.B.Keller, *Elliptic boundary value problems suggested by nonlinear diffusion process*, Arch.Rational.Mech.Anal. 35 (1969), pp.363-370.

[5] M.N.Le Roux, *Semidiscretization in time of nonlinear parabolic equations with blowup of the solutions*, SIAM J.Numer.Anal., 31, No2(1994).

[6] M.N.Le Roux, *Resolution numerique d'un probleme de reaction-diffusion avec blow-up de la solution* , Publication interne No9303, Univ. Bordeaux I.

[7] M.N.Le Roux, H.Wilhelmsson, *External boundary effects on simultaneous diffusion and reaction processes*, Physica Scripta, Vol.40, pp.674-681.

[8] P.E.Sacks, *Global behavior for a class of nonlinear evolution equations*, SIAM J.Math Anal., Vol.16, No2, (1985).

Address of the Author:
CeReMaB
Universite Bordeaux I
33405 - TALENCE Cedex,
FRANCE

A Bounding Numerical Approach to Seismic Interaction between Adjacent Buildings

Liolios, A. A.

Abstract

The paper deals with a numerical procedure, which provides upper and lower bounds for solution quantities, either global or local, for dynamic inequality problems of structural mechanics, when the time-integration method of Houbolt is used.

1 Introduction

Upper and lower solution bounds can be used either to check the numerical results or to predict an interval of values for a solution quantity in practical engineering problems, which in most cases can be solved numerically only. Methods of deriving such two-sided solution bounds for elliptic linear equality boundary value problems of mathematical physics and especially of elastostatics are already well-known, see e.g. [1,2]. A first generalization of these bounding procedures for elliptic, inequality problems of unilateral elastostatics has been recently obtained in [3].

The aim of the present paper is to extend further these methods to hyperbolic inequality problems of structural elastomechanics using time-discretization. So, a procedure is presented, which provides upper and lower bounds, either global or local, for solution quantities of the general convex, piece-wise linearized problem of unilateral elastodynamics, when this problem is discretized in time by the Houbolt method. Finally, the proposed numerical approach is applied to a civil engineering problem concerning the seismic interaction between adjacent buildings.

2 Problem Conditions

The problem formulation of [3] is followed herein and extended to unilateral elastodynamics. So the problem is discussed for simplicity in matrix operator notation. First the structure is discretized into finite elements of the "natural" type [7] and the unilateral behaviour is considered as piece-wise linearized. Thus the constitutive conditions for the unassembled structure are

$$\mathbf{e} = \varepsilon + \theta + \mu; \quad \sigma = \mathbf{E}\varepsilon \text{ or } \varepsilon = \mathbf{E}^{-1}\sigma, \qquad (1);(2a,b)$$

$$\mu = \mathbf{Bv}, \quad \mathbf{t} = \mathbf{B}^T\sigma - \mathbf{Av} - \mathbf{r}, \quad \mathbf{r} \geq \mathbf{0}, \qquad (3),(4a,b)$$

$$\mathbf{t} \leq \mathbf{0}, \quad \mathbf{v} \geq \mathbf{0}, \quad \mathbf{v}^T\mathbf{t} = \mathbf{0}, \qquad (5a,b,c)$$

where $\mathbf{e}, \varepsilon, \theta, \mu$ are the total, elastic, imposed and unilateral strain vectors, respectively; σ is the stress vector; $\mathbf{t}, \mathbf{v}, \mathbf{r}$ are the stress, deformation and resistence vectors of the unilateral constraints, respectively; $\mathbf{B}$ is the unilateral-transformation matrix; and $\mathbf{E}, \mathbf{A}$ are the "natural" stiffness and unilateral-interaction matrices, respectively, both symmetric and positive definite.

Further, compatibility and equilibrium for the assembled structure are expressed by the following conditions

$$\mathbf{e} = \mathbf{G}^T\mathbf{u}, \qquad (6)$$

$$\mathbf{G}\sigma = \mathbf{p} - \mathbf{M}\ddot{\mathbf{u}} - \mathbf{C}\dot{\mathbf{u}}. \qquad (7)$$

Here $\mathbf{G}$ is the rectangular equilibrium matrix and $\mathbf{G}^T$ its transposed; $\mathbf{M}$ and $\mathbf{C}$ are the symmetric and positive definite mass and damping matrices, respectively; $\mathbf{p}$ is the load vector; and $\mathbf{u}$ is the displacement vector. As usually, dots over symbols denote time derivatives. Finally, the initial conditions are

$$\mathbf{u}(t=0) = \mathbf{u}_0, \quad \dot{\mathbf{u}}(t=0) = \dot{\mathbf{u}}_0. \qquad (8a,b)$$

If $\mathbf{E}, \mathbf{G}, \mathbf{A}, \mathbf{B}, \mathbf{M}$ and $\mathbf{C}$ are known, then the so-formulated problem consists in finding for the time interval $[0,T)$ the time dependent set $\{\mathbf{u}, \mathbf{v}; \sigma, \mathbf{t}; \mathbf{e}, \varepsilon, \mu\}$ that corresponds to the given history $\{\mathbf{p}, \theta\}$ and initial $\{\mathbf{u}_0, \dot{\mathbf{u}}_0\}$ sets and satisfies (1)-(8). Solution existence, uniqueness and regularity questions for this hyperbolic problem can be treated by the variational inequality approach [5]. Here we assume that a unique solution exists, for which two-sided solution bounds are required.

3 Houbolt Time-discretization

Next we discretize the problem (1)-(8) in time by the step-by-step Houbolt method [4]. So, supposing that the solution has been found at the time $t - \Delta t$, where Δt is the time-step, and substituting into (7)

$$\ddot{\mathbf{u}}(t) = \frac{1}{(\Delta t)^2}[2\mathbf{u}(t) - 5\mathbf{u}(t-\Delta t) + 4\mathbf{u}(t-2\Delta t) - \mathbf{u}(t-3\Delta t)], \qquad (9a)$$

$$\dot{\mathbf{u}}(t) = \frac{1}{6\Delta t}[11\mathbf{u}(t) - 18\mathbf{u}(t-\Delta t) + 9\mathbf{u}(t-2\Delta t) - 2\mathbf{u}(t-3\Delta t)], \qquad (9b)$$

we obtain for the time t the equilibrium condition

$$\mathbf{G}\sigma = \mathbf{q} - \mathbf{D}\mathbf{u}, \qquad (10a)$$

where

$$\mathbf{D} = \frac{2}{(\Delta t)^2}\mathbf{M} + \frac{11}{(6\Delta t)}\mathbf{C} \qquad (10b)$$

and

$$\begin{aligned}\mathbf{q} = \; & [\frac{5}{\Delta t^2}\mathbf{M} + \frac{3}{\Delta t}\mathbf{C}]\mathbf{u}(t-\Delta t) - [\frac{4}{\Delta t^2}\mathbf{M} + \frac{3}{2\Delta t}\mathbf{C}]\mathbf{u}(t-2\Delta t) + \\ & [\frac{1}{\Delta t^2}\mathbf{M} + \frac{1}{3\Delta t}\mathbf{C}]\mathbf{u}(t-3\Delta t) + \mathbf{p}.\end{aligned} \qquad (10c)$$

Thus, the problem conditions at the time t are the relations (1)-(6) and (10a), where the latter has replaced (7). For every time-step these relations constitute now an elliptic inequality problem and consequentely the procedures proposed in [3] and [6] can be applied, as it will be shown in the following sections.

4 The Hilbert Space of the Structure states and Some Basic inequalities

The ordered set $\{\sigma, \mathbf{t}; \mathbf{u}, \mathbf{v}\}$ is said to be the state $\mathbf{S}$ of the structure at the time t. From relations (1)-(6) and (10) it follows that such a structure state can be fully determined by either the pair $\{\sigma, \mathbf{t}\}$ or the pair $\{\mathbf{u}, \mathbf{v}\}$. Further, with each state $\mathbf{S}$ we associate an element of a real Hilbert space $\mathcal{H}$. For this space we define as scalar product of two states $\mathbf{S}_1$ and $\mathbf{S}_2$ the bilinear form

$$\mathbf{S}_1 \cdot \mathbf{S}_2 = \varepsilon_1^T \mathbf{E} \varepsilon_2 + \mathbf{v}_1^T \mathbf{A} \mathbf{v}_2 + \mathbf{u}_1^T \mathbf{D} \mathbf{u}_2. \qquad (11)$$

This symmetric form is positive definite, because the matrices $\mathbf{E}, \mathbf{A}$ and $\mathbf{D}$ are so. According to definition (11), the squared norm $||\mathbf{S}||^2$ is the double of the generalized energy of the structure at the time t.

Now we introduce two sets of admissible states. A state $\mathbf{S}_\sigma$ determined by the stress pair $\{\sigma_\sigma, \mathbf{t}_\sigma\}$ is said to be a stress admissible state, iff this state satisfies (2)-(4), (5a) and (10). On the other hand, we define the kinematically admissible states $\mathbf{S}_k$ as those which are determined by the pairs $\{\mathbf{u}_k, \mathbf{v}_k\}$ and satisfy (1)-(4), (5b) and (6). Taking into account these definitions of $\mathbf{S}_\sigma$ and $\mathbf{S}_k$ and the rels. (1)-(6),(10) and (11), we have for the solution state $\mathbf{S}_0$ and any pair $\{\mathbf{S}_\sigma, \mathbf{S}_k\}$

$$\mathbf{v}_0^T\mathbf{t}_\sigma \leq 0, \quad \mathbf{v}_k^T\mathbf{t}_0 \leq 0, \quad \alpha = -\mathbf{v}_k^T\mathbf{t}_\sigma \geq 0 \tag{12a, b, c}$$

$$(\mathbf{S}_0 - \mathbf{S}_\sigma)\cdot(\mathbf{S}_0 - \mathbf{S}_k) = +\mathbf{v}_k^T\mathbf{t}_0 + \mathbf{v}_0^T\mathbf{t}_\sigma - \mathbf{v}_k^T\mathbf{t}_\sigma. \tag{13}$$

Due to (12), equality (13) is transformed to the following basic inequalities:

$$(\mathbf{S}_0 - \mathbf{S}_\sigma)\cdot(\mathbf{S}_0 - \mathbf{S}_k) \leq \alpha \tag{14}$$

$$(\mathbf{S}_0 - \mathbf{S}_\sigma)^2 + (\mathbf{S}_0 - \mathbf{S}_k)^2 \leq -2\mathbf{v}_k^T\mathbf{t}_\sigma + (\mathbf{S}_\sigma - \mathbf{S}_k)^2 + 2\mathbf{v}_0^T\mathbf{t}_\sigma, \tag{15}$$

$$(\mathbf{S}_0 - \mathbf{S}_\sigma)^2 + (\mathbf{S}_0 - \mathbf{S}_k)^2 \leq -2\mathbf{v}_k^T\mathbf{t}_\sigma + (\mathbf{S}_\sigma - \mathbf{S}_k)^2 + 2\mathbf{v}_0^T\mathbf{t}_0, \tag{16}$$

5 Global Two-sided Solution Bounds

Applying now the procedures of [3] and [6], we derive from inequalities (14)-(16) the following upper and lower global bounds for the solution state $\mathbf{S}_0$:

$$[\mathbf{S}_0 - \mathbf{C}_0]^2 \leq R_0^2; \quad (||\mathbf{C}_0|| - R_0) \leq ||\mathbf{S}_0|| \leq (||\mathbf{C}_0|| + R_0), \tag{17a, b}$$

$$-\Pi_c(\mathbf{S}_\sigma) \leq -\Pi_c(\mathbf{S}_0) = \Pi_p(\mathbf{S}_0) \leq \Pi_p(\mathbf{S}_k), \tag{18}$$

where

$$\mathbf{C}_0 = \frac{1}{2}(\mathbf{S}_\sigma + \mathbf{S}_k); \quad R_0^2 = \alpha + \frac{||\mathbf{S}_\sigma - \mathbf{S}_k||^2}{4} = [a + \Pi_p(\mathbf{S}_k) + \Pi_c(\mathbf{S}_\sigma)]/2, \tag{19a, b}$$

$$\Pi_p(\mathbf{S}) = \Pi_p(u, v) = \frac{1}{2}[\varepsilon^T\mathbf{E}\varepsilon + \mathbf{v}^T\mathbf{A}\mathbf{v} + \mathbf{u}^T\mathbf{D}\mathbf{u}] - \mathbf{q}^T\mathbf{u} + \mathbf{r}^T\mathbf{v}, \tag{20}$$

$$\Pi_c(\mathbf{S}) = \Pi_c(\sigma, \mathbf{t}) = \frac{1}{2}[\sigma^T\mathbf{E}^{-1}\sigma + \mathbf{v}^T\mathbf{A}\mathbf{v} + \mathbf{u}^T\mathbf{D}\mathbf{u}] + \sigma^T\theta. \tag{21}$$

6 Local Two-sided Solution Bounds

Further, to obtain local bounds for a solution quantity $F(\xi)$ at the point ξ of the structure, we introduce [3] a suitable Green state $\check{\mathbf{S}}$ with corresponding stress, $\check{\mathbf{S}}_\sigma$, and kinematically, $\check{\mathbf{S}}_k$, admissible states. So we obtain finally

$$|F(\xi) - F_\alpha(\xi)| \leq R_0\check{D}, \tag{22a}$$

where

$$F_\alpha(\xi) = (\mathbf{q}^T\check{\mathbf{u}}_k - \check{\sigma}_\sigma^T\theta) + \frac{1}{2}(\mathbf{S}_\sigma + \mathbf{S}_k)\cdot(\check{\mathbf{S}}_\sigma - \check{\mathbf{S}}_k), \tag{22b}$$

$$\check{D} = ||\check{\mathbf{S}}_\sigma - \check{\mathbf{S}}_k||. \tag{22c}$$

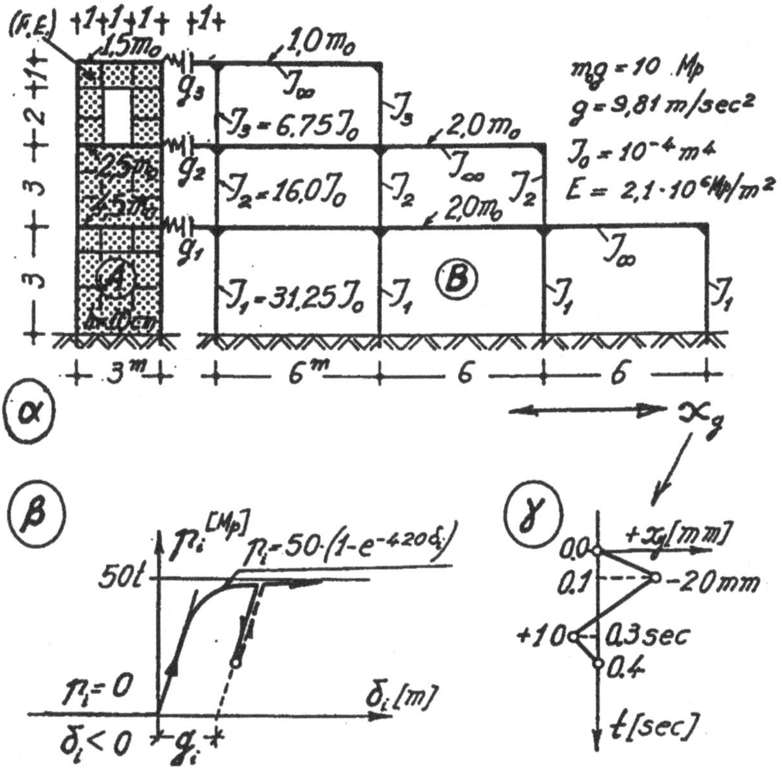

Figure 1: The Numerical Example

7 Numerical Realization of the Bounds

Two-sided bounds in terms of the admissible states $\mathbf{S}_\sigma$, $\mathbf{S}_k$, $\check{\mathbf{S}}_\sigma$ and $\check{\mathbf{S}}_k$ are provided by (17), (18) and (22) for the solution state $\mathbf{S}_0$ at time t. Improvements of these bounds can be obtained by minimizing the bounds-gaps R_0 and $\check{D}$. This minimization leads to quadratic programming problems. Moreover, (18) expresses two dual minimum principles, the first one of the generalized potential energy (20) and the second of the generalized complementary energy (21). Thus, the required admissible states $\mathbf{S}_\sigma$ and $\check{\mathbf{S}}_\sigma$ (resp. $\mathbf{S}_k$ and $\check{\mathbf{S}}_k$) can be numerically constructed as solutions of convex optimization problems by using stress diffusive (resp. displacement conforming) finite element models [8]. Finally, the bounding procedure presented herein can be applied in connection with other time-discretization schemes to various hyperbolic problems of structural elastomechanics.

8 Numerical Example

The developed procedure is now applied to an earthquake engineering unilateral problem. This problem concerns the seismic interaction between the two adjacent structures (A) and (B) of reinforced concrete shown in Fig. 1a.

The elastoplastic behaviour $(p-\delta)$ of the three unilateral constraints and the given seismic ground displacement $x_g(t)$ are depicted in Fig. 1b and 1c, respectively. The shear wall (A) is discretized into 25 quadratic finite elements of the compatible displacemenmt (resp. diffusive equilibrium) type [8] for the kinematically (resp. stress) admissible states. Due to elastoplastic unilateral contact behaviour allowing for unloading-reloading processes, the bounding procedure is applied in incremental form. Various aspects concerning the seismic interaction between adjacent buildings can be found in [9,10].

From the results obtained by applying the herein procedure are shown indicatively in Fig. 2 those concerning the upper and lower bounds for the unilateral contact forces $r_i(i=1,2,3)$ in the time interval $(0.3-0.4)$ sec. The range of the bounds gaps is about $8\%-26\%$ in respect to lower bounds. These relatively poor bounds can be improved by increasing the number of the finite elements and consequentely the calculation time.

9 Concluding Remarks

Upper and lower solution bounds for the convex unilateral problem in earthquake engineering have been obtained for the case where the Houbolt method for time-integration is used. For the derivation of these bounds, some functional analysis concepts have been used and interpreted from the structural mechanics point of view. The numerical realization of the bounds requires the use of stress and kinematically admissible states. These are constructed numerically by the finite element method and quadratic programming. Thus, by using available computer programms the herein presented procedure can be used either to control numerical results or to predict an interval of values for a design quantity in practical engineering problems.

References

[1] Synge, J. L., *The Hypercircle in Mathematical Physics*, Cambridge University Press, 1957.

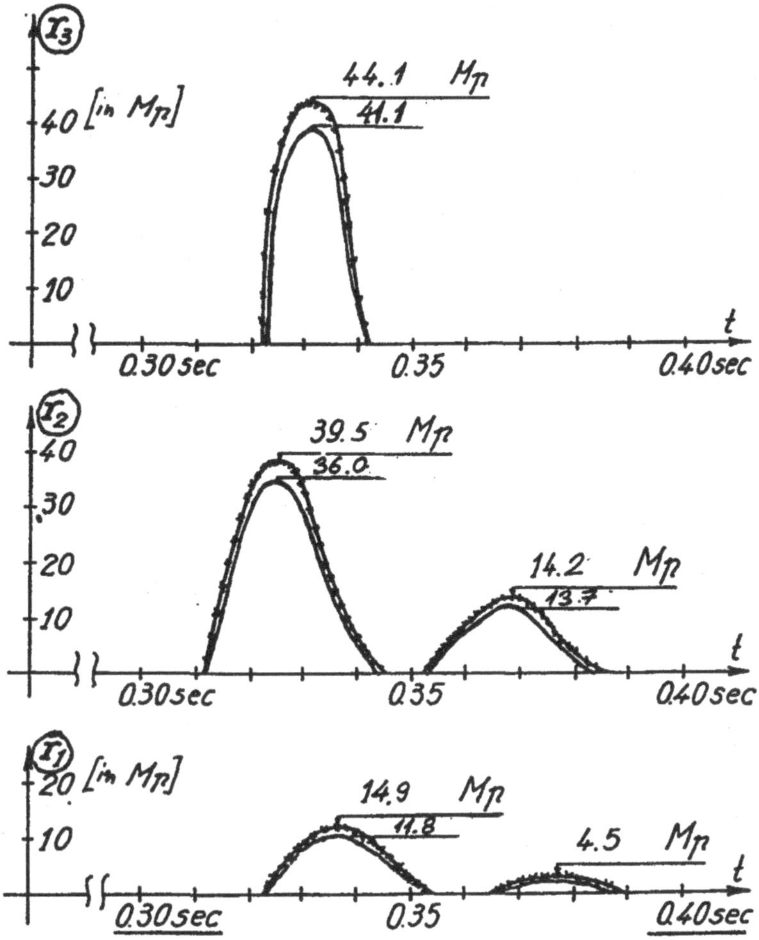

Figure 2: Bounds for the contact forces

[2] Stumpf, H., *Eingrenzungsverfahren in der Elastomechanik*, Westdeutscher Verlag, Köln, 1970.

[3] Liolios, A. A., *Upper and Lower Bounds in Unilateral and Nonlinear Structural Elastostatics*, Jnl Appl. Math. Mech. (ZAMM), **62** (1982), T138-T139.

[4] Bathe, K.-J., *Finite Element Procedures in Engineering Analysis*, Prentice-Hall, Inc., New Jersey, 1982.

[5] P. D. Panagiotopoulos, *Inequality Problems in Mechanics and Applications. Convex and Nonconvex Energy Functions*, Birkhäuser Verlag, Boston,Basel 1985 (Russian translation: MIR Publ. Moscow 1989)

[6] Liolios, A. A., *A Direct Formulation of Dual Extremum Principles in Unilateral Elastostatics by the Generalization of the Hypercircle Method*, Jnl Appl. Math. Mech. (ZAMM), **65** (1985), T348-T350.

[7] Maier, G., *Incremental Elastoplastic Analysis in the Presence of Large Displacements and Physical Instabilizing Effects*, Int. Jnl Solids and Structures **7** (1971), 345-372.

[8] Fraeijs de Veubeke, B., *Displacement and Equilibrium Models in the Finite Element Method*, pp. 145–197, in: Zienkiewicz, O. C., and Holister, G. S., (Eds.), *Stress Analysis*, Wiley, 1985.

[9] Liolios, A. A., *A Finite-Element Central-Difference Approach to the Dynamic Problem of Nonconvex Unilateral Contact between Structures*, pp. 394-401 in: B. Sendov, R. Lazarov and P. Vasilevski (Eds.), *Numerical Methods and Applications*, Proc. 1st Int. Conf., BAS, Sofia, 1984.

[10] Liolios, A. A., *A linear complementarity approach to the nonconvex dynamic problem of unilateral contact with friction between adjacent structures*, Jnl Appl. Math. Mech. (ZAMM), **69** (1989), T420-T422.

Democritus-University of Thrace
Dept. of Civil Engineering
GR–67100 Xanthi
Greece

Numerical Simulation of Wind-Driven Water Flows

D.T. Marinov S.P. Miladinova K.B. Radev

Abstract

The work deals with modelling of vertical structure of wind-driven flows. The model includes turbulent boundary layer approximation, the air-sea interaction and the fluxes caused by the slopping bottom. The finite difference algorithm for calculation of tide in estuaries and flows in shelf regions is used. The model is evaluated against experimental data for wind-driven flow in a channel and with the improved model an example of circulation in Bulgarian shelf is demonstrated.

1 Introduction

The flow patterns in coastal marine zones are caused by combined influences of many and different factors. However, the main impact for nearshore regions without strong tide flows is the wind forcing. Most of the existing numerical models of coastal and shelf seas [2, 5, 6] deal with general circulation in larger areas. In spite of including many forcing factors these models can not describe well the vertical structure of the sea flows in nearshore regions.

In the present work a two-dimensional mathematical model for the vertical structure of the water circulation is developed. The model involves the turbulent boundary layer approximation, which takes into account the fluxes across the air-sea interface described by wind surface stress and fluxes caused by the shelf bottom slope and roughness. The calculations are based on finite difference algorithm for computations of tide and surge in estuaries and the flows in elongated channels and shallow continental shelf regions [3, 11]. First, the model is tested against the experimental data for wind-driven flow in channel with uniform depth and with slope bottom. In this case the influence of different model parameters is investigated. Further, an example of the application of proposed model for simulation of water circulation in Bulgarian Black Sea shelf is shown.

2 Physical and Mathematical Model

For describing the coastal flow field, a right handed Cartesian coordinate system (x, y) is used with the x-axis aligned perpendicular to the shore and positive onshore and the y-axis vertical. The origin is far from the coast at the mean water level. The corresponding mean velocity components are u, v and u', v' are fluctuation ones.

The development of the basic hydrodynamical model is based on several assumptions: *i)* the wind driven transport is either onshore or offshore, so that, the sea motion can be regarded as two-dimensional, i.e. as having negligible gradients in velocity in the alongshore direction (further the x-axis is called longitudinal); *ii)* the maximum shelf depth h_m is rather small than the shelf width L; *iii)* the water level is variable and the sea free surface shape can be changed with time and onshore distance $f = f(t, x)$; *iv)* the fluid is incompressible and homogeneous i.e. the fluid motion is independent of the density stratification; *v)* the bottom shape h is changed only in longitudinal direction, i.e. $h = h(x)$; *vi)* the role of waves in the momentum transfer process from air to water is not considered.

The equations describing the mean flow in domain $\Omega = \{0 \le x \le L, -h \le y \le f, t \ge 0\}$ are based on the hydrostatic approximation and can be written as [11]:

$$\frac{\partial u}{\partial t} + u\frac{\partial u}{\partial x} + v\frac{\partial u}{\partial y} = -g\frac{\partial f}{\partial x} + \frac{\partial}{\partial y}[(\nu + \nu_t)\frac{\partial u}{\partial y}] + K_x\frac{\partial^2 u}{\partial x^2} \tag{1}$$

$$\frac{\partial u}{\partial x} + \frac{\partial v}{\partial y} = 0, \qquad \frac{\partial f}{\partial t} + u|_{y=f}\frac{\partial f}{\partial x} = v|_{y=f} \tag{2}$$

where g is gravity acceleration and ν is molecular kinematic viscosity. The two last terms in the first equation include the Reynolds stresses: $-\overline{u'v'} = \nu_t(\partial u/\partial y)$ and $-\overline{u'^2} = K_x(\partial u/\partial x)$ where the coefficients ν_t and K_x characterize the vertical and longitudinal turbulent transport, respectively. For closing the system it is necessary to know ν_t and K_x. The most important role for turbulent energy growth is played by the vertical gradient of velocity, therefore, one has to be cautions when choosing the appropriate turbulent model determined ν_t. In this study the classic Prandtl's hypothesis is used:

$$\nu_t = l^2\left|\frac{\partial u}{\partial y}\right|, \qquad l = \frac{\kappa}{h+f}(h + y + y_b)(f - y + y_s) \tag{3}$$

where l is the Prandtl's mixing length determining here by the Monthgomery relation (see in [11]). It takes into account the influence of the boundary layers - near the bottom and near the air-sea interface. In (3) $\kappa \simeq 0.4$ is the Von Karman constant; y_b and y_s are the roughness' lengths of the bottom and the sea surface, respectively.

For wind-driven flows the longitudinal turbulent transport is an important factor. The coefficient of horizontal turbulent diffusion K_x is assumed constant as usually in oceanology but it can be a function of the characteristic length scale L [7, 11].

The zero initial conditions for velocities and free surface lavel are used. The boundary condition for the mean velocity on the coast is $u = 0$ but at the open sea boundary u tends to a fixed value. At the solid bottom no slip boundary conditions for velocity components are supposed, while at the air-sea boundary the shear stress balance is assumped. The stress is represented by the friction velocity u_*, which is usually given by the relation: $u_*^2 = c_h U^2$ [4, 11] where c_h is a friction coefficient and U is the wind velocity at $10m$ above the mean water level. The friction coefficient is an empirical constant and its value will be described below.

The more important element for the sea turbulent flow is the choice of the length and velocity scales. In the present study the characteristic length in vertical and longitudinal direction are respectively the maximum sea depth h_m and the shelf width L. The wind velocity U is chosen as a characteristic velocity. In accordance with these scales the nondimensional parameters of the problem are:

$$\epsilon = \frac{h_m}{L} \quad K = \frac{K_x}{\nu} \quad \lambda = \frac{\rho_{water}}{\rho_{air}} \quad Re = \frac{Uh_m}{\nu} \quad Fr = \frac{U^2}{gL} \tag{4}$$

where Re is the Reynolds number and Fr is the Froude number.

For the purpose of transforming the area Ω in a rectangular one, which provides the possibility of using a constant number of grid points in vertical direction, the following transformation is made: $\tau = t, \xi = x, \eta = (y+h)/(h+f)$, introduced into meteorology by *Phillips* (see in [3]). In new area $\Omega_* = \{0 \le \xi \le 1, 0 \le \eta \le 1, \tau \ge 0\}$, the governing equations involving only the main part of the longitudinal diffusion, the initial conditions and the boundary conditions are, respectively:

$$\frac{\partial u}{\partial \tau} + u\frac{\partial u}{\partial \xi} + V\frac{\partial u}{\partial \eta} = -\frac{1}{Fr}\frac{\partial f}{\partial \xi} + \frac{1}{\epsilon Re}\frac{1}{H^2}\frac{\partial}{\partial \eta}[(1+\frac{\nu_t}{\nu})\frac{\partial u}{\partial \eta}] + \frac{\epsilon K}{Re}\frac{\partial^2 u}{\partial \xi^2} \tag{5}$$

$$\frac{\partial f}{\partial \tau} = -[H\int_0^1 u d\eta] \tag{6}$$

$$H = h + f, \quad \frac{\nu_t}{\nu} = \kappa^2 Re H(1 - \eta + \frac{y_s}{H})^2(\eta + \frac{y_b}{H})^2\left|\frac{\partial u}{\partial \eta}\right|$$

$$V(\xi,\eta) = \frac{1}{H}[\eta\int_0^1 \frac{\partial (Hu)}{\partial \xi}d\eta - \int_0^\eta \frac{\partial (Hu)}{\partial \xi}d\eta].$$

The initial conditions at $\tau = 0$

$$u = v = f = 0 \quad \text{at} \quad 0 \le \xi \le 1, \ 0 \le \eta \le 1. \tag{7}$$

The boundary conditions for $\tau > 0$

$$\frac{\partial u}{\partial \xi} = 0 \ \text{ at } \ \xi = 0, \ \ 0 \le \eta \le 1 \quad u = 0 \ \text{ at } \ \xi = 1, 0 \le \eta \le 1$$

$$u = v = 0 \quad \text{at} \quad \eta = 0, \ \ 0 \le \xi \le 1 \tag{8}$$

$$\left[\frac{1}{H} + \kappa^2 Re(1+\frac{y_b}{H})^2(\frac{y_s}{H})^2\left|\frac{\partial u}{\partial \eta}\right|\right]\frac{\partial u}{\partial \eta} = \lambda c_h Re \quad \text{at } \eta = 1, 0 \le \xi \le 1.$$

The vertical velocity V is found from the continuity equation. The numerical technique for solving the system (5)-(8) is given further.

3 Method of numerical solution

The described system (5)-(8) will be solved on a finite-difference grid in domain Ω_*. An uniform mesh in ξ-direction is defined with a step $\Delta\xi = L/(N-1)$, N is a number of points in longitudinal direction. For a good simulation of the both boundary layers, it is important that the finite-difference analogue of eq. (5) be correctly formulated and that the mesh size be smaller near the free surface at $\eta = 0$ and near the sea bottom at $\eta = 1$. So, in η-direction an irregular mesh with M points is defined:

$$\eta_m = \overline{\eta_m} = 2\left[\frac{m-1}{M-1}\right]^2, \quad m = 1, 2, \ldots, \left[\frac{M+1}{2}\right];$$

$$\eta_m = 1 - \overline{\eta_{M+1-m}}, \quad m = \left[\frac{m+1}{2}\right] + 1, \ldots, M.$$

It provides the better accuracy of solution in the areas close to the boundaries. A time-step $\Delta\tau$ is constant and its magnitude is determined in accordance to the condition for the scheme stability [8].

The discretization of eq. (5) is based on the method MAC [8] for time- dependent viscous incompressible flow with free surface. In [3] this method is developed for numerical computations of sea flows. Here a version of the method is used [11]. The grid-point values of dependent variables are denoted by a conventional subscript and for time-level is used a superscript. Therefore, the finite-difference analogue of eq. (5) has the form:

$$\begin{aligned}\frac{u_{n,m}^{k+1,j+1} - u_{n,m}^{k}}{\Delta\tau} &= R_{n,m}^{k+1,j} + \frac{1}{\epsilon Re}\frac{2}{H_n(\eta_{m+1} - \eta_{m-1})}\times \\ &\left[c_{n,m+1/2}^{k+1,j}\frac{(u_{n,m+1}^{k+1,j+1} - u_{n,m}^{k+1,j+1})}{\eta_{m+1} - \eta_m} - c_{n,m-1/2}^{k+1,j}\frac{(u_{n,m}^{k+1,j+1} - u_{n,m-1}^{k+1,j+1})}{\eta_m - \eta_{m-1}}\right]\end{aligned} \tag{9}$$

for $n = 3, 5, \ldots, N-2$ (N is an odd number); $m = 2, \ldots, M-1; k = 1, 2, \ldots, j = 1, 2, \ldots$, where $c = (1 + \nu_t/\nu)/H$. The term R incorporates the effects of vertical and horizontal advection and horizontal diffusion. All terms in $R_{n,m}^{k+1,j}$ are approximated with formal second order accuracy in ξ and η directions. The vertical velocity V is taken at the lower time-level. The value of the coefficient $c_{n,m+1/2}^{k+1,j}$ (resp. $c_{n,m-1/2}^{k+1,j}$) is computed by its average values at both adjacent η-points.

The following finite-difference analogue of the boundary conditions are used: $u_{1,m}^{k+1} = u_{3,m}^{k+1}$, at $\xi = 0$; $\quad u_{N,m}^{k+1} = 0$, at $\xi = 1$ where: $m = 1, \ldots, M; k = 1, 2, \ldots$. The boundary conditions at $\eta = 0$ and $\eta = 1$ add the system of algebraic equations with equation for $m = 1$: $u_{n,1}^{k+1,j+1} = 0$ and for $m = M$:

$$\left[1 + \frac{\nu_t}{\nu}\right]_{n,M-\frac{1}{2}}^{k+1,j} \frac{\left(u_{n,M}^{k+1,j+1} - u_{n,M-1}^{k+1,j+1}\right)}{\eta_M - \eta_{M-1}} = H\lambda c_h Re.$$

There are used two kinds of superscribes: k is related to time-level and j is related to new additional iterations in the borders of one time step. The additional iterations are needed because of the nonlinear terms in eq. (13) describing the vertical turbulent diffusion, horizontal and vertical advection and the horizontal diffusion term. If the variable $u_{n,m}^k$ is known the calculations for the new time level start with $u_{n,m}^{k+1,1} = u_{n,m}^{k+1}$. The equation for free surface level is approximated with second order accuracy and it is included in the iteration procedure. The free surface is calculated for even values of index n while the momentun equation - for uneven ones. The solution of the full algebraic system is based on Gauss elimination method with the main element choice [10]. The system is solved while the following condition

$$max \left| \frac{(u^{j+1} - u^j)}{u^j} \right|_{n,m}^{k+1} \leq \sigma \quad (\sigma = 0.01) \tag{10}$$

is satisfied. Than the obtained solution will be the next: $u_{n,m}^{k+1} = u_{n,m}^{k+1,j+1}$; $n = 1, \ldots, N$; $m = 1, \ldots, M$.

4 Results and Discussion

In [1] an experimental investigation of the wind-driven water currents is described. A series of experiments are conducted in which the stationary mean flow and the turbulence properties are measured in a channel of uniform depth and in a channel with a slopping bottom. The experimental channel has a $0.9m$ by $0.9m$ cross section and is $9m$ long. The calculations based on the present numerical technique are compared with their experimental data for $Re = 1.19 \times 10^6$, $Re = 1.19 \times 10^6$ and corresponding $Fr = 5.9$, $Fr = 12$. The coefficient of horizontal turbulent diffusion and the roughness parameters y_s and y_b are not measured in the experiments. The appropriate values of these parameters are found during the calibration of the model. The model is run long enough to establish a stationary flow in the channel.

On Fig.1 the vertical distribution of the velocity u/u_* is plotted for the central part $\xi = 0.5$ in the channel of uniform depth and wind shear stress extracted from the experiment $\lambda c_h = 1 \times 10^{-3}$. The agreement with observations is quite good. Only for η about 0.7 the calculated u/u_* are not so close to the experimental data. This phenomena may be due to the fact that the experimental volume flux in the direction of the wind is approximately 70 per cent of the return flow. For this simple situation the influence of unknown parameters is studied. The dimensionless coefficient of

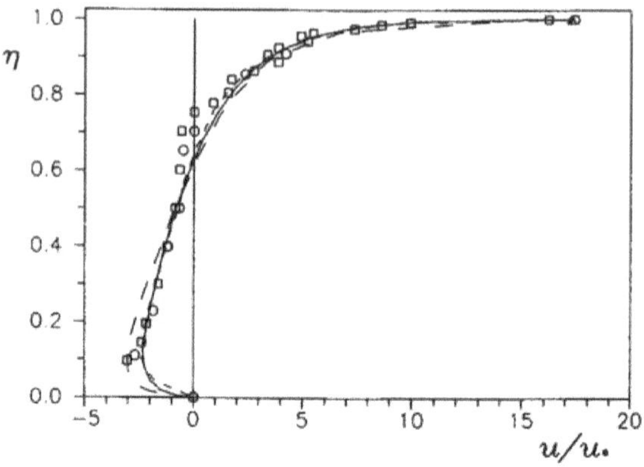

Figure 1: Velocity profiles in the middle section of the channel of uniform depth:
△ Keulegan [1], $Re = 1.5 \times 10^6$;
○ Baines and Knapp [1], $Re = 1.19 \times 10^6$;
□ Baines and Knapp [1], $Re = 1.8 \times 10^6$;
-- Reid's results [1];
--- Svensson's model [8];
— this work

horizontal turbulent transport K is varied from 10^{-2} to 10^2. These values correspond to the experimental observations of [7] for this range of characteristic lengths. It is not found strong effect of numerical results from this parameter which can be explained by the small length of the experimental channel. So, it is chosen to work with $K = 1$. From the accumulated experiences in the numerical tests one can conclude that the relative roughness y_b is responsible for the maximum value of u near to the bottom. In the same way y_s has important role for determination of free surface velocity. During of calibration of the numerical model these dimensionless parameters have been changed in the intervals: $0.0001 \leq y_b \leq 0.01$ and $0.0001 \leq y_s \leq 0.001$. Good correspondence with the experimental observations is obtained when $y_b = 0.002$ and $y_s = 0.0005$. The values of K, y_b and y_s are kept the same in the next calculations for the channel. In preliminary tests different mesh sizes are tried - from $N \times M = 51 \times 101$ to $N \times M = 101 \times 201$ - and it is not found dependance of results from the numerical mesh.

On Fig.1 also a comparison is made with the numerical results of Svensson (see in [9]) and with the analytical solution for velocity derived by Reid (see in [1]). Svensson's model is based on traditional $k - E$ model and considered the free surface as moving sheet. On the sheet the shear stress is applied obtained by the experiment. Nearly identical results have been obtained by the $k - E$ turbulent model and the simple turbulent model described in the present study.

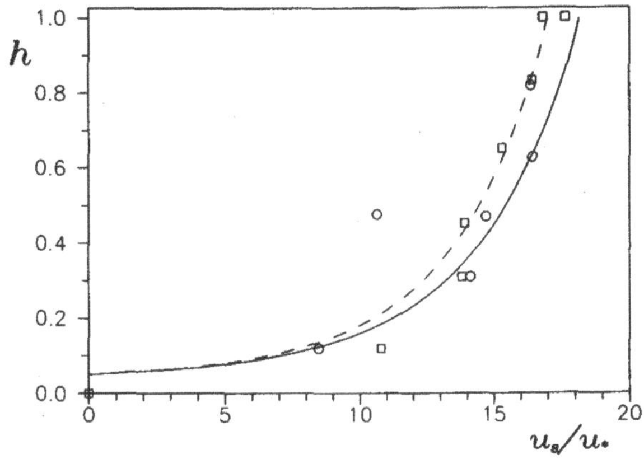

Figure 2: Surface velocity over the beach of slope 1/15:
○ Baines and Knapp [1], $Re = 1.19 \times 10^6$;
□ Baines and Knapp [1], $Re = 1.8 \times 10^6$;
- - - this work, $Re = 1.19 \times 10^6$;
— this work, $Re = 1.8 \times 10^6$

The variations of surface velocity u_s/u_* in the channel with a beach of slope 1/15 as a function of the channel depth, for both Reynolds numbers are shown on Fig.2. Some care must be taken in comparing the present numerical results to the experimental data. There is not a sufficient number of experimental observations near the free surface and there are relatively big deviations of the experimental points. In spite of that, one can conclude that the calculated surface velocities agree closely with the experimental data. For a large range of nondimensional depth, from 1 to 0.5, the surface velocity changes a little. As the depth become smaller, the surface velocity decreases rapidly. Here it has to be pointed out that in these calculations it is not possibly the depth of the slopping channel to be equal to zero at $\xi = 1$. For this reason the end of the computational domain is placed at the point where $h = 0.05$.

The International experiment "*Kamchia* 77 – 79" is carried out on the stationary platforms near the beach of Bulgarian coastal zone. The first platform is placed on the beach and the last one at about $1km$ offshore. The air velocity, the friction velocity and the surface roughness under changing wind directions are obtained at these stations. The average values of these parameters for a day are used in this study [4]. The investigated shelf region can be represented by a domain $2km$ long with a depth varying from $0.5m$ at the coast to $19m$ at $2km$ offshore. It is assumed that the wind blows perpendicular to the coast - onshore or offshore, i.e. regarding the sea flow as two-dimensional. The typical flow patterns of this region are demonstrated on Fig.3, where the stream lines are plotted for wind speed $U = 5m/s$ onshore. For the

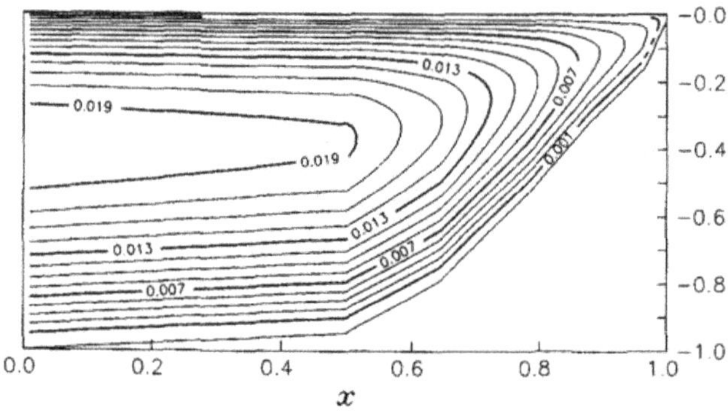

Figure 3: Stream lines of the Kamchia region of the Black Sea shelf: $U = 5m/s$ onshore

nondimensional time equal to about 50 the stationary flow is obtained. The results show a very similar structure of the flow for the opposite wind direction. Due to the large length scale in the vertical direction it is not possible to see on Fig.3 the free surface development. The maximum free surface elevation for $U = 2m/s$ onshore is about 0.3 per cent from the maximum shelf depth and for $U = 8m/s$ - about 0.7 per cent.

References

[1] Baines, W.D. and Knapp, D.J., *Wind driven water currents*, J. Hydraulics Division, ASCE, Vol. 91, 1965, pp. 205-221.

[2] Blumberg, A.F. and Mellor, G.L., A coastal ocean numerical model, In: J. Sunderman, K.P. Holz (Editors), *Mathematical modelling in estuarine physics*, Springer-Verlag, Berlin, 1980, pp. 203-219

[3] Johns, B., *The modelling of tidal flow in a channel using a turbulent energy closure scheme*, J. Phys. Oceanogr., Vol. 8, 1978, pp. 1042-1049

[4] Kuznetsov, O., Changes in dynamic characteristics of the flow upon changing in wind direction. In: Z. Belberov, V. Zahariev, O. Kuznetsov, N.Pykhov, B. Filyushkin and M. Zaslavski (Editors), *Interaction of the atmosphere, hydrosphere and litosphere in the nearshore zone*, Publ. House of BAS, Sofia, 1983, pp. 54-59

[5] Marchuk, G.I. and Sarkisyan, A.S., *Numerical modelling of the ocean circulation*, Springer-Verlag, Berlin, 1988

[6] Nihoul, J.C.J., Deleersnijder, E. and Djenidi, S., *Modelling the general circulation of shelf seas by 3-D $k-\epsilon$ models*, Earth-Science Rev., vol.26, 1989, pp. 163-189

[7] Ozmidov, P., *Diffusion of admixtures in the ocean*,Gidro- metereoizdat, Leningrad, 1986, 280 pp.

[8] Roache, P.J., *Computational fluid dynamics*, Hermosa Publishers, Albuquerque, 1976, 367 pp.

[9] Rodi, W., Turbulence models for environmental problems, In: W.Kollmann (Editor), *Prediction methods for turbulent flows*, Hemisphere Publ. Corp., Washington, D.C., 1980, pp. 227-315

[10] Samarsky, A. and Nikolaev, E., *Methods for solving the finite-difference equations*, Nauka, Moscow, 1978, 592 pp.

[11] Voltzinger, N., Klevanny, K. and Pelinovsky, E., *Long-wave dynamics of the coastal zone*, Gidrometereoizdat, Leningrad, 1989, 271 pp.

Address for correspondence:
Dr. Dimitar Marinov
Institute of Water Problems
1 Academician G. Bonchev Street
Sofia 1113, Bulgaria
Mrs. S. Miladinova, Dr. K. Radev
Institute of Mechanics
4 Academician G. Bonchev Street
Sofia 1113, Bulgaria

Numerical Approach to Ideal Separation on Airfoils. Application of Deformed Quasi-Parabolic Coordinates

M. D. Todorov C. I. Christov*

Abstract

The notion of discontinuous ideal flow consisting of potential streaming and stagnation zone (Helmholtz-Kirchhoff scheme) is applied for modelling the flow around airfoils. Deformed quasi-parabolic coordinates are introduced and difference schemes are devised for Laplace equation and for the equation for shape function of the free boundary. An iterative procedure is employed for defining the separation point. The performance of the numerical technique is displayed on NACA2412 airfoil. Results are presented for two angles of attack (8° and 16°). The pressure distributions on the lower and upper surfaces of airfoil compare quantitatively very well with the experimental data. The predictions for the drag and lifting forces are also in very good agreement (about 95%) with the experimentally measured values.

1 Introduction

Despite the numerous publications (see, e.g., [9] for the classical works), the investigation of inviscid flows with separation not only retains its actuality, but gathers some new impetus because of the quest for the limiting solutions of Navier–Stokes equations for large Reynolds numbers ([14, 12]). The problem is that $Re \to \infty$ yields singular asymptotic problem and the so-called "inner solution" can not be matched to the potential solution for the ideal flows around bluff bodies. In these flows, the boundary layer separates and violates the main assumptions of "inner in asymptotic sense" solution. The singular asymptotic nature of Navier–Stokes equations for vanishing viscosity (large Reynolds numbers) dictates that the "outer" solution must be of more complex nature than the mere potential flow.

*On leave from the National Institute of Meteorology and Hydrology, Bulgarian Academy of Sciences, Sofia, Bulgaria

Another candidate for an "outer" solution appears to be the Helmholtz–Kirchhoff separated inviscid flow introduced first by Helmholtz [10]. The Helmholtz model consists of potential and stagnation zones that match at prior unknown free stream line (tangential discontinuity). The free stream line originates at the body in a point called "detachment" or "separation" point. When the body has sharp edges, the position of detachment point is one of these edges and then the method of hodograph, as employed by Kirchhoff [11], was higly efficient for solving the separation problem. For bluff bodies, however, the problem of detachment point hindered the development of the techniques for solution. A condition [17] for smooth separation (called "condition of Brillouin–Villat") was established to determine the position of separation. Brodetsky [2] was the first to find an approximate solution to the problem of circular cylinder with Brillouin–Villat condition acknowledged.

The problem with Brodetsky's solution is that it predicts a very early separation of the flow (around 55° from the leading stagnation point), while the experiments show a value of order of 80°. In our opinion, this unsatisfactory result can be explained by the rough approximation of the Brodetsky method. He used only three terms in the Lorain series for approximate representation of the conformal mapping in the framework of hodograph method. Recently the authors [3, 4, 5] revisited numerically the Helmholtz–Kirchhoff–Brodetsky problem (the separated flow around circular cylinder). A finite–differences method was used with a specially devised algorithm for iterative determination of detachment point. It was found that the separation point is at 78° which differed significantly from Brodetsky's result being much closer to the experimental measurements and to the high-*Re* direct numerical simulations [7, 8] in which the separation angle was 95°. In [15] another solution to the same problem was developed, based on the integral–equations method and with similar iterative algorithm for defining the detachment point. The integral–equations method confirmed quantitatively the results of the finite–difference method.

The finite–difference algorithm was also applied to the ideal separated flow around airfoils [16, 6] and encouraging results were obtained for airfoils from the series of NACA. The purpose of the present work is to enhance the efficiency of the said numerical scheme by means of introducing a specially devised quasi–parabolic coordinates that are topologically most suited to the geometry of the flow around airfoils at considerable angle of attack.

2 Quasi–Parabolic Coordinates

In our previous works concerned with the circular–cylinder flow, we used parabolic coordinates since they are better suited to describe stagnation zones extending to infinity. In [16] we applied the parabolic coordinates to the separated flow around airfoil, as well, but part of the advantages of the parabolic coordinates were lost when the flow had non-trivial angle of attack. Then one faces the choice to orient the parabolic coordinates either along the chord of profile (the horizontal part of

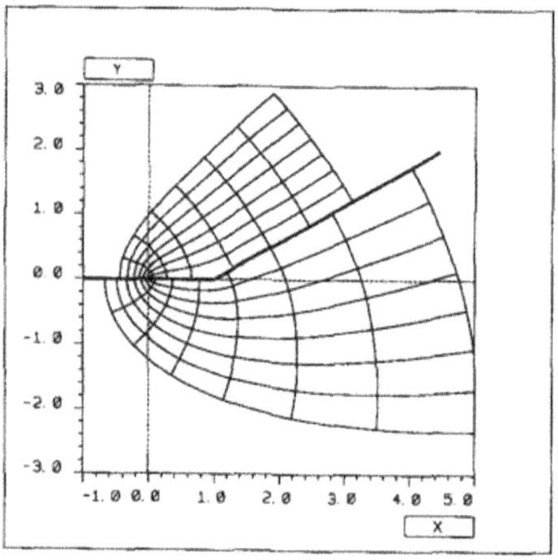

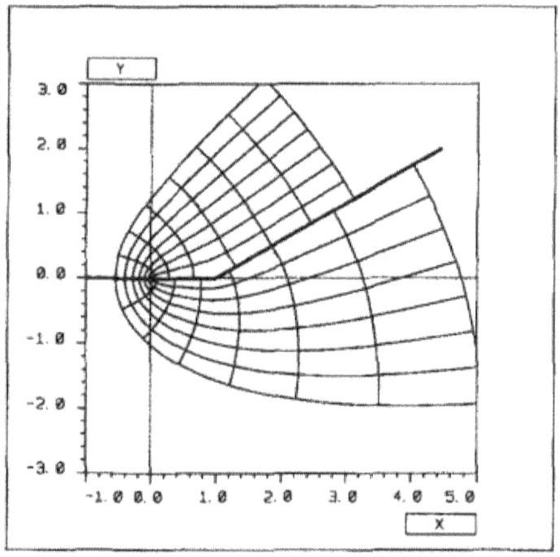

Figure 1: Left: the parabolic coordinates in the upper and lower regions. Right: quasi-parabolic coordinates in the whole region.

the thick line in Fig. 1) or along the direction of the flow (the inclined part of the thick line in Fig. 1). The former case yields a naturally dense mesh in the vicinity of the leading point of the airfoil where the good resolution is badly needed but then the asymptotic behaviour of the free lines is not the one of a coordinate line. The latter case is attractive because in it the asymptotic behaviour of separation lines is conveniently expressed as approaching some constants at infinity. When the constants are different from zero one has a parabolically expanding zone (Kirchhoff's solution). When both the asymptotic constants are equal to zero – a cusped zone is modelled. Due to the specifics of our algorithmic implementation we had to opt in [16] for the parabolic coordinates that were oriented with the direction of the flow. This forced us to introduce significantly nonuniform mesh in the vicinity of the leading point of the airfoil in order to capture the behaviour of the flow there. Here we combine the advantages of the two approaches by means of quasi-parabolic coordinate system.

Consider the upper and lower regions (above and bellow the thick line in Fig. 1, respectively) into which the whole region is divided by the chord and the direction of flow. These regions are subjected to conformal mappings rendering them into the upper and lower half-planes of the transformed region. Thus the thick line becomes a straight line. Note that the upper region in the physical plane is bounded by an angle lesser than 2π, while the lower region – by an angle larger than 2π. Let us denote now by $\omega_{\mp}$ the respective half-plane in the transformed region and by $z_{\mp}$ – the original region(s), i.e., $x_{\mp} = \mathbf{Re}[z_{\mp}]$, $y = \mathbf{Im}[z_{\mp}]$. The sign "−" refers to the upper half-plane, and the sign "+" – to the lower. Then for the inverse transformation of the two conformal mappings we have

$$z_{\mp} = r_{\mp}^{\frac{\pi \mp \alpha}{\pi}} \exp\left[i \left(\frac{\pi \mp \alpha}{\pi} \theta_{\mp} + \alpha \right) \right] , \qquad r_{\mp} = \sqrt{\xi_{\mp}^2 + \zeta_{\mp}^2} , \quad \theta_{\mp} = \arctan \frac{\zeta_{\mp}}{\xi_{\mp}} . \tag{1}$$

Here $\xi_{\mp}$ and $\zeta_{\mp}$ are the variables in the transformed region(s).

It is interesting to note that the above defined coordinate system ξ, ζ is orthogonal and its two metric coefficients are equal, say $Q(\xi,\zeta) \equiv g^{11} = g^{22}$. The Laplace equation for the stream function and the modulus of the velocity adopt the following simple forms

$$\frac{1}{Q(\xi,\zeta)}(\psi_{\xi\xi} + \psi_{\zeta\zeta}) = 0 . \qquad |\nabla \psi|^2 \equiv \frac{1}{Q(\xi,\zeta)}(\psi_{\xi}^2 + \psi_{\zeta}^2) . \tag{2}$$

Now we introduce a local parabolic coordinate system $O\sigma_{\mp}\tau_{\mp}$ in each of the half–planes of the transformed plane, namely

$$\xi_{\mp} = \frac{1}{2}(\tau_{\mp}^2 - \sigma_{\mp}^2) - X_0 , \quad \sigma_{\mp} \geq 0 ; \qquad \zeta_{\mp} = \sigma_{\mp}\tau_{\mp} , \quad -\infty < \tau_{+} < 0 , \ \ 0 < \tau_{-} < \infty . \tag{3}$$

Here X_0 is the focal point of the parabolic coordinates. Without this parameter the focal point of parabolic coordinates would have been at the vertex of the angle between the chord and direction of the flow. Without loosing the generality we can set it equal to unity, since we use it as the characteristic scale when rendering the problem dimensionless. Practically speaking, X_0 is the focal point of the parabola fitting best the leading edge of the airfoil under consideration, i.e., the characteristic length X_0 is slightly shorter than the chord L.

The two parabolic coordinate systems defined by (3) are different in the two half-planes (see the left part of Fig. 1). However, they can be matched along the line $\tau_{\mp} = 0$ (or which is the same for $-\infty < \xi_{\mp} < -1$). If a new σ–coordinate common for the two systems is introduced, namely

$$\frac{\sigma_{\mp}^4}{4} + \sigma_{\mp}^2 + 1 = \left(\frac{\sigma^4}{4} + \sigma^2 + 1 \right)^{\frac{\pi}{\pi \mp \alpha}} \quad \text{or} \quad \sigma_{\mp} = \sqrt{\left| \sigma^2 + 2 \right|^{\frac{\pi}{\pi \mp \alpha}} - 2} \tag{4}$$

If one defines the new τ–coordinate in the upper and lower half–plane respectively as $\tau \equiv \tau_{-}$ or $\tau \equiv \tau_{+}$ one gets the coordinate system, depicted in the right part of Fig. 1. Note that after the introduction of the new σ-coordinate the two parabolic coordinate systems do not match neither on the chord nor on the cut behind the body and that is why the quasi– parabolic coordinates are discontinuous (across the thick line in Fig. 1. This is not a problem, because the stream function is not supposed to be continuous when crossing the thick line. Physically, the flows in the two regions do not interact with each other. In a sense they are screened by the chord and by the cut (the separation lines).

In the practical implementation of the quasi-parabolic coordinates QPC we prefer not to introduce the transformation (4) into the governing equations since it would lead us to variable coefficient in the operator in Laplace equation responsible for the σ second-derivative. The other way around is to keep in mind when defining the difference function approximating the stream function that when crossing the line $\tau_{\mp} = 0$ we have to match difference functions with different σ number. Of course, this is possible only because both parabolic systems that comprise our QPC system are orthogonal to the said τ-line. Moreover, even after the transformation (4) they remain orthogonal.

Furthermore we omit the indices $\mp$ of ξ and η or σ and τ. They will be explicitly referred to only in the places where confusion may arise.

3 Posing the problem

In order to render the computational domain to a region with fixed boundaries we introduce scaled coordinates

$$\eta = \frac{\sigma - S}{S_\infty - S}, \quad \tau = \tau; \qquad S_\infty = \frac{\sqrt{2}x_\infty}{\sqrt{x_\infty} + \tau} \tag{5}$$

where x_∞ is an appropriately chosen parameter called "actual infinity". Transformation (5) renders the physical region of the flow into the strip $[0 \le \eta \le 1\ ,\ -\infty < \tau < \infty]$ of the computational domain. The advantage of (5) over the transformations with $S_\infty = const$ is that for $\tau \to \infty$, it gives $S_\infty \to 0$, i.e., the outer boundary of the region shrinks at infinity behind the body. Thus the excessive coarseness of the η-grid at large distances behind the body can be avoided. In terms of the scaled variables (η, τ), the Laplace equation for the stream function adopts the form

$$a\psi_{\eta\eta} - 2b\psi_{\eta\tau} + \psi_{\tau\tau} + d\psi_\eta = 0\ ,$$

$$a = \frac{1 + (\eta(S' - S'_\infty) - S')}{(S_\infty - S)^2}, \quad b = \frac{S' - \eta(S' - S'_\infty)}{S_\infty - S}, \tag{6}$$

$$d = \frac{-S'' + \eta(S'' - S''_\infty)}{S_\infty - S} + \frac{[S' - \eta(S' - S_\infty)](S'_\infty - S')}{(S_\infty - S)^2}.$$

This equation is solved with the boundary conditions

$$\psi|_{\eta=0} = 0\ , \quad \psi|_{\tau\to\pm\infty} \approx U_\infty [S + \eta(S_\infty - S)]\,\tau \tag{7}$$

The first one acknowledges the fact that boundary of the composite region "body / stagnation zone" is a stream line. The second one gives the asymptotic behaviour at infinity of the undisturbed flow.

The dynamic condition (Bernoulli integral) now becomes an explicit equation for the shape function $S(\tau$ of the free boundary and reads

$$\frac{1}{(S_\infty - S)^2(S^2+\tau^2)}(1+S'^2)\left[\frac{\partial\psi}{\partial\eta}\Big|_{\eta=0}\right]^2 = 1\ , \quad \text{for} \quad \tau_u^* \le \tau\ , \tau_l^* \ge \tau\ . \qquad (8)$$

where τ_l^* and τ_u^* are the coordinates of the detachment points, respectively on the lower and upper surfaces of airfoil. Thus the b.v.p. to be solved is completed.

For the drag and lifting-force coefficients one has

$$C_x = \int_{-\tau_1}^{\tau_1} p(S+S'\tau)\,d\tau\ , \qquad C_y = \int_{-\tau_1}^{\tau_1} p(SS'-\tau)\,d\tau$$

where p is the dimensionless pressure.

4 Numerical Implementation

We use in τ-direction $N_1 = N + 2N_W$ nodes and introduce a mesh which is non-uniform both over the chord (the first N points) of profile and over the two "shores" of the cut (the other $2N_W$ points).

Here we do not discuss in detail the difference approximation to Laplace equation for stream function. It is similar to [3, 4, 16] and makes use of splitting method as an iterative procedure. The scheme for the Bernoulli integral is of second order of approximation. In the upper half-plane of the flow the approximation is as follows.

$$\hat{S}_{j+1}^{k+1} - \hat{S}_j = \pm g_{j+1}\sqrt{\frac{\left[2S_\infty^{j+\frac{1}{2}} - \left(\hat{S}_j + S_{j+1}^{\alpha,k}\right)\right]^2\left[\left(\hat{S}_j + \hat{S}_{j+1}^k\right)^2 + (\tau_j+\tau_{j+1})^2\right]}{4\left(T_j^\alpha + T_{j+1}^\alpha\right)^2} - 1}\ , \quad (9)$$

for $j = j_u^*, \ldots, N$. It is similar in the lower part but the calculations are conducted for $j = j_l^*, \ldots, 1$. j_u and j_l are the indices of the detachment points on the upper and lower airfoil surface, respectively. For the sake of brevity is denoted

$$S_\infty^{j\pm\frac{1}{2}} = \frac{2\sqrt{2}x_\infty}{2\sqrt{x_\infty} + (\tau_j + \tau_{j\pm1})}$$

The scheme (9) requires inlet condition $\hat{S}_{j_u^*} = S_{j_u^*}^B$, where $S^B(\tau)$ is the known shape function of the rigid boundary. Then at each point internal iterations are conducted (referred by the index k) until convergence and thus the value $\hat{S}_j$ is obtained as a solution of the nonlinear difference equation.

The algorithm requires also a global iteration. Let us denote by S^α the shape function on the current iterative stage. Having S^α prescribed we solve the Laplace equation. Thus the "new" values for $\psi^{\alpha+1}$ are obtained. Then we calculate the quantity $\psi_\eta|_{\eta=0}$ and then solve the difference approximation of (9) and obtain a prediction for the shape function $\hat{S}(\tau)$. The strongly coupled character of the model requires some measures for preventing instability. The simplest one is to introduce relaxation for S according to the standard formula with the only difference that the relaxation parameter ω_j is taken here to be a function of the point, namely

$$S_j^{\alpha+1} = \omega_j \hat{S}_j + (1-\omega_j) S_j^\alpha \,,$$

$$\omega_j \equiv \frac{\tau_2}{\tau_2 + \epsilon_{rj}} \,, \quad \epsilon_{rj} = \frac{\hat{S}_j - S_j^\alpha}{\hat{S}_j} \quad \text{for} \quad j = j_u^*, \ldots, N_1 \,; \quad j = 1, \ldots, j_l^* \tag{10}$$

Our numerical experiments show that the optimal values for the "overall" relaxation parameter τ_2 lie in the interval $0.01 \le \tau_2 \le 0.05$. The criterion for terminating the global iteration is $\max\left|(\hat{S}_j - S_j^\alpha)/\hat{S}_j\right| \le 10^{-4}$

5 Results and Discussion

We have performed numerical tests with different angles of attack for the airfoil NACA2412 for whose shape there exists analytical representation, and for which comprehensive experimental data can be found in the literature, as well.

For $x_\infty = 20$, the pressure on the free surface is smaller than 10^{-3}, i.e. on such distances the calculated flow velocity is virtually equal to unity and that is the value for the "actual infinity" used in the present work. The number of nodes along the chord of the airfoil was taken in different experiments between 23 and 49. For the particular airfoil under consideration, its optimal value turned out to be 43. The number of nodes N_W alongside the the cut was taken between 15 and 40. For $x_\infty = 7$ the optimal number appeared to be 20. For $N_W > 20$ the result did not change significantly in comparison with $N_W = 20$, while smaller N_W led to unacceptably rough results.

We have found the solution for both the signs of the radical in eq.(8), but the solution for the positive sign corresponds to the so-called laminar separation and has not practical significance for airfoils. Hence, in the present work we resort to the other case – the negative sign. Following the terminology adopted in our previous works, we call it *solution I*. In Figs. 2 are shown the shapes of the two free boundaries of the separation zone (*solution I*) for angles of attack $\alpha = 8^\circ$ and $\alpha = 16^\circ$

Figs. 3 presents the chord-wise distribution of the surface pressure for the same two cases. For comparison, the experimental data of [13] for $Re = 2.7.10^6$ for the nearest available angle of attack is juxtaposed in Figs. 3. The agreement between the numerical solution and experimental measurements is fully satisfactory.

An important follow-up of our model are the predictions for the forces exerted from the discontinuous flow upon the body (airfoil). For $\alpha = 8^\circ$ we obtained values

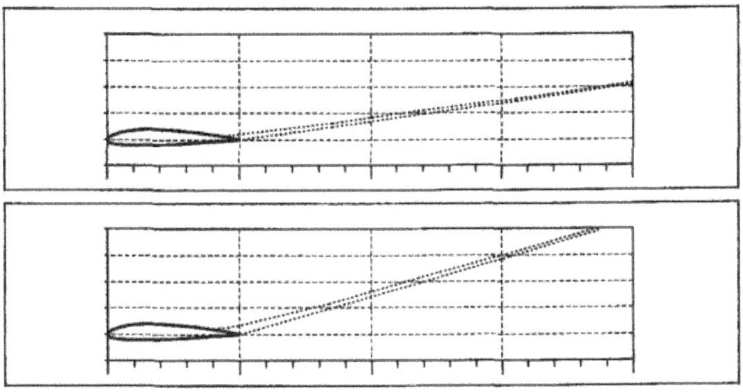

Figure 2: Shape of the separated zone for NACA2412. Up: $\alpha = 8°$; down: $\alpha = 16°$

$C_x = 0.017$ for the drag coefficient and $C_y = 0.9354$ – for the lifting-force coefficient. These results are in 95% agreement with the experimental results [1]. The important point here is that our results were obtained without any fudging parameters and without postulating the presence of a circulation around the airfoil (so-called Kutta-Joukowski postulate). In addition to the delightful prediction for the lifting force, we obtain the drag coefficient, the latter being zero by definition for the models employing the circulation postulate.

Acknowledgment A Sabbatical Fellowship form the Spanish Ministry of Science and Education is gratefully acknowledged by CIC. CIC also acknowledges the hospitality of the Lab. de Mecanique, Université de Caen during the preparation of the manuscript.

References

[1] I. E. Abott and A. E. Doenhoff. *Theory of wing sections.* Dover, 1959.

[2] S. Brodetsky. Discontinuous fluid motion past circular and elliptic cylinders. *Proc. Roy. Soc., London A*, 718:542–553, 1923.

[3] C. I. Christov and M. D. Todorov. Numerical investigation of separated or cavitating inviscid flows. In *Proc. Int. Conf. Num. Methods and Applications, Sofia 1984*, pages 216–233, 1985.

[4] C. I. Christov and M. D. Todorov. On the determination of the shape of stagnation zone in separated inviscid flows around blunt bodies. In *Proc. XV Jubilee Session on Ship Hydrodynamics*, page paper 10, Varna, 1986. BSHC.

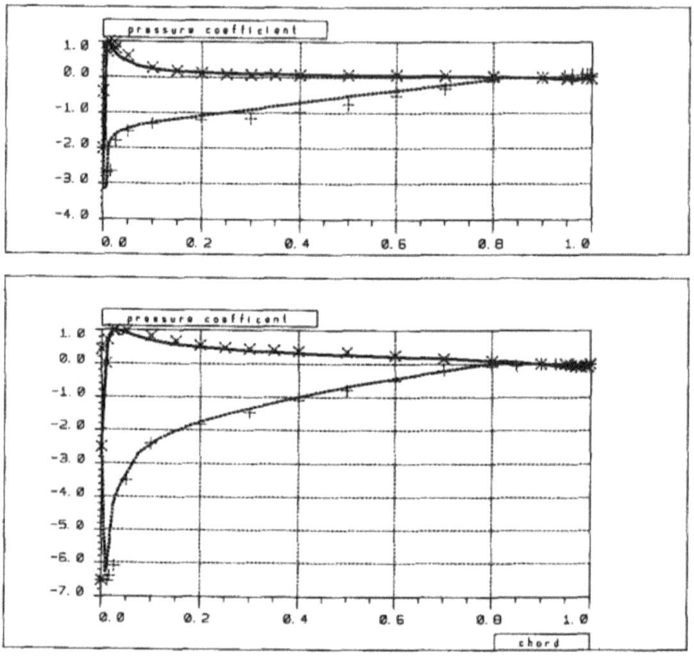

Figure 3: Pressure distribution on the lower and upper surfaces for NACA2412. Up: $\alpha = 8°$; down: $\alpha = 16°$ Lines – the present solution; symbols - $Re = 2.7.10^6$ (Riegels)

[5] C. I. Christov and M. D. Todorov. An inviscid model of flow separation around blunt bodies. *Compt. Rend. Acad. Bulg. Sci.*, 7:43–46, 1987.

[6] C. I. Christov and M. D. Todorov. Numerical solution of 2d Helmholtz problem for blunt bodies and airfoils. In V. V. Kozlov and A. V. Dovgal, editors, *Separated Flows and Jets*, pages 175–180, Berlin, 1990. Springer.

[7] B. Fornberg. A numerical study of steady viscous flow past a circular cylinder. *J. Fluid Mech.*, 98:819–855, 1980.

[8] B. Fornberg. Steady viscous flow past a circular cylinder up to Reynolds number 600. *J. Comput. Phys.*, 61:297–320, 1985.

[9] M. I. Gurevich. *Theory of Jet Flows of Ideal Fluids.* Nauka, Moscow, 1979. in Russian.

[10] H. Helmholtz. Uber discontinuirliche flussigkeitsbewegnungen. *Monatsbericht. d. Akad. d. Wiss.*, (Berlin):215–228, 1868.

[11] G. Kirchhoff. Zur theorie freier flussigkeitsstrahlen. *J. Reine Angew. Math.*, 70:289–298, 1869.

[12] D. H. Peregrine. A note on the steady high-Reynolds-number flow about a circular cylinder. *J. Fluid Mech.*, 157:493–500, 1985.

[13] F. Riegels. *Aerodynamische Profile.* Oldenbourg, Munchen, 1958.

[14] F. T. Smith. A structure for laminar flow past a bluff body at high Reynolds number. *J. Fluid Mech.*, 155:175–191, 1985.

[15] M. D. Todorov. Christov's algorithm for Helmholtz problem: an integral-equations implementation. In *Proc.XVIII National Summer School "Application of Mathematics in Technology", Varna, 25.8.-2.9.1992*, pages 191–193, Sofia, 1993.

[16] M. D. Todorov and C. I. Christov. Application of the model of inviscid separation for investigation of flows around airfoils. *Ann. Univ. Sof., Fac. Math. & Inform.*, 82 (Book 2 - Mechanics):193–217, 1988. In Russian.

[17] H. Villat. *Sur la resistance des fluides, Apercus theoriques.* Number No 38. Gauthier-Villars, Paris, 1920.

M. D. Todorov
Dept. Diff. Equations, Institute of Applied Mathematics & Informatics,
Technical University, Sofia 1156, BULGARIA

C. I. Christov
Instituto Pluridisciplinar, Universidad Complutense,
Paseo Juan XXIII, No 1, 28040, Madrid, SPAIN

Numerical solution of hyperbolic equations with semilinear dynamical conditions

Krasimira Z. Zlateva

Abstract

A semidiscrete scheme for numerical solution of one-dimensional hyperbolic equations with semilinear dynamical boundary conditions is introduced. For temporal discretization extrapolation and imbedded Nyström methods are implemented and tested. Global and blow-up (in time) solutions are numerically examined.

Key words and phrases: hyperbolic equations, nonlinear boundary conditions, Galerkin method, ordinary differential and differential-algebraic equations.

AMS(MOS) subject classification. 65M60

1 Introduction

We study the numerical approximation to solutions of the problems:

$$\ddot{u} - u'' = f(x,t,u) \text{ in } (0,1) \times R_+ \times R \tag{1}$$

$$\ddot{u}(0,t) - u'(0,t) = g_0(t, u(0,t)), \quad t \in R_+ \tag{2}$$

$$\ddot{u}(1,t) + u'(1,t) = g_1(t, u(1,t)), \quad t \in R_+ \tag{3}$$

$$u(x,0) = \varphi(x), \quad \dot{u}(x,0) = \psi(x), \qquad x \in (0,1) \tag{4}$$

and

$$-u'' = f(x,t,u) \text{ in } (0,1) \times R_+ \times R \tag{5}$$

with (2), (3) and

$$u(0,0) = \varphi_0, \quad \dot{u}(0,0) = \psi_0, \tag{6}$$

$$u(1,0) = \varphi_1, \quad \dot{u}(1,0) = \psi_1, \tag{7}$$

By dots and prime we denote time and space derivatives respectively.

Problems of this type occurs in connection with gravity waves [1, Chapter 9], [2]. They can be used as models to describe the motion of a fluid in a container or to describe the displacements of a fluid in a medium without gravity (artificial satellite).

Existence of local (in time) solutions for problems related to the above is proved in [3]. Blow-up criteria (in the sense of the following Definition) are established in [4,5].

Definition A point $x \in [0,1]$ is a blow-up point if there exist $\{(x_k, t_k)\}$ such that $t_k \to T_{max}$, $x_k \to x$ and $|u(x_k, t_k)| \to +\infty$, where T_{max} is the maximal life-span time of the local solution.

The aim of the present paper is twofold. First, in Section 2,3 numerical methods for solution of a specific class of partial differential equations, namely (1)-(4) and (5), (2),(3), (6), (7) are proposed. We use Galerkin semidiscretization to obtain systems of differential (or algebraic differential) equations for the nodal functions. Then Nyström methods for their solution are implemented. Second, as the global existence of solutions for (1)-(4) and (5),(2),(3),(6),(7) is still an open problem numerical experiments presented in Section 4 throw light on the global behaviour of these solutions.

Convergence of the Galerkin and fully discrete methods will be presented in next papers.

2 Galerkin method

The interval $[0,1]$ is divided into $(n+1)$ elements of equal length $h = 1/(n+1)$ by means of mesh $x_i = ih, \quad i = 0, 1, \ldots, n+1$. We consider the usual space S_h of piecewise linear functions. $\Phi_i \in S_h, \quad i = 0, 1, \ldots, n+1, \quad \Phi_i(x_i) = 1, \quad \Phi_i(x_j) = 0$, $i \neq j$. The Galerkin approximation of the solution of the problem (1)-(4) is a mapping $t \to Y(t)$, $Y(t) \in S_h$ defined by [6],

$$\begin{aligned} &(\ddot{Y}, \Phi) + \ddot{Y}(0,t)\Phi(0) + \ddot{Y}(1,t)\Phi(1) = \\ &= g_0(t, Y(0,t))\Phi(0) + g_1(t, Y(1,t))\Phi(1) + \\ &-(Y', \Phi') + (f, \Phi), \quad \Phi \in S_h, \end{aligned} \tag{8}$$

$$Y(0) - \varphi(x), \quad \dot{Y}(0) - \psi(x), \quad \text{"small"}$$

We denote by $(.,.)$, $\|.\|$ the inner product and norm in $L^2(0,1)$, respectively.

We carry out the following approximation in S_h:

$$u(x,t) \cong Y(x,t) = \sum_{i=0}^{n+1} y_i(t)\Phi_i(x) \tag{9}$$

$$f(x,t,Y) \cong \sum_{i=0}^{n+1} f(x,t,y_i(t))\Phi_i(x) \tag{10}$$

The last approximation is called "product approximation". Conditions (8)-(10) are already seen to be equivalent to the following system of ordinary differential equations (ODEs) for the nodal values:

$$M\ddot{Y} = NY + Pf(t) + Q \tag{11}$$

$$Y(t=0) = \varphi(ih), \quad \dot{Y}(t=0) = \psi(ih)$$

where we set

$$Y(t) = (y_0(t), ..., y_{n+1}(t))^T, \quad f(t) = \Big(f(t,x_0,y_0(t)), \cdots, f(t,x_{n+1},y_{n+1}(t))\Big)^T$$

$$M = \frac{h}{6}\begin{pmatrix} \frac{6}{h}+2 & 1 & & & \\ 1 & 4 & 1 & & \\ \cdots & \cdots & \cdots & & \cdots \\ & 1 & 4 & 1 & \\ & & 1 & \frac{6}{h}+2 & \end{pmatrix}, \quad N = \frac{1}{h}\begin{pmatrix} 1 & -1 & & & \\ -1 & 2 & -1 & & \\ \cdots & \cdots & \cdots & \cdots & \cdots \\ & & -1 & 2 & -1 \\ & & & -1 & 1 \end{pmatrix},$$

$$P = \frac{h}{6}\begin{pmatrix} 2 & 1 & & & \\ 1 & 4 & 1 & & \\ \cdots & \cdots & \cdots & \cdots & \cdots \\ & & 1 & 4 & 1 \\ & & & 1 & 4 \end{pmatrix}, \qquad Q = \begin{pmatrix} g_0(0,t,y_0) \\ 0 \\ \vdots \\ 0 \\ g_1(1,t,y_{n+1}) \end{pmatrix}.$$

Therefore, solving system (1)-(4) boils down to the calculating the solution of an ODEs of $n+2$ nonlinear equations.

Applying the Galerkin method to the problem (5), (2), (3), (6), (7) we get the algebraic differential equations system:

$$-y_{i-1}(t) + 2y_i(t) - y_{i+1}(t) = h^2 f(x_i,t,y_i(t)), \quad i = 1, \ldots, n \tag{12}$$

$$\ddot{y}_0 = \frac{y_1 - y_0}{h} + g_0(t, y_0)$$
$$\ddot{y}_{n+1} = -\frac{y_{n+1} - y_n}{h} + g_1(t, y_{n+1}) \tag{13}$$

$$y_0(0) = \varphi_0 \qquad \dot{y}_1(0) = \psi_0 \tag{14}$$
$$y_{n+1}(0) = \varphi_1 \qquad \dot{y}_{n+1}(0) = \psi_1 \tag{15}$$

The local accuracy of the interior equations of (11) and of (12) is $O(h^2)$ and that of exterior of (11) and of (13) — $O(h)$. The last local errors on the solution $u(x,t)$ will be of order $O(h^2)$ if instead of (13) we take

$$\ddot{y}_0 = \frac{y_1 - y_0}{h} + 0.5hf(0, t, y_0) + g_0(0, t, y_0),$$
$$\ddot{y}_{n+1} = -\frac{y_{n+1} - y_n}{h} + 0.5hf(1, t, y_{n+1}) + g_1(1, t, y_{n+1}) \tag{16}$$

3 Integration in time

To obtain the fully-discrete approximations we shall discretize the system (11) in time by Nyström methods described in [7] on the problem

$$\ddot{y} = z(t, y), \quad y(t_0) = y_0 \quad \dot{y}(t_0) = \dot{y}_0$$

Such system can be transformed into first-order ODEs of doubled dimension

$$\begin{pmatrix} y \\ \dot{y} \end{pmatrix} = \begin{pmatrix} \dot{y} \\ z(t, y) \end{pmatrix}, \qquad \begin{matrix} y(t_0) = y_0 \\ \dot{y}(t_0) = \dot{y}_0 \end{matrix} \tag{17}$$

E.J.Nyström (1925) proposed a method of the form

$$\begin{aligned} k_i &= z(t_0 + c_i\tau, y_0 + c_i\tau\dot{y}_0 + \tau^2 \sum_j a_{ij}k_j) \\ y_1 &= y_0 + \tau\dot{y}_0 + \tau^2 \sum_i \bar{b}_i k_i \\ \dot{y}_1 &= \dot{y}_0 + \tau \sum_i b_i k_i, \end{aligned} \tag{18}$$

where the coefficients do not necessary satisfy the relation $\bar{b}_i = \sum_j b_j a_{ji}$. A Nyström method (18) has order p if for sufficiently smooth problems (17)

$$y(t_0+\tau) - y_1 = O(\tau^{p+1}), \qquad \dot{y}(t_0+\tau) - \dot{y}_1 = O(\tau^{p+1})$$

Order conditions for Nyström methods are derived in [7,8].

The numerical experiments reported in the next section use subroutines ODEX2 and DOPRIN. These subroutines are part of codes for nonstiff, stiff and differential-algebraic problems which were sent to us by E. Hairer and we highly appreciate his kindness.

ODEX2 is based on an extrapolation algorithm using explicit midpoint rule (with stepsize control and order selection), cf [7, section II.13].

DOPRIN is based on the Nyström method of order (6)7 due to DORMAND & PRINCE (with stepsize control), cf [7, section II.13].

For other numerical results regarding the order of these methods, their efficiency and the calculation of various errors, see [7,8] and the references therein.

In this paper we shall treat (12)-(15) when f is a linear function with respect to u, i.e. $f = f_0(x,t) + f_1(t)u$. Then we solve (12) explicitly for $Y_n \equiv (y_1, \dots, y_n)^T$

$$Y_n = R^{-1}F$$

where

$$R = \begin{pmatrix} 2+h^2 f_1(t) & -1 & & & \\ -1 & 2+h^2 f_1(t) & -1 & & \\ \dots & \dots & \dots & \dots & \\ & -1 & 2+h^2 f_1(t) & -1 & \\ & & -1 & 2+h^2 f_1(t) \end{pmatrix},$$

$$F = \begin{pmatrix} h^2 f_0(x_1,t) + y_0 \\ h^2 f_0(x_2,t) \\ \vdots \\ h^2 f_0(x_n,t) \\ h^2 f_0(x_n,t) + y_{n+1} \end{pmatrix}.$$

It's easy to express in closed form the elements of the inverse matrix of R: $R^{-1} = \|b_{ij}\|_{i,j=1}^{n}$, cf[9]. Then

$$y_i = b_{i1}(h^2 f_0(x_1,t) + y_0) + b_{i2}h^2 f_0(x_2,t) + \dots + b_{in}(h^2 f_0(x_{n+1},t) + y_{n+1}, \qquad i = 1,n.$$

We insert y_1, y_n in (16) to obtain the system of two ODEs:

$$\begin{aligned}\ddot{y}_0 &= \left(\frac{b_{11}-1}{h} + 0.5hf_1(t)\right) y_0 + g_0(0,t,y_0) + b_{1n}y_{n+1} + \\ &+ h\left(b_{11}f_0(x_1,t) + \ldots + b_{1n}f_0(x_n,t)\right) \\ \ddot{y}_{n+1} &= b_{n1}y_0 + h\left(b_{n1}f_0(x,t) + \ldots + b_{nn}f_0(x_n,t)\right) + \\ &+ \left(\frac{b_{nn}-1}{h} + 0.5hf_1(t)\right) y_{n+1} + g_1(1,t,y_{n+1})\end{aligned}$$

with initial conditions (16).

4 Numerical experiments.

All computations are performed on the IBM/PC 486 computer without coprocessor.

The degree of accuracy is evaluated by calculating the relative error E_r of the approximated solutions $Y(x,t)$ as follows

$$E_r(t) = sup_i \left|\frac{u(x_i,t) - y_i(t)}{u(x_i,t)}\right|$$

First example.

We choose in (1)-(3) functions f, g_0, g_1 as:

$$f \equiv 0, \quad g_0 = -\cos t, \quad g_1 = -(\cos 1 + \sin 1)\cos t$$

The following solution of (1):

$$u(x,t) = \cos x \cos t$$

corresponds to the initial conditions (4)

$$u(x,0) = \varphi(x) = \cos x, \quad u_t(x,0) = \psi(x) \equiv 0.$$

We record the results as shown in Table 1.

Second example.

We choose in (5), (2), (3) functions f, g_0, g_1 as:

$$f \equiv 0, \quad g_0 \equiv 0, \quad g_1 = u(u^2+1)$$

The following solution of (5):

	n=11	n=30
T=1	$0.3876842292\times10^{-3}$	$0.4622432373\times10^{-4}$
T=2	$0.1446319735\times10^{-2}$	$0.1721677486\times10^{-3}$
T=3	$0.9176665630\times10^{-3}$	$0.1089311261\times10^{-3}$

Table 1: Relative error E_r

	x_5	x_{10}	x_{15}
T= 1.41	0.7065972796×10^{2}	0.1589843879×10^{3}	0.2473090479×10^{3}
	0.7065972796×10^{2}	0.1589843879×10^{3}	0.2473090479×10^{3}
T= 1.41421	$0.1360492229\times10^{13}$	$0.3061107516\times10^{13}$	$0.4761722802\times10^{13}$
	$0.16955890216\times10^{13}$	$0.3680075297\times10^{13}$	$0.5724561574\times10^{13}$

Table 2: Exact and numerical solution ($n = 20$)

$$u(x,t) = \frac{\sqrt{2}x}{\sqrt{2}-t}$$

corresponds to the initial conditions (6) and (7)

$$u(0,0) = \varphi_0 = 0, \quad \dot{u}(0,0) = \psi_0 = 0$$

$$u(1,0) = \varphi_1 = 1, \quad \dot{u}(1,0) = \psi_1 = \frac{\sqrt{2}}{2}$$

We record the results as shown in Table 2, 3.

Third example The functions f, g_0, g_1 in (5), (2), (3) are chosen as:

$$f \equiv 0, \quad g_0 = u^2, \quad g_1 = u^2$$

and the initial conditions (6) and (7) are:

$$u(0,0) = \varphi_0 = 10^{-1}, \quad \dot{u}(0,0) = \psi_0 = 10$$

	x_{10}	x_{20}	x_{30}
T= 1.41	0.7745331901×10^{2}	0.1635125624×10^{3}	0.2495718057×10^{3}
	0.7745331901×10^{2}	0.1635125623×10^{3}	0.2495718057×10^{3}
T= 1.41421	$0.1491308790\times10^{13}$	$0.3148318556\times10^{13}$	$0.4805328322\times10^{13}$
	$0.17922857196\times10^{13}$	$0.3784920748\times10^{13}$	0.577698429×10^{13}

Table 3: Exact and numerical solution ($n = 40$)

x_1	x_{10}	x_{20}	x_{30}
0.1130793338×10^8	0.7815366313×10^7	0.3998069574×10^7	0.1507728352×10^6

Table 4: Numerical solution near before the blow-up time.

$$u(1,0) = \varphi_1 = 10^{-2}, \quad \dot{u}(1,0) = \psi_1 = 10$$

The computed blow-up time is in the interval (1.481, 1.482) and the numerical solution at time 1.481 and $n = 30$ nods is represented in four of them in Table 4.

5 Concluding remarks

In this paper we have discussed the application of Galerkin method combined with extrapolation and imbedded Nyström methods to hyperbolic equations with dynamical boundary conditions. Three example problems are used to illustrate the numerical method. We consider these results to be very satisfactory. However, we feel that testing on more difficult (nonlinear) problems needs to be done in order to study the nature of blow-up solutions. Then mesh-refinement strategy can be developed.

References

[1] H. Lamb, *Hydrodynamics,* OGIS, Moscow, (1947),(Russian).

[2] A. Friedman, M. Shinbrot, *The initial value problem for the linearized equations of water waves*, J. Math. Mech. 17(1968), pp. 107-180.

[3] T. Hinterman, *Evolution equations with dynamic boundary conditions*, Proc. Roy. Soc. Einburgh, Sect.A, 113A (1989), pp. 43-60.

[4] M. Kirane, *Blow-up for some equations with semilinear dynamical boundary conditions of parabolic and hyperbolic type*, Hokkaido Math. J. 21 (1992), pp. 221–229.

[5] L. Vulkov, *Nonexistence of global solutions for hyperbolic equations with dynamical nonlinear boundary conditions*, preprint, Technical University-Rousse (1994).

[6] C. A. J. Fletcher, *Computational Galerkin methods*, Springer, N.Y. (1984).

[7] E. Hairer, S. P. Norsett, G. Wanner, *Solving ordinary differential equations I*, Springer, N.Y. (1980).

[8] E. Hairer, S. P. Norsett, G. Wanner,*Solving ordinary differential equations II*, Springer, N.Y. (1991).

[9] J. Lewy, *Inversion of tridiagonal matrices,* Numer.Math. 38(1982), pp. 333-345.

Address of Author:
Techn. Univ., Dept. of Math.
Rousse, BULGARIA

Author Index

www.ingramcontent.com/pod-product-compliance
Lightning Source LLC
LaVergne TN
LVHW071223011026
849280LV00007B/34
* 9 7 8 9 8 1 0 2 1 9 2 6 0 *